RETINAL DEGENERATIONS

OPHTHALMOLOGY RESEARCH

JOYCE TOMBRAN-TINK, PhD, AND COLIN J. BARNSTABLE, DPhil
SERIES EDITORS

Retinal Degenerations: *Biology, Diagnostics, and Therapeutics,* edited by *Joyce Tombran-Tink, PhD, and Colin J. Barnstable, DPhil, 2007*

Ocular Angiogenesis: *Diseases, Mechanisms, and Therapeutics,* edited by *Joyce Tombran-Tink, PhD, and Colin J. Barnstable, DPhil, 2006*

RETINAL DEGENERATIONS

BIOLOGY, DIAGNOSTICS, AND THERAPEUTICS

Edited by

JOYCE TOMBRAN-TINK, PhD

and

COLIN J. BARNSTABLE, DPhil

Department of Ophthalmology and Visual Science,
Yale University School of Medicine,
New Haven, CT

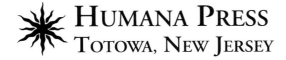
HUMANA PRESS
TOTOWA, NEW JERSEY

© 2007 Humana Press Inc.
999 Riverview Drive, Suite 208
Totowa, New Jersey 07512

www.humanapress.com

Production Editor: Christina Thomas

Cover design by Donna Niethe

For additional copies, pricing for bulk purchases, and/or information about other Humana titles, contact Humana at the above address or at any of the following numbers: Tel.: 973-256-1699; Fax: 973-256-8341, E-mail: orders@humanapr.com; or visit our Website: www.humanapress.com

This publication is printed on acid-free paper. ∞
ANSI Z39.48-1984 (American National Standards Institute) Permanence of Paper for Printed Library Materials.

Photocopy Authorization Policy:
Authorization to photocopy items for internal or personal use, or the internal or personal use of specific clients, is granted by Humana Press Inc., provided that the base fee of US $30.00 is paid directly to the Copyright Clearance Center at 222 Rosewood Drive, Danvers, MA 01923. For those organizations that have been granted a photocopy license from the CCC, a separate system of payment has been arranged and is acceptable to Humana Press Inc. The fee code for users of the Transactional Reporting Service is: [1-58829-620-2/07 $30.00].

Printed in the United States of America.   10 9 8 7 6 5 4 3 2 1

eISBN 1-59745-186-X

Library of Congress Cataloging-in-Publication Data

Retinal degenerations : biology, diagnostics, and therapeutics / edited by Joyce Tombran-Tink and Colin J. Barnstable.
       p. ; cm. -- (Ophthalmology research)
     Includes index.
     ISBN 1-58829-620-2 (alk. paper)
     1. Retinal degeneration.   I. Tombran-Tink, Joyce. II. Barnstable, Colin J. III. Series.
     [DNLM: 1. Retinal Degeneration.   WW 270 R438239 2007]
     RE661.D3R482 2007
     617.7'35--dc22
                                                                                     2006024144

PREFACE

For centuries, humans have tried to explain the complex process of vision and find effective treatments for eye diseases. Perhaps the oldest surviving record of ancient ophthalmic practices is the Babylonian code of Hammurabi that over 4000 years ago, mentioned fees for eye surgery—and penalties for unsuccessful operations that led to loss of the eye. Babylonian medicine was controlled by priests who directed the work of skilled surgeons. The earliest records of Egyptian medicine date from almost the same time. The Ebers Papyrus, dating back to more than 3500 years ago is a superbly preserved document in which a section outlines a relatively advanced system of diagnosis and treatment of various ocular pathologies. The text reveals that ancient Greek and Egyptian physicians prescribed "liver juice" for night blindness. This was obtained from roasted and crushed ox liver. We now know that their prescription contained a remarkable amount of vitamin A. It was only within the last century, however, that we have recognized the importance of vitamin A to the function of photoreceptors and visual acuity and that its deficiency can result in night blindness.

Egyptian ophthalmological practices were held in high esteem in the ancient world and so were their medical institutes, called "peri-ankh," which existed since the first dynasty. Herodotus, the fifth century BC Greek historian, comments on the specialization of the physicians: "Each physician treats just one disease. Some treat the eye, some the teeth, some of what belongs to the abdomen and other internal diseases." The profession was so organized that there were even specific titles to describe the physicians: "swnm" for the Lay Physician, "imy-r swnw" for the Overseer of Physicians, "wr swnw" for the Chief Physician, and "shd swnw" for the Inspector of Physicians. Qualified female physicians were also popular at the time. Peseshet (imyt-r swnwt), the first documented female physician in history, practiced during the fourth dynasty and was given the title "Lady Overseer of the lady physicians."

Many ocular diseases and their treatment were also known in Asia as well. During the golden age of his reign, the Yellow Emperor (2696–2598 BC) of China composed his Neijing Suwen or Basic Questions of Internal Medicine, also known as the Huangdi Neijing. Within this text are descriptions of eye diseases including descriptions of floaters within the eye, small eyes, corneal diseases as well as the use of needle penetrations to alleviate some diseases. Modern scholarly opinion holds that the extant text of this Treatise was compiled by an eponymous scholar between the Chou and Han dynasties more than two thousand years later than tradition reports, although some parts of the extant work may have originated as early as 1000 BC.

Although medical specialties were highly developed in Egypt and in other parts of the world, progress in the field remained static for centuries because ancient medical practices were bound by galling fetters of supernatural beliefs, and rigid notions that the disease originated from hostile spirits and angry gods stifled innovative observations.

Nevertheless, it is still fascinating to examine a few of the ancient theories and to compare them to our modern knowledge. Interest in how the human visual system works dates back at least to the time of Aristotle, the prominent fourth century BC philosopher. His explanation for the mechanism underlying human vision was that the object being viewed altered the "medium" between the object and the eye of the viewer. The object is then perceived when the altered transparent medium propagates to the eye, suggesting that the object itself had innate properties that allowed vision. Democritus (425 BC) proposed that the emissions from the object viewed entered the eye and formed an image, but could not explain how a single object could generate enough emissions to create sight if many people were viewing it simultaneously. Vigorous debates ensued and the intromission perception of vision, accepted by scholars such as Democritus and Epicurus (342–270 BC), was refuted and replaced by an opposing view that the eyes sent out emissions to the object in view and that those rays promoted vision. Plato (427–347 BC) introduced the concept of "ocular beams" projecting from the eye at great speed and in a straight line to interact with objects in view. The objects deflected the beams back to the eye, which, in turn, transmitted these rays along hollow tubes connecting the eye with the brain to create sight. The extromission model of visual cognition then became a widely accepted scientific dogma by many scholars and persisted until the early 1600s, although not without challenge and modifications from theorists who had difficulties integrating this proposal with prior beliefs. Aristotle pointed out that it was unreasonable to think that a ray from the eye could reach as far as the stars and Galen (AD 129–AD 216) argued that larger images could not fit through the tiny pupils of the eye. Alexandrian vision theorists considered the lens to be the seat of vision, a view that Galen promoted when he found the retina lining the posterior aspect of the lens in a strategic position to serve as a mirror to reflect the object viewed. Needless to say, speculation was rife and sparked much controversy and the schism was too wide to be easily bridged. Although inaccurate, these early concepts have spawned our interest in developing experimental scientific methods from which we have gained our understanding of the visual process today.

It must also be noted that during the Dark Ages of Western civilization, knowledge of the eye and eye diseases was maintained and developed in the Arab world. Ophthalmic departments were important components of hospitals and many surgical procedures perfected. The earliest anatomical drawings of the eye are in Hunayn ibn Ishâq's *Book of the Ten Treatises on the Eye*. After the fall of the Roman Empire, Arabic philosophers carried on Greek science and mathematics. One of the greatest of these Arabians was Alhazen (965–1040), who defended the intromission concept of vision. He was the first to propose that the eye is a receiver, a dark chamber through which light enters carrying information from the outside world. He argued that if the air and the eye are transparent, light from objects would reach and enter the eye; therefore, visual rays are unnecessary to explain vision (Alhazen's *De aspectibus*). Translations of Arab ophthalmic texts into Latin promoted the revival and development of ophthalmology that occurred in the renaissance period in Europe.

By the 1600s, Leonardo da Vinci (1452–1519) and Felix Plater (1536–1614) largely promoted a historical shift in thinking by endorsing the concept of a "camera obscura" in vision. Felix Plater's *De corporis humani structura et usu* of 1583 is the first published work in which the retina is identified as the target of light—"the retina and not the lens was the receiving plate of the eye." Alessandro Achillini (1466–1533) may have been the first to challenge the idea of the crystalline lens as the main organ of sight. In 1543,

Andreas Vesalius' *De humani corporis fabrica* alluded to the retina as the "seat of vision" as well, but it was not until the 17th century that modern theories of vision and the role of the retina in this process was born as the focus shifted from the lens and the cornea to the retina as that structure of the eye required for the detection of light. Johannes Kepler of Germany and René Descartes of France, both avant-garde physicists of their time, applied the physical concepts of light rays and geometric optics to the visual process. As a result Kepler, in *Ad Vitellionem paralipomena* (1604), proposed that the lens re-focused intromitted rays on the retina, creating a real optical image which he called "pictura." This proposal was quite controversial and stirred up yet another debate in the scientific community since it had long been accepted that if rays were crossed within the eye, the image created would be reversed and inverted causing us to see the world upside down. This dilemma of vision was recognized since antiquity, and scholars such as Fabri de Peiresc (1580–1637) and Pierre Gassendi (1592–1655) of France set out to refute Kepler's retinal inversion hypothesis. Their experiments eventually culminated in a proposal for the existence of a "retina mirror" that would upright the image on the retina by reflecting it back toward the center of the eye. Others such as Christopher Scheiner were among the first to embrace Kepler's optical analogy of the eye and the camera obscura and, in 1619, he provided the first direct observation of image formation on the retina. The theory that the retinal image is inverted was eventually confirmed by the landmark experiment of Descartes in 1637, which showed the first direct evidence that visual images were inverted as a result of being focused onto the retina by the lens. What Descartes did was quite amazing. He surgically removed an eye from an ox, scraped the sclera from the back of it to make the orb transparent, then he placed the eye on the ledge of a window as if the ox were looking out of the window. He then looked at the back of the eye and saw an inverted image of the scenery outside.

At the end of the 18th century, however, our knowledge of the retina was still rudimentary. Although Briggs described fibers in the retina in 1678 and Mariotte identified the blind spot in 1681, it was not until 1819 that the astute Irish physician, Arthur Jacob, provided us with the first anatomically detailed description of the retina. Jacob described the retina as "the most beautiful specimen of a delicate tissue which the human body affords." He described a layer in the eye consisting of three sublayers: a limiting, a nervous, and a choroid layer. The nervous layer, which became known for a time as "Jacob's Membrane," consists of the rods and cones.

Identification of the rods and cones, however, should probably be credited to Leeuwenhoek, pioneer of the microbe world, who was the first to perform microscopic examination of the retina and noted images of these cells during his studies in 1684. These highly specialized photoreceptors were rediscovered in 1834 by Treviranus and so named because of their microscopic appearance. Perhaps one of the greatest achievements in cell biology that promoted his findings was the development of the compound microscope. A new realm of observation was ushered in with the increasing technical sophistication of this instrument and largely brought about the dawn of the cellular theory and our knowledge of the retina.

The pace quickened in the last 200 years, and our understanding of the visual process has increased dramatically since Aristotle, Plato, and Galen. What followed was a series of significant advances in retinal biology. Additional research showed that the rod and cone cells were responsive to light. Max Schultze (1825–1874) discovered that the cones are the color receptors of the eye and the rod cells are not sensitive to color but very sensitive to light at low levels. Helmholtz invented the ophthalmoscope in 1851. Then in 1854, Mueller proved that photoreception occurs in the rods and cones. In 1866, Holmgren discovered the electroretinogram. In 1893, Ramon y Cajal's "La retine des vertebras" was the first complete description of retinal neuroanatomy as revealed by Golgi stain. This work, which carefully classified retinal cell types by anatomical criteria, remains the benchmark of retinal anatomy and in the century following its publication, we have come to appreciate that almost all of Ramon y Cajal's retinal cell types can also be identified by unique sets of molecular and physiological properties. In 1925, Holm demonstrated that vitamin A deficiency causes night blindness. The chemical structure of vitamin A and its precursor, β-carotene, was unraveled in 1930 by Swiss researchers. In 1933, Wald found vitamin A in rhodopsin, Stiles and Crawford demonstrated directional sensitivity of rods and cones, and Cooper, Creed, and Granit demonstrated the first electronically amplified human electroretinogram. In 1938, Selig Hecht showed the exquisite sensitivity of rod cells by demonstrating that a single photon can initiate a response in a rod cell.

Identification and understanding of retinal diseases has lagged behind that of diseases of the cornea and lens, but closely followed the increasing knowledge of retinal anatomy. Color blindness was first described in detail by the English chemist and physicist John Dalton (1766–1844). Many of the degenerative diseases were carefully described only toward the end of the 19th century through the first half of the 20th Century. For example, Retinitis pigmentosa was first described clinically in 1853 by van Trigt (1), X-linked forms of the disease by Usher in 1935 (2), and X-linked retinoschisis by Haas in 1898 (3). Hutchinson and Tay (1875), and Robert Walter Doyne (1899), were the first to describe whitish spots (drusen) in the macula, a condition frequently leading to age-related macular degeneration.

Discoveries made in the last two centuries have increased our understanding of how the retina works at the biochemical level and therapeutic strategies that should be developed to slow its dysfunction. We have come a long way since the discovery of the rods and cones. In the last few years, we have seen convincing results from experiments conducted in the laboratory and in clinical trials with growth factors, micronutrient supplements, antibodies, and small molecules that offer hope for retinal degenerations. We have made advances in nanotechnology, cell and molecular biology, ocular genetics, gene manipulation and delivery methods, and cellular transplantation, which now have wide-range implications for the future of blinding eye diseases.

The early pioneers who preceded us in the field were at the vanguard of science during their time and their investigative approach laid the foundation for mature scientific discovery and for the development of new therapies to combat retinal degenerative diseases. Regrettably, despite these advances, few therapies lived up to their alleged claims or to the hope that they aroused. Clearly, there are numerous difficulties encountered in developing treatments and we cannot predict which therapy will result in the best outcomes, but it is important that we continue to foster collaborative efforts to pursue several distinct therapeutic initiatives. It is our goal to continue at the frontline of science with the same passion, persistence, and vision as our forerunners. We hope that our contemporaries in the lab, the clinic, and the industry will be challenged by the recent advances described in *Retinal Degenerations: Biology, Diagnostics, and Therapeutics* by respected leaders and by the potential to develop innovative strategies to translate investigative research into viable therapeutics for retinal diseases.

Joyce Tombran-Tink, PhD
Colin J. Barnstable, DPhil

References

1. van Trigt, A.C. (1852–1853) De oogspiegel. *Nederlandisch Lancet*, third series, Utrecht, 2d, 417–509.
2. Usher, C.H. (1935) On a few hereditary eye affections. *Trans. Ophthal. Soc. UK* **55,** 164–245.
3. Haas, J. (1898) Ueber das Zusammenvorkommen von Veraenderungen der Retina und Choroidea. *Arch. Augenheilkd.* **37,** 343–348

CONTENTS

CONTRIBUTORS

UMER F. AHMAD, MD • *Edward S. Harkness Eye Institute, Columbia University, New York, NY*

RANDO ALLIKMETS, PhD • *Ophthalmology, Columbia University Eye Institute Research, New York, NY*

ROBERT B. ARAMANT, PhD • *Department of Anatomical Sciences and Neurobiology, University of Louisville, Ocular Transplantation LLC, Louisville, KY*

COLIN J. BARNSTABLE, DPhil • *Ophthalmology and Visual Science, Yale University School of Medicine, New Haven, CT*

JAMES K. CAPE • *Republic of South Africa*

BO CHANG, PhD • *The Jackson Laboratory, Bar Harbor, ME*

GORDON F. COUSINS, MD • *Republic of South Africa*

MURIEL T. DAVISSON, PhD • *The Jackson Laboratory, Bar Harbor, ME*

MALIA M. EDWARDS, PhD • *The Jackson Laboratory, Bar Harbor, ME*

G. JANE FARRAR, PhD • *Ocular Genetics Unit, Department of Genetics, Trinity College Dublin, Dublin, Republic of Ireland*

JACQUIE GREENBERG, PhD • *Division of Human Genetics, University of Cape Town, Cape Town, South Africa*

NORMAN L. HAWES • *The Jackson Laboratory, Bar Harbor, ME*

J. R. HECKENLIVELY, MD • *W. K. Kellogg Eye Center, The University of Michigan, Ann Arbor, MI*

PETER HUMPHRIES, PhD • *Ocular Genetics Unit, Department of Genetics, Trinity College Dublin, Dublin, Republic of Ireland*

ULRICH KELLNER, MD • *Eye Centre Siegburg, Siegburg, Germany*

FIONA KERNAN, PhD • *Ocular Genetics Unit, Department of Genetics, Trinity College Dublin, Dublin, Republic of Ireland*

HENRY KLASSEN, MD, PhD • *Children's Hospital of Orange County, University of California, Irvine, Orange, CA*

RONALD KLEIN, MD, MPH • *Ophthalmology and Visual Science, University of Wisconsin-Madison, Madison, WI*

ROBERT K. KOENEKOOP, MD, PhD • *McGill Ocular Genetics Laboratory, Montreal Children's Hospital, Montreal, Quebec, Canada*

ALEJANDRO J. LAVAQUE, MD • *New England Retina Research and Education, Spring Glen Medical Center, Hamden, CT*

PETER E. LIGGETT, MD • *New England Retina Research and Education, Spring Glen Medical Center, Hamden, CT*

RAYMOND D. LUND, PhD • *Casey Eye Institute, Oregon Health Sciences University, Portland, OR*

DENNIS M. MADDOX, PhD • *The Jackson Laboratory, Bar Harbor, ME*

GERALD MCGWIN, JR., MS, PhD • *Department of Ophthalmology, School of Medicine, University of Alabama at Birmingham, Birmingham, AL*

ALEX G. MCKEE, PhD • *Ocular Genetics Unit, Department of Genetics, Trinity College Dublin, Dublin, Republic of Ireland*

PAUL MITCHELL, MD, PhD • *Centre for Vision Research, Department of Ophthalmology, Westmead Millennium Institute, University of Sydney, Westmead Hospital, Westmead, N.S.W., Australia*

JÜRGEN K. NAGGERT, PhD • *The Jackson Laboratory, Bar Harbor, ME*

KRISTINA NARFSTRÖM, DVM, PhD • *University of Missouri-Columbia, Columbia, MO*

PATSY M. NISHINA, PhD • *The Jackson Laboratory, Bar Harbor, ME*

CYNTHIA OWSLEY, MSPH, PhD • *Department of Ophthalmology, School of Medicine, University of Alabama at Birmingham, Birmingham, AL*

YOU-WEI PENG, PhD • *Department of Ophthalmology, Duke University Medical Center, Durham, NC*

NORMAN D. RADTKE, MD • *Retina Vitreous Resource Center, Norton Audubon Hospital, Department of Ophthalmology, University of Louisville, Louisville, KY*

RAJKUMAR RAMESAR, PhD • *Division of Human Genetics, University of Cape Town, Cape Town, Republic of South Africa*

GEORGE REBELLO, PhD • *Division of Human Genetics, University of Cape Town, Cape Town, Republic of South Africa*

MATHIAS SEELIGER, MD, PhD • *Department of Pathophysiology of Vision and Neuroophthalmology, University Eye Hospital, Tubingen, Germany*

MAGDALENE J. SEILER, PhD • *Ophthalmology, Keck School of Medicine, University of Southern California, Los Angeles, CA*

XINHUA SHU, PhD • *MRC Human Genetics Unit, Western General Hospital, Edinburgh, UK*

R. THEODORE SMITH, MD, PhD • *Edward S. Harkness Eye Institute, Columbia University, New York, NY*

SYLVIA B. SMITH, PhD • *Medical College of Georgia, Department of Cellular Biology and Anatomy, Augusta, GA*

JANET R. SPARROW, PhD • *Department of Ophthalmology, Columbia University, New York, NY*

MAKOTO TAMAI, MD, PhD • *Department of Ophthalmology, Tohoku University, School of Medicine, Sendai, Japan*

WENG TAO, MD, PhD • *Neurotech USA, Lincoln, RI*

JOYCE TOMBRAN-TINK, PhD • *Ophthalmology and Visual Science, Yale University School of Medicine, New Haven, CT*

GREGORY E. TULLIS, PhD • *Molecular Microbiology and Immunology, Life Sciences Center, University of Missouri-Columbia, Columbia, MO*

JIE JIN WANG, MMED, PhD • *Centre for Vision Research, Department of Ophthalmology, Westmead Millennium Institute, University of Sydney, Westmead Hospital, Westmead, N.S.W., Australia*

SHAOMEI WANG, PhD • *Casey Eye Institute, Oregon Health Sciences University, Portland, OR*

BERNHARD H. F. WEBER, PhD • *University of Regensburg, Institute of Human Genetics, Regensburg, Germany*

RONG WEN, MD, PhD • *Department of Ophthalmology, University of Pennsylvania, School of Medicine, Philadelphia, PA*

DAVID S. WILLIAMS, PhD • *Departments of Pharmacology and Neurosciences, University of California, San Diego School of Medicine, La Jolla, CA*

JUNGYEON WON, PhD • *The Jackson Laboratory, Bar Harbor, ME*

FULTON WONG, PhD • *Department of Ophthalmology, Duke University Medical Center, Durham, NC*

ALAN F. WRIGHT, PhD • *MRC Human Genetics Unit, Western General Hospital, Edinburgh, Scotland, UK*

JOANNE YAU • *The Schepens Eye Research Institute, Department of Ophthalmology, Harvard Medical School, Boston, MA*

MICHAEL J. YOUNG, PhD • *The Schepens Eye Research Institute, Department of Ophthalmology, Harvard Medical School, Boston, MA*

TASNEEM ZAHIR, PhD • *The Schepens Eye Research Institute, Department of Ophthalmology, Harvard Medical School, Boston, MA*

Companion CD ROM

All illustrations, both black and white and color, are contained on the accomapnying CD ROM.

I
LIVING WITH RETINAL DEGENERATION

1

Coping With Retinal Degeneration
A Patient's View

Gordon Cousins and James Cape

CONTENTS

INTRODUCTION

We have been asked to write a chapter on coping with retinal degeneration (RD) from the point of view of the patient. James Cape, my co-author, has macular degeneration (MD) and I have retinitis pigmentosa ([RP]; dominant inheritance RP13). Because our two different retinal conditions create different experiences and viewpoints, we decided to use a common structure, but to write it from our own perspective. Our viewpoints are different because a picture always looks different if you can only see the middle bits (RP) or only the part around the outside (MD). However, as James and I often jokingly say, together we believe we have 20/20 vision. In this chapter, we will attempt to provide an all-round 20/20 view of the problems of coping with RD. I list here the broad subjects that we have tried to cover. I think you will find that we have not necessarily covered them in sequence, and that both of us sees these subjects from a different view. However, we hope that you will see a common thread running through the chapter.

CHAPTER STRUCTURE

After discussing the most important symptoms of our conditions and how we were first diagnosed, we attempt to deal with the difficult problem of the impact of continuous, slow vision loss. Thereafter, we reveal a number of situations we experienced as young people: with doctors, parents, children, siblings, and our partners. We also divulge our experiences at school, university, and in our careers. We discuss our challenges while participating in sports, social occasions, and travel. Finally, we highlight

From: *Ophthalmology Research: Retinal Degenerations: Biology, Diagnostics, and Therapeutics*
Edited by: J. Tombran-Tink and C. J. Barnstable © Humana Press Inc., Totowa, NJ

how the situation today is different, especially because of the efforts of Retina International and all of our member countries in raising awareness of our condition. Furthermore, given the tremendous advances in technology to aid partially sighted people, and legislative and environmental improvements, we conclude with a view toward the future. We hope you enjoy our contribution.

COPING WITH RETINAL DEGENERATION: *JAMES CAPE*

Born in 1951, as the youngest of three children, I enjoyed a normal and happy childhood. In 1958, my sister, the eldest of us siblings who was 12 at the time, was diagnosed with MD. As little was known at the time, my brother—13 months older than me—and I were duly examined but were found to have normal vision.

During the ensuing years, I adapted to life with a visually impaired family member, but found it strange that there did not appear to be any magical cure. Throughout my teens and early adult years, I participated in all levels of ball sports, representing my region at the highest level.

At the age of 27 and at the pinnacle of my sporting career and duly making progress in my career as a bank official, I was called up for a military camp as part of the country's national defence force. Subject to a full medical examination, the military optometrist could not fully understand why I was having problems reading the normal optometrist chart. Consequently, I was simply overlooked on the basis that I was endeavouring to be exempt. Knowing my sister's condition, I immediately identified my shortcoming and the feeling of dismay remained with me for the duration of my 3-mo call up.

Immediately on my dismissal, I visited my ophthalmologist for a full assessment. After a full diagnosis, he referred me to a colleague for a second opinion. Both concluded that I had MD and were somewhat bewildered by the fact that I had been examined on an annual basis and that the condition had only reared its head at this later stage of my life. At the time, it had only been diagnosed in patients under the age of 25 and after the age of 55 this is known as age-related MD (AMD).

As a young adult enjoying the fruits of life with ambitions and aspirations, the news came as a devastating blow. My immediate reaction was one of anger and bewilderment. I had many questions that needed to be answered by the medical fraternity, my parents, and my faith. This forced me into a state of denial and I almost became an introvert, not sharing or prepared to divulge my inner feelings and hurt. Part of the thought process I experienced was perhaps that there was no light at the end of this tunnel and, therefore, giving up the aspects of living seemed to be the only alternative.

Because my parents had experienced my sister's condition, they found themselves in an awkward position, as they still did not have what seemed to be the right answers or solutions. Dealing with the emotions and the reality of the situation is not easy to describe as my loss of sight was gradual at the time. But what it did do was start to place an emphasis on the day-to-day aspects with which I was having difficulty. I noticed that I was perhaps not as competitive as I started to focus on the negative aspects of my sporting activities rather than the positive.

At the workplace (I wore spectacles being far sighted in one eye and near sighted in the other for clerical work), I had experienced problems with certain printed material and had merely put it down to an adjustment required to the lenses of my spectacles.

At the time, this did not bother me, but, rather, the daunting thoughts of the future did, based on what my sister had experienced. Slowly, I started to restrict my sporting activities and denied my problems at the workplace. It was during this time that I perhaps received the best advice from my local physician and one that I continue to share with others: Do not give up on your eyes, but let your eyes give you up. What he meant by this was simply, do not give up your day-to-day activities, sports, and work, but rather let them give you up and let logic prevail. With this in mind and a fresh outlook, I approached my day-to-day activities in a more optimistic manner, focusing on the positive aspects rather than the negative. I continued with my sporting activities, taking the opportunity to use my experience and skills rather than relying totally on my sight. In the working environment, it was merely a matter of adjusting to conditions and looking at a smarter approach rather than the norm. I continued to enjoy and play sports and have only recently given up playing indoor hockey. The first of the sports that I retired from was softball and baseball, which I found became more and more difficult. Because the ball spends much of its time airborne and blends in with the background, I was unable to see it and position myself accordingly. Furthermore, the expectations of your teammates continue to be in line with one's performance and they could not understand or appreciate why my performance had declined. It is worth mentioning that I normally batted at third or fourth in the line up and was considered to be one of the leading run scorers. This was indeed a sad moment.

My other passion was that of squash, which I had played actively from my school days and thoroughly enjoyed playing men's league squash. With the deterioration of my sight, I found the need to adjust my style of play and particularly focused on keeping the ball at a low to middle projection, which seemed to fit in with my field of vision. The high drop shots caused me the most problem and more than often I would forfeit the point as I was unable to track the ball. At the same time, I was finding that I was being hit by the ball, especially when the ball came off the back wall. Those who know the game will appreciate how much this can hurt, leaving one with bruise marks. My friends typically took advantage of the situation despite the fact that they had empathy for me. I continued to play squash until the early 1990s when it became apparent that I was no longer competitive and perhaps was endangering myself.

I have continued to play field hockey and now have the pleasure of having my two sons on my team. I must mention that four or five of the players have played hockey with me since 1972 and they continuously prompt me throughout the game. This has often lead to confusion with those at the back prompting me to move left whereas those facing me are prompting me to move to the right. I have become known in the league as that deaf player as my teammates were in the early years congratulated for accommodating a deaf player because of the shouting/calling that took place. So you are probably wondering how (or do) I see the ball. With my instinct for the game and my ability to see movement and duly assisted by the prompting, I am able to follow the play of the game. It is only when play is within 1 m or so that I am able to "pick up" the ball provided the ball is in a good, clean condition. When the ball is traveling at a speed, I am unable to "pick it" up resulting in a couple of bumps and bruises. Any free hit taken close to the goal mouth, I am kindly asked to move out of the area.

The team always calls on me to take free hits as I have a natural ability to hit the ball extremely hard. As you can appreciate in field hockey, the ball can be lifted fairly easy depending on the condition of the field and this, combined with me not being able to distinguish between my team and the opposition, depending on the team colors, has unfortunately meant that I often hit players with the ball. My reputation in this regard has certainly exceeded me, resulting in players reluctant to stand too close to me when free hits are awarded. On the subject of the team colors, I have often been generous in passing the ball to the opposition, which often has my team in a tizz.

From the time that I was diagnosed, I immediately informed my employer of the situation and it was mutually agreed that we would address the situation on a year-to-year basis. I have now worked for the same company for 34 yr and have gone through many trials and tribulations.

Owing to the deterioration of my sight, I have tried many forms of glasses. At one stage, the one lens of my glasses was almost 1-cm thick and my colleagues used to comment jokingly that I should not sit with the sun behind me as it could cause the documents I was working on to catch on fire. It certainly does require an adjustment to the many comments from customers and alike regarding how closely I read and that perhaps I should change my optometrist. Naturally, this restricted me in my career development because of the limitation of such matters as transport. I also learned that if you silently teach people to adapt to your needs, they do so willingly and without realizing it. A good example of this is that when staff extract fax messages, they will automatically read the fax and not just merely place it in my in tray. They will ensure that work presented to me is formatted and printed in a manner in which they know I will be able to read. To assist me with computer work, I make use of a software package called ZOOMTEXT, which has a variety of features, such as magnification and text reading. I also make use of a micro reader, which facilitates the reading of documents and books. Despite the fact that I have been overlooked for certain positions, I was recently asked to visit certain African countries to consult on operational issues. This was certainly a challenge and not only from a consultancy aspect, but also the whole aspect of dealing with mobility and obtaining appropriate assistance from people within these countries.

Let me share with you some of my trials and tribulations. Shortly after my diagnosis, I realized that I had to read with my nose closer to the print. As mentioned, my first set of spectacles had one lens that was extremely thick and of course very heavy to the extent that I would have marks on my nose from the frames. As the spectacles were for reading only and the fact that I would have to take them on and off, I often misplaced the glasses resulting in such comments as: "you will have to get a pair of spectacles to find your spectacles."

Ensuring that the light conditions were favorable became a priority. Too much light meant a reflection off the documents, resulting in strain on my eyes; simultaneously, too little light resulted in difficulty in reading the material. It also became difficult to write, taking into account the closeness of my face to the writing material. To accommodate this, I naturally adjusted my head to a more suitable position with the added consequence of my handwriting deteriorating. My two sons often criticized my handwriting, stating that it was similar to that of a medical physician's and they would always suggest I scribe letters to the school.

At about this time, I started to experience problems when driving my motor vehicle, depending on a host of conditions such as the time of day, the condition of traffic signs, the condition of the façade, and the amount of traffic on the road. I found that I was not adhering to road rules, going through red lights and not stopping at stop signs. In order to compensate, I found it in my interest to travel routes that I was familiar with and would follow the vehicle in front of me on the basis that they were adhering to the road rules. If there were no vehicles, the cautionary action taken would be to turn left at a traffic light rather than proceed through. (Do not forget we travel on the left side of the road.)

My family and I were preparing to leave on a short vacation and as we left our abode, the motor vehicle stalled and we were unable to get the vehicle started. As it was our intention to call on our dentist on the way because my eldest son was experiencing toothache, my wife decided to proceed to the dentist using our second vehicle, leaving me to sort out the problems. Shortly after my wife departed, I was able to get the car started and, with my youngest son as a guide, made my way to the nearest service center (which was luckily up the road). The motor mechanic requested that I pull the motor vehicle into the garage in order to undertake an examination of the problem. Despite my appeal, the mechanic insisted I drive the vehicle, which I duly did under guidance from my son seated in the back seat. The workshop was a hive of activity and took some careful maneuvering to reach the allocated service bay. It did not take the mechanic all that long to attend to the problem. After which, he suggested we take the car for a test drive. Once again he insisted that I drive, so with the mechanic in the passenger seat and my son in the back seat, off we set. My son was quite clear on his instructions, which went something like—slow down you are about 5 m from the stop street, okay you should now stop, it looks clear you can proceed, you are coming towards a pedestrian crossing, etc. This prompted the mechanic to reply that it was most encouraging to hear a youngster so involved and knowledgeable on our local road ordinance. When we returned to the workshop, the only means of payment was by cheque, which my son duly completed on my behalf, requesting me to sign in the appropriate space. The owner of the workshop was somewhat bewildered and also commented on how nice it was to see a youngster being taught at such an early age.

After some minor accidents and having nearly written the motor car off, I decided it was time to stop driving because I realized that I was not only becoming a danger to myself but the consequences of my actions could be severe. This news I knew would delight my ophthalmologist, who on each visit cautioned me not to drive. The loss of one's independence certainly takes a huge amount of adjustment, especially in a country like ours where public transportation is almost nonexistent. Part of the adjustment entailed re-organizing getting to and from work, visiting customers, attending conferences, and even on the domestic front. At the time, it was difficult to engage the full impact, but this certainly placed an additional burden on my wife and friends. To this day, this remains one of my stumbling blocks of life, as it does not matter what arrangements I make, I always need to take into account the whole transportation matter. Strange but true, I have a good sense of direction and, once I have been on a particular route, I have the ability to memorize it. As a result, I am invariably requested to direct people from point A to B.

The nature of my job has required that I undertake trips to outlying areas, necessitating the use of air travel. I do not experience difficulty in getting around airports as one only needs to ask for directions in order to get to the various areas requiring departure, e.g., obtaining boarding passes, security, and access to airlines. I do experience problems when boarding the aircraft. The location of my designated seat is indeed a challenge and has often resulted in confusion with fellow passengers.

When arriving at new destinations, it is always important for me to familiarize myself with the lie of the land. What I mean by this is finding out where the dining room, toilets, lifts, and reception desk are located. This action also extends to hotel rooms—making sure that I fully appreciate all the small things. I identify which bottle has the shampoo vs hand cream. To make my life easier, the better I understand or familiarize myself with the environment, the easier it is to adjust. On a recent visit, I exited the hotel carrying a briefcase and, as the stairs leading to the pavement were not clearly marked, I ended up sprawled all over the pavement. Pedestrians passing by were to my amazement reluctant to assist me and I could sense from their remarks that they considered whether or not I was intoxicated.

Living with a disabled person I do believe takes quite an adjustment and does impact on all family members, from time to time. My two sons grew up with my condition from the onset and to them this was the way of life. When playing with their toy cars, the lady always did the driving and it was only when they went to school that they realized that there was in fact a difference. However, being the ages they were, this did not deter them and they played on. They would automatically do certain things differently, e.g., they would look for me after school or school events rather than me looking for them. Without hesitation, they would guide me up or down stairs and describe my immediate surroundings. When walking through the school grounds, they would politely advise me of any teacher that was approaching in order that I could formally greet them. Of course, it was difficult for me to help with school homework, particularly reading.

I would, from time to time, participate with my sons at certain scout outings like day trips, camps, and hiking. Without their help and assistance, these outings would not have been possible. During an overnight hike through some unforeseen circumstances, the boys got split into two groups and for all in tense and purposes were lost. I was in charge of the group and the hike itself was in a nature reserve, which stretched over many kilometres. It took me quite a while to overcome the feeling of absolute helplessness and gather my wits and attend to the immediate problem. The area was covered in tall grass, which compounded the problem. Through applying some good logic and some lady luck, we were able to regroup. As this was on the first day of the hike, it taught me just how observant and careful I had to be.

This reminds me of one incident when my wife was at work one Saturday morning. The boys asked me to read them a story and, not wanting to disappoint them, I chose a book that I was familiar with, not knowing it happened to be a favorite of theirs, which my wife had read on several occasions to them. With my sons sitting on either side of me, I pretended to read the book by telling them the story. It was not long before one of them notified me that I was deviating from the story as read by their mother.

I took the liberty of cleansing my sons inside by default. This came about when my wife went out and I was mowing the lawn and I was left instructions as to provide them with cool drinks and biscuits as and when the need arose. Youngsters being who they are, it was not long before they requested that I pour a cool drink. I went into the kitchen where the cool drink was kept and unfortunately on this occasion the dishwashing liquid stood next to it (both were green). I took the dishwashing liquid by mistake and poured it into their mugs, added water, and told them to drink up. Naturally the liquid created a burning sensation and, in order to resolve this quickly, I dosed them with milk. I asked them not to repeat this to their mom on her return, but this was to no avail. You can imagine the first words they spoke to their mother and in what detail!

My wife has adapted to the situation and has been a pillar of support throughout the years. She was fully aware of my eye condition at the time of our marriage, but I do not believe anybody can anticipate the impact and the personal sacrifices that are associated with a partner having this condition. With the loss of my sight being gradual, my wife subconsciously takes on more menial day-to-day tasks. I am unable, for example, to read the price of items when shopping. Although I can memorize where the items are in a shop, the condition and otherwise of the article requires assistance of a sighted person. When eating at a restaurant, menus remain a stumbling block and always require somebody in your company to assist you. It can be quite daunting to be told by restaurant staff that the bathroom "is over there," followed by pointing in the general direction. It does not end there: when you locate the bathroom, the signage is invariably small and difficult to decipher. In order to return to your table, one has to memorize the environment.

Attending to domestic accounts and such matters can also be difficult, especially without the assistance of your partner. I have been fortunate that my family has adapted so well, ensuring that small furniture items are not placed haphazardly in the room. Drinking glasses are stacked in manner that I am able to see them and take them out of the appropriate cupboard without any damage being caused. The watching of television is somewhat unusual as my family sits in their normal positions on the couch and I sit to the side of the television and in close proximity, i.e., almost on top of it. Sliding doors are clearly marked.

The family has become almost experts at describing the countryside when we travel on location. On arrival, I get a full description of the lie of the land, which certainly helps me to orientate myself and enjoy and share their enthusiasm. This is done invariably without having to prompt them and seems to come naturally from them. Typically, this type of behavior has been adopted by friends and family, who invariably are more than willing to assist without making me feel disabled.

Once a year, I visit my local ophthalmologist for a full eye examination. Despite the fact that there is no cure or treatment available, it is important to ensure that the rest of my eye is healthy. During the examination my vision acuity is measured. I am actively involved with Retina South Africa, which gives me the opportunity to keep abreast of the latest developments both on a local and international basis. This includes research treatment and technology enhancements. Public awareness remains an ongoing challenge and, in particular, in our country. This has a distinct disadvantage in that it is difficult to explain to people that you have sight but cannot see. The yellow cane concept, which was muted some years ago, (as a means of indicating that the individual is partially sighted as opposed to the white cane, which is interpreted that the person is blind), still carries merit and should be reconsidered. Because of the lack of awareness, employers are also unaware of the various conditions and are hesitant to employ people who are visually impaired and, in many cases, will terminate employment normally through unethical means. What has become important to meet the day-to-day challenges is the use of technology. Certainly, computers allow one to compete on a compatible basis. The use of ZOOMTEXT has given me the access to the computer environment and, with such features as text reading, I am able to keep up with the latest developments and stay actively in communication with friends and customers. I also make full use of a micro viewer, which allows me to enlarge print, thus giving me the ability to peruse documents and books. Although it is a slow means of reading, I can read. Other gadgets, such as talking watches and alarm clocks, make life simpler. I have tried to obtain speech text for my mobile phone, but this technology is new and expensive at the moment. As most aids for visually impaired people are imported, the costs are extremely high.

To maintain my standing at the workplace is very demanding. As there is no doubt that it does take longer than a normal-sighed individual to maintain the required levels of efficiency, I work extended hours and often take work home. This is in order to spend time with the family. When attending courses, my first approach is to liaise with the lecturer, explaining my condition so that he and I are not placed in an embarrassing position. Work colleagues normally prepare a set of enlarged print slides when inviting me to their presentation. I have experienced that, because I have taken the time to explain my condition to colleagues and peers, they are accommodating and understanding and as such have made my life that much easier. In conclusion, what has become important to me in the way forward is Stephen Covey's first habit in his book, "The 7 Habits of Highly Effective People," Covey, S. R. (2004). The 7 Habits of Highly Effective People. New York: Free Press. The gist of which is to concentrate on that which we can influence rather than that which we cannot.

COPING WITH RETINAL DEGENERATION: *GORDON COUSINS*

Introduction

My objective in writing this chapter is to attempt to help people who suffer from RD conditions and people associated in some ways with RD sufferers. I have attempted to cover the various aspects of life and to highlight my experiences, my learning, and my comments on coping in these different personal conditions. All of the situations referred to are extracts from my own experiences and, as such, represent a very personal view of how to cope with a retinal disorder.

With RP, as you will see from other chapters in this book, the nature of inheritance has a significant impact on the date of onset, the severity, and the rate of degeneration of the condition. I am a dominant inheritance RP, which tends to be the mildest form of RP, in that, it implies early onset but with reasonably slow degeneration, particularly visual acuity, resulting in usable central vision, often until the RP sufferer is quite old. I, at 57 years of age, have lost all of my peripheral vision, but still have about a 15-degree field of central vision. This implies that I am still able to continue my career as an international lecturer and trainer, running my own training business. Coping with RP has therefore, over the years, become not only a personal, but also a professional necessity and in order to understand how to cope with RP, it is first necessary to understand the major symptoms of RP that are experienced by an RP sufferer and how these symptoms play out in different situations.

With reference to the symptoms of RP, there are five particular symptoms that cause problems for an RP sufferer. They are tunnel vision, contrast sensitivity, night blindness, slow dark adaptation, and loss of visual acuity. It is important for the reader to understand how these symptoms are experienced and what they present to an RP sufferer. In all of the situations that are covered in the forthcoming pages, we are referring to these symptoms time and again. Therefore, it is worth spending some time on a brief discussion of each of these symptoms and how they impact on an RP sufferer.

1. Loss of peripheral vision (tunnel vision). In early years, this takes the form of a ring scotoma, which does not significantly impact on mobility or vision for a classic RP. This is especially so in my case, as I am a dominant RP and therefore had onset at an earlier age, but a much slower rate of degeneration. However, once the condition is more advanced and the peripheral vision is lost completely, then the impact of tunnel vision can be more severe. I have a 15-degree field, which presents certain mobility challenges in that the inability to see left, right, up, and down poses some difficulties. One of the biggest difficulties posed is the fact that I appear to have absolutely normal vision and yet, will often bump into people, fall over objects at floor level, etc. This often results in people assuming that I have been consuming alcohol in large quantities.

2. Contrast sensitivity. The major problem for a patient with RP caused by lack of contrast sensitivity is that vision becomes variable, depending on the location of the light source and the position of shadow. With low-contrast sensitivity, the shading and shadow in a sight situation becomes exceedingly important, as can be seen in the Figs. 1 and 2, which show a set of steps leading onto the beach in Zanzibar, taken from two different angles.
   As can be seen, the lack of shadows from one angle makes the stairs more difficult to see, if you are an RP. This presents some interesting challenges.

3. Night blindness. Night blindness or the inability to see in low light situations, is one of the first symptoms experienced by an RP and this tends to be one of the most debilitating aspects of the condition, as will be covered extensively in the discussion of various situations later on.

4. Dark adaptation. The problem for an RP sufferer is that the time required to adjust to differences in light intensity is much slower than for a person with normal vision. This presents certain problems when one changes from light to dark situations.

5. Visual acuity. In general, for most young patients with RP, visual acuity is not experienced as a significant problem. However, as one gets into later years and, depending on the condition itself, the loss of visual acuity has significant problems.

6. The final difficulty that an RP sufferer has to deal with is the fact that the visual loss is experienced over an extended period of time and, therefore, psychologically one has to deal with the process of loss as a continuous lifestyle. The implication of this is that one gets used to the thought of loss, but certain key activities or capabilities are lost at particular points in one's life and it is these focal points that bring up the whole psychological problem of loss of vision, again and again. For many people with RP, it is the fact that the sense of loss is experienced time and time again, which is a particularly challenging and difficult situation to deal with.

First Diagnosis

I was first diagnosed with RP at the age of 6. However, in many ways I was fortunate because, being from a dominant inheritance family, RP was already in the family and,

Fig. 1.

therefore, perhaps, there was some, if limited, realization of what the implications of the condition were. I first experienced night blindness as a very young boy, when playing out at night with a friend, I ran into a tree, which my friend saw, but I did not. A few similar experiences eventually resulted in my parents taking me to an ophthalmologist, who diagnosed RP. The first diagnosis of RP is a particular challenge for most parents and for young RP sufferers. Our advice would be not to underestimate the child's ability to understand the condition and to deal with the truth of the condition. When I went for my first eye examination, drops were placed in my eyes to dilate the pupils so that the retina could be examined. Unfortunately, nobody told me that this dilation % would cause an inability to focus and also nobody told me how long it would take for the dilation to disappear. I can still remember, vividly, returning home from the doctor's rooms, at 11 o'clock in the morning and for the rest of the day, being unable to read and unable to see clearly. It was an extremely traumatic experience, which could have easily been avoided, had the eye-care professional simply taken the time to explain to this 7-yr-old boy that he would find some blurredness in his vision for the rest of the day. My experience is that most professionals tend to err on the side of too little information to young patients with RP rather than too much. Being diagnosed with RP in the 1950s, there was very little information around on how the disease would progress, how the symptoms would be experienced, and what methods were available to cope with RP. This is a common experience of young patients with RP, particularly in the case of patients

Fig. 2.

with RP who are diagnosed from a recessive inheritance condition. In this situation, there is no experience of RP in the family and, therefore, no knowledge of how to cope with the condition. Given the extent of knowledge that exists on the condition today, it should be possible to assist a young patient with RP to cope much more adequately with the condition. However, it appears that because of the lack of awareness and lack of sensitivity amongst both parents and eye-care professionals, not enough work is being done to help the young patients with RP to understand that they will need to cope with contrast sensitivity, night blindness, loss of peripheral vision, slow dark adaptation, and eventual loss of visual acuity. How did I experience these situations as I grew up?

Youth

As a young person, one desperately wants not to be different from any of one's friends and, therefore, the usual behavior of a young person with RP is to pretend that there is nothing wrong with his or her vision. This inevitably results in the young person with RP not wanting to talk about it to his or her friends and not wanting to ask for assistance. The net result is that many young people with RP withdraw from social and personal contact with his or her peers, often initially in low light situations and then subsequently in sporting and other situations. The biggest gift that one can give to a young person with RP is to encourage him or her to recognize that he or she has a condition and to feel free to talk about the condition and the implications of the condition.

I, as a youth with RP, went through a similar process of trying to avoid contact, particularly in low light situations. The hit song, "You Will Always Find Him in the Kitchen at Parties" could well have been written for an RP sufferer because, at most parties, the lights were always on in the kitchen and were always off in the dance room. This created some significant difficulties in trying to find a girlfriend in a party situation, as the pretty girls were always in the dance room, waiting to be asked to dance!

The inability to share one's disability and the implications of the symptoms with one's friends are probably the most significant debilitating factors for a young person with RP If this can be overcome at an early age, it can make a significant difference to the life of a young person with RP.

Personal Situations

In this section of the chapter, I cover the problems of coping with RP with respect to one's parents, partners, siblings, friends, and the question of children. With regard to parents, the problem of a retinal degenerative condition is that it is inherited and, therefore, there is implied guilt on the parents for having given birth to a child suffering from a degenerative eye disease. This inevitably results in the parents wanting to do more for the visually impaired child than is either necessary or healthy and often results in parents increasing the sense of handicap of the child. In many cases, we have seen that parents (because of their deep desire to help) are quick to consider putting the child into special schooling facilities and other assistance programs, which I would not recommend. In fact, I do not feel they are necessary for the child. My recommendation is to allow the child to be as normal as he possibly can be and to not over compensate for the child's handicap. I was very fortunate in that I came from a family of RPs and my mother had her hands full dealing with my father and his RP. Therefore, she did not have the time to concern herself excessively with my condition and handicap. She also had the wisdom to understand that leaving me to cope with my condition was effectively teaching me coping skills and equipping me for the conditions I would experience in the rest of my life. In almost all of the retina associations around the world, the problem of trying to get parents to deal with their guilt for having given birth to a child with a degenerative eye condition and to get them to deal with their guilt within themselves and not over compensate in terms of the way they behave with the child is a common challenge.

When it comes to partners, someone with RP needs again to be as open, honest, and direct as possible, with regard to what condition they have and what it implies for their partner. In many situations, we have noticed the tendency for people with RP to not involve their partners in their condition and in the implications in the attempt not to "burden" their partner. This behavior generally results in continuing problems and, therefore, should be avoided. The problem with partners is that the question of children soon arises and, in this situation, we would recommend competent genetic counseling to ensure that both parties understand the risks that are involved in having children when one of the parties has a retinal degenerative condition. In my case, being of dominant inheritance, there was a 50/50 chance of the condition being passed down to my children and my daughter suffers from RP. This is a situation that still presents challenges to both my wife and me in terms of how to deal with a child with RP. The last subject in this section is that of siblings and friends. In the case of siblings, there is often a tendency on the part of RP sufferers to ignore the feelings of ones brothers and sisters, dismissing them with a view that they are, in fact, not suffering from the condition. In many cases, however, there is also guilt on the siblings part and an inability from them to know how far to go to assist their brother or sister with RP. The same situation extends to friends and it is very common for a person with RP not to know how to graciously accept the assistance that is offered by friends. One of the problems of a person with RP is that one's vision is variable, particularly because of the impact of contrast sensitivity, which results in the person with RP being able to see and not see in very similar light conditions, depending on the position of shadows and the extent of contrast. This makes it very difficult for sighted people, because in some situations, the RP sufferer is definitely in need of assistance and in other situations, which appear very similar, the RP is absolutely fine with regard to their mobility. It is this variability of vision that causes most of the interpersonal problems with friends and siblings. It is also this apparent variability that sometimes causes problems, even with the public. On many occasions, we have had complaints from the public that they have witnessed a person walk down a street using a cane, and then sit down at a bus stop, take out the bus time table, and read the time table. It is this incongruous situation that presents challenges both to the visually impaired person and to the public.

Doctors and Professionals

As has been referred to earlier, the biggest problem for a person with RP in coping with doctors and other professional people is the instant assumption on behalf of many professionals that the patients with RP can see nothing and, second, that these patients should not be burdened with too much technical information regarding their condition. My experience in Retina International is that the best way to assist patients with RP in coping with their condition is to give them as much information as possible and to assist them as far as possible to understand their condition. Many of our members have complained that the eye care professionals they consult do not fully comprehend the variability of the vision of the RP sufferer and are not able to give clear advice with regard to how to deal with the condition. This, of course, is quite natural given the wide number of inherited forms of RP that exist, and the variability of visual loss and disease progression that exists. My advice to the professionals, however, would be to give your patient as much information as possible, almost irrespective of their age, as patients with RP usually have become quite adept at dealing with the difficulties that their vision presents to them. Providing them with insights as to how the condition may develop, in fact, empowers them to deal with the condition more effectively.

Study

When it comes to education, school, and university, the RP symptoms of contrast sensitivity and dark adaptation often present some significant difficulties. This especially occurs with regard to seeing information on a chalkboard or overhead projector, and it is very common for young people with RP, in an attempt to disguise their condition, not to disclose that they have a problem in the accessing of this information. It is highly recommended that parents of a child with RP spend some time with their child's teachers and lecturers to explain the implications of lack of contrast sensitivity, slow dark adaptation, and, in some cases, the loss of visual acuity. Also, take the time to explain how this will impact on the child in a classroom situation. Probably the most important thing that one could do for a child with RP in the study situation would be to attempt to provide information to the rest of the students and to the teachers on the symptoms of the condition: how they will be experienced by the child with RP, how people withoutRP will experience an RP child, and what can be done to assist in that situation. The problem about this is that it first requires permission from your child to communicate to their peers that they are an RP sufferer. Frequently, in attempting to assist your child to cope with RP, you can create other difficulties and embarrassments, which present more of a problem to your child than the RP does. My advice always is to consult with your child and to not make the condition overtly public, unless your child is comfortable with that strategy. However, I would encourage you to continuously encourage your child to "come out of the closet" and declare his or her visual handicap. The most significant thing that one can do for an RP is to inform people around them as to how the condition will manifest.

Work and Career

The problem with RP is that the rate of degeneration is not known and cannot be accurately predicted for any particular form of RP. Therefore, the extent to which vision loss is going to impact on one's career cannot be predicted. One comment that I would have, however, is to recognize that in this day and age, one need not have a career for life and, therefore, if your child is intent on following a particular career direction, and they feel passionate about it, then encourage them to follow their heart. I have been self-employed for the last 27 yr and have benefited from being able to control my work environment to suit myself. James, however, has been employed in corporate positions throughout his life and has also managed to structure his work environment to suit himself. The most important thing that I do today, being an international consultant and lecturer, is to openly disclose my RP and the implications that this will have on delegates to my seminars. This public announcement of my condition has significantly assisted me to continue to perform my professional activities without significant restrictions caused by my RP. The same message pervades throughout all of what we are saying. The best way of coping with the condition is to understand the symptoms oneself and to ensure that everybody around understands those symptoms as well.

Sport

Sport is probably the one area where most people with RP experience the most significant challenges and have their most humorous experiences with the condition. It seems that people with RP continuously wish to ignore their condition in their choice of sport and many become involved in sport that requires a fairly high degree of visual acuity to be successful. There are very few sports that involve a ball, which do not require a certain amount of visual acuity and yet people with RP seem to disregard this fact totally in their choice of this sport. This "stubbornness" is highly commendable and should be encouraged. It is in their sport that people with RP have the chance to express their desire to be able to behave like fully sighted people. With age and maturity comes a small amount of wisdom. In my older years, I have learned that certain sports are, in fact, much easier to undertake as a visually impaired person, but I still insist on challenging myself with sport that require some visual skills, including golf, snow ski-ing, some cross country running, and similar activities. Even in these situations, how-ever, coping skills can assist significantly. I have not yet gotten involved in snow skiing with radio-assisted support, but I typically ski following my wife or a friend and, there-fore, mostly manage to make it from the top of the slope to the bottom unscathed. My challenge, however, is to recognize exactly when to turn on the piste and I have often gone slightly wider than would be recommended, therefore, disappearing into some soft powder snow. It is these experiences that remind an RP suffers of their condition and yet allows them to enjoy the fact that they are able to undertake sports, which, gen-erally speaking, perhaps would not be recommended for them, without allowing their handicap to restrict them in any way. My biggest challenge in scuba diving has been to keep my buddy in some degree of visual contact and, therefore, ensure that I am fol-lowing the correct group. When diving off the Barrier Reef, where there were a large number of other divers, most of whom were wearing black wetsuits with yellow tanks on their backs, often caused me to follow the wrong group in totally the wrong direc-tion. In that situation, my son, who does not have RP, had his work cut out to ensure that he kept me roughly in line with where the dive was intended to go—a proverbial underwater sheepdog! You can imagine the look of surprise on another diver's face when someone he does not know, suddenly attaches himself to him as a dive buddy. One benefit is that if you can explain it away, you do get to meet new people, but this is a little difficult to do 20 m below the surface. Still, my recommendation would always be to any visually impaired person, not to allow your visual impairment to restrict you from doing anything that you think you can do. Only ensure that you inform everyone around you of your condition and what the implications are for both you and them, so that you can try and ensure your own safety, as well as the safety of the other people with whom you are enjoying your pastime. On climbing Kilimanjaro last year, I expe-rienced the challenge of having to set out for the summit at 11 PM at night and climb 7 h in the dark. Although this was a particularly challenging experience, the exhilaration of having completed the climb and reaching the summit of the highest peak in Africa was worth all of the difficulties en route. However, my Tanzanian mountain guide, who struggled to get me across the lava rocks without injury, might not agree!

Sports and recreation are probably the most important way in which a visually impaired person can experience their own handicap and their own ability to conquer it. Use the people and support around you wisely, and anything is possible.

Travel and Transport

Travel and transport probably represents the most significant challenge to a visually impaired person. It is the area in which we are often reliant on other people and, fre-quently, these people are not people that are friends of, or connected with us. We are reliant on the general public and mostly in conditions in which we do not have the time or opportunity to explain our condition. In areas where there are no efficient forms of public transport, the RP sufferer is particularly handicapped and this is probably the most difficult handicap to deal with. In these situations, most people with RP often have to employ drivers or make extensive use of taxis, which have significant cost implications. Even using public transportation, such as trains and planes, have significant challenges to someone with RP. My biggest concern in using trains is if one is traveling at night and needs to leave the train at a particular station, then the ability to see the station name becomes a core competence with which I have some difficulty. In this situation, one can only resort to requesting the help of your fellow travellers. Also in situations of travel, railway stations and airports present particular challenges to someone with RP. Someone with RP may have a small visual field (in my case about 15 degrees) and have good central vision for objects that are further away, yet is blinded to objects that are very near or outside the field of vision. As a result, the person with RP tends to be quite "fixed" on a visual point somewhere in the distance and will often move quite speedily to that point, ignoring the fact that there are people and objects to the left and to the right of the visual field that may well come into play. I have been known to trip more people in airports and railway stations than is probably good for them or myself, and yet I travel extensively throughout the world, frequently on my own, and am able to cope quite effectively. The biggest coping problem in these situations is often the remarks of people that one encounters (or trips) along the way. To these people, the RP sufferer appears to be a perfectly sighted person, who appears to act extremely rudely in that he will often crowd this person or bump this person without realizing that it is happening. This often results in some disparaging remarks from strangers. The important coping skill for the RP sufferer, and for his or her partner if they are traveling together, is to not allow these remarks be upsetting, but simply to recognize that this is part of the cost of suffering from a retinal degenerative condition.

What tips can I offer for coping in a transport situation? Once again, my recommendation on key coping skills is, at every opportunity, to inform the person of the implications that the symptoms of the condition will cause. I remember a most enjoyable plane trip where, quite early on in the flight, I managed to explain to the person sitting next to me the implications of tunnel vision, lack of contrast sensitivity, and night blindness. The passenger next to me then took on the task during the rest of the flight of informing me as to when a crewmember was wishing to attract my attention to pass me some food, hand me some travel documents, etc. Again, it is bringing people into our world that allows us to cope with the conditions in our world. I am fortunate that, these days, I travel mostly with my wife and, therefore, do not have the same challenges with regard to travel and transport that I used to have.

Social: Including Dinners, Parties, Dating, etc.

Some comments earlier, with regard to youth, have already referred to the problems that are experienced at parties, or when dating, and of course, those comments are not restricted to young people with RP, but are relevant to sufferers of any age. The biggest difficulty in this situation is when one is exposed to dinner guests, who are not close friends and where one has not necessarily had the opportunity or the time to enlighten the people as to one's visual condition. This is particularly so in the case of business dinner engagements where, on meeting a customer for the first time, it is not necessarily appropriate to talk of one's visual condition immediately. The risk of not doing so, however, is that one may end up creating some difficulties, such as knocking one's drink over into your business associate's lap and then subsequently trying to explain that this was because of a visual condition. This does not usually help to ease the embarrassment. For most young people with RP, however, they inevitably tend to err on the side of not informing people of their condition, until it is too late. Finding ways to inform people at a very early opportunity of the implications of having RP is the best way of coping with the resulting difficult situations. Failing that opportunity, you need to be creative with finding ways to cope. I remember as a young boy I used to go to the movies with my father who also had RP. I always wanted to leave my seat during intermission to buy popcorn, but the challenge was how to find my seat in the dark. My father and I worked out a solution. He would sit on the aisle seat with his arm extended out into the aisle. On my way back, I would simply walk next to the seats until I bumped into his arm, and found my seat. The sense of relief and accomplishment at finding my seat was tremendous! However, I never stopped to think of the verbal abuse that my father probably endured for his "inconsiderate" behavior. I guess one of the key coping mechanisms for someone with RP is to develop a "thick skin."

Changes Today

The situation today creates new challenges and also offers new opportunities for people in coping with a retinal degenerative condition. First, technology now provides significant assistance for visually impaired people and there is significantly more public awareness in many parts of the world of the retinal degenerative conditions and how they affect the people who suffer from them. This public awareness, together with the activities of Retinal Associations and other support structures around the world, are making it easier for suffers to talk openly about their condition, in a way that helps people to understand the situation. In many parts of the world, governments are being sensitized to the physical needs of visually impaired people and are adapting infrastructures to make it easier for us all to cope. However, the most significant progress that needs to be made in dealing with the condition is the acceptance of the condition inside one's own head and, therefore, to have the maturity and courage to admit and talk openly to suffering from a retinal degenerative condition.

Conclusion

In conclusion, what can I say about coping with retinal degenerative conditions? I think that the best tips, with regard to suffering from RP, all center around "I." The first advice on "I" is to stop thinking that the world revolves around "I," that is, me. Many people with RP and MD become totally preoccupied with their condition, their struggle to accept their condition, and their struggle to live with their condition. They lose sight of the big picture.

The second "I" relates to information and informing. One should ensure that one acquires as much information as possible about one's condition, including the amazing developments that are happening on the scientific front and the progressive movement towards treatments occurring across the world. It is then important to inform everybody that one comes into contact with about the condition, the symptoms, how they affect a sufferer, how they will impact on other people, and how other people can be of assistance to someone with the condition.

The next "I" refers to "involve." I recommend that RP sufferers involve themselves in the scientific and other activities of the Retinal Association in their country. Involve the people around you: friends, family, and business associates in your condition and in ways to assist you in coping with your condition.

The next "I" involves the word "initiate." With a retinal degenerative condition, you should continually ensure that you are initiating situations that test your ability to cope with the condition. Exposing yourself to new challenges and finding new ways of coping with those challenges helps you cope with the retinal condition in general and assists others in realizing how they can help you with the condition.

The final "I" refers to the word "inhibit." Do not, in any way, inhibit yourself from taking on any challenges, experiences, and new situations because of your retinal condition. The more you are able to treat this condition as simply a fact of life that you need to deal with rather than a restrictive condition that will stop you from reaching your goals and dreams, the better you are able to cope with the condition. Coping is all in the head!

II

DEGENERATIVE DISEASES OF THE RETINA

2

Epidemiology of Age-Related Macular Degeneration Early in the 21st Century

Jie Jin Wang, MMed, PhD, Paul Mitchell MD, PhD, and Ronald Klein MD, MPh

CONTENTS

INTRODUCTION
CURRENT "GOLD STANDARD" FOR IDENTIFYING AMD PHENOTYPES
PREVALENCE OF AMD
INCIDENCE AND PROGRESSION OF AMD
RISK FACTORS ASSOCIATED WITH AMD PREVALENCE OR INCIDENCE
IMPACT OF AMD
SUMMARY
REFERENCES

INTRODUCTION

Age-related macular degeneration (AMD) is the leading cause of irreversible blindness and moderate visual impairment in older, white persons *(1–11)* and will remain a major threat to vision in coming decades *(12,13)*. AMD research has progressed substantially in the last quarter of the 20th century and provides clues for future research directions *(12,14–27)*. Although the exact etiology of AMD remains uncertain *(12)*, we now have a considerably better understanding of this condition than 30 yr ago. The purpose of this chapter is to summarize progress on understanding the epidemiology of AMD in the last two to three decades.

CURRENT "GOLD-STANDARD" FOR IDENTIFYING AMD PHENOTYPES

The establishment of a standard retinal photographic grading method and the development of the Wisconsin Age-Related Maculopathy (ARM) Grading System *(28)*, followed by description of the International Classification and Grading System for ARM and AMD *(29)* was an important milestone in the epidemiological study of AMD. Photographic documentation of AMD lesions permits validation and thus is a highly reliable diagnostic method with a high level of clinical accuracy in identifying the AMD phenotype *(17)*. The almost uniform employment of The Wisconsin Grading System (with or without

From: *Ophthalmology Research: Retinal Degenerations: Biology, Diagnostics, and Therapeutics*
Edited by: J. Tombran-Tink and C. J. Barnstable © Humana Press Inc., Totowa, NJ

modifications) to grade retinal photographs, or the International Classification System *(29)* by investigators of the most recent large population-based studies *(30–40)* has not only enhanced the comparability of findings across studies, but also has permitted data pooling *(41–43)* and meta-analysis *(44)*. As a result, data obtained from such recent population-based studies have been extrapolated to other similar populations *(44)*.

Various AMD severity scales based on the Wisconsin AMD Grading and Classification System *(28)* have been proposed and used in the Beaver Dam Eye Study *(45,46)*, the Age-Related Eye Disease Study (AREDS) system *(47,48)* and the Rotterdam Study team *(49,50)*.

With the advance of imaging technology, the replacement of stereoscopic retinal photography by digital imaging is inevitable and already now underway. The detection of late AMD lesions (geographic atrophy and apparent neovascular lesions) from either stereoscopic or nonstereoscopic digital images has been found to be reasonably comparable to detection using stereoscopic slides *(51,52)*. Agreement on the detection of early AMD lesions, drusen, or pigmentary changes (hyper- and hypopigmentation), however, has been found to vary substantially *(51,52)* depending on grader experience and lesion appearance. The detection of retinal hyperpigmentation associated with AMD from digital images has been found to have the lowest agreement compared with assessment from stereoscopic photographic film-based grading *(52)*. This would be expected as stereoscopic photography provides multiple fields (enabling viewing from different angles) that can enhance the detection sensitivity for this lesion.

PREVALENCE OF AMD

Whites

Prior to the mid 1980s, a number of studies were conducted to assess the prevalence of AMD, most with relatively small sample sizes *(17,53–57)*. As a result of the wide variations in the methods, criteria, and different observers used in ascertaining AMD phenotype by these studies, comparison of findings between these early studies is limited *(17)*. Since the mid 1980s, a growing number of large population-based studies have been conducted, including the Baltimore Eye Survey *(30)*, the Beaver Dam Eye Study (BDES) *(31)*, and the Salisbury Eye Evaluation (SEE) project *(58,59)* in the United States, the Rotterdam Study (RS) *(33)* in the Netherlands, and the Blue Mountains Eye Study (BMES) *(34)* and Visual Impairment Project (VIP) *(36)* in Australia. All of these studies uniformly employed the "gold standard" of retinal photographic documentation with subsequent masked grading to ascertain AMD phenotypes and have provided robust estimates of the prevalence of AMD for white populations aged 40 yr or older. The meta-analysis conducted by the Eye Disease Prevalence Research Group *(44)* showed relatively high consistency in the age-specific prevalence of late AMD (Fig. 1) and large soft drusen (≥125 μm in diameter) (Fig. 2) for whites across these different populations *(44)*. In all of these studies, the prevalence of late AMD (geographic atrophy and neovascular AMD) increased exponentially with increasing age, rising from approximately less than 0.5% at age 60 yr to around 10% at more than age 80 yr *(44)* (Fig. 1). The prevalence of large drusen shows a similar age-related increasing trend but the rising curve associated with increasing age is more linear than exponential *(44)* (Fig. 2). Possible misclassification

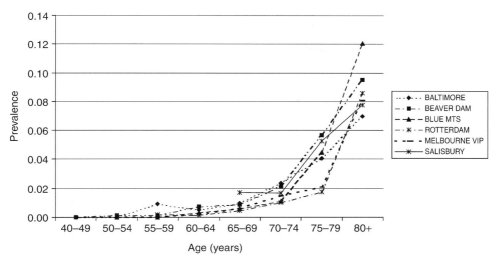

Fig. 1. Prevalence of late AMD by age in whites (reproduced from the Eye Disease Prevalence Research Group report *[44]*).

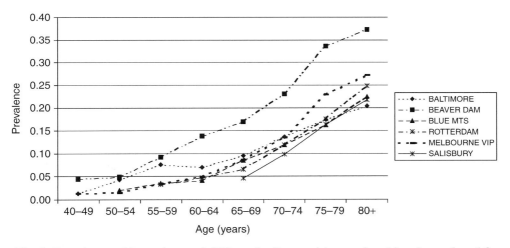

Fig. 2. Prevalence of large drusen (≥125 μm in diameter) by age in whites (reproduced from the Eye Disease Prevalence Research Group report *[44]*).

in the detection of drusen, together with the age-related natural course of AMD (involving the disappearance of large drusen associated with the progression to a more advanced AMD stage with aging *(60)*, could explain the observed age-related linear pattern, rather than an exponential pattern, in the prevalence of large drusen.

Recent reports of the third National Health and Nutrition Examination Survey (NHANES III) *(61,62)*, the Atherosclerosis Risk in Communities (ARIC) study *(35)*, the Cardiovascular Health Study (CHS) *(38)*, and the Reykjavik Eye Study *(39)* have provided further AMD prevalence data from different populations. The NHANES III

(61,62) (conducted in a multiracial US population aged >40 yr), the ARIC study *(35)* (conducted in persons aged 45–64 yr), and the CHS *(38)* (conducted in persons aged 69–97 yr) took photographs from only one eye of each subject. This could explain, in part, the lower overall late AMD prevalence in non-Hispanic whites (0.5%) *(62)* found in the NHANES III compared to the BDES (1.6%) *(31)*, BMES (1.9%) *(34)*, and RS (1.7%) *(33)* studies in which two eyes were photographed then graded for AMD and defined from the worse eye. The same reasons may apply for the finding of lower AMD prevalence in the ARIC study *(35)* compared to studies in which both eyes were photographed and graded *(31,33,34)*. On the other hand, the CHS found similar age- and gender-specific AMD prevalence in white participants compared to these three aforementioned studies (BDES, RS, and BMES) *(31,33,34)*. The relatively older age range of the CHS population may partly explain the similarity in findings between the CHS and the earlier studies, as AMD is a bilateral condition *(63,64)* and bilateral involvement increases with age *(64)*. The Reykjavik Eye Study *(39)* found a higher prevalence of geographic atrophy in Iceland than that found in other white populations *(31,34,44)*, including another European population (the Netherlands) *(33)*. A genetic influence for all these geographic atrophy cases has been suggested *(39)*.

Other Ethnicities

Ethnic differences in AMD prevalence have been shown in the NHANES III *(61,62)* and the Colorado-Wisconsin Study of ARM *(65)*. Non-Hispanic blacks were found to have a lower AMD prevalence than non-Hispanic whites. The Baltimore Eye Survey *(30)*, the SEE project *(58,59)*, and particularly the Barbados Eye Study *(32)*, which examined a more racially pure sample, have provided AMD prevalence data for black populations. Consistent findings suggest that the prevalence of AMD is lower in blacks than in whites *(30,32,35,38,44,59,61,62,66)*, although possible selection bias may have occurred in black participants because of the low participation rate *(35)*, a high proportion with ungradable retinal photographs *(30,35,38)*, or poor survival in the eldest old group *(30)*. However, the possibility that protection against AMD, particularly against late-stage AMD, from increased ocular pigmentation or gene variation in black populations is also a likely explanation of such differences, warranting further investigation *(30)*.

Mexican Americans (Hispanic whites) were also found to have a relatively lower prevalence of late but not early AMD than non-Hispanic whites in previous studies *(10,61,62,65)*, now confirmed by the recently completed Los Angeles Latino Eye Study (LALES) *(40)* with a much larger population-based sample. Survival bias could partly account for this finding in Hispanic populations, as they have much higher proportions of younger (aged <60 yr) than older (aged >70 yr) participants, seen in both the Proyecto Vision Evaluation and Research (VER) *(10)* and the LALES *(40)*, compared to other white populations such as the BMES *(7)*. The observed ethnic differences in AMD prevalence rates *(30,32,61,62)* could also be related to differences in the prevalence of many other risk factors, including genetic influences.

Relatively well-conducted population-based studies of AMD prevalence in Asian countries are currently emerging *(37)*.

INCIDENCE AND PROGRESSION OF AMD

Whites

Incidence

Prior to 1997, when the BDES study team published the first population-based finding on the 5-yr cumulative incidence of AMD *(67)*, there were no population-based data available. The Chesapeake Bay Waterman Study *(68)* conducted in a small sample of watermen aged 30 or more years provided some information on the incidence of early AMD lesions. Late AMD incidence was previously estimated either using prevalence data *(69)* or blindness registry data *(3,4)*.

After the turn of this century, a number of population-based studies, following the BDES *(67)*, have provided AMD incidence in older white populations *(50,60,70–72)* or from Medicare claims databases *(73)*. The cumulative 5-yr, person-specific incidence of early and late AMD is very similar between the BDES (8.2 and 0.9%) *(67)*, the RS (7.9 and 0.9%) *(50)*, and the BMES (8.7 and 1.1%) *(60)*, all of which used similar study protocols in AMD phenotype ascertainment *(41)*, although the baseline age range is slightly younger in the BDES (43–86 yr) *(31)* and slightly older in the RS (55–98 yr) *(74)* than the BMES population (49–97 yr) *(41)*. The VIP *(71)* reported slightly lower person-specific incident rates of early (5.4%, using the BMES definition) and late AMD (0.5%) over a 5-yr period, partially reflecting the younger age group included in this study population (aged 40–102 yr) *(36)*. In the Reykjavik Eye Study population *(72)*, the 5-yr person-specific incidence of geographic atrophy was 0.85% (7 out of 846), which was higher than the incidence reported by the RS (0.4%) *(50)*, but similar to the incidence found in the BMES (0.8%) *(60)*. An overall high 5-yr incidence of early AMD (14.8%) found in this Icelandic population could perhaps have been because of a different definition used to classify early AMD (intermediate drusen were included in the early AMD category) *(39,72)*, but the low reported incidence of neovascular AMD (0%) *(72)* is difficult to interpret, apart from an effect of the relatively small sample size.

Similar to AMD prevalence, the incidence of this condition is strongly age-related *(50,60,67,71)*, showing an exponential curve with increasing age (*see* Fig. 2 of the RS report *[50]*).

A substantially higher incidence of neovascular AMD (varying from 12 to 26% over 5 yr) has been observed in the second eye of clinic patients with unilateral neovascular AMD *(75–82)*. In a population-based case sample (BMES) *(60)*, a similar incidence of second eye unilateral late-AMD cases was observed over 5 yr: 19% of second eyes developed either of the two late-AMD lesions (neovascular AMD or geographic atrophy with or without involvement of the fovea) in subjects with either late lesion in the first eye; 29% of second eyes developed end-stage AMD (either neovascular AMD or geographic atrophy involving the fovea) in subjects with unilateral end-stage AMD *(60)*.

To date, only the BDES *(45)* has reported 10-yr cumulative incidence and progression of AMD data. The person-specific incidence was 12.1% for early and 2.1% for late AMD over 10 yr *(45)*, including 1.4% for incident neovascular AMD and 0.8% for incident geographic atrophy. In addition to age, the incidence of late AMD was also strongly related to the severity of early-AMD lesions at baseline *(45)*.

Progression

Progression of AMD from soft or reticular drusen, or retinal pigmentary abnormalities to geographic atrophy, or neovascular AMD has been well documented in a number of clinical case series *(81,83–86)*. In the BDES and the BMES, early ARM lesion characteristics have been studied in detail in terms of the distribution, location, size, lesion type, and area involved in relation to the risk of subsequent development of late AMD *(45,87–89)*. Large soft drusen (≥125 μm in diameter), soft drusen with indistinct margins (indistinct soft or reticular drusen), or pigmentary changes involving large macular areas plus close proximity to the fovea indicate a much higher risk of subsequent late AMD development *(45,88)*.

Late AMD lesions do not usually regress, apart from retinal pigment epithelial (RPE) detachment *(45)*. Although large soft drusen may disappear without signs of progression, such disappearance is often associated with progression of the disease to a more advanced stage *(60)*.

Other Ethnicities

Currently, there are no AMD incidence data available for other ethnicities except blacks. The Barbados Eye Study recently reported the 4-yr incidence of macular changes in a black population aged 40 to 84 yr at baseline *(90)*. Incident neovascular AMD was observed in 1 of the 2362 persons at risk (0.04%) and incident geographic atrophy in none of the 2419 at risk (0%) over the 4-yr period. The incidence of early AMD lesions, including intermediate-sized drusen, was also relatively low (5.2%, 60 out of 1160) *(90)*. The proportion of participants in this study with ungradable retinal photographs, however, was relatively high (19.3%, 616 out of 3193) *(90)*.

RISK FACTORS ASSOCIATED WITH AMD PREVALENCE OR INCIDENCE

Genetic Influences

All observational studies, including clinic-based case-control studies *(91)* and population-based surveys *(92–95)*, have shown that family history is a consistent, strong risk factor for AMD with a risk ratio around 3 or higher for subjects with an AMD family history compared to those without. Familial aggregation in AMD prevalence *(93,96–100)* and incidence *(94)* has been confirmed in different study populations. Studies in twins *(101–104)* have provided further evidence supporting a genetic basis for AMD by showing that the concordance of AMD phenotypes is much higher in monozygotic twins than in dizygotic twins *(103,104)*, spouses *(102)*, or in the general population *(101)*. A strong genetic component in the etiology of AMD is thus beyond doubt *(19,24,103,105)*.

The search for AMD-related genes, however, has been far from conclusive *(24,106)*. Much research effort has been put into the search for AMD genes *(24,46,106–142)*. To date, inconsistent findings from this research suggest that multiple genes are likely to be responsible for the AMD susceptibility of affected individuals *(24,105,138,143)*. Although inconsistent findings have been reported from different study populations *(110)*, particularly with respect to exact gene loci *(24)*, two AMD-related chromosome regions have been identified in more than one study population *(24)*; 1q25–31 *(107,108,138)* and 10q26 *(108,137–139)*. Other chromosome regions found to have highly significant associations with AMD, or Hlod scores* of at least 3, from single study populations include 17q25 *(108)*, 6q14 *(109)*, and 15q21 *(138)*.

Candidate gene approaches and single nucleotide polymorphisms in the detection of gene variations have been widely used in the search for AMD genes *(24)*. Candidate genes for AMD are causal genes responsible for monogenic degenerative retinal conditions. One of the genes studied has been found to be possibly associated with AMD in case-control studies *(24)*. The *ABCR*, or *ABCA4*, a retinal-specific adenosine triphosphate-binding cassette transporter gene, known to be causally responsible for autosomal recessive juvenile-onset Stargardt macular dystrophy *(144)*, was found associated with AMD susceptibility in more than one study population *(111–114)*. These studies all shared a contribution from the same principal investigator, with one conducted in a large pooled sample *(113)*. The *ABCR*–AMD association, however, was not able to be confirmed in other study populations *(115–120,145,146)*.

The cholesterol transporter gene apolipoprotein E (*APOE*) appeared to be a promising AMD-related gene. In 1998, Souied et al. *(121)* reported a lower frequency of *APOE* ε4 allele carriers in 116 patients with neovascular AMD compared to 186 age- and sex-matched controls. The RS team *(122)* also reported that the *APOE* ε4 allele was associated with a lower risk of AMD and that the ε2 allele might be associated with a higher risk of AMD in a population-based case-control study. Since then, *APOE* polymorphisms have been attracting considerable research attention *(123–129)*. The protective association between the *APOE* ε4 allele and AMD has now been confirmed in four different studies *(124,125,128,129)*. Findings from one of these four studies suggest that the *APOE* ε4 allele–AMD association may exist only in familial cases but not in sporadic cases *(124)*, whereas another study did not find an association in familial cases *(127)*. In an analytic study which pooled data from four case-control study populations (three from the United States and another from the RS) *(126)*, a significant protective effect from the *APOE* ε4 allele in both men and women and a possible increased risk from the *APOE* ε2 allele in men, but not in women, was replicated, although the pooled study samples and the study investigators were not completely independent of previous reports *(122,124)*. Another study in a Chinese population, however, could not confirm the *APOE*–AMD association *(123)*.

A recent meta-analysis of findings on the associations between *APOE* polymorphisms and AMD suggests a risk effect of up to 20% for the *APOE* ε2 allele and a protective effect of up to 40% for the *APOE* ε4 allele *(351)*. In early 2005, four independent groups, Klein et al. *(140)*, Edwards et al. *(141)*. Haines et al. *(142)* and Hageman et al. *(352)*, simultaneously indentified a polymorphism in the *CFH* gene on chromosome region 1q25-32, a tyrosine-to-histidine substitution at amino acid 402 (Y402H), strongly associated with AMD. A second susceptibility gene marker at *LOC387715*, on chromosome region 10q26, has also been documented in many different independent studies *(353–355)*.

---

*lod is an acronym for logarithm of odds, and the Hlod parameter takes into account both lod score likelihood estimations and heterogeneity *(24)*.

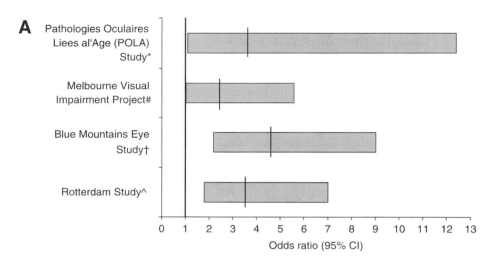

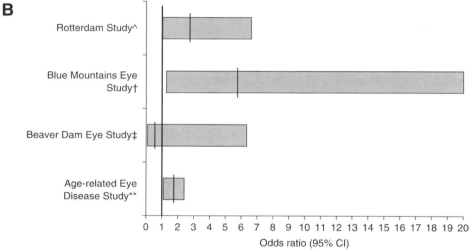

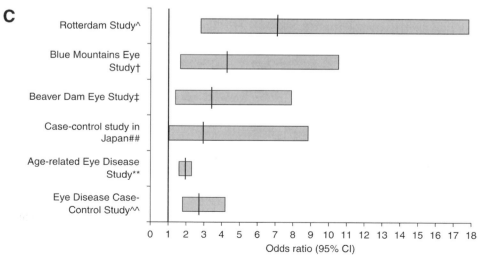

The complexity of AMD genetics, including heterogeneity in both phenotypes and genotypes, unknown correlations between specific genotypes and phenotypes, low penetrance, and a high susceptibility to nongenetic factors, is well summarized by Gorin et al. *(143)* and Tuo et al. *(24)*. Despite these difficulties, the recent replication of associations of AMD with gene variants in the *CFH, LOC387715 (353–355)* and *C2/BF (356)* genes is encouraging and will accelerate revealing the pathogenesis of AMD. Using a quantitative trait linkage approach to increase power in detecting linkage signals from genome-wide screening may help to partially overcome some of these difficulties *(24,46,147)*.

Lifestyle and Environmental Factors

Smoking

To date, smoking is the only modifiable AMD risk factor consistently found across different study populations *(148–159)*. This association is also consistent in cross-sectional analytic studies of clinical case-control study samples *(148,157–159)*, in population-based samples *(41,149–154,160)* (Fig. 3A–C), and in longitudinal analytic studies of large samples *(42,155,156,161–163)* (Fig. 4). In the BDES population, people who smoked at baseline were more likely to develop early AMD lesions 5 yr *(162)* and 10 yr later *(163)*, though a significant association between smoking and 10-yr incident neovascular AMD was evident only in men *(163)*.The magnitude of the risk for late AMD in current vs noncurrent smokers, or in ever smokers vs never smokers, is between two- and sixfold higher (Fig. 4). In the BMES population *(161,164)*, baseline current smokers developed late AMD at a mean age of 67 yr, whereas past smokers developed this at a mean age of 73 yr and never smokers at a mean age of 77 yr. This 10-yr earlier development of late AMD among current smokers implies a substantial increase in the burden on affected individuals, their families and aged/disability care.

In an experimental choroidal neovascularization (CNV) mouse model, Suner et al. *(165)* demonstrated that nicotine increased both the size and vascularity of CNV and that this effect can be blocked by subconjunctival injection of hexamethonium, a nonspecific nicotinic receptor antagonist.

Based on current, available evidence from epidemiological studies *(15,16,19,21–23, 25,164,166–168)* and experiments in animal models *(165)*, smoking is likely to have a causal role (as trigger and/or promoter) in the course of the development of neovascular AMD *(164,168)*. Currently, eye health practitioners give too little weight to smoking cessation and tobacco control, and public health specialists insufficient attention to

---

Fig. 3. *(Opposite page)* **(A)** Smoking and prevalence of late AMD. **(B)** Smoking and prevalence of geographic atrophy. **(C)** Smoking and prevalence of neovascular AMD. *Pathologies Oculaires Liees al'Age (POLA) study $N = 2196$, aged 60+ yr, risk for current smokers *(154)*; #Melbourne Visual Impairment Project $N = 4744$, aged 40+ yr, risk for smoking more than 40 yr *(153)*; †Blue Mountains Eye Study $N = 3654$, aged 49–97 yr, risk for current smokers *(41,152)*; ^Rotterdam Study $N = 7983$, aged 55–106 yr, risk for current smokers *(41,160)*; ‡Beaver Dam Eye Study $N = 4756$, aged 43–86 yr, risk for current smokers *(41)*; **Age-related Eye Disease Study $N = 4757$, aged 60–80 yr, risk for past and current smokers *(158)*; ##Case-Control Study in Japan $N = 138$, aged 50–69 yr, risk for current smokers *(157)*; ^^Eye Disease Case-Control Study $N = 1036$, aged 55–80 yr, risk for current smokers *(148)*.

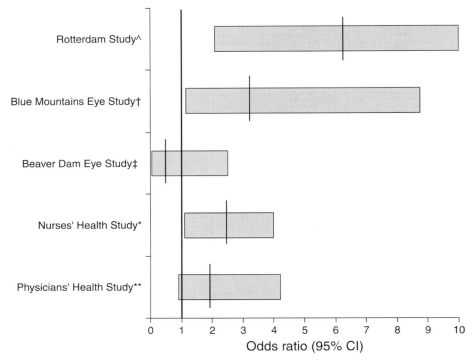

Fig. 4. Smoking and incidence of late AMD. ^Rotterdam Study N = 7983, aged 55+ yr, follow-up for 6 yr, risk for current smokers at baseline *(42)*; †Blue Mountains Eye Study N = 3654, aged 49+ yr, follow-up for 5 yr, risk for current smokers at baseline *(42)*; ‡Beaver Dam Eye Study N = 4926, aged 43–86 yr, follow-up for 5 yr, risk for current smokers at baseline *(42)*; *Nurses' Health Study N = 31,843, aged 50–59 yr, follow-up for 12 yr, current smokers who smoke ≥25 cigarettes/d *(156)*; **Physicians' Health Study N = 21,157, aged 40–84 yr, follow-up for 12.2 yr, risk for current smokers at baseline *(155)*.

eye disease *(164)*. The ocular hazards of smoking should be publicized more, and appropriate smoking cessation support should be offered in eye and medical services *(164,168,169)*.

Alcohol Consumption

In a subsample of the NHANES I study population *(170)*, persons who reported moderate wine consumption were significantly less likely to have AMD than nondrinkers. The subsample (*n* = 3072 out of 10,127) was chosen based on the age criterion (45–74 yr) and the availability of alcohol consumption data. The proportion of participants who responded to the alcohol questionnaire, however, was not stated *(170)*. AMD was defined during a clinical examination *(170)* and so could not be validated. This association between wine and AMD, however, was not confirmed in either cross-sectional *(171,172)* or longitudinal data *(173–175)* obtained from other large population-based studies. The BDES 5-yr follow-up data showed that beer drinking in men was associated with an increased risk of soft indistinct drusen *(173)*. The BDES 10-yr follow-up data also showed that people who reported being heavy drinkers at baseline were more likely to develop late AMD *(163)*. This was not found in either the BMES or RS (unpublished data). To date, accumulated evidence on the association between alcohol consumption and AMD appears to suggest a U-shape association: moderate alcohol consumption associated with a reduced risk, and no or heavy consumption associated with an increased risk.

Diet/Supplement Intake

ANTIOXIDANT NUTRIENTS (CAROTENOIDS/LUTEIN/ZEAXANTHIN)

The hypothesis that oxidative damage may be involved in the pathogenesis of AMD *(176,177)* is supported by a number of factors: (1) a high concentration of polyunsaturated fatty acids is present in the outer segments of photoreceptors in the retina; (2) there is high photo-oxidative stress *(178)* and relatively high oxygen tension in this region; and (3) the well-known susceptibility of polyunsaturated fatty acids to undergo oxidation in the presence of oxygen or oxygen-derived radical species *(177)*. Therefore, antioxidant nutrients may have a protective effect on AMD *(177,179–182)*. Findings from most case-control studies *(148,179,183–185)*, except one *(186)*, support this hypothesis. Observations from large study samples suggest that consumption of fruits and vegetables is associated with a reduced likelihood of AMD *(187–189)*. Whether the beneficial effect from consumption of fruits and vegetables is caused by carotenoids (a precursor of vitamin A), including lutein and zeaxanthin, is less conclusive *(190,191)*. An inverse association between AMD prevalence and serum carotenoids was suggested by one study *(192)*, but has not been evident in the majority of population-based nested case-control studies *(193–195)*. Nor has there been an inverse association evident between AMD prevalence and dietary carotenoid intake *(195–197)* or the use of vitamin and zinc supplements *(192,198)* in cross-sectional analytical studies of population-based data. Positive findings that were only observed in stratified analyses in the NHANES III *(195)* could have been because of chance, resulting from multiple comparisons *(195)*. In the Pathologies Oculaires Liees a l'Age (POLA) study population, plasma α-tocopherol (vitamin E) level was associated with a reduced AMD prevalence *(199)*, whereas glutathione peroxidase was significantly associated with an increased late-AMD prevalence *(200)*. Further, apart from an inverse association between intakes of pro-vitamin A carotenoids, dietary vitamin E, and the 5-yr incidence of large drusen observed in the BDES population *(201)*, findings from most population-based longitudinal studies have not provided convincing evidence supporting an inverse association between antioxidant or zinc intake (diet or supplements) and the subsequent development of AMD *(189,201–204)*. Potential misclassification in phenotype identification could have played a role in the negative findings from some of these studies *(189,202,203)*, as self-reported study outcomes with confirmation from doctors' records were used.

Direct evidence supporting a link between antioxidant nutrient intake and a protective effect on AMD needs to come from intervention trials *(191)*, a number of which have been conducted in recent years *(48,205–210)*. Currently, evidence as to the effectiveness of intervention with antioxidant vitamin and mineral supplementation on slowing AMD progression is dominated by the findings from one large randomized clinical trial, the AREDS *(48,205)*. In this study, a modest beneficial effect (25% reduction) from antioxidant and zinc supplementation on progression to late AMD was observed in persons with moderate to relatively advanced AMD *(48)*. Although the majority of intervention trials conducted so far were underpowered to detect small differences

(206–210), there is limited evidence at present suggesting that people with early AMD lesions should take supplements *(190,205)*, despite commercial advertisements and preparations already available in the market. AREDS findings should only be applied with caution to appropriate at-risk patients *(211,212)*.

Coffee and caffeine consumption have not been found to be associated with AMD risk *(213)*.

FATTY ACIDS AND LIPIDS

The hypothesis behind the postulated association between cholesterol, dietary fat intake, and AMD is that AMD and cardiovascular disease may share some common risk factors *(214)* or share a similar pathogenesis like atherosclerosis and arteriosclerosis *(215–218)*. In an experimental animal model, young transgenic mice fed a high-fat diet to achieve elevated plasma triglyceride and cholesterol levels, then exposed to nonphototoxic levels of blue-green light, showed a high frequency of developing basal laminar deposits (BLD) *(219)*, an early degenerative change in RPE cells associated with risk of CNV *(220)*.

An early case-control study conducted in the United Kingdom *(221)* found no difference between 65 AMD cases and 65 controls in plasma polyunsaturated acid content and erythrocyte phospholipids. In the Eye Disease Case-Control Study *(148)*, elevated serum total cholesterol level was associated with an increased likelihood of neovascular AMD. Cross-sectional population-based data showed that a high intake of saturated fat and cholesterol was associated with an increased prevalence of early AMD in the BDES population *(215)*, and that high intake of cholesterol was associated with an increased prevalence of late AMD in the BMES population *(222)*. However, in the NHANES III *(223)* and the CHS *(38)* populations, no significant association between dietary fat intake and AMD prevalence was found.

Recently, in the older subsamples of the Nurses' Health Study (NHS) and Health Professionals Follow-Up Study (HPFS) populations, Cho et al. *(217)* reported that a high total fat intake, including saturated, mono-unsaturated, and trans-unsaturated fats, were all associated with a modest, marginally significant increase in AMD risk. In two hospital-based samples, Seddon et al. *(214,224)* reported an increased risk for AMD progression in subjects with high intake of fat, including vegetable, mono- and poly-unsaturated fats. However, these associations could not be replicated in 5-yr incident analyses using BMES population data (unpublished data).

High intake of omega-3 fatty acids was associated with a reduced AMD risk in the Eye Disease Case-Control Study population *(224)*. Regular consumption of fish has been consistently found to be associated with a reduced AMD prevalence *(222)* and incidence (unpublished data) in the BMES population, an observation consistent across different populations *(214,217,224)*.

Use of cholesterol-lowering medications (statins) appeared to have an inverse association with AMD *(225,226)* in hospital-based clinical samples, but no significant protective effect on AMD prevalence *(227)* or incidence *(43,228–230)* was evident in analyses of large population-based samples.

Interpretation of findings from hospital-based case-control populations *(214,224)* and the NHS and HPFS *(217)* needs to consider possible indication bias, with changes in health behavior and diet occurring after the diagnosis of early AMD. Furthermore, people with high total fat dietary intakes are also more likely to have high intakes of all fat subtypes.

Iris Color, Skin Color, and Sunlight Exposure

Light-colored irides were found associated with modest increased odds of AMD in case-control studies *(91,231,232)* and in patients with unilateral neovascular AMD *(233)*. Population-based, cross-sectional findings, however, have been inconsistent *(234,235)*. Population-based longitudinal data also provide conflicting findings of either no association between iris color and 5-yr incident AMD *(236,237)* or a significantly decreased 10-yr risk of retinal pigmentary abnormalies in brown compared to blue eyes *(238)*. Persons with very fair skin appeared to have an increased risk of geographic atrophy compared to those with fair skin in the BMES population *(237)*. Sensitivity to glare or susceptibility to sun burn in persons with very fair skin could be markers for AMD susceptibility; however, sun-avoidance behavior in these persons (observed in Australian populations *[235,239]*) is likely to have confounded the association between these markers and AMD *(239)*. Macular pigment and melanin have not been found to be associated with AMD in a subsample of the RS *(240)*.

Although biologically plausible, the hypothesis of an association between sunlight exposure and AMD has proved difficult to investigate, because of a lack of precise, quantitative measures on life-time exposure and confounding by sun-avoidance behavior in subjects with very fair skin who are more susceptible to sun-related skin damage *(239)*. Surrogate factors (questionnaires, sun-related skin damage, or presence of pterygium—a sunlight-related conjunctival degenerative condition) have been used in epidemiological studies *(158,235,239,241–244)*. The Chesapeake Bay Waterman Study *(245)* reported no statistically significant association between ultraviolet UV-B or UV-A exposure and AMD, but a significant association between exposure to blue/visible light in the previous 20 yr and late AMD later in life (odds ratio [OR] 1.4, 95% confidence interval [CI] 1.0–1.9) *(241)*. There were only eight cases of late AMD in the Waterman Study population *(241)* and recall bias would be likely to occur during questioning for past exposure. In the BDES population, some consistency can be revealed from baseline cross-sectional *(242)*, 5-yr *(246)*, and 10-yr *(247)* longitudinal data. Time spent outdoors in summer, particularly early in life (teens to thirties), was associated with an increased risk of early AMD later in life *(246,247)*, whereas the use of hats or sunglasses appeared to provide some protection against development of early AMD lesions later in life *(246,247)*. Using pterygium as a surrogate factor for life-time sunlight exposure, the BMES team *(244)* reported that persons presenting with pterygium or reporting a past history of pterygium surgery at baseline had an increased risk of 5-yr incident late (adjusted OR 3.3, 95% CI 1.1–10.3) and early AMD (adjusted OR 1.8, 95% CI 1.1–2.9). Findings are not consistent across different study populations *(243)*, likely resulting from the difficulty in measurement. Although these combined data suggest the possibility of a sunlight exposure–AMD link, the impact from sunlight exposure is likely to be only modest *(21,25)*.

Demographic and Socio-Economic Factors

Female Gender and Sex Hormones

Gender differences have been observed in the prevalence of some AMD lesions *(31,34,36)*, but these differences were not consistent across different populations *(19,32,33,35,38–41,248)* and were not statistically significant. No significant gender difference has been found for AMD incidence *(45,50,67,71)*, although in the BMES population, women were (nonsignificantly) more likely to develop both early and late AMD *(60)*, and in the BDES population, a similar trend was observed in the oldest old group (aged >75 yr) *(67)*. Greater longevity in women could be a possible explanation for these observed, nonsignificant gender differences *(19)*. Interestingly, in the LALES population *(40,95)*, Hispanic men (12.3%) were significantly more likely than women (7.8%) to have early AMD, independent of other AMD risk factors *(95)*. In the Hisayama study population *(249)*, Japanese men (1.0%) were also much more likely to have late AMD than women (0.1%). The participation rate in the Hisayama Study, however, was relatively low (60.4%), so that a possible differential distribution of potential risk factors between participants and nonparticipants could have led to the huge observed gender difference in late AMD prevalence found in this study.

An association between exogenous estrogen exposure and AMD was postulated by investigators of the Eye Disease Case-Control Study Group based on findings from a hospital-based clinic sample *(148)*. This hypothesis of an association between female hormones and AMD has been investigated in basic research *(250,251)* and supported by some, but not all, observational studies *(252,253)*. Estrogen receptors have been found in human RPE cells *(250)*. In laser-induced CNV mice models, estrogen supplementation appeared to increase the severity of CNV and was found to interact with macrophages *(251)*. An inverse association between the prevalence of AMD and the duration of endogenous estrogen exposure (number of years between menarche and menopause) was observed in a nested case-control study of the baseline RS population *(252)* and in a cross-sectional analytic study of baseline BMES data *(253)*, but not in the BDES *(254)* or in pooled data from the BDES, RS, and BMES studies *(41)*. In the POLA study *(255)*, no association was evident between hormone replacement therapy (HRT) and AMD prevalence in 1451 postmenopausal women aged 60 yr or older.

There have also been no longitudinal associations from population-based studies in support of the estrogen–AMD hypothesis *(42,256)*. Findings from case-control studies in postmenopausal women *(257,258)* share these inconsistent findings. One study *(257)* lent support to the hypothesis that longer lifetime exposure to endogenous estrogen, or that exposure to exogenous estrogen (HRT) after menopause, is associated with a reduced likelihood of having advanced AMD, defined as grades 4 and 5 (level 4) using the AREDS classification system *(47)*. Another study found no evidence in support of this hypothesis *(258)*. An influence from selection and recall biases on these case-control study findings, however, cannot be ruled out. On the other hand, a possible role for estrogen interacting with other factors during the development of AMD also cannot be ruled out. A recent report from the Women's Health Initiative (WHI) Study *(357)* showed that women who received conjugated equine estrogens combined with progestin for an average of 5 yr had a reduced prevalence of soft drusen and neovascular AMD.

Socio-Economic Factors

No biologically plausible mechanism supports a causal association between socio-economic factors and AMD. However, socio-economic factors can serve as surrogates for other causal or noncausal factors. The AREDS reported a cross-sectional association between lower education level and AMD prevalence in this large, multisite clinic-based study sample *(259)*. In the BDES population, lower socio-economic status was not related to AMD in a cross-sectional analytic study of baseline data *(260)*, but nonprofessional occupation was found to be related to the 5-yr incidence of early AMD *(261)*. The inconsistency of these findings suggests that the positive associations found are likely to be the result of either chance, selection bias, or residual confounding effects from measured and unmeasured risk factors *(95,158,187,261)*.

Systemic Factors

Cardiovascular Disease and Risk Factors

Many cardiovascular disease risk factors have also been suggested as potential risk factors for AMD. These include age *(41,249,262)*, smoking *(41,148–154,157–159)*, atherosclerosis *(263)*, elevated blood pressure *(153,158,249,264,265)*, elevated total cholesterol level *(148)*, high body mass index (BMI) *(158,266,267)*, alcohol consumption *(170)*, and past history of cardiovascular diseases *(91)*. These observations have led to the proposal that AMD and cardiovascular disease may share common etiologic pathways *(19,148,218)*. Negative associations between AMD and cardiovascular risk factors from cross-sectional analytical studies have also been reported *(38,159,267, 268)*. Longitudinal observational studies have provided support for a link between AMD progression or development and smoking *(42,155,156,161,162)*, heavy drinking *(163)*, high BMI *(269,270)*, high pulse pressure *(271–273)* or elevated systolic blood pressure *(272,273)*, atherosclerosis *(273)*, elevated serum total cholesterol level *(42)*, and hypertensive retinal vessel wall signs *(274)*. However, longitudinal analytical findings of an association between high levels of serum high-density lipoprotein and increased AMD risk *(272,275)*, and an inverse association between serum total cholesterol and incident neovascular AMD *(42)* run counter to the cardiovascular disease associations with these lipids. These observed lipid–AMD associations could also be caused by genetic influences or interaction with susceptibility gene variations *(275)*. An association between AMD and cardiovascular mortality has also been reported *(276,277)*.

Use of antihypertensive or cholesterol-lowering medications and low-dose aspirin has not been found to have a significant protective effect on AMD prevalence *(227,278)* or incidence *(43,228–230,279)* in large population-based samples.

Inflammatory Markers

Leukocytes were observed in CNV tissues from post mortem eyes with neovascular AMD *(280)*. In laser-induced CNV mice models, macrophages have been shown to have a role in relation to the extent of choroidal neovascularization *(281)*. The possibility that infectious agents may play a role in the pathogenesis of vascular diseases and AMD is partially attributed to the recognition that atherosclerosis is an inflammatory disease initiated by endothelial injury *(282,283)*. Some inflammatory markers, such as high leukocyte count *(268,284)* and high plasma fibrinogen level *(266)*, were found to be associated with AMD prevalence *(266,268)* and incidence *(284)* in the BDES and BMES. C-reactive protein was found to be significantly higher among patients with advanced AMD compared to controls in the AREDS *(285)*. Recent reports *(286,287)* also suggest possible links between AMD and *Chlamydia pneumoniae* infection *(286)*, neovascular AMD and prior cytomegalovirus infection *(287)*, and neovascular AMD and monocyte/macrophage activation status *(288)*. Neither the BDES *(358)* nor BMES studies (unpublished data) could confirm the association between *Chlamydia pneumoniae* infection and AMD. These findings need further confirmation *(283,289)*. Clarification is also needed regarding whether: (1) the inflammatory process or the infection is specific or nonspecific *(283)* in terms of the link to AMD etiology, as these infectious agents are known to cause chronic, persistent inflammation *(283)* or (2) the inflammation is an antecedent or a consequential phenomenon in the course of neovascular AMD.

Use of anti-inflammatory medications was not found to have a beneficial effect on AMD prevalence or incidence *(279,290)*.

Hyperglycemia or Diabetes

There has been little evidence supporting an association between glycosylated hemoglobin, impaired fasting glucose, or presence of diabetes and AMD *(276,291,292)*.

Ocular Factors

Cataract and AMD

Aside from the fact that cataract and AMD are both strongly age related, findings have been inconsistent for an association between cataract and AMD in case-control studies *(158,232,248)* and in population-based cross-sectional *(245,293–296)* or longitudinal studies *(236)*, and no definite link can be confirmed between these two conditions. The positive associations between cataract and AMD found in case-control studies are likely the result of selection bias, a residual confounding effect or a chance finding *(158)*.

Cataract Surgery and the Risk of AMD

Whether the risk of late AMD is increased in eyes after cataract surgery is a long-standing but unresolved clinical question. Early clinical case series reports suggested a link between AMD and cataract surgery *(297–299)*. A report from post mortem eyes also suggested that neovascular AMD was more frequently observed in pseudophakic than phakic eyes *(300)*. These case series, however, cannot exclude the possibility that subtle, unrecognized, new vessels existed before surgery.

In a small number of cases, Pollack et al. *(301,302)* carefully documented that the risk of AMD increased within 6 to 12 mo after surgery in cataract patients with bilateral, symmetric early AMD. Each patient had only one eye operated, where the fellow eye served as a control, and fluorescein angiography was performed pre- and postoperatively. In contrast, Armbrecht et al. *(303)* could not confirm such an observation between surgical and nonsurgical cataract patients. In addition to small sample size and possible selection bias, shortcomings of clinical case series also include that no randomization procedure was used to decide which patient or which eye should be operated. Doctors' decisions could have been biased towards eyes or patients with worse vision. Hence the comparability of the operated and the nonoperated groups in these studies is likely to be low.

A recent report using pooled cross-sectional data from the SEE Project, the Proyecto VER, and the Baltimore Eye Survey *(59)* supported an association between prior cataract surgery (performed before baseline) and an increased prevalence of AMD. The link was stronger among persons who had cataract surgery at least 5 yr earlier. The BDES longitudinal data have consistently indicated a link between cataract surgery prior to the baseline examination and the incidence of AMD after 5 *(236)* and 10 yr *(304)* (Fig. 5A,B). The increased AMD risk was around fourfold in pseudophakic compared to phakic eyes (Fig. 5B). Presence of AMD lesions at baseline was not significantly related to 5-yr incident cataract surgery in the BDES population *(305)*.

Both the BDES and the BMES followed similar study protocols in diagnosing AMD, with all incident late-AMD cases confirmed using side-by-side grading and mutual cross-checking by investigators of the two studies. Pooled, longitudinal data from these two studies have shown an approx 10 times higher crude incidence of late AMD in operated compared to nonoperated eyes *(306)*. The age-adjusted relative risk was around 4; after multivariable adjustment the increased odds for subsequent development of late AMD was between 3 to 5, a magnitude similar to the effect of smoking on AMD (Fig. 5A). Higher odds were observed in models that also adjusted for baseline early AMD status *(306)*, suggesting that nonphakic eyes that developed late AMD did not have as much advanced early AMD at baseline as phakic eyes that developed late AMD.

Cataract surgery is currently the most commonly performed and most successful ophthalmic surgical procedure worldwide, with case numbers continuously increasing. In the BMES examination of two cross-sectional surveys, prevalence of cataract surgery increased by one-third over a mean 7-yr interval, from 6 to nearly 8%, in cross-sectional population samples with a similar age range *(307)*. The increase in surgical procedures was predominantly for the eldest old group (aged 80+ yr) and for second eyes *(307)*. If the link between prior cataract surgery and higher risk of subsequent late AMD is confirmed, it is likely that AMD incidence will increase further, over and above the increase from population aging alone. Longitudinal cohort studies on cataract surgical patients are needed to answer this important, but still unresolved, clinical question. A few such studies are currently underway. Discussion of a possible higher risk of progression with patients who already have risk signs (unilateral late AMD or significant early signs bilaterally) by their doctors prior to cataract surgery is potentially important.

Refractive Errors

Previous case series *(308)*, case-control studies *(91,148,158,232,309,310)*, and population-based cross-sectional analyses *(187,311,312)* have indicated a possible link between hyperopia and AMD. However, apart from the RS *(312)*, the majority of population-based longitudinal data do not support a link between refractive error and incident AMD *(236,313,314)*. Without a biologically plausible mechanism *(311,312)*, the previous positive findings of weak association could have been as a result of selection bias or chance findings *(158)*.

IMPACT OF AMD

Vision-Related Quality of Life, Depression, and Other Disabilities

Links between AMD and visual impairment *(1–11,315,316)* and decreased scores on measures of vision-related quality of life *(315,317,318)* via impaired vision *(317)* are

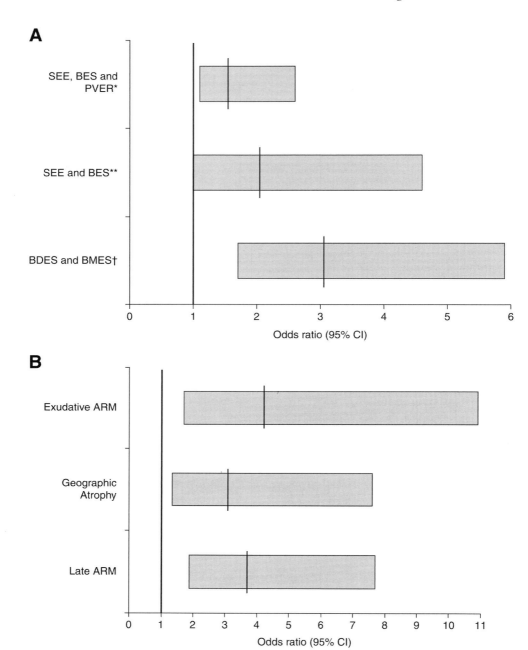

Fig. 5. (A) Prior cataract surgery and the risk of subsequent development of late AMD. *Salisbury Eye Evaluation (SEE), Baltimore Eye Survey (BES), and Proyecto VER (PVER) pooled cross-sectional data *(59)*: prior cataract surgery and late AMD detected at the surveys; **SEE and BES pooled cross-sectional data *(59)*: cataract surgery ≥5 yr ago and late AMD detected at the surveys; ⁺Beaver Dam Eye Study (BDES) and Blue Mountains Eye Study (BMES) pooled incidence data *(306)*: cataract surgery prior to baseline and 5-yr incidence of late AMD. **(B)** Cataract surgery prior to baseline and risk of 10-yr incidence of AMD, the BDES *(304)*.

obvious. Many studies have shown that visual impairment decreases the ability of older people to perform usual activities of daily living *(58,319–328)*. Visual impairment, a direct consequence of AMD, also links with other conditions and disabilities in older people including falls *(329,330)*, fractures *(331,332)*, depression *(333–335)*, hearing loss *(336)*, Alzheimer's disease *(337)*, cognitive function decline *(338)*, and low self-rated health *(339)*. A possible contribution from visual impairment to subsequent premature admission to nursing homes *(340)* and to reduced survival *(276,277,319,341–345)* have also been reported.

The Burden From AMD

It was estimated that 3.5% of the UK population aged 75 yr or older in 2003 (214,000 persons) were visually impaired as a result of AMD *(346)*. The number of persons with geographic atrophy and neovascular AMD in the United Kingdom was estimated at 172,000 and 245,000, respectively *(346)*. In the United States, the estimated proportion and number of persons aged 40 yr or older with AMD using the 2000 US Census population data was 1.5% and 1.75 million, respectively *(44)*. The projected numbers with AMD 10 to 20 yr in the future will be higher, owing to population aging *(44,346)*. The economic impact of visual impairment from AMD and the cost utility of screening and treatment of early AMD with antioxidants has been explored recently *(347–349)*, but remains under-researched.

SUMMARY

Early in the 21st century, our knowledge of the epidemiology of AMD has grown substantially. Reliable data on prevalence and incidence is available for projecting the magnitude of this disease *(44)* and to estimate the societal burden it causes *(44,346)*. Investigations of etiological factors for AMD have led to the identification of a number of definite or probable risk factors—particularly age, susceptibility genes, and smoking—and a number of possible risk factors such as light exposure, antioxidant status, inflammatory processes, cataract surgery or aphakic/pseudophakic status, and cardiovascular-related pathogenesis including atherosclerosis and hypertension. At present, our understanding of AMD etiology can be summarized thus: it appears that AMD is caused by environmental factors triggering disease in genetically susceptible individuals *(106)*. If AMD-related genes are identified as susceptibility factors that interact with environmental factors during the aging process, then multiple genes and environmental factors, together with all possible combinations for the possible interactions between each of these genes and environmental factors will result in a much more complex picture of the pathogenesis of AMD *(350)*. Research in animal models has already shed some light on the complexity of possible interactions *(219)*. A speculative etiology model proposed by the authors is that interactions of susceptibility genes with various environmental factors result in multiple pathogenetic pathways leading to the same disease. An example supporting this possible model is that elevated blood pressure, high serum cholesterol, elevated blood glucose, and high BMI can all lead to coronary heart disease (CHD). It is also possible that multiple pathogenetic pathways overlap between different diseases and so the same pathogenetic pathway can lead to different diseases.

An example supporting the latter is that high BMI can lead to a number of chronic conditions like diabetes, hypertension, and CHD. This hypothetical AMD etiology model may mirror etiologies of other diseases, such as cardiovascular disease and cancer. The birth cohort effect on AMD prevalence observed in the BDES population *(262)* also suggests that unmeasured risk factors and/or unknown interactions between susceptibility genes and environmental factors may exist. The elucidation of the genetic basis of AMD and the gene–environmental interactions responsible for the development of AMD in some but not all aged individuals *(19)*, is one of the challenges of ophthalmic research within the next decade or two *(143)*. With the collaboration between epidemiology and genetics research teams, and use of currently advancing genetic *(46,108,138,147)* and statistical techniques, together with basic scientists' experiments in vivo and in vitro, this goal is apparently achievable *(350)*. Preventive and therapeutic strategies should follow.

ACKNOWLEDGMENTS

The authors would like to thank Ms. Kirsten Jakobsen and Ms. Bronwen Taylor for editing and proof reading, and Ms. Ava Tan for preparing Figs. 3, 4, and 5.

REFERENCES

1. Ferris FL 3rd, Fine SL, Hyman L. Age-related macular degeneration and blindness due to neovascular maculopathy. Arch Ophthalmol 1984;102:1640–1642.
2. Sommer A, Tielsch JM, Katz J, et al. Racial differences in the cause-specific prevalence of blindness in east Baltimore [see comments]. N Engl J Med 1991;325:1412–1417.
3. Rosenberg T, Klie F. The incidence of registered blindness caused by age-related macular degeneration. Acta Ophthalmol Scand 1996;74:399–402.
4. Evans J, Wormald R. Is the incidence of registrable age-related macular degeneration increasing? Br J Ophthalmol 1996;80:9–14.
5. Rahmani B, Tielsch JM, Katz J, et al. The cause-specific prevalence of visual impairment in an urban population. The Baltimore Eye Survey. Ophthalmology 1996;103:1721–1726.
6. Klein R, Wang Q, Klein BE, Moss SE, Meuer SM. The relationship of age-related maculopathy, cataract, and glaucoma to visual acuity. Invest Ophthalmol Vis Sci 1995;36:182–191.
7. Attebo K, Mitchell P, Smith W. Visual acuity and the causes of visual loss in Australia. The Blue Mountains Eye Study. Ophthalmology 1996;103:357–364.
8. Wang JJ, Foran S, Mitchell P. Age-specific prevalence and causes of bilateral and unilateral visual impairment in older Australians: the Blue Mountains Eye Study. Clin Experiment Ophthalmol 2000;28:268–273.
9. Klaver CC, Wolfs RC, Vingerling JR, Hofman A, de Jong PT. Age–specific prevalence and causes of blindness and visual impairment in an older population: the Rotterdam Study. Arch Ophthalmol 1998;116:653–658.
10. Rodriguez J, Sanchez R, Munoz B, et al. Causes of blindness and visual impairment in a population-based sample of U.S. Hispanics. Ophthalmology 2002;109:737–743.
11. Evans JR, Fletcher AE, Wormald RP. Age-related macular degeneration causing visual impairment in people 75 years or older in Britain: an add-on study to the Medical Research Council Trial of Assessment and Management of Older People in the Community. Ophthalmology 2004;111:513–517.
12. Ambati J, Ambati BK, Yoo SH, Ianchulev S, Adamis AP. Age-related macular degeneration: etiology, pathogenesis, and therapeutic strategies. Surv Ophthalmol 2003;48:257–293.

13. Fine SL. Age-related macular degeneration 1969–2004: a 35-year personal perspective. Am J Ophthalmol 2005;139:405–420.
14. Hyman L. Epidemiology of eye disease in the elderly. Eye 1987;1:330–341.
15. Vingerling JR, Klaver CC, Hofman A, de Jong PT. Epidemiology of age-related maculopathy. Epidemiol Rev 1995;17:347–360.
16. Hawkins BS, Bird A, Klein R, West SK. Epidemiology of age-related macular degeneration. Mol Vis 1999;5:26.
17. Klein R, Klein BE, Cruickshanks KJ. The prevalence of age-related maculopathy by geographic region and ethnicity. Prog Retin Eye Res 1999;18:371–389.
18. Klein R. Epidemiology. In: Berger JW, Fine SL, and Maguire MG, eds. Age-related Macular Degeneration. St. Louis: Mosby, 1999:31–56.
19. Evans JR. Risk factors for age-related macular degeneration. Prog Retin Eye Res 2001;20:227–253.
20. Gottlieb JL. Age-related macular degeneration. JAMA 2002;288:2233–2236.
21. Hyman L, Neborsky R. Risk factors for age-related macular degeneration: an update. Curr Opin Ophthalmol 2002;13:171–175.
22. van Leeuwen R, Klaver CC, Vingerling JR, Hofman A, de Jong PT. Epidemiology of age-related maculopathy: a review. Eur J Epidemiol 2003;18:845–854.
23. Klein R, Peto T, Bird A, VanNewkirk MR. The epidemiology of age-related macular degeneration. Am J Ophthalmol 2004;137:486–495.
24. Tuo J, Bojanowski CM, Chan CC. Genetic factors of age-related macular degeneration. Prog Retin Eye Res 2004;23:229–249.
25. Seddon JM, Chen CA. The epidemiology of age-related macular degeneration. Int Ophthalmol Clin 2004;44:17–39.
26. Fine SL, Berger JW, Maguire MG, Ho AC. Age-related macular degeneration. N Engl J Med 2002;342:483–492.
27. Klein R. Epidemiology of Age-related macular degeneration. In: Penfold PL and Provis JM, eds. Macular Degeneration. Springer-Verlag GmbH Co, Berlin 2004:79–102.
28. Klein R, Davis M, Magli Y, Segal P, Klein B, Hubbard L. The Wisconsin age-related maculopathy grading system. Ophthalmology 1991;98:1128–1134.
29. Bird AC, Bressler NM, Bressler SB, et al. An international classification and grading system for age- related maculopathy and age-related macular degeneration. The International ARM Epidemiological Study Group. Surv Ophthalmol 1995;39:367–374.
30. Friedman DS, Katz J, Bressler NM, Rahmani B, Tielsch JM. Racial differences in the prevalence of age-related macular degeneration: the Baltimore Eye Survey. Ophthalmology 1999;106:1049–1055.
31. Klein R, Klein BE, Linton KL. Prevalence of age-related maculopathy. The Beaver Dam Eye Study. Ophthalmology 1992;99:933–943.
32. Schachat AP, Hyman L, Leske MC, Connell AM, Wu SY. Features of age–related macular degeneration in a black population. The Barbados Eye Study Group. Arch Ophthalmol 1995;113:728–735.
33. Vingerling JR, Dielemans I, Hofman A, et al. The prevalence of age-related maculopathy in the Rotterdam Study. Ophthalmology 1995;102:205–210.
34. Mitchell P, Smith W, Attebo K, Wang JJ. Prevalence of age-related maculopathy in Australia. The Blue Mountains Eye Study. Ophthalmology 1995;102:1450–1460.
35. Klein R, Clegg L, Cooper LS, et al. Prevalence of age-related maculopathy in the Atherosclerosis Risk in Communities Study. Arch Ophthalmol 1999;117:1203–1210.
36. VanNewkirk MR, Nanjan MB, Wang JJ, Mitchell P, Taylor HR, McCarty CA. The prevalence of age-related maculopathy: the visual impairment project. Ophthalmology 2000;107:1593–1600.

37. Oshima Y, Ishibashi T, Murata T, Tahara Y, Kiyohara Y, Kubota T. Prevalence of age related maculopathy in a representative Japanese population: the Hisayama study. Br J Ophthalmol 2001;85:1153–1157.

38. Klein R, Klein BE, Marino EK, et al. Early age-related maculopathy in the cardiovascular health study. Ophthalmology 2003;110:25–33.

39. Jonasson F, Arnarsson A, Sasaki H, Peto T, Sasaki K, Bird AC. The prevalence of age–related maculopathy in iceland: Reykjavik eye study. Arch Ophthalmol 2003;121: 379–385.

40. Varma R, Fraser-Bell S, Tan S, Klein R, Azen SP. Prevalence of age-related macular degeneration in Latinos: the Los Angeles Latino eye study. Ophthalmology 2004;111: 1288–1297.

41. Smith W, Assink J, Klein R, et al. Risk factors for age-related macular degeneration: Pooled findings from three continents. Ophthalmology 2001;108:697–704.

42. Tomany SC, Wang JJ, van Leeuwen R, et al. Risk factors for incident age-related macular degeneration: pooled findings from 3 continents. Ophthalmology 2004;111:1280–1287.

43. van Leeuwen R, Tomany SC, Wang JJ, et al. Is medication use associated with the incidence of early age-related maculopathy? Pooled findings from 3 continents. Ophthalmology 2004;111:1169–1175.

44. Friedman DS, O'Colmain BJ, Munoz B, et al. Prevalence of age-related macular degeneration in the United States. Arch Ophthalmol 2004;122:564–572.

45. Klein R, Klein BE, Tomany SC, Meuer SM, Huang GH. Ten-year incidence and progression of age-related maculopathy: The Beaver Dam eye study. Ophthalmology 2002; 109:1767–1779.

46. Schick JH, Iyengar SK, Klein BE, et al. A whole-genome screen of a quantitative trait of age-related maculopathy in sibships from the Beaver Dam Eye Study. Am J Hum Genet 2003;72:1412–1424.

47. Age-Related Eye Disease Study Research Group. The Age-Related Eye Disease Study system for classifying age-related macular degeneration from stereoscopic color fundus photographs: the Age-Related Eye Disease Study Report Number 6. Am J Ophthalmol 2001;132:668–681.

48. Age-Related Eye Disease Study Research Group. A Randomized, Placebo-Controlled, Clinical Trial of High-Dose Supplementation With Vitamins C and E, Beta Carotene, and Zinc for Age-Related Macular Degeneration and Vision Loss. AREDS Report Number 8. Arch Ophthalmol 2001;119:1417–1436.

49. Klaver CC, Assink JJ, van Leeuwen R, et al. Incidence and progression rates of age-related maculopathy: the Rotterdam Study. Invest Ophthalmol Vis Sci 2001;42:2237–2241.

50. van Leeuwen R, Klaver CC, Vingerling JR, Hofman A, de Jong PT. The risk and natural course of age-related maculopathy: follow-up at 6 1/2 years in the Rotterdam study. Arch Ophthalmol 2003;121:519–526.

51. van Leeuwen R, Chakravarthy U, Vingerling JR, et al. Grading of age-related maculopathy for epidemiological studies: is digital imaging as good as 35-mm film? Ophthalmology 2003;110:1540–1544.

52. Scholl HP, Dandekar SS, Peto T, et al. What is lost by digitizing stereoscopic fundus color slides for macular grading in age-related maculopathy and degeneration? Ophthalmology 2004;111:125–132.

53. Sperduto RD, Seigel D. Senile lens and senile macular changes in a population-based sample. Am J Ophthalmol 1980;90:86–91.

54. Bressler NM, Bressler SB, West SK, Fine SL, Taylor HR. The grading and prevalence of macular degeneration in Chesapeake Bay watermen. Arch Ophthalmol 1989;107:847–852.

55. Mitchell RA. Prevalence of age related macular degeneration in persons aged 50 years and over resident in Australia. J Epidemiol Community Health 1993;47:42–45.

56. Wu L. Study of aging macular degeneration in China. Jpn J Ophthalmol 1987;31:349–367.
57. Vinding T. Age-related macular degeneration. Macular changes, prevalence and sex ratio. An epidemiological study of 1000 aged individuals. Acta Ophthalmol Copenh 1989;67:609–616.
58. West SK, Munoz B, Rubin GS, et al. Function and visual impairment in a population-based study of older adults. The SEE project. Salisbury Eye Evaluation. Invest. Ophthalmol Vis Sci 1997;38:72–82.
59. Freeman EE, Munoz B, West SK, Tielsch JM, Schein OD. Is there an association between cataract surgery and age-related macular degeneration? Data from three population-based studies. Am J Ophthalmol 2003;135:849–856.
60. Mitchell P, Wang JJ, Foran S, Smith W. Five-year incidence of age-related maculopathy lesions: the blue mountains eye study. Ophthalmology 2002;109:1092–1097.
61. Klein R, Rowland ML, Harris MI. Racial/ethnic differences in age-related maculopathy. Third National Health and Nutrition Examination Survey. Ophthalmology 1995; 102:371–381.
62. Klein R, Klein BE, Jensen SC, Mares-Perlman JA, Cruickshanks KJ, Palta M. Age-related maculopathy in a multiracial United States population: the National Health and Nutrition Examination Survey III. Ophthalmology 1999;106:1056–1065.
63. Lavin MJ, Eldem B, Gregor ZJ. Symmetry of disciform scars in bilateral age-related macular degeneration. Br J Ophthalmol 1991;75:133–136.
64. Wang JJ, Mitchell P, Smith W, Cumming RG. Bilateral involvement by age related maculopathy lesions in a population. Br J Ophthalmol 1998;82:743–747.
65. Cruickshanks KJ, Hamman RF, Klein R, Nondahl DM, Shetterly SM. The prevalence of age-related maculopathy by geographic region and ethnicity. The Colorado-Wisconsin Study of Age-Related Maculopathy. Arch Ophthalmol 1997;115:242–250.
66. Frank RN, Puklin JE, Stock C, Canter LA. Race, iris color, and age-related macular degeneration. Trans. Am Ophthalmol Soc 2000;98:109–115.
67. Klein R, Klein BE, Jensen SC, Meuer SM. The five-year incidence and progression of age-related maculopathy: the Beaver Dam Eye Study. Ophthalmology 1997;104:7–21.
68. Bressler NM, Munoz B, Maguire MG, et al. Five-year incidence and disappearance of drusen and retinal pigment epithelial abnormalities. Waterman study. Arch Ophthalmol 1995;113:301–308.
69. Podgor MJ, Leske MC, Ederer F. Incidence estimates for lens changes, macular changes, open- angle glaucoma and diabetic retinopathy. Am J Epidemiol 1983;118:206–212.
70. Klaver CC, Assink JJ, van Leeuwen R, et al. Incidence and progression rates of age-related maculopathy: the Rotterdam Study. Invest Ophthalmol Vis Sci 2001;42:2237–2241.
71. Mukesh BN, Dimitrov PN, Leikin S, et al. Five-year incidence of age-related maculopathy: the Visual Impairment Project. Ophthalmology 2004;111:1176–1182.
72. Jonasson F, Arnarsson A, Peto T, Sasaki H, Sasaki K, Bird AC. Five-year incidence of age-related maculopathy in the Reykjavik Eye Study. Ophthalmology 2005;112:132–138.
73. Javitt JC, Zhou Z, Maguire MG, Fine SL, Willke RJ. Incidence of exudative age-related macular degeneration among elderly Americans. Ophthalmology 2003;110:1534–1539.
74. Jaakkola A, Anttila PM, Immonen I. Interferon alpha–2a in the treatment of exudative senile macular degeneration. Acta Ophthalmol Copenh 1994;72:545–549.
75. Gregor Z, Bird AC, Chisholm IH. Senile disciform macular degeneration in the second eye. Br J Ophthalmol 1997;61:141–147.
76. Strahlman ER, Fine SL, Hillis A. The second eye of patients with senile macular degeneration. Arch Ophthalmol 1983;101:1191–1193.
77. Roy M, Kaiser-Kupfer M. Second eye involvement in age-related macular degeneration: a four-year prospective study. Eye 1990;4:813–818.
78. Macular Photocoagulation Study Group. Five-year follow-up of fellow eyes of patients with age-related macular degeneration and unilateral extrafoveal choroidal neovascularization. Macular Photocoagulation Study Group. Arch Ophthalmol 1993;111:1189–1199.

79. Chang B, Yannuzzi LA, Ladas ID, Guyer DR, Slakter JS, Sorenson JA. Choroidal neovascularization in second eyes of patients with unilateral exudative age-related macular degeneration. Ophthalmology 1995;102:1380–1386.

80. Sandberg MA, Weiner A, Miller S, Gaudio AR. High-risk characteristics of fellow eyes of patients with unilateral neovascular age-related macular degeneration. Ophthalmology 1998;105:441–447.

81. Pieramici DJ, Bressler SB. Age-related macular degeneration and risk factors for the development of choroidal neovascularization in the fellow eye. Curr Opin Ophthalmol 1998;9:38–46.

82. Uyama M, Takahashi K, Ida N, et al. The second eye of Japanese patients with unilateral exudative age related macular degeneration. Br J Ophthalmol 2000;84:1018–1023.

83. Smiddy WE, Fine SL. Prognosis of patients with bilateral macular drusen. Ophthalmology 1984;91:271–277.

84. Bressler NM, Bressler SB, Seddon JM, Gragoudas ES, Jacobson LP. Drusen characteristics in patients with exudative versus non- exudative age-related macular degeneration. Retina 1988;8:109–114.

85. Holz FG, Wolfensberger TJ, Piguet B, et al. Bilateral macular drusen in age-related macular degeneration. Prognosis and risk factors. Ophthalmology 1994;101:1522–1528.

86. Arnold JJ, Sarks SH, Killingsworth MC, Sarks JP. Reticular pseudodrusen. A risk factor in age-related maculopathy. Retina 1995;15:183–191.

87. Wang Q, Chappell RJ, Klein R, et al. Pattern of age-related maculopathy in the macular area. The Beaver Dam Eye Study. Invest. Ophthalmol Vis Sci 1996;37:2234–2242.

88. Wang JJ, Foran S, Smith W, Mitchell P. Risk of age-related macular degeneration in eyes with macular drusen or hyperpigmentation: the Blue Mountains Eye Study cohort. Arch Ophthalmol 2003;121:658–663.

89. Knudtson MD, Klein R, Klein BE, Lee KE, Meuer SM, Tomany SC. Location of lesions associated with age-related maculopathy over a 10-year period: the Beaver Dam Eye Study. Invest Ophthalmol Vis Sci 2004;45:2135–2142.

90. Leske MC, Wu SY, Hyman L, Hennis A, Nemesure B, Schachat AP. Four-year incidence of macular changes in the Barbados Eye Studies. Ophthalmology 2004;111:706–711.

91. Hyman LG, Lilienfeld AM, Ferris FL 3rd, Fine SL. Senile macular degeneration: a case-control study. Am J Epidemiol 1983;118:213–227.

92. Smith W, Mitchell P. Family history and age-related maculopathy: the Blue Mountains Eye Study. Aust NZ J Ophthalmol 1998;26:203–206.

93. Klaver CC, Wolfs RC, Assink JJ, van Duijn CM, Hofman A, de Jong PT. Genetic risk of age-related maculopathy. Population-based familial aggregation study. Arch Ophthalmol 1998;116:1646–1651.

94. Klein BE, Klein R, Lee KE, Moore EL, Danforth L. Risk of incident age-related eye diseases in people with an affected sibling: The Beaver Dam Eye Study. Am J Epidemiol 2001;154:207–211.

95. Fraser-Bell S, Donofrio J, Wu J, Klein R, Azen SP, Varma R. Sociodemographic factors and age-related macular degeneration in Latinos: The Los Angeles Latino Eye Study. Am J Ophthalmol 2005;139:30–38.

96. Heiba IM, Elston RC, Klein BE, Klein R. Sibling correlations and segregation analysis of age-related maculopathy: the Beaver Dam Eye Study. Genet Epidemiol 1994;11:51–67.

97. Silvestri G, Johnston PB, Hughes AE. Is genetic predisposition an important risk factor in age-related macular degeneration? Eye 1994;8:564–568.

98. Seddon JM, Ajani UA, Mitchell BD. Familial aggregation of age-related maculopathy. Am J Ophthalmol 1997;123:199–206.

99. Keverline MR, Mah TS, Keverline PO, Gorin MB. A practice-based survey of familial age-related maculopathy. Ophthalmic Genet 1998;19:19–26.

100. Yoshida A, Yoshida M, Yoshida S, Shiose S, Hiroish G, Ishibashi T. Familial cases with age-related macular degeneration. Jpn J Ophthalmol 2000;44:290–295.

101. Klein ML, Mauldin WM, Stoumbos VD. Heredity and age-related macular degeneration. Observations in monozygotic twins. Arch Ophthalmol 1994;112:932–937.

102. Gottfredsdottir MS, Sverrisson T, Musch DC, Stefansson E. Age related macular degeneration in monozygotic twins and their spouses in Iceland. Acta Ophthalmol Scand 1999; 77:422–425.

103. Hammond CJ, Webster AR, Snieder H, Bird AC, Gilbert CE, Spector TD. Genetic influence on early age-related maculopathy: a twin study. Ophthalmology 2002;109:730–736.

104. Grizzard SW, Arnett D, Haag SL. Twin study of age-related macular degeneration. Ophthalmic Epidemiol 2003;10:315–322.

105. de Jong PT, Bergen AA, Klaver CC, van Duijn CM, Assink JM. Age-related maculopathy: its genetic basis. Eye 2001;15:396–400.

106. Yates JR, Moore AT. Genetic susceptibility to age related macular degeneration. J Med Genet 2000;37:83–87.

107. Klein ML, Schultz DW, Edwards A, et al. Age-related macular degeneration. Clinical features in a large family and linkage to chromosome 1q. Arch Ophthalmol 1998;116: 1082–1088.

108. Weeks DE, Conley YP, Tsai HJ, et al. Age-related maculopathy: an expanded genome-wide scan with evidence of susceptibility loci within the 1q31 and 17q25 regions. Am J Ophthalmol 2001;132:682–692.

109. Kniazeva M, Traboulsi EI, Yu Z, et al. A new locus for dominant drusen and macular degeneration maps to chromosome 6q14. Am J Ophthalmol 2000;130:197–202.

110. Hayashi M, Merriam JE, Klaver CC, et al. Evaluation of the ARMD1 locus on 1q25–31 in patients with age-related maculopathy: genetic variation in laminin genes and in exon 104 of HEMICENTIN–1. Ophthalmic Genet 2004;25:111–119.

111. Allikmets R, Shroyer NF, Singh N, et al. Mutation of the Stargardt disease gene (ABCR) in age-related macular degeneration. Science 1997;277:1805–1807.

112. Zhang K, Kniazeva M, Hutchinson A, Han M, Dean M, Allikmets R. The ABCR gene in recessive and dominant Stargardt diseases: a genetic pathway in macular degeneration. Genomics 1999;60:234–237.

113. Allikmets R. Further evidence for an association of ABCR alleles with age-related macular degeneration. The International ABCR Screening Consortium. Am J Hum Genet 2000;67:487–491.

114. Shroyer NF, Lewis RA, Yatsenko AN, Wensel TG, Lupski JR. Cosegregation and functional analysis of mutant ABCR (ABCA4) alleles in families that manifest both Stargardt disease and age-related macular degeneration. Hum Mol Genet 2001;10: 2671–2678.

115. De La Paz MA, Guy VK, Abou-Donia S, et al. Analysis of the Stargardt disease gene (ABCR) in age-related macular degeneration. Ophthalmology 1999;106:1531–1536.

116. Fuse N, Suzuki T, Wada Y, et al. Molecular genetic analysis of ABCR gene in Japanese dry form age-related macular degeneration. Jpn. J Ophthalmol 2000;44:245–249.

117. Souied EH, Ducroq D, Rozet JM, et al. ABCR gene analysis in familial exudative age-related macular degeneration. Invest Ophthalmol Vis Sci 2000;41:244–247.

118. Guymer RH, Heon E, Lotery AJ, et al. Variation of codons 1961 and 2177 of the Stargardt disease gene is not associated with age-related macular degeneration. Arch Ophthalmol 2001;119:745–751.

119. Webster AR, Heon E, Lotery AJ, et al. An analysis of allelic variation in the ABCA4 gene. Invest Ophthalmol Vis Sci 2001;42:1179–1189.

120. Schmidt S, Postel EA, Agarwal A, et al. Detailed analysis of allelic variation in the ABCA4 gene in age-related maculopathy. Invest Ophthalmol Vis Sci 2003;44:2868–2875.

121. Souied EH, Benlian P, Amouyel P, et al. The epsilon4 allele of the apolipoprotein E gene as a potential protective factor for exudative age-related macular degeneration. Am J Ophthalmol 1998;125:353–359.

122. Klaver CC, Kliffen M, van Duijn CM, et al. Genetic association of apolipoprotein E with age-related macular degeneration. Am J Hum Genet 1998;63:200–206.

123. Pang CP, Baum L, Chan WM, Lau TC, Poon PM, Lam DS. The apolipoprotein E epsilon4 allele is unlikely to be a major risk factor of age-related macular degeneration in Chinese. Ophthalmologica 2000;214:289–291.

124. Schmidt S, Saunders AM, De La Paz MA, et al. Association of the apolipoprotein E gene with age-related macular degeneration: possible effect modification by family history, age, and gender. Mol Vis 2000;6:287–293.

125. Simonelli F, Margaglione M, Testa F, et al. Apolipoprotein E polymorphisms in age-related macular degeneration in an Italian population. Ophthalmic Res 2001;33:325–328.

126. Schmidt S, Klaver C, Saunders A, et al. A pooled case-control study of the apolipoprotein E (APOE) gene in age-related maculopathy. Ophthalmic Genet 2002;23:209–223.

127. Schultz DW, Klein ML, Humpert A, et al. Lack of an association of apolipoprotein E gene polymorphisms with familial age-related macular degeneration. Arch Ophthalmol 2003;121:679–683.

128. Zareparsi S, Reddick AC, Branham KE, et al. Association of apolipoprotein E alleles with susceptibility to age-related macular degeneration in a large cohort from a single center. Invest Ophthalmol Vis Sci 2004;45:1306–1310.

129. Baird PN, Guida E, Chu DT, Vu HT, Guymer RH. The epsilon2 and epsilon4 alleles of the apolipoprotein gene are associated with age-related macular degeneration. Invest Ophthalmol Vis Sci 2004;45:1311–1315.

130. Kimura K, Isashiki Y, Sonoda S, Kakiuchi-Matsumoto T, Ohba N. Genetic association of manganese superoxide dismutase with exudative age-related macular degeneration. Am J Ophthalmol 2000;130:769–773.

131. Ikeda T, Obayashi H, Hasegawa G, et al. Paraoxonase gene polymorphisms and plasma oxidized low-density lipoprotein level as possible risk factors for exudative age-related macular degeneration. Am J Ophthalmol 2001;132:191–195.

132. Guymer RH, McNeil R, Cain M, et al. Analysis of the Arg345Trp disease-associated allele of the EFEMP1 gene in individuals with early onset drusen or familial age-related macular degeneration. Clin. Experiment. Ophthalmol 2002;30:419–423.

133. Baird PN, Chu D, Guida E, Vu HT, Guymer R. Association of the M55L and Q192R paraoxonase gene polymorphisms with age-related macular degeneration. Am J Ophthalmol 2004;138:665–666.

134. Sauer CG, White K, Stohr H, et al. Evaluation of the G protein coupled receptor-75 (GPR75) in age related macular degeneration. Br J Ophthalmol 2001;85:969–975.

135. Zurdel J, Finckh U, Menzer G, Nitsch RM, Richard G. CST3 genotype associated with exudative age related macular degeneration. Br J Ophthalmol 2002;86:214–219.

136. Stone EM, Braun TA, Russell SR, et al. Missense variations in the fibulin 5 gene and age-related macular degeneration. N Engl J Med 2004;351:346–353.

137. Kenealy SJ, Schmidt S, Agarwal et al. Linkage analysis for age-related macular degeneration supports a gene on chromosome 10q26. Mol Vis 2004;10:57–61.

138. Iyengar SK, Song D, Klein BE, et al. Dissection of genomewide-scan data in extended families reveals a major locus and oligogenic susceptibility for age-related macular degeneration. Am J Hum Genet 2004;74:20–39.

139. Majewski J, Schultz DW, Weleber RG, et al. Age-related macular degeneration—a genome scan in extended families. Am J Hum Genet 2003;73:540–550.

140. Klein RJ, Zeiss C, Chew EY, et al. Complement Factor H Polymorphism in Age-Related Macular Degeneration. Science 2005;308:362–364.

141. Edwards AO, Ritter IR, Abel KJ, Manning A, Panhuysen C, Farrer LA. Complement Factor H Polymorphism and Age-Related Macular Degeneration. Science 2005;308:421–424.

142. Haines JL, Hauser MA, Schmidt S, et al. Complement Factor H Variant Increases the Risk of Age-Related Macular Degeneration. Science 2005;308:419–421.

143. Gorin MB, Breitner JC, de Jong PT, et al. The genetics of age-related macular degeneration. Mol Vis 1999;5:29.

144. Allikmets R, Singh N, Sun H, et al. A photoreceptor cell-specific ATP-binding transporter gene (ABCR) is mutated in recessive Stargardt macular dystrophy. Nat Genet 1997;15:236–246.

145. Stone EM, Webster AR, Vandenburgh K, et al. Allelic variation in ABCR associated with Stargardt disease but not age-related macular degeneration [letter]. Nat Genet 1998;20:328–329.

146. Rivera A, White K, Stohr H, et al. A comprehensive survey of sequence variation in the ABCA4 (ABCR) gene in Stargardt disease and age-related macular degeneration. Am J Hum Genet 2000;67:800–813.

147. Daiger SP. Was the Human Genome Project Worth the Effort? Science 2005;308:362–364.

148. Eye Disease Case-Control Study Group. Risk factors for neovascular age-related macular degeneration. Arch Ophthalmol 1992;110:1701–1708.

149. Vinding T, Appleyard M, Nyboe J, Jensen G. Risk factor analysis for atrophic and exudative age-related macular degeneration. An epidemiological study of 1000 aged individuals. Acta Ophthalmol Copenh 1992;70:66–72.

150. Klein R, Klein BE, Linton KL, DeMets DL. The Beaver Dam Eye Study: the relation of age-related maculopathy to smoking. Am J Epidemiol 1993:137:190–200.

151. Vingerling JR, Hofman A, Grobbee DE, de Jong PT. Age-related macular degeneration and smoking. The Rotterdam Study. Arch Ophthalmol 1996:114:1193–1196.

152. Smith W, Mitchell P, Leeder SR. Smoking and age-related maculopathy. The Blue Mountains Eye Study. Arch Ophthalmol 1996;114:1518–1523.

153. McCarty CA, Mukesh BN, Fu CL, Mitchell P, Wang JJ, Taylor HR. Risk factors for age-related maculopathy: the Visual Impairment Project. Arch Ophthalmol 2001;119:1455–1462.

154. Delcourt C, Diaz JL, Ponton Sanchez A, Papoz L. Smoking and age-related macular degeneration. The POLA Study. Pathologies Oculaires Liees al'Age. Arch Ophthalmol 1998;116:1031–1035.

155. Christen WG, Glynn RJ, Manson JE, Ajani UA, Buring JE. A prospective study of cigarette smoking and risk of age-related macular degeneration in men. JAMA 1996;276:1147–1151.

156. Seddon JM, Willett WC, Speizer FE, Hankinson SE. A prospective study of cigarette smoking and age-related macular degeneration in women. JAMA 1996;276:1141–1146.

157. Tamakoshi A, Yuzawa M, Matsui M, Uyama M, Fujiwara NK, Ohno Y. (1997). Smoking and neovascular form of age related macular degeneration in late middle aged males: findings from a case-control study in Japan. Research Committee on Chorioretinal Degenerations. Br J Ophthalmol 1997;81:901–904.

158. Age-Related Eye Disease Study Research Group. Risk factors associated with age-related macular degeneration. A case-control study in the age-related eye disease study: age-related eye disease study report number 3. Ophthalmology 2000;107:2224–2232.

159. DeAngelis MM, Lane AM, Shah CP, Ott J, Dryja TP, Miller JW. Extremely discordant sib-pair study design to determine risk factors for neovascular age-related macular degeneration. Arch Ophthalmol 2004;122:575–580.

160. Klaver CC, Assink JJ, Vingerling JR, Hofman A, de Jong PT. Smoking is also associated with age-related macular degeneration in persons aged 85 years and older: The Rotterdam Study [letter]. Arch Ophthalmol 1997;115:945.

161. Mitchell P, Wang JJ, Smith W, Leeder SR. Smoking and the 5-year incidence of age-related maculopathy: the Blue Mountains Eye Study. Arch Ophthalmol 2002;120: 1357–1363.

162. Klein R, Klein BE, Moss SE. Relation of smoking to the incidence of age-related maculopathy. The Beaver Dam Eye Study. Am J Epidemiol 1998;147:103–110.

163. Klein R, Klein BE, Tomany SC, Moss SE. Ten-year incidence of age-related maculopathy and smoking and drinking: the Beaver Dam Eye Study. Am J Epidemiol 2002;156: 589–598.

164. Kelly SP, Edwards R, Elton P, Mitchell P. Age related macular degeneration: smoking entails major risk of blindness. BMJ 2003;326:1458–1459.

165. Suner IJ, Espinosa-Heidmann DG, Marin-Castano ME, Hernandez EP, Pereira-Simon S, Cousins SW. Nicotine increases size and severity of experimental choroidal neovascularization. Invest Ophthalmol Vis Sci 2004;45:311–317.

166. Chan D. Cigarette smoking and age-related macular degeneration. Optom. Vis Sci 1998;75:476–484.

167. DeBlack SS. Cigarette smoking as a risk factor for cataract and age-related macular degeneration: a review of the literature. Optometry 2003;74:99–110.

168. Mitchell P, Chapman S, Smith W. "Smoking is a major cause of blindness": a new cigarette pack warning? Med J Aust 1999;171:173–174.

169. Schwartz D. The Beaver Dam Eye Study: the relation of age-related maculopathy to smoking. Surv Ophthalmol 1994;39:84–85.

170. Obisesan TO, Hirsch R, Kosoko O, Carlson L, Parrott M. Moderate wine consumption is associated with decreased odds of developing age-related macular degeneration in NHANES–1. J Am Geriatr Soc 1998;46:1–7.

171. Ritter LL, Klein R, Klein BE, Mares Perlman JA, Jensen SC. Alcohol use and age-related maculopathy in the Beaver Dam Eye Study. Am J Ophthalmol 1995;120:190–196.

172. Smith W. Mitchell P. Alcohol intake and age-related maculopathy. Am J Ophthalmol 1996;122:743–745.

173. Moss SE, Klein R, Klein BE, Jensen SC, Meuer SM. Alcohol consumption and the 5-year incidence of age-related maculopathy: the Beaver Dam eye study. Ophthalmology 1998;105:789–794.

174. Ajani UA, Christen WG, Manson JE, et al. A prospective study of alcohol consumption and the risk of age-related macular degeneration. Ann Epidemiol 1999;9:172–177.

175. Cho E, Hankinson SE, Willett WC, et al. Prospective study of alcohol consumption and the risk of age-related macular degeneration. Arch Ophthalmol 2000;118:681–688.

176. Sarma U, Brunner E, Evans J, Wormald R. Nutrition and the epidemiology of cataract and age-related maculopathy. Eur J Clin Nutr 1994;48:1–8.

177. Winkler BS, Boulton ME, Gottsch JD, Sternberg P. Oxidative damage and age-related macular degeneration. Mol Vis 1999;5:32.

178. Gu X, Meer SG, Miyagi M, et al. Carboxyethylpyrrole protein adducts and autoantibodies, biomarkers for age-related macular degeneration. J Biol Chem 2003;278:42,027–42,035.

179. Seddon JM, Ajani UA, Sperduto RD, et al. Dietary carotenoids, vitamins A, C, and E, and advanced age-related macular degeneration. Eye Disease Case-Control Study Group [see comments]. J Am Med Assoc 1994;272:1413–1420.

180. Pratt S. Dietary prevention of age-related macular degeneration. J Am Optom Assoc 1999;70:39–47.

181. Beatty S, Murray IJ, Henson DB, Carden D, Koh H, Boulton ME. Macular pigment and risk for age-related macular degeneration in subjects from a Northern European population. Invest Ophthalmol Vis Sci 2001;42:439–446.

182. Horowitz S. Lutein update: protecting the eyes and the heart with the yellow carotenoid. Alt Comp Ther 2000;6:272–277.

183. Eye Disease Case-Control Study Group. Antioxidant status and neovascular age-related macular degeneration. Arch Ophthalmol 1993;111:104–109.

184. Snellen EL, Verbeek AL, Van Den Hoogen GW, Cruysberg JR, Hoyng CB. Neovascular age-related macular degeneration and its relationship to antioxidant intake. Acta Ophthalmol Scand 2002;80:368–371.

185. Gale CR, Hall NF, Phillips DI, Martyn CN. Lutein and zeaxanthin status and risk of age-related macular degeneration. Invest Ophthalmol Vis Sci 2003;44:2461–2465.

186. De La Paz MA, Zhang J, Fridovich I. Red blood cell antioxidant enzymes in age-related macular degeneration. Br J Ophthalmol 1996;80:445–450.

187. Goldberg J, Flowerdew G, Smith E, Brody J, Tso M. Factors associated with age-related macular degeneration. An analysis of data from the first National Health and Nutrition Survey. Am J Epidemiol 1988;128:700–710.

188. Vaicaitiene R, Luksiene DK, Paunksnis A, Cerniauskiene LR, Domarkiene S, Cimbalas A. Age-related maculopathy and consumption of fresh vegetables and fruits in urban elderly. Medicina (Kaunas) 2003;39:1231–1236.

189. Cho E, Seddon JM, Rosner B, Willett WC, Hankinson SE. Prospective study of intake of fruits, vegetables, vitamins, and carotenoids and risk of age-related maculopathy. Arch Ophthalmol 2004;122:883–892.

190. Seddon JM, Hennekens CH. Vitamins, minerals, and macular degeneration. Promising but unproven hypotheses [editorial]. Arch Ophthalmol 1994;112:176–179.

191. Cooper DA, Eldridge AL, Peters JC. Dietary carotenoids and certain cancers, heart disease, and age-related macular degeneration: a review of recent research. Nutr Rev 1999;57:201–214.

192. West S, Vitale S, Hallfrisc J, et al. Are antioxidants or supplements protective for age-related macular degeneration? Arch Ophthalmol 1994;112:222–227.

193. Mares-Perlman JA, Brady WE, Klein R, et al. Serum antioxidants and age-related macular degeneration in a population-based case-control study. Arch Ophthalmol 1995;113:1518–1523.

194. Smith W, Mitchell P, Rochester C. Serum beta carotene, alpha tocopherol, and age-related maculopathy: the Blue Mountains Eye Study. Am J Ophthalmol 1997;124:838–840.

195. Mares-Perlman JA, Fisher AI, Klein R, et al. Lutein and Zeaxanthin in the Diet and Serum and Their Relation to Age-Related Maculopathy in the Third National Health and Nutrition Examination Survey. Am J Epidemiol 2001;153:424–432.

196. Mares-Perlman JA, Klein R, Klein BE, et al. Association of zinc and antioxidant nutrients with age-related maculopathy. Arch Ophthalmol 1996;114:991–997.

197. Smith W, Mitchell P, Webb K, Leeder SR. Dietary antioxidants and age-related maculopathy: the Blue Mountains Eye Study. Ophthalmology 1999;106:761–767.

198. Kuzniarz M, Mitchell P, Flood VM, Wang JJ. Use of vitamin and zinc supplements and age-related maculopathy: the Blue Mountains Eye Study. Ophthalmic Epidemiol 2002;9:283–295.

199. Delcourt C, Cristol JP, Tessier F, Leger CL, Descomps B, Papoz L. Age-related macular degeneration and antioxidant status in the POLA study. POLA Study Group. Pathologies Oculaires Liees al'Age. Arch Ophthalmol 1999;117:1384–1390.

200. Delcourt C, Cristol JP, Leger CL, Descomps B, Papoz L. Associations of antioxidant enzymes with cataract and age-related macular degeneration. The POLA Study. Pathologies Oculaires Liees al'Age. Ophthalmology 1999;106:215–222.

201. VandenLangenberg GM, Mares Perlman JA, Klein R, Klein BE, Brady WE, Palta M. Associations between antioxidant and zinc intake and the 5-year incidence of early age-related maculopathy in the Beaver Dam Eye Study. Am J Epidemiol 1998;148:204–214.

202. Christen WG, Ajani UA, Glynn RJ, et al. Prospective cohort study of antioxidant vitamin supplement use and the risk of age-related maculopathy. Am J Epidemiol 1999;149:476–484.

203. Cho E, Stampfer MJ, Seddon JM, et al. Prospective study of zinc intake and the risk of age-related macular degeneration. Ann Epidemiol 2001;11:328–336.
204. Flood V, Smith W, Wang JJ, Manzi F, Webb K, Mitchell P. Dietary antioxidant intake and incidence of early age-related maculopathy: the Blue Mountains Eye Study. Ophthalmology 2002;109:2272–2278.
205. Evans JR. Antioxidant vitamin and mineral supplements for age-related macular degeneration. Cochrane Database Syst Rev 2002;volume 2.
206. Richer S. Multicenter ophthalmic and nutritional age-related macular degeneration study—part 2: antioxidant intervention and conclusions. J Am Optom Assoc 1996;67:30–49.
207. Stur M, Tittl M, Reitner A, Meisinger V. Oral zinc and the second eye in age-related macular degeneration. Invest Ophthalmol Vis Sci 1996;37:1225–1235.
208. Teikari JM, Laatikainen L, Virtamo J, et al. Six-year supplementation with alpha-tocopherol and beta-carotene and age-related maculopathy. Acta Ophthalmol Scand 1998; 76:224–229.
209. Richer S, Stiles W, Statkute L, et al. Double-masked, placebo-controlled, randomized trial of lutein and antioxidant supplementation in the intervention of atrophic age-related macular degeneration: the Veterans LAST study (Lutein Antioxidant Supplementation Trial). Optometry 2004;75:216–230.
210. Koh HH, Murray IJ, Nolan D, Carden D, Feather J, Beatty S. Plasma and macular responses to lutein supplement in subjects with and without age-related maculopathy: a pilot study. Exp Eye Res 2004;79:21–27.
211. Seigel D. AREDS investigators distort findings. Arch Ophthalmol 2002;120:100–101.
212. Ambati J, Ambati BK. Age-related eye disease study caveats. Arch Ophthalmol 2002;120:997–999.
213. Tomany SC, Klein R, Klein BE. The relation of coffee and caffeine to the 5-year incidence of early age-related maculopathy: the Beaver Dam Eye Study. Am J Ophthalmol 2001;132:271–273.
214. Seddon JM, Cote J, Rosner B. Progression of age-related macular degeneration: association with dietary fat, transunsaturated fat, nuts, and fish intake. Arch Ophthalmol 2003;121:1728–1737.
215. Mares-Perlman JA, Brady WE, Klein R, VandenLangenberg GM, Klein BE, Palta M. Dietary fat and age-related maculopathy. Arch Ophthalmol 1995;113:743–748.
216. Friedman E. Dietary fat and age-related maculopathy [letter]. Arch Ophthalmol 1996;114:235–236.
217. Cho E, Hung S, Willett WC, Spiegelman D, et al. Prospective study of dietary fat and the risk of age-related macular degeneration. Am J Clin Nutr 2001;73:209–218.
218. Snow KK, Seddon JM. Do age-related macular degeneration and cardiovascular disease share common antecedents? Ophthalmic Epidemiol 1999;6:125–143.
219. Espinosa-Heidmann DG, Sall J, Hernandez EP, Cousins SW. Basal laminar deposit formation in APO B100 transgenic mice: complex interactions between dietary fat, blue light, and vitamin E. Invest Ophthalmol Vis Sci 2004;45:260–266.
220. Sarks JP, Sarks SH, Killingsworth MC. Evolution of soft drusen in age-related macular degeneration. Eye 1994;8:269–283.
221. Sanders TA, Haines AP, Wormald R, Wright LA, Obeid O. Essential fatty acids, plasma cholesterol, and fat-soluble vitamins in subjects with age-related maculopathy and matched control subjects. Am J Clin Nutr 1993;57:428–433.
222. Smith W, Mitchell P, Leeder SR. Dietary fat and fish intake and age-related maculopathy. Arch Ophthalmol 2000;118:401–404.
223. Heuberger RA, Mares-Perlman JA, Klein R, Klein BE, Millen AE, Palta M. Relationship of dietary fat to age-related maculopathy in the Third National Health and Nutrition Examination Survey. Arch Ophthalmol 2001;119:1833–1838.

224. Seddon JM, Rosner B, Sperduto RD, et al. Dietary fat and risk for advanced age-related macular degeneration. Arch Ophthalmol 2001;119:1191–1199.
225. McGwin G Jr., Owsley C, Curcio CA, Crain RJ. The association between statin use and age related maculopathy. Br J Ophthalmol 2003;87:1121–1125.
226. Wilson HL, Schwartz DM, Bhatt HR, McCulloch CE, Duncan JL. Statin and aspirin therapy are associated with decreased rates of choroidal neovascularization among patients with age-related macular degeneration. Am J Ophthalmol 2004;137:615–624.
227. McCarty CA, Mukesh BN, Guymer RH, Baird PN, Taylor HR. Cholesterol-lowering medications reduce the risk of age-related maculopathy progression. Med J Aust 2001; 175:340.
228. Klein R, Klein BE, Tomany SC, Danforth LG, Cruickshanks KJ. Relation of statin use to the 5-year incidence and progression of age-related maculopathy. Arch Ophthalmol 2003;121:1151–1155.
229. Klein R, Klein BE, Jensen SC, et al. Medication use and the five-year incidence of early age-related maculopathy: the Beaver Dam Eye Study. Arch Ophthalmol 2001; 119:1354–1359.
230. van Leeuwen R, Vingerling JR, Hofman A, de Jong PT, Stricker BH. Cholesterol lowering drugs and risk of age related maculopathy: prospective cohort study with cumulative exposure measurement. BMJ 2003;326:255–256.
231. Holz FG, Piguet B, Minassian DC, Bird AC, Weale RA. Decreasing stromal iris pigmentation as a risk factor for age- related macular degeneration. Am J Ophthalmol 1994;117:19–23.
232. Chaine G, Hullo A, Sahel J, et al. Case-control study of the risk factors for age related macular degeneration. France-DMLA Study Group. Br J Ophthalmol 1998;82:996–1002.
233. Sandberg MA, Gaudio AR, Miller S, Weiner A. Iris pigmentation and extent of disease in patients with neovascular age-related macular degeneration. Invest Ophthalmol Vis Sci 1994;35:2734–2740.
234. Vinding T. Pigmentation of the eye and hair in relation to age-related macular degeneration. An epidemiological study of 1000 aged individuals. Acta Ophthalmol Copenh 1990;6:53–58.
235. Mitchell P, Smith W, Wang JJ. Iris color, skin sun sensitivity, and age-related maculopathy: the Blue Mountains Eye Study. Ophthalmology 1998;105:1359–1363.
236. Klein R, Klein BE, Jensen SC, Cruickshanks KJ. The relationship of ocular factors to the incidence and progression of age-related maculopathy. Arch Ophthalmol 1998;116:506–513.
237. Wang JJ, Jakobsen K, Smith W, Mitchell P. Five-year incidence of age-related maculopathy in relation to iris, skin or hair colour, and skin sun sensitivity: the Blue Mountains Eye Study. Clin Experiment Ophthalmol 2003;31:317–321.
238. Tomany SC, Klein R, Klein BE. The relationship between iris color, hair color, and skin sun sensitivity and the 10-year incidence of age-related maculopathy: the Beaver Dam Eye Study. Ophthalmology 2003;110:1526–1533.
239. Darzins P, Mitchell P, Heller RF. Sun exposure and age-related macular degeneration. An Australian case-control study. Ophthalmology 1997;104:770–776.
240. Berendschot T, Willemse-Assink JJ, Bastiaanse M, de Jong PT, van Norren D. Macular pigment and melanin in age-related maculopathy in a general population. Invest Ophthalmol Vis Sci 2002;43:1928–1932.
241. Taylor HR, Munoz B, West S, Bressler NM, Bressler SB, Rosenthal FS. Visible light and risk of age-related macular degeneration. Trans Am Ophthalmol Soc 1990;88:163–173.
242. Cruickshanks KJ, Klein R, Klein BE. Sunlight and age-related macular degeneration: the Beaver Dam Eye Study. Arch Ophthalmol 1993;111:514–518.
243. Delcourt C, Carriere I, Ponton-Sanchez A, Fourrey S, Lacroux A, Papoz L. Light exposure and the risk of age-related macular degeneration: the Pathologies Oculaires Liees al'Age (POLA) study. Arch Ophthalmol 2001;119:1463–1468.

244. Pham TQ, Wang JJ, Rochtchina E, Mitchell P. Pterygium/pinguecula and the five-year incidence of age-related maculopathy. Am J Ophthalmol 2005;139:536–537.
245. West SK, Rosenthal FS, Bressler NM, et al. Exposure to sunlight and other risk factors for age-related macular degeneration. Arch Ophthalmol 1989;107:875–879.
246. Cruickshanks KJ, Klein R, Klein BE, Nondahl DM. Sunlight and the 5-year incidence of early age-related maculopathy: the Beaver Dam Eye Study. Arch Ophthalmol 2001; 119:246–250.
247. Tomany SC, Cruickshanks KJ, Klein R, Klein BE, Knudtson MD. Sunlight and the 10-year incidence of age-related maculopathy: the Beaver Dam Eye Study. Arch Ophthalmol 2004;122:750–757.
248. Hirvela H, Luukinen H, Laara E, Sc L, Laatikainen L. Risk factors of age-related maculopathy in a population 70 years of age or older. Ophthalmology 1996;103:871–877.
249. Miyazaki M, Nakamura H, Kubo M, et al. Risk factors for age related maculopathy in a Japanese population: the Hisayama study. Br J Ophthalmol 2003;87:469–472.
250. Marin-Castano ME, Elliot SJ, Potier M, et al. Regulation of estrogen receptors and MMP–2 expression by estrogens in human retinal pigment epithelium. Invest Ophthalmol Vis Sci 2003;44:50–59.
251. Espinosa-Heidmann DG, Marin-Castano ME, Pereira-Simon S, Hernandez EP, Elliot S, Cousins SW. Gender and estrogen supplementation increases severity of experimental choroidal neovascularization. Exp Eye Res 2005;80:413–423.
252. Vingerling JR, Dielemans I, Witteman JC, Hofman A, Grobbee DE, de Jong PT. Macular degeneration and early menopause: a case-control study. BMJ 1995;310: 1570–1571.
253. Smith W, Mitchell P, Wang JJ. Gender, oestrogen, hormone replacement and age-related macular degeneration: results from the Blue Mountains Eye Study. Aust New Zealand J Ophthalmol 1997;25 Suppl 1:S13–S15.
254. Klein BE, Klein R, Jensen SC, Ritter LL. Are sex hormones associated with age-related maculopathy in women? The Beaver Dam Eye Study. Trans Am Ophthalmol Soc 1994;92:289–295.
255. Defay R, Pinchinat S, Lumbroso S, Sutan C, Delcourt C. Sex steroids and age-related macular degeneration in older French women: the POLA study. Ann Epidemiol 2004;14:202–208.
256. Klein BE, Klein R, Lee KE. Reproductive exposures, incident age-related cataracts, and age-related maculopathy in women: the beaver dam eye study. Am J Ophthalmol 2000; 130:322–326.
257. Snow KK, Cote J, Yang W, Davis NJ, Seddon JM. Association between reproductive and hormonal factors and age-related maculopathy in postmenopausal women. Am J Ophthalmol 2002;134:842–848.
258. Abramov Y, Borik S, Yahalom C, et al. The effect of hormone therapy on the risk for age-related maculopathy in postmenopausal women. Menopause 2004;11:62–68.
259. Rulon LL, Robertson JD, Lovell MA, Deibel MA, Ehmann WD, Markesber WR. Serum zinc levels and Alzheimer's disease. Biol Trace Elem Res 2000;75:79–85.
260. Klein R, Klein BE, Jensen SC, Moss SE, Cruickshanks KJ. The relation of socioeconomic factors to age-related cataract, maculopathy, and impaired vision: the Beaver Dam Eye Study. Ophthalmology 1994;101:1969–1979.
261. Klein R, Klein BE, Jensen SC, Moss SE. The relation of socioeconomic factors to the incidence of early age-related maculopathy: the Beaver Dam Eye Study. Am J Ophthalmol 2001;132:128–131.
262. Huang GH, Klein R, Klein BE, Tomany SC. Birth cohort effect on prevalence of age-related maculopathy in the Beaver Dam Eye Study. Am J Epidemiol 2003;157: 721–729.

263. Vingerling JR, Dielemans I, Bots ML, Hofman A, Grobbee DE, de Jong PT. Age-related macular degeneration is associated with atherosclerosis. The Rotterdam Study. Am J Epidemiol 1995;142:404–409.

264. Sperduto RD, Hiller R. Systemic hypertension and age-related maculopathy in the Framingham Study. Arch Ophthalmol 1986;104:216–219.

265. Hyman L, Schachat AP, He Q, Leske MC. Hypertension, cardiovascular disease, and age-related macular degeneration. Age-Related Macular Degeneration Risk Factors Study Group. Arch Ophthalmol 2000;118:351–358.

266. Smith W, Mitchell P, Leeder SR, Wang JJ. Plasma fibrinogen levels, other cardiovascular risk factors, and age-related maculopathy: the Blue Mountains Eye Study. Arch Ophthalmol 1998;116:583–587.

267. Delcourt C, Michel F, Colvez A, Lacroux A, Delage M, Vernet MH. Associations of cardiovascular disease and its risk factors with age-related macular degeneration: the POLA study. Ophthalmic Epidemiol 2001;8:237–249.

268. Klein R, Klein BE, Franke T. The relationship of cardiovascular disease and its risk factors to age-related maculopathy: the Beaver Dam Eye Study. Ophthalmology 1993;100:406–414.

269. Schaumberg DA, Christen WG, Hankinson SE, Glynn RJ. Body mass index and the incidence of visually significant age-related maculopathy in men. Arch Ophthalmol 2001;119:1259–1265.

270. Seddon JM, Cote J, Davis N, Rosner B. Progression of age-related macular degeneration: association with body mass index, waist circumference, and waist-hip ratio. Arch Ophthalmol 2003;121:785–792.

271. Klein R, Klein BE, Jensen SC. The relation of cardiovascular disease and its risk factors to the 5-year incidence of age-related maculopathy: the Beaver Dam Eye Study. Ophthalmology 1997;104:1804–1812.

272. Klein R, Klein BEK, Tomany SC, Cruickshanks KJ. The association of cardiovascular disease with the long-term incidence of age-related maculopathy: the Beaver Dam Eye Study. Ophthalmology 2003;110:1273–1280.

273. van Leeuwen R, Ikram MK, Vingerling JR, Witteman JC, Hofman A, de Jong PT. Blood pressure, atherosclerosis, and the incidence of age-related maculopathy: the Rotterdam Study. Invest Ophthalmol Vis Sci 2003;44:3771–3777.

274. Wang JJ, Mitchell P, Rochtchina E, Tan AG, Wong TY, Klein R. Retinal vessel wall signs and the 5-year incidence of age related maculopathy: the Blue Mountains Eye Study. Br J Ophthalmol 2004;88:104–109.

275. van Leeuwen R, Klaver CC, Vingerling JR, et al. Cholesterol and age-related macular degeneration: is there a link? Am J Ophthalmol 2004;137:750–752.

276. Voutilainen-Kaunisto RM, Terasvirta ME, Uusitupa MI, Niskanen LK. Age-related macular degeneration in newly diagnosed type 2 diabetic patients and control subjects: a 10-year follow-up on evolution, risk factors, and prognostic significance. Diabetes Care 2000;23:1672–1678.

277. Clemons TE, Kurinij N, Sperduto RD. Associations of mortality with ocular disorders and an intervention of high-dose antioxidants and zinc in the Age-Related Eye Disease Study: AREDS Report No. 13. Arch Ophthalmol 2004;122:716–726.

278. Wu KH, Wang JJ, Rochtchina E, Foran S, Ng MK, Mitchell P. Angiotensin-converting enzyme inhibitors (ACEIs) and age-related maculopathy (ARM): cross-sectional findings from the Blue Mountains Eye Study. Acta Ophthalmol Scand 2004;82:298–303.

279. Christen WG, Glynn RJ, Ajani UA, et al. Age-related maculopathy in a randomized trial of low-dose aspirin among US physicians. Arch Ophthalmol 2001;119:1143–1149.

280. Penfold PL, Provis JM, Billson FA. Age-related macular degeneration: ultrastructural studies of the relationship of leucocytes to angiogenesis. Graefes Arch Clin Exp Ophthalmol 1987;225:70–76.

281. Espinosa-Heidmann DG, Suner IJ, Hernandez EP, Monroy D, Csaky KG, Cousins SW. Macrophage depletion diminishes lesion size and severity in experimental choroidal neovascularization. Invest Ophthalmol Vis Sci 2003;44:3586–3592.

282. Ross R. Atherosclerosis—an inflammatory disease. N Engl J Med 1999;340:115–126.

283. Margolis TP, Lietman T, Strauss E. Infectious agents and ARMD: a connection? Am J Ophthalmol 2004;138:468–470.

284. Klein R, Klein BE, Tomany SC, Cruickshanks J. Association of emphysema, gout, and inflammatory markers with long-term incidence of age-related maculopathy. Arch Ophthalmol 2003;121:674–678.

285. Seddon JM, Gensler G, Milton RC, Klein ML, Rifai N. Association between C-reactive protein and age-related macular degeneration. JAMA 2004;291:704–710.

286. Kalayoglu MV, Galvan C, Mahdi OS, Byrne GI, Mansour S. Serological association between Chlamydia pneumoniae infection and age-related macular degeneration. Arch Ophthalmol 2003;121:478–482.

287. Miller DM, Espinosa-Heidmann DG, Legra J, et al. The association of prior cytomegalovirus infection with neovascular age-related macular degeneration. Am J Ophthalmol 2004;138:323–328.

288. Cousins SW, Espinosa-Heidmann DG, Csaky KG. Monocyte activation in patients with age-related macular degeneration: a biomarker of risk for choroidal neovascularization? Arch Ophthalmol 2004;122:1013–1018.

289. Ishida O, Oku H, Ikeda T, Nishimura M, Kawagoe K, Nakamura K. Is Chlamydia pneumoniae infection a risk factor for age related macular degeneration? Br J Ophthalmol 2003;87:523–524.

290. Wang JJ, Mitchell P, Smith W, Gillies M, Billson F. Systemic use of anti-inflammatory medications and age-related maculopathy: the Blue Mountains Eye Study. Ophthalmic Epidemiol 2003;10:37–48.

291. Klein R, Klein BE, Moss SE. Diabetes, hyperglycemia, and age-related maculopathy: the Beaver Dam Eye Study. Ophthalmology 1992;99:1527–1534.

292. Mitchell P, Wang JJ. Diabetes, fasting blood glucose and age-related maculopathy: The Blue Mountains Eye Study. Aust New Zealand J Ophthalmol 1999;27:197–199.

293. Sperduto RD, Hiller R, Seigel D. Lens opacities and senile maculopathy. Arch Ophthalmol 1981;99:1004–1008.

294. Liu IY, White L, LaCroix AZ. The association of age-related macular degeneration and lens opacities in the aged. Am J Public Health 1989;79:765–769.

295. Klein R, Klein BE, Wang Q, Moss SE. Is age-related maculopathy associated with cataracts? Arch Ophthalmol 1994;112:191–196.

296. Wang JJ, Mitchell PG, Cumming RG, Lim R. Cataract and age-related maculopathy: the Blue Mountains Eye Study. Ophthalmic Epidemiol 1999;6:317–326.

297. Oliver M. Posterior pole changes after cataract extraction in elderly subjects. Am J Ophthalmol 1966;62:1145–1148.

298. Blair CJ, Ferguson J, Jr. Exacerbation of senile macular degeneration following cataract extraction. Am J Ophthalmol 1979;87:77–83.

299. Klein BE, Klein R, Moss SE. Exogenous estrogen exposures and changes in diabetic retinopathy. The Wisconsin Epidemiologic Study of Diabetic Retinopathy. Diabetes Care 1999;22:1984–1987.

300. van der Schaft TL, Mooy CM, de Bruijn WC, Mulder PG, Pameyer JH, de Jong PT. Increased prevalence of disciform macular degeneration after cataract extraction with implantation of an intraocular lens. Br J Ophthalmol 1994;78:441–445.

301. Pollack A, Marcovich A, Bukelman A, Oliver M. Age-related macular degeneration after extracapsular cataract extraction with intraocular lens implantation. Ophthalmology 1996;103:1546–1554.

302. Pollack A, Marcovich A, Bukelman A, Zalish M, Oliver M. Development of exudative age-related macular degeneration after cataract surgery. Eye 1997;11:523–530.
303. Armbrecht AM, Findlay C, Aspinall PA, Hill AR, Dhillon B. Cataract surgery in patients with age-related macular degeneration: one-year outcomes. J Cataract Refract Surg 2003;29:686–693.
304. Klein R, Klein BE, Wong TY, Tomany SC, Cruickshanks KJ. The association of cataract and cataract surgery with the long-term incidence of age-related maculopathy: the Beaver Dam eye study. Arch Ophthalmol 2002;120:1551–1558.
305. Klein BE, Klein R, Moss SE. Incident cataract surgery: the Beaver Dam Eye Study. Ophthalmology 1997;104:573–580.
306. Wang JJ, Klein R, Smith W, Klein BEK, Tomany S, Mitchell P. Cataract surgery and the 5-year incidence of age-related macular degeneration: Pooled findings from the Beaver Dam and Blue Mountains Eye Studies. Ophthalmology 2003;110:1960–1967.
307. Tan AG, Wang JJ, Rochtchina E, Jakobsen K, Mitchell P. Increase in cataract surgery prevalence from 1992–1994 to 1997–2000: analysis of two population cross-sections. Clin Experiment Ophthalmol 2004;32:284–288.
308. Boker T, Fang T, Steinmetz R. Refractive error and choroidal perfusion characteristics in patients with choroidal neovascularization and age-related macular degeneration. Ger J Ophthalmol 1993;2:10–13.
309. Maltzman BA, Mulvihill MN, Greenbaum A. Senile macular degeneration and risk factors: a case-control study. Ann Ophthalmol 1979;11:1197–1201.
310. Sandberg MA, Tolentino MJ, Miller S, Berson EL, Gaudio AR. Hyperopia and neovascularization in age-related macular degeneration. Ophthalmology 1993;100:1009–1013.
311. Wang JJ, Mitchell P, Smith W. Refractive error and age-related maculopathy: the Blue Mountains Eye Study. Invest. Ophthalmol Vis Sci 1998;39:2167–2171.
312. Ikram MK, van Leeuwen R, Vingerling JR, Hofman A, de Jong PT. Relationship between refraction and prevalent as well as incident age-related maculopathy: the Rotterdam Study. Invest Ophthalmol Vis Sci 2003;44:3778–3782.
313. Wong TY, Klein R, Klein BE, Tomany SC. Refractive errors and 10-year incidence of age-related maculopathy. Invest Ophthalmol Vis Sci 2002;43:2869–2873.
314. Wang JJ, Jakobsen KB, Smith W, Mitchell P. Refractive status and the 5-year incidence of age-related maculopathy: the Blue Mountains Eye Study. Clin Experiment Ophthalmol 2004;32:255–258.
315. Mangione CM, Gutierrez PR, Lowe G, Orav EJ, Seddon JM. Influence of age-related maculopathy on visual functioning and health-related quality of life. Am J Ophthalmol 1999;128:45–53.
316. Foran S, Wang JJ, Mitchell P. Causes of incident visual impairment: the blue mountains eye study. Arch Ophthalmol 2002;120:613–619.
317. Chia EM, Wang JJ, Rochtchina E, Smith W, Cumming RG, Mitchell P. Impact of Bilateral Visual Impairment on Health-Related Quality of Life: the Blue Mountains Eye Study. Invest Ophthalmol Vis Sci 2004;45:71–76.
318. Maguire M. Baseline characteristics, the 25-Item National Eye Institute Visual Functioning Questionnaire, and their associations in the Complications of Age-Related Macular Degeneration Prevention Trial (CAPT). Ophthalmology 2004;111:1307–1316.
319. La Forge RG, Spector WD, Sternberg J. The relationship of vision and hearing impairment to one-year mortality and functional decline. J Aging Health 1992;4:126–148.
320. Rudberg MA, Furner SE, Dunn JE, Cassel CK. The relationship of visual and hearing impairments to disability: an analysis using the longitudinal study of aging. J Gerontol 1993;48:M261–M265.
321. Carabellese C, Appollonio I, Rozzini R, et al. Sensory impairment and quality of life in a community elderly population. J Am Geriatr Soc 1993;41:401–407.

322. Salive ME, Guralnik J, Glynn RJ, Christen W, Wallace RB, Ostfeld AM. Association of visual impairment with mobility and physical function. J Am Geriatr Soc 1994;42: 287–292.

323. Rubin GS, Roche KB, Prasada Rao P, Fried LP. Visual impairment and disability in older adults. Optom Vis Sci 1994;71:750–760.

324. Dargent Molina P, Hays M, Breart G. Sensory impairments and physical disability in aged women living at home. Int J Epidemiol 1996;25:621–629.

325. Lee PP, Spritzer K, Hays RD. The impact of blurred vision on functioning and well-being. Ophthalmology 1997;104:390–396.

326. Lee P, Smith JP, Kington R. The relationship of self-rated vision and hearing to functional status and well-being among seniors 70 years and older. Am J Ophthalmol 1999;127: 447–452.

327. Reuben DB, Mui S, Damesyn M, Moore AA, Greendale GA. The prognostic value of sensory impairment in older persons. J Am Geriatr Soc 1999;47:930–935.

328. Harries U, Landes R, Popay J. Visual disability among older people: a case study in assessing needs and examining services. J Public Health Med 1994;16:211–218.

329. Ivers RQ, Cumming RG, Mitchell P, Attebo K. Visual impairment and falls in older adults: the Blue Mountains Eye Study. J Am Geriatr Soc 1998;46:58–64.

330. Ivers R, Cumming R, Mitchell P. Poor vision and risk of falls and fractures in older Australians: the Blue Mountains Eye Study. NSW Public Health Bull 2002;13:8–10.

331. Ivers RQ, Norton R, Cumming RG, Butler M, Campbell AJ. Visual impairment and risk of hip fracture. Am J Epidemiol 2000;152:633–639.

332. Ivers RQ, Optom B, Cumming RG, Mitchell P, Simpson JM, Peduto AJ. Visual risk factors for hip fracture in older people. J Am Geriatr Soc 2003;51:356–363.

333. Brody BL, Gamst AC, Williams RA, et al. Depression, visual acuity, comorbidity, and disability associated with age-related macular degeneration. Ophthalmology 2001;108: 1893–1900.

334. Rovner BW, Casten RJ, Tasman WS. Effect of depression on vision function in age-related macular degeneration. Arch Ophthalmol 2002;120:1041–1044.

335. Casten RJ, Rovner BW, Tasman W. Age-related macular degeneration and depression: a review of recent research. Curr Opin Ophthalmol 2004;15:181–183.

336. Klein R, Cruickshanks KJ, Klein BE, Nondahl DM, Wiley T. Is age-related maculopathy related to hearing loss? Arch Ophthalmol 1998;116:360–365.

337. Klaver CC, Ott A, Hofman A, Assink JJ, Breteler MM, de Jong PT. Is age-related maculopathy associated with Alzheimer's disease? The Rotterdam Study. Am J Epidemiol. 1999;150:963–968.

338. Wong TY, Klein R, Nieto FJ, et al. Is early age-related maculopathy related to cognitive function? The Atherosclerosis Risk in Communities Study. Am J Ophthalmol 2002;134:828–835.

339. Wang JJ, Mitchell P, Smith W. Vision and low self-rated health: the Blue Mountains Eye Study. Invest Ophthalmol Vis Sci 2000;41:49–54.

340. Wang JJ, Mitchell P, Cumming RG, Smith W. Visual impairment and nursing home placement in older Australians: the Blue Mountains Eye Study. Ophthal Epidemiol 2002;10: 3–13.

341. Kleiner RC, Enger C, Alexander MF, Fine SL. Contrast sensitivity in age-related macular degeneration. Arch Ophthalmol 1988;106:55–57.

342. Klein R, Klein BE, Moss SE. Age-related eye disease and survival: the Beaver Dam Eye Study. Arch Ophthalmol 1995;113:333–339.

343. Wang JJ, Mitchell P, Simpson JM, Cumming RG, Smith W. Visual impairment, age-related cataract, and mortality. Arch Ophthalmol 2001;119:1186–1190.

344. Taylor HR, McCarty CA, Nanjan MB. Vision impairment predicts five-year mortality. Trans Am Ophthalmol Soc 2000;98:91–96.

345. Buch H, Vinding T, la Cour M, Jensen GB, Prause JU, Nielsen NV. Age-related maculopathy: a risk indicator for poorer survival in women: the Copenhagen City Eye Study. Ophthalmology 2005;112:305–312.

346. Owen CG, Fletcher AE, Donoghue M, Rudnicka AR. How big is the burden of visual loss caused by age related macular degeneration in the United Kingdom? Br J Ophthalmol 2003;87:312–317.

347. Hopley C, Carter R, Mitchell P. Measurement of the economic impact of visual impairment from age-related macular degeneration in Australia. Clin Experiment Ophthalmol 2003;31:522–529.

348. Lanchoney DM, Maguire MG, Fine SL. A model of the incidence and consequences of choroidal neovascularization secondary to age-related macular degeneration. Comparative effects of current treatment and potential prophylaxis on visual outcomes in high-risk patients. Arch Ophthalmol 1998;116:1045–1052.

349. Hopley C, Salkeld G, Wang JJ, Mitchell P. Cost utility of screening and treatment for early age-related macular dageneration with zinc and antioxidants. Br J Ophthalmol 2004;88:450–454.

350. Wright AF. Strategies for mapping susceptibility genes in age-related maculopathy. Eye 2001;15:401–406.

351. Thakkinstian A, Bowe S, McEvoy M, Smith W, Attia J. Association between apolipoprotein E polymorphisms and age-related macular degeneration: A HuGE review and meta-analysis. Am J Epidemiol 2006 (in press).

352. Hageman GS, Anderson DH, Johnson LV, et al. A common haplotype in the complement regulatory gene factor H (HF1/CFH) predisposes individuals to age-related macular degeneration. Proc Natl Acad Sci USA 2005;102:7227–7232.

353. Jakobsdottir J, Conely YP, Weeks DE, Mah TS, Ferrell RE, Gorin MB. Susceptibility genes for age-related maculopathy on chromosome 10q26. Am J Hum Genet 2005;77: 389–407.

354. Rivera A, Fisher SA, Fritsche LG et al. Hypothetical LOC387715 is a second major susceptibility gene for age-related macular degeneration, contributing independently of complement factor H to disease risk. Hum Mol Genet 2005;14:3227–3236.

355. Schmidt S, Hauser MA, Scott WK et al. Cigarette smoking strongly modifies the association of LOC387715 and age-related macular degeneration. Am J Hum Genet 2006;78:852–864.

356. Gold B, Merriam JE, Zernant J, et al. Variation in factor B (BF) and complement component 2 (C2) genes is associated with age-related macular degeneration. Nat Genet 2006;38: 458–462.

357. Haan MN, Klein R, Klein BE, et al. Hormone therapy and age-related macular degeneration: the Women's Health Initiative Sight Exam Study. Arch Ophthalmol 2006;124:988–992.

358. Klein R, Klein BE, Knudrson MD, Wong TY, Shankar A, Tsai MY. Systemic markers of inflammation, endothelial dysfunction, and age-related maculopathy. Am J Ophthalmol 2005;140:35–44.

3

Leber Congenital Amaurosis

A Hereditary Childhood Form of Blindness and a Model to Elucidate Retinal Physiology and Development

Robert K. Koenekoop, MD, PhD

CONTENTS

INTRODUCTION

Leber congenital amaurosis ([LCA], MIM 204000) is an important, currently untreatable congenital retinal dystrophy that inexorably leads to blindness. Its importance is twofold and lies in the fact that it creates a tremendous burden on the affected child, the family, and society, as the blindness is life long and commences at birth. Also, LCA gene discoveries have led to an increased understanding of the molecular determinants of retinal physiology and retinal development by identifying new biochemical and cellular pathways. Therefore, LCA serves as a model for all human retinal dystrophies and human retinal development and physiology. LCA has a worldwide prevalence of 3 in 100,000 newborns and accounts for 5% or more of all inherited retinopathies and approx 20% of children attending schools for the blind *(1)*. We estimate that 180,000 patients are affected worldwide *(2)*. Leber defined LCA in 1869 as a congenital form of retinitis pigmentosa (RP) with profound visual loss at birth, nystagmus, amaurotic pupils, and a pigmentary retinopathy *(3)*. A severely reduced electroretinogram (ERG) was added to the definition as this distinguishes it from a complex set of overlapping retinal dystrophies *(4)*.

LCA is both genetically and clinically heterogeneous, and since 1996, 10 genes/loci participating in a wide variety of retinal functional pathways have been implicated in the disease process. LCA associated proteins participate in phototransduction (GUCY2D) *(5)*, vitamin A metabolism (retinal pigment epithelium [RPE]65, Retinal dehydrogenase 12 [RDH12] *(6,7)*, photoreceptor development (cone-rod homeobox [CRX]) *(8)*,

From: *Ophthalmology Research: Retinal Degenerations: Biology, Diagnostics, and Therapeutics*
Edited by: J. Tombran-Tink and C. J. Barnstable © Humana Press Inc., Totowa, NJ

photoreceptor cell development and structure (Crumbs homologue 1 [CRB1]) *(9)*, biosynthesis of cGMP phosphodiesterase (PDE) (aryl hydrocarbon interacting protein like-1 [AIPL1]) *(10)*, and protein trafficking (RP GTPase Regulator Interacting Protein-1 [RPGRIP1]) *(11)*. The LCA genes at chromosomal loci 6q11 *(12)*, 14q24 *(13)*, and 1p36 *(14)* remain to be identified. The seven currently known LCA genes account for approx 40% of the cases, whereas genes underlying the remaining 60% of cases await discovery. Histopathological variability of LCA cases has also been reported and three possible groups may be distinguished *(2)*. In group I, extensive retinal degeneration was reported *(15–21)*; in group II, the retinal architectural appearance is essentially normal, suggesting a dysfunction *(22–24)*; and in group III, primitive abnormal cells are reported in the outer nuclear layer (ONL) of the retina, suggesting an aplasia or hypoplasia *(22,25,26)*. This is unlike the pathology reported for the later-onset RP *(27–29)*. Based on visual function, it is also possible to group patients into three categories *(2)*. In group A, stable function was documented; in group B, declining function was noted; and in group C, small improvements were found *(30–33)*. It is also now evident that LCA and RP animal models respond to gene replacement and drug therapy *(34–43)*. The purpose of this chapter is to summarize currently known biological, diagnostic, and therapeutic concepts as they relate to LCA.

DIFFERENTIAL DIAGNOSIS OF LCA

The differential diagnosis of congenital blindness with nystagmus (which includes LCA) can be difficult, as at least seven ocular phenotypes overlap with LCA. In a retrospective study of 75 "LCA patients," 30 patients had a different diagnosis on later analysis, including congenital stationary nightblindness, achromatopsia, joubert syndrome, zellweger syndrome, infantile refsum disease, and juvenile RP *(33)*. Differentiating these various entities is important for counseling parents and children regarding exact diagnosis, prognosis, and management: (1) certain types of congenital blindness may be associated with systemic disease; (2) some forms are stationary, others progressive; (3) future potential therapies with gene replacement, cell transplants, or drugs will be disease specific; and (4) genetic counseling. A study by Weiss and Bierhoff in 1979 *(44)* revealed that patients with congenital blindness visit many ophthalmologists before the final diagnosis is made. This final diagnosis is usually made in the teenage years. However, with carefully chosen diagnostic tests, LCA could be diagnosed before the age of 1 yr. The differential diagnosis of LCA includes: complete achromatopsia, incomplete achromatopsia, complete congenital stationary night blindness (cCSNB), incomplete congenital stationary night blindness (iCSNB), albinism, optic nerve hypoplasia, and aniridia.

Patients with complete achromatopsia, also known as rod monochromatism, present with striking photoaversion and blepharospasm in the light; whereas, in the dark, their eyes open and their visual functioning appears to improve. Typically, patients remain stable throughout life with 20/200 acuities, and they have absent color vision on D15, FM100, and HRR testing. The ERG is diagnostic and reveals absent or severely reduced cone function with essentially normal rod function. Retinal appearance may be completely normal although mild foveal abnormalities have been reported. Complete achromatopsia is an autosomal recessive (AR) condition and mutations in three genes encoding phototransduction proteins including both the α and β subunits of the cGMP-gated channel proteins (CNGA3, CNGB3) *(45,46)* and the α subunit of cone transducin (GNAT1) have been reported *(47)*. Incomplete achromatopsia or blue cone monochromacy (BCM) occurs only in males and also presents with photoaversion. It is an X-linked recessive (XR) condition and is called incomplete achromatopsia because blue cones function normally in BCM, whereas in the complete form of achromatopsia they do not. A D-15 test will reveal protan and deutan axes of confusion in the BCM patient but the tritan axis is intact, whereas in complete achromatopsia, there is no color vision. This feature can be tested during an eye examination and used to distinguish the two in an older child by color vision discrimination. Acuities in BCM range from 20/50 to 20/200 and a prominent maculopathy may develop, unlike complete achromatopsia. The molecular biology of BCM is characterized by a loss of the 5′ locus control center of the X-linked red and green genes, which occur in tandem and lead to inactivation of both genes. Alternatively unequal homologous recombination and a point mutation in the remaining gene have been found as a cause of BCM.

CSNB is an important, common and heterogeneous group of retinal dysfunctions, with significant clinical overlaps with LCA. Several reports have found that blind infants initially diagnosed with LCA were reclassified as CSNB following repeat ERG testing *(48)*. CSNB may be inherited as an autosomal dominant (AD), autosomal recessive (AR) or X linked recessive (XR) condition. An electronegative ERG is the hallmark finding in CSNB, which identifies an ERG where the b-wave amplitude is less than the a-wave amplitude of the scotopic ERG. An electronegative ERG is not diagnostic for CSNB, as this type of ERG may also be found in retinoschisis and central retinal artery occlusion and other disease entities. Two types of ERGs are further found in CSNB, the Riggs type ERG, with an absent a-wave, and the Schubert-Bornschein ERG, in which the a-wave is present. Patients with CSNB with Riggs type ERG abnormalities have gene defects encoding phototransduction proteins; "Rhodopsin type" CSNB harbors heterozygous rhodopsin mutations, "Nougaret type" CSNB patients have heterozygous α rod transducin mutations lesions, whereas "Rambusch type" were found to have heterozygous mutations in the β subunit of rod cGMP PDE gene *(49)*. Children with CSNB usually see better in the day, although they usually report very significant night blindness, and their visual course is usually stable, and acuities range from 20/20 to 20/200. AR patients with CSNB with the Oguchi phenotype harbour mutations in phototransduction genes as well, both arrestin and rhodopsin kinase (RK) mutations have been described *(49)*. Another AR CSNB type of disease, fundus albipunctatus, was found to harbour mutations in 11-*cis* retinal dehydrogenase *(49)*. X-linked patients with CSNB have Schubert-Bornschein ERGs (in which the a-wave is clearly present) and are divided into two types: complete CSNB (cCSNB) with relatively normal cone waves on photopic ERG testing, and incomplete CSNB (iCSNB) in which these cone waves are absent. cCSNB is associated with lesions in nyctalopin (NYX), whereas iCSNB results from mutations in the voltage-gated L-type α subunit of the calcium channel protein (CACNA1F) *(49)*. Patients with albinism present with photophobia and nystagmus, and can usually be easily distinguished from LCA in that the acuities may range from normal to 20/200, their hypopigmented retinal appearance with foveal hypoplasia, by increased decussation of the ganglion cells axons (detectable by visual evoked potentials), and normal or supernormal ERGs. Many gene-encoding proteins involved in melanin production have been mutated in the different forms of albinism. Optic nerve hypoplasia (ONH) is an important and relatively common developmental abnormality of the optic nerve and mimics LCA, as it presents with poor visual fixation and nystagmus and the abnormal optic nerve appearance may not be obvious. ONH may be unilateral, bilateral, total, or segmental and may be part of the De Morsier syndrome (also known as septo-optic dysplasia), which consists of absent corpus callosum, absent septum pellucidum, and pan-hypopituitarism. Because of the decrease in the number of ganglion cells, the foveal mound appears shallow, and the optic nerve appears small and may appear pale within a normal-sized scleral canal, creating the "double-ring" size. Acuities are variable and range from 20/20 to no light perception. Distinguishing features with LCA are the optic nerve appearance and a normal ERG. Aniridia is a dominantly inherited pan-ocular disorder and presents with nystagmus and poor fixation. The iris is always abnormal, but this ranges from subtle changes to almost complete loss of iris tissue. In the former instance, aniridia may be difficult to distinguish from LCA. Distinguishing features are the foveal hypoplasia and normal ERG of aniridia.

LCA must also be distinguished from an LCA ocular phenotype, which may be part of a systemic disease. Nephronophtisis (NPHP) is a cystic kidney disorder which may be present with congenital blindness as part of the Senior-Loken syndrome (MIM 266900), whereas skeletal anomalies or osteopetrosis may be found with congenital blindness in the Saldino-Mainzer syndrome (cone-shaped epiphyses of the hand bones and ataxia; MIM 266920). Cerebellar hyperplasia with blindness is known as the l'Hermitt-Duclos syndrome. The Joubert syndrome consists of congenital blindness with cerebellar vermis hypoplasia, oculomotor difficulties, and respiratory problems (MIM 243910). Other systemic disorders with an LCA-type phenotype that must be distinguished from LCA are the Bardet-Biedl syndrome, Alstrom syndrome, abetalipoproteinemia, the peroxisomal disorders (such as infantile Refsum disease, neonatal adrenoleuko-dystrophy, and Zellweger disease), and Batten disease (also known as neuronal ceroid lipofuscinosis).

BIOLOGY OF THE LCA GENES

AIPL1

AIPL1 was discovered and cloned by Sohocki et al. *(10)*, localized to 17p13.1, found to contain eight exons, and is expressed in retina *(10)*. This gene is mutated in approx 7% of LCA patients and in an unknown percentage of patients with RP and cone-rod degeneration (CRD) *(50)*. The AIPL1 protein shows 70% similarity to AIP and both possess three consecutive tetratricopeptide (TPR) repeats. TPR domains are sites of protein–protein interactions, in particular with the C terminus of partner proteins. Sequence comparisons place AIPL1 in a family of proteins (FK-506-binding protein family) with molecular chaperone functions *(51)*. The chaperone function of this family of proteins is not at the initial step of polypeptide folding, but rather in the more specialized steps of maturation, subunit assembly, transport, or degradation *(51)*. Many of the client proteins are components of signal transduction pathways such as phototransduction in photoreceptors. In a yeast two hybrid study by Akey et al, using a bovine retinal cDNA library and *AIPL1* as bait, AIPL1 was found to interact with Nedd8-ultimate buster 1 (NUB1), a protein thought to play an important role in the Wnt/Frizzled pathway that is critical in regulating cell cycle progression and cell fate specification in the developing retina *(52)*. *AIPL1* mutations from patients with LCA were expressed and their interactions with *NUB1* were tested in vitro. Most, but not all, *AIPL1* mutations significantly reduced the interactions between the two proteins. Both *AIPL1* and *NUB1* were found to be expressed in retinoblastoma Y79 cells, a known retinal progenitor cell. *In situ* hybridization showed expression of both *NUB1* and *AIPL1* in inner segments (IS) of photoreceptors in both embryonic and post natal mice *(52)*. *AIPL1* is predominantly cytoplasmic, but *NUB1* is primarily nuclear *(53)*. Van der Spuy demonstrated that *AIPL1* modulates the nuclear translocation of *NUB1* *(53)*. Kanaya et al. *(54)* determined that *AIPL1* directly interacts with *NUB1* through AIPL1's amino acid residues 181–330, and then showed that I206N, G262S, R302L, and P376S *AIPL1* mutations are able to interact directly with *NUB1*, but *AIPL1* mutations W278X, A197P, and C239R do not *(54)*. This suggests that the former mutations cause LCA, not through an aberrant interaction with NUB1, but through a different mechanism *(54)*.

In a second yeast two hybrid study by Ramamurthy et al., *AIPL1* was found to interact with farnesylated proteins *(51)*. Farnesylation is a type of prenylation, which represents a posttranslational modification of the C terminal of several proteins. Several retinal proteins involved in phototransduction are known to be farnesylated *(51)*, including cGMP PDE, transducin, and RK. Prenylation of proteins is a three-step process *(55)*. Step 1 involves the attachment of a farnesyl group to the cysteine residue of the CAAX box-containing protein, catalyzed by farnesyltransferase. During step 2, the newly farnesylated proteins are targeted to the endoplasmic reticulum (ER), where the last three amino acids are removed. In step 3, the proteins are carboxymethylated in the ER at the newly created farnesyl cysteine groups. Finally, once the posttranslational modifications are completed, the newly farnesylated proteins are transported to their target membranes *(51,55)*. Ramamurthy et al. *(51)* showed that *AIPL1* interacts specifically with farnesylated proteins, and that in cells lacking farnesyl transferase (which catalyzes the farnesylation process), the farnesylated proteins fail to interact with *AIPL1*. They also showed that *AIPL1* enhances this process by 2.5-fold. Three classes of *AIPL1* mutations have been associated with various human retinal dystrophies *(51)*. Class I mutations include missense mutations in the N-terminal part of *AIPL1*, whereas class II mutations are missense mutations in the TPR domains, and nonsense mutations that lack one or more TPR domains. Class III mutations are deletion type mutations in the C terminal of *AIPL1*. Class I and II are associated with LCA, whereas class III are associated with juvenile RP and CRD. Experiments with class I and II mutations (M79T, A197P, C239R, and W278X) by Ramamurthy et al. revealed that the mutant *AIPL1* does not interact with farnesylated proteins, and the most common mutations W278X (a class II mutation) causes complete loss of AIPL1. Surprisingly, the well-known R302L mutation interacts normally with the farnesylated proteins, suggesting that it is not a disease causing variation. They found that *AIPL1* is expressed in both the IS and the synaptic terminals of the outer plexiform layer (OPL) *(51)*. Also, both A197P and C239R forms of AIPL1 were unable to interact with the farnesylated proteins *(51)*, but maintained the ability to interact with NUB1 *(52)*.

The divergent results of the two studies *(51,52)* would suggest that there are two possibilities for the essential roles of *AIPL1* in the retina and, therefore, two potential causes for the development of LCA as a result of aberrant *AIPL1* functioning. One is that because of *AIPL1's* interaction with *NUB1*, the essential role of *AIPL1* may be to control photoreceptor proliferation and/or differentiation. Based on this interaction, the severity of LCA caused by *AIPL1* defects, and the early expression of both *AIPL1* and *NUB1* in the embryological retina, it is reasonable to suppose that the essential role of *AIPL1* may be in the initial development of photoreceptors. Secondly, because of *AIPL1's* interaction with farnestylated retinal proteins, *AIPL1's* role may be to enhance farnestylation of all or one of the known farnesylated proteins (PDE, RK, and or transducin). Because prenylation has been shown to maintain retinal architecture, photoreceptor structure, stability of cGMP *PDE*, and *PDE's* ability to associate with its target membrane, *AIPL1's* role is then suggested to be photoreceptor maintenance.

The development of three separate *AIPL1* knockout (KO) mice *(19,56,57)* has led to new insights into the essential role of *AIPL1* in retinal development and functioning, and resolved the divergent results of the two-yeast hybrid studies. The *AIPL1* KO mouse shows that rods and cones develop normally, but degenerate quickly *(19,56,57)*, starting on day 12, is complete in week 4 and proceeds centrifugally (from the center to the periphery) *(19)*. Ramamurthy et al. found that ERG responses were not present in the KO at any age, and the photoreceptors were disorganized and the IS contained vacuolar inclusions *(19)*. Although rhodopsin, guanylyl cyclase, transducin, and recoverin protein levels were normal, there was a striking reduction of all three subunits of *PDE*, which corresponded with a 7- to 9-fold increase in the levels of cGMP. Immunoblot analysis showed that the α, β, and γ subunits of PDE protein levels were reduced by 90%, whereas mRNA levels were normal *(19)*, indicating a posttranslational abnormality. The link between rapid retinal degeneration and elevated cGMP is well known from a well-studied mouse retinal degeneration model rd/rd, in which there are null mutations in the *PDE* gene, and the cGMP levels are thought to be toxic *(54,58,60)*. Therefore, cGMP elevation may be a common disease pathway in retinal degeneration *(19)*. Therefore, in the absence of normal *AIPL1*, it is possible that *PDE* is not properly or efficiently farnesylated, and becomes unstable and is degraded *(19)*. The result of mutated *AIPL1* in LCA may therefore be a toxic elevation of cGMP, and consequently elevated intracellular calcium levels and rapid apoptosis. An alternative but similar interpretation may be that *AIPL1* acts as a chaperone that promotes folding and the assembly of *PDE* and that the cGMP phenotype results from *PDE* subunit destabilization *(57)*. In their *AIPL1* KO, Liu et al. *(57)* also found a progressive photoreceptor degeneration after apparent normal development of both rods and cones. They also found a lower abundance of rod *PDE*, but not of other retinal proteins, including farnesylated proteins such as transducin and RK. cGMP levels and calcium levels were not significantly elevated which was unexpected *(57)*. Dyer et al. *(56)* created a KO *AIPL1* mouse and also found a near total loss of photoreceptors. In addition, they found abnormal ERGs in the heterozygote, which correlates with our findings in the heterozygous human parents of LCA offspring with *AIPL1* mutations *(61–63)*. Immunostaining showed a reduced number of cones, a possible expansion of bipolar cells, a marked increase in Müller cell reactive gliosis, and significant disruption of the ribbon synapse formation in the OPL *(56)*. Microarray analysis confirmed the marked reduction of *PDE*, but also found significant reductions of two *Wnt* genes, important in retinal proliferation and cell fate determination. Real time reverse-transcriptase polymerase chain reaction (RT-PCR) for markers of retinal cell proliferation and cell fate at several different and key developmental stages demonstrated normal expression of 25 probes (including *Rb, Nr2e3, Nrl, Rpr1*, and *GFAP*). Only the expression of the secreted frizzled-2 gene was markedly increased *(56)*. These findings suggest that, despite *AIPL1's* known interaction with a cell cycle regulatory protein, *NUB1, AIPL1* is not required for retinal proliferation or retinal cell fate determination, but that *AIPL1* is involved in a crucial photoreceptor function and OPL synapse formation. This function relates to the phototransduction protein PDE, which becomes unstable in the absence of *AIPL1*, either because of an abnormality in farnesylation of PDE *(19)*, or interruption of the chaperone function of *AIPL1* upon PDE *(57)*. Interestingly, *AIPL1* is expressed only in mature rods (not in mature cones) *(64)*, but is expressed in both developing rods and cones *(65)*. Because of *PDE's* early expression in the retina (cones and rods in early development, but only rods later in life), a possible role of *PDE* in the *Wnt-frizzled* signalling pathway, coupled with the significant reductions of two *Wnt* genes and an increase in frizzled-2 in the *AIPL1* KO mouse, is that *AIPL1* performs a dual role in retinal function *(56)*.

The clinical phenotype of human patients with *AIPL1* lesions is particularly severe as most patients have hand motion, light perception or no light perception vision, and the retinal appearances reveal extensive degenerative changes including macular colobomas *(61,66)*. We have delineated the disease phenotype of *AIPL1* type LCA patients, by screening 303 patients with LCA and identifying 26 patients who carried two *AIPL1* mutations (17 were homozygotes, 9 were compound heterozygotes) *(61)*. The patients with the *AIPL1* phenotype all presented with poor or no visual fixation at or shortly after birth, wandering nystagmus and amaurotic pupils, whereas the ERGs, which were done in the majority of cases, were nondetectable. Compared to the published clinical phenotypes of the other five LCA types, we found overlapping features but also found the following: in our cohort *(61)*, the phenotype included an atrophic or pigmentary maculopathy in 100% of the cases, varying degrees of midperipheral chorioretinal atrophy with intraretinal pigment migration in all patients over age 6 yr, keratoconus and cataracts in 33%, and optic disc pallor in 100% after age 6. We also established for the first time that *AIPL1* heterozygotes have significant rod ERG abnormalities (but normal cone function), which correlates with the predominant rod-expression profile of the adult *(64)*. *AIPL1* carriers may also develop patchy peripheral retinal atrophy and older *AIPL1* carriers may develop cone and rod ERG abnormalities *(62,67)*.

In conclusion, *AIPL1* lesions lead to a particularly severe form of LCA often associated with a maculopathy, keratoconus, cataracts, and optic disc pallor in the affected patients *(61)* and subclinical but recognizable ERG and retinal abnormalities in the obligate carrier parents *(61,62)*. The molecular mechanism appears to be a toxic elevation of cGMP, through a defect in the chaperone function *(57)* or defective farnesylation *(19)* of the critical phototransduction protein cGMP phosphodiesterase (Table 1).

CRX

The *CRX* gene *(68)* resides on 19q13.3, has three exons, and encodes a protein of 299 amino acids. *CRX* is mutated in a wide variety of retinal dystrophies, including LCA,

Table 1
Proposal for a Molecular Classification of LCA

Molecular defect	Presumed mechanism	Evidence (ref.)
1. cGMP elevation		
AIPL1	Defective biosynthesis of cGMP PDE	57,19
CRX	Lowered expression of rhodopsin and γ subunit of cGMP PDE	71,72
RPGRIP1	Mislocalization of rhodopsin, normal localization of cGMP PDE	146, no direct evidence
2. cGMP depression		
GUCY2D	Lowered cGMP production	105–107
RPE65	Constitutively active opsin, with lowering of cGMP	129
RDH12	Constitutively active opsin, with lowering of cGMP	No direct evidence
3. OLM fragmentation and ectopic photoreceptor formation		
CRB1	Loss of polarity and loss of zonula adherens	84,86

CRD, and RP. The gene encodes a paired-type homeobox transcription factor that belongs to the orthodenticle (OTX) family and is expressed by both photoreceptors. Members of the OTX family of proteins participate in regulating early neuronal development. The expression of *CRX* commences very early in retinal development, at the time when the photoreceptors first appear. The CRX protein binds to specific upstream DNA sequence elements in vitro and transactivates several important photoreceptor genes, such as rhodopsin, the β subunit of rod cGMP *PDE*, arrestin, and interphotoreceptor retinoid binding protein. *CRX* also interacts synergistically with *NRL*, a transcription factor expressed both in retina and the brain. *CRX* controls outer segment (OS) formation *(69,70)*, and mutant *CRX* is associated with maldevelopment of the OS of both rods and cones *(70)*.

The *CRX* KO mouse was developed by Furukawa et al. *(69)*. When comparing retinal sections of wild-type (WT) *CRX* ⁻/⁻ and *CRX*⁻/⁺, they found no differences between the retinas at postnatal day 10 (P10) (this is before OS form). At P14, however, OS were normal in WT, not present in the KO, and shorter than normal in the heterozygote. ERGs were recorded in 1-mo-old mice, and WT was normal, in the KO both rod and cone responses were abnormal, and in the heterozygote, the rod signals were abnormally decreased, whereas the cone ERG was not yet present. At 6 mo, the rod and cone ERG of the heterozygote was normal. At P10, the expression of several key photoreceptor genes, including rhodopsin, cone opsins, rod transducin, arrestin, and recoverin were significantly decreased. *CRX* was not found to be essential for cell fate determination as markers for cell types were found to be normal *(69)*.

Further analyses of the *CRX* KO were done to understand the changes of the inner retina. Using immunocytochemical techniques on *CRX* KO mice retinas, Pignatelli et al. *(71)* showed some very unexpected inner retinal changes, which are of major significance to our understanding of LCA retinal pathology and our expectations that future therapy may be feasible. Rod bipolar cells are normally connected to rod spherules by large dendritic arbors in the OPL, and to the inner plexiform layer (IPL) by axonal arbors, and their postnatal development is complete at around 1 mo. In the *CRX* KO, two important changes were found, first the axonal endings to the IPL were abnormally small, and later the dendritic arborisation to the OPL was completely retracted. Similar but later changes were found in the cone bipolars. In the horizontal cells, the authors found intense neurite sprouting that increased in size and number during the mouse's lifetime. Ganglion cell, and both cholinergic and dopaminergic amacrine cell morphology, were surprisingly normal. The histological similarity with the rd mouse (in which cGMP PDE is mutated and leads to cGMP elevated levels of cGMP) *(60)* is striking and provides indirect evidence that the molecular mechanisms of *CRX* defects and rd defects may be similar *(71)*. These aforementioned morphological changes are remarkably similar to a completely different RP animal model, namely the rd/rd mouse, in which there is a complete absence of the phototransduction enzyme the β subunit of rod phosphodiesterase, and a presumed high level of cGMP, which leads to a very rapid photoreceptor degeneration *(60)*. Despite the marked differences in genes, and presumed mechanisms, the histological picture is surprisingly similar, leading the authors to postulate that second-order neuron modification and remodeling in the retina depends on photoreceptor degeneration itself, not on the genetic mutation that causes the retina to degenerate in the first place. Furthermore, microarray analysis of $CRX^{+/+}$ (WT) and $CRX^{-/-}$ (KO) showed differential expression of 16 retinal genes, 5 of which encode phototranduction pathway proteins *(72)*. Livesey et al. *(72)* showed that rhodopsin, rod transducin (α subunit), recoverin, rod cGMP phosphodiesterase (γ subunit), and arrestin, providing indirect evidence that cGMP may be elevated. Tsang et al. *(73)* showed that KO of the γ subunit of cGMP *PDE* leads to a rapid retinal degeneration with elevated levels of cGMP, similar to the rd mouse (with mutant cGMP *PDE*) *(60)*.

In an attempt to elucidate the biological actions of *CRX* mutations and to understand whether they act in a dominant negative fashion vs a loss of function, Rivolta et al. *(74)* divided the reported *CRX* mutations into two categories; the "missense" group consists of 13 missense mutations, and short inframe deletions which preferentially affect residues of the homeodomain, whereas the 9 mutations in the "frameshift" group consists of insertions and deletions of 1, 2, or 4 bp are all in the terminal exon 3 and affect the OTX domain of the protein. Because frameshift mutations lead to premature termination codons, they are usually assumed to create null alleles. Because *CRX's* frameshifts occur in the terminal exon, this is probably not the case. Although recent new knowledge about mRNA decay mechanisms indicates that premature termination codon (PTC) lead to rapid mRNA decay (and therefore a null allele), PTC mutations in the terminal exon of a gene escape this detection mechanism. Therefore, Rivolta et al. *(74)* predict that both the "missense" and "frameshift" groups of *CRX* mutations will lead to a stable RNA message. They also predict that the "missense" group in the homeodomain will interfere with *CRX's* ability to interact with NRL or with DNA binding, whereas the "frameshift" group of mutations will produce truncated *CRX* molecules that fail to transactivate the promotors of targetted retinal photoreceptor genes *(74)*.

In humans with LCA, both heterozygous and homozygous *CRX* mutations have been found. Swaroop et al. *(75)* showed that recessive R90W mutations in the homeodomain of the CRX protein cause the LCA phenotype. They also showed that the CRX R90W protein shows reduced binding to DNA sequence elements, and the R90W mutation also reduces CRX mediated transactivation of the rhodopsin promotor in vitro experiments. The carrier parents with the heterozygous R90W *CRX* mutations developed a mild CRD, which consisted of photoaversion, dark adaptation difficulties, subtle ERG abnormalities of both rod and cone systems, color vision abnormalities (measured by Ischihara plates), and perifoveal punctate changes of the retina *(75)*. One frameshift *CRX* mutation was found in both an affected LCA child and her normal-vision father, suggesting mosaicism in the father or recessive inheritance in the affected child *(67)*. *CRX* is implicated in both dominant LCA *(74)* and recessive LCA *(75)*, but also in dominant cone-rod dystrophy (CORD) *(68)*, and dominant RP *(76)*. Most authors report a severe phenotype for the *CRX* genotype *(76)* with a profound maculopathy, indicating abnormal foveal development, whereas we have reported *(77)* one patient with LCA with marked improvement in acuity, visual field, and cone ERG, when measured over a period of 11 yr. A patient with LCA with a heterozygous Ala 177 (1bp del) frameshift mutation was diagnosed with LCA at age 6 wk. Repeat ERGs showed a nondetectable rod and cone signal. At age 6, we measured acuities at 20/900 and this improved to 20/150 at age 11. An ERG at age 11 showed an electronegative cone signal with a small a- and b-wave, whereas the visual field was measurable by Goldman perimetry. We provide clinical evidence that in some patients with *CRX*, postnatal OS lengthening and improved vision may take place after the initial insult of the *CRX* lesion *(77)*.

In conclusion, the *CRX* defects lead to a highly variable phenotypic spectrum of disease. *CRX* mutations may lead to a dominant CORD *(68)* or a dominant or recessive form of LCA *(74)*. *CRX* defects appear to be a rare cause of LCA *(78,79)*. Patients affected with LCA have moderately severe visual loss, often associated with a macular lesion, although one report noted improvements in several visual function parameters *(61)*. The obligate carrier parents appear to have a mild cone-rod dysfunction (on ERG) *(75)*. The histological phenotype is very similar to the rd mouse (with a cGMP *PDE* defect and elevated cGMP levels) *(71)*, and the expression of the γ subunit of cGMP *PDE* is markedly decreased *(72)*, suggesting that the molecular mechanism of *CRX* lesions, may be cGMP elevation (Table 1).

CRB1

CRB1 was cloned by den Hollander et al. in the Netherlands *(80)*. The gene resides on 1q31, has 12 exons, and encodes a protein of 1376 amino acids and harbors mutations that have been found in a wide variety of retinal dystrophies *(9,80–83)*. These include AR RP (ARRP) of the original RP12 family, which has a severe phenotype with hyperopia, nystagmus, and preservation of the para-arteriolar RPE (also known as PPRPE type of RP), ARRP without the PPRPE, RP with the Coats-like exudative vasculopathy, and LCA with and without the PPRPE type of retinal changes.

The CRB1 protein shows 35% structural similarity to the Drosophila Crumbs *(CRB)* gene. The similarity with the CRB protein suggests a role for CRB1 in cell–cell interaction and possibly in the maintainance of epithelial cell polarity of ectodermally derived cells. The human CRB1 protein contain a signal peptide, 19 epidermal growth factor-like domains, three laminin A Globular-like domains, a transmembrane domain, and a highly conserved 37 amino acid cytoplasmic domain with a C-terminal ERL1 motif *(82)*. The cytoplasmic domain of the Drosophila CRB and human CRB1 proteins domains, when overexpressed in Drosophila embryos, rescue the Crumbs phenotype to a large degree, suggesting a critical function for this small domain *(82)*.

The architecture of epithelial cells depends on the distribution of cellular junctions and other membrane associated protein complexes. A core component of these complexes is the transmembrane protein CRB. Human photoreceptors are packed together in the ONL of the retina together with processes of Müller glial cells for structural and metabolic support. The establishment and maintenance of apical–basal polarization and cell adhesion is important for the photoreceptors. The apical part of photoreceptors is the OS (which abuts the RPE), whereas the basal part represents the synapse of the photoreceptor with bipolar cells. At the apical site of photoreceptors an adhesion belt named the outer limiting membrane (OLM) contains multiple adherence junctions which are present between photoreceptors and Müller cells. These adherence junctions consist of multiprotein complexes and are linked to the cytoskeleton of the cell for cell shape. The adhesion belt runs along the photoreceptor cells at the division of the IS and the cell-body and nucleus. The potential space between the OLM and the RPE represents the subretinal space and the photoreceptor IS and OS normally project and "float" in this space. Cells lacking CRB fail to organize a continuous zonular adherens and fail to maintain cell polarity. Pellikka et al. *(83)* showed that CRB1 localizes to a subdomain of the photoreceptor apical plasma membrane, namely the IS. They proposed that CRB is a central component of a molecular scaffold that controls zonula adherens assembly. The role of mammalian CRB1, however was still unclear.

Clues about the function of CRB1 came initially from a natural *CRB1* mutant mouse, called retinal degeneration 8 (rd8). This mouse model produces a secreted truncated CRB1 protein of 1207 amino acids, which lacks both the transmembrane domain and the intracellular domain. The rd8 mouse develops striking retinal abnormalities, including focal photoreceptor degeneration and irregularities of the OLM *(84)*. Mehalow et al. *(84)*, reported that *CRB1* localizes to both Müller cells and photoreceptor IS. They found clinically that the rd8 mice developed irregular-shaped large white subretinal spots, more heavily concentrated in the inferonasal quadrant of the retina. These spots correspond histopathologically to regions with retinal folds and pseudorosettes. They also found OLM fragmentation, OS shortening but normal IS. By 5 mo, the OS have virtually disappeared, the IS are swollen, and the Müller cell processes are unusually prominent. The retinal degeneration was focal in appearance, with nearly normal retina present at the edge of a region with severe degeneration *(85)*. The photoreceptor dysplasia and degeneration reported by Mehalow et al. in the rd8 mouse with the *CRB1* mutations strongly vary with the genetic background, suggesting modifier effects from other retinal genes.

The rd8 mutation is probably not a null allele, therefore van de Pavert et al. *(86)* inactivated both *CRB1* alleles (*CRB1*$^{-/-}$) to produce a complete null and examined the resulting mouse retina. There were marked differences between the *CRB1*$^{-/-}$ and the rd8 retinal findings. Two-week- nor 2-mo-old *CRB1*$^{-/-}$ mice had retinal abnormalities. At 3 mo, however, the KO mice developed focal areas of retinal degeneration. The OLM was ruptured and there was protrusion of single or multiple photoreceptors both into the subretinal space and into the inner nuclear layer. One of the most strikng histological findings were double photoreceptors layers (half rosettes). These rosettes developed normal inner segments and a full OLM, very much unlike the rd8 model. This finding suggests that *CRB1* is not essential for the formation of junctional complexes and OLM, but rather for the maintenance of these structures. In 6-mo-old *CRB1*$^{-/-}$ mice large ectopic photoreceptor layers were identified, which were so large that they resembled a "funnel" abutting the ganglion cell and inner limiting membrane. They suggest that the initial insult of the *CRB1* mutation is the loss of the photoreceptor to Müller glial cell adhesion in retinal foci. Light exposure experiments revealed a significant increase in retinal degeneration in the *CRB1*$^{-/-}$ mouse especially inferotemporally; therefore, light enhances the retinal degeneration in the *CRB1*$^{-/-}$ retinas *(86)*. This may be a correlate of the thickened retinas that lacked the distinct layers of the normal retina, found by Jacobson et al. *(87)*, who used in vivo high-resolution microscopy (also known as optical coherence tomography [OCT 3]) in patients with LCA with known *CRB1* mutations. Comparable retinas from patients with LCA with *RPE65* mutations were thinned when examined by the same methodology *(87)*.

Because of the severity of the PPRPE type RP, *CRB1* was postulated also to cause LCA. Den Hollander et al. *(9)* and Lotery et al. *(85)* indeed found that 13% of cases with LCA can be explained by mutations in the *CRB1* gene, making it a common and important gene for LCA. In addition, *CRB1* mutations were identified in five of the nine patients with RP and Coats-like exudative vasculopathy, a severe complication of RP *(9)*. Patients with LCA with *CRB1* mutations in three of seven cases also showed the PPRPE picture *(9)*. An overview of the *CRB1* mutation spectrum can be found in den Hollander et al. *(83)*.

The phenotype of patients with LCA with *CRB1* lesions may be distinct as the following constellation of findings may point to a *CRB1* defect. Visual acuities range from 20/40 to light perception *(61,85)*. Many patients have hyperopic refractions and the retinal appearance, although variable and overlapping with other LCA phenotypes may be distinct when PPRPE or white dots are found. Many but not all patients with LCA with *CRB1* mutations have one of these two features. Thick, abnormally laminated retinas by OCT imaging may be a pathognomonic feature of LCA retinas with degeneration caused by *CRB1* mutations *(87)*. Pigmented paravenous chorioretinal atrophy, a dominant form of RP characterized by paravenous pigmentary deposits, was found to be associated with a heterozygous Val162Met *CRB1* missense mutation in a large RP family *(88)*. Obligate heterozygous parents of offspring with LCA and *CRB1* mutations have an ERG and a multifocal ERG phenotype that may be distinct from the phenotype of obligate heterozygotes with other LCA gene defects *(62,89)*. Cremers hypothesized that LCA may be associated with complete loss of function of CRB1, whereas patients with RP (early-onset RP with and without PPRPE, and RP with Coats-like exudative vasculopathy) in whom the visual loss is more gradual and later in life, may have residual CRB1 function *(90)*. This hypothesis is supported by the facts that 37% of LCA CRB1 alleles are presumed null alleles, whereas only 19% of RP CRB1 alleles appear to be null ($p = 0.01$ by Fisher's exact test) *(90)*.

In conclusion, *CRB1* defects are an important and frequent cause of LCA *(9,85)*. Both the affected status and obligate heterozygotic parents may be recognizable clinically *(61,62)*. The affected *CRB1* phenotype may be relatively mild and distinct, as some patients have PPRPE and relatively good visual function *(61)*. The partial *(84)* and complete *(86)* KO mice models of *CRB1* suggest that the pathology involves fragmentation of the adhesion belt, the OLM, and absence of the zonula adherens, which results in the loss of polarity of the photoreceptors, ectopic development of photoreceptors, and initially a focal retinal degeneration.

GUCY2D

The gene for retinal guanylyl cyclase *GUCY2D* was cloned by Shyjan et al. in 1992 *(91)* and mapped to 17p13.1 by Oliviera et al. *(92)*. In 1995, Camuzat et al. *(93,94)* mapped a gene for LCA to this region (17p13.1) by a homozygosity-mapping strategy using consanguineous LCA families of North African origin. Perrault et al. *(5)* subsequently reported missense and frameshift mutations in *GUCY2D* in four unrelated LCA1 probands of these same families. The human gene contains 20 exons, and has thus far been implicated in AR LCA and, surprisingly, also the later-onset AD CRD *(95)* (also known as CORD6). *GUCY2D* is expressed in the plasma membrane of photoreceptor OS, but at higher levels in cones than in rods *(96,97)*. The protein of *GUCY2D* is GC and is a transmembrane protein that serves a key function in photoreceptor physiology as it synthesizes the intracellular transmitter of photoexcitation guanosine 3′,5′-cyclic monophosphate (cGMP). A unique feature of GC is its regulation by three small Ca^{2+}-binding proteins called GC activating proteins (GCAPs) *(98–101)*. These regulatory proteins sense changes in the cytoplasmic Ca^{2+}-concentration (Ca^{2+}) during illumination and activate GCs when the (Ca^{2+}) decreases below the value in a dark-adapted cell of 500–600 n*M (102)*. GC is responsible for the production of second messenger cGMP after it has been hydrolysed by cGMP *PDE* (for a complete review, *see* ref. *103*). cGMP *PDE* becomes activated after light stimulation and depletes cGMP, resulting in closure of photoreceptor cGMP gated channels and subsequent decreased levels of calcium, which leads to hyperpolarization of the membrane. The drop in calcium stimulates GCAP, to stimulate GC to produce cGMP. GC is an enzyme that is composed of an extracellular ligand-binding N-terminal segment, a transmembrane domain, an internal protein kinase homology region, and a C-terminal catalytic domain. In an attempt to identify the significance of the extracellular domain (ECD), Laura et al. *(104)* created deletion mutants of GC and removed the ECD and showed normal GCAP stimulated cGMP production by the mutant GC, concluding that GCAP interacts with GC through the intracytoplasmic portion. Duda et al. *(105)* showed that a Phe514Ser mutation (found in a patient with LCA) from the kinase homology domain of the protein severely compromised the ability to produce cGMP *(105)*, whereas Rozet et al. *(106)* tested the basal functional capacities of catalytic domain and ECD mutations and found that the catalytic mutations almost completely abolished basal cGMP production. The extracellular domain mutations had no effect on basal cGMP production. Tucker et al. *(107)* then showed that catalytic domain mutations severely compromise the ability of GCAP to stimulate cGMP production, whereas ECD mutations cause only about 50% decrease in catalytic ability *(108)*. Tucker et al. also showed that recessive *GUCY2D* mutations in heterozygous state found in humans with LCA cause dominant negative effects on the WT allele *(108)*, and this correlates with significantly abnormal cone electroretinographic responses in parents who are obligate heterozygotes for these same mutations *(109)*, who have children with LCA.

We were puzzled by the observations that mutations in the same gene, *GUCY2D*, can cause two different diseases with very different onsets and severities; i.e., dominant *GUCY2D* mutations (in the dimerization domain of the protein) are associated with cone-rod dystrophy, whereas recessive *GUCY2D* mutations (found in all other protein domains) are well known causes of the much more severe LCA. These facts prompted us to hypothesize that parents of children with LCA with *GUCY2D* mutations must be obligate heterozygotes for the same mutations and may therefore exhibit a mild cone-rod disease *(109,107)*. We tested this hypothesis by performing detailed ERGs on parents with known *GUCY2D* lesions (P858S, L954P) and documented repeatable and significant cone dysfunctions in four subjects *(109)*. To address the question of whether abnormalities in the heterozygous state were caused by haploinsufficiency or dominant negative effects, we tested the same mutations in an in vitro expression system *(107)*. We constructed the mutations in *GUCY2D* cDNA by site-directed mutagenesis *(108)* and tested the ability of the mutant protein to produce cGMP. First, we showed *(107,110)* that mutations in the catalytic domain dramatically compromise the ability of GCAP to stimulate cGMP production *(107)*. Membrane GCs are thought to exist in a dimeric state *(111,112)* and, therefore, we tested whether the P858S and L954P mutants affect WT RetGC-1 activity when co-expressed. We co-transfected HEK 293 cells with 2.5 µg WT RetGC-1 and 2.5 or 5 µg of either pRc-cytomegalovirus (CMV) (as a control), P858S, or L954P. Western blots showed that the total amount of WT and mutant protein transfected correlates with the amount of protein used in the assays. The addition of increasing amounts of either P858S or L954P reduced GCAP-2 stimulated activity by upto 55%. These results showing a significant decrease in wild-type RetGC-1 activity when L954P or P858S are coexpressed suggest that any heterodimers formed are inactive or poorly active *(107)*. We then expressed mutations from the extracellular domain of RetGC-1 (C105Y and L325P) to investigate their effects on cGMP production and found that these only reduce RetGC1 activity by 50%.

Disruption of the retinal GC gene in mice leads to normal development and normal numbers of rod and cone photoreceptors *(113)*. However, by 5 wk, the number of cones in the mouse KO has been reduced dramatically and rapidly in the null mice compared to age-matched controls *(113)*. The degeneration of cones is not matched by a rod decline. Both numbers and morphology of the rod photoreceptors remains normal. However, the rod responses are also dramatically decreased despite their normal appearance *(113)*. The GC KO in the chicken also leads to rapid retinal degeneration after the photoreceptors appear to have developed normally *(114)*. Light-driven translocation of photoreceptor proteins between inner and outer segment plays an important role in adaptation of photoreceptors to light and this process is disrupted in the mouse GC KO *(115)*. The α cone transducin molecule, which is normally found in the OS, was found in IS and synapses of cones in the GC KO, whereas arrestin, which is normally shifted from the OS to IS during light exposure of the WT mice, was found only in the outer segments of the GC KO *(115)*. This may indicate that the absence of GC may be the biochemical equivalent of light exposure and the sequestration of transducin in the inner segmet may be a protective mechanism in the face of chronic light adaptation and chronic hyperpolarization of the cell membrane *(115)*. The molecular mechanism of photoreceptor cell death in LCA retinas with *GUCY2D* defects may relate to the abnormally low cGMP levels, low calcium levels, chronic hyperpolarization, all consequences of chronic light adaptation. The link between the former biochemical changes and apoptosis still needs to clarified.

In some populations, *GUCY2D* may be the most common LCA gene, with 20% of patients with LCA carrying this genotype *(79,116)*. Others, including our own, found that 8% of LCA is caused by this gene *(78)*. The phenotype of *GUCY2D*-type LCA is distinct *(78,117)*. Dharmaraj et al. found a severe phenotype for patients with LCA with *GUCY2D* defects, but the visual course was stable (over a 20-yr period). Marked hyperopia, photo-aversion, and an essentially normal retinal appearance were also noted *(78)*. Obligate heterozygous parents with *GUCY2D* mutations were found to have mild cone ERG abnormalities *(109)*, and may be recognizable clinically, reducing the time and effort of the molecular diagnostic process *(109)*. Silva et al. determined that a *GUCY2D* mutation modified the phenotype of a patient with LCA with *RPE65* defects *(118)*. Histological studies of a patient with LCA with *GUCY2D* mutations revealed the surprizing presence of cone photoreceptors *(24)*. In this study, Milam et al. *(24)* found a Arg660Gln mutation and a deletion adjacent to the *GUCY2D* gene in an 11-yr-old patient with LCA with light perception vision, +4.00 D hyperopia, and a subtle retinal pigmentary mottling. OCT showed a thinned retina, whereas further histological study showed OS loss, a normal bipolar layer, and a thinned ganglion cell layer *(24)*. Autofluorescence studies (which measure the active lipofuscin deposition in the RPE, and indicates viability of the photoreceptor–RPE complex) showed normal autofluorescence patterns in patients with LCA with *GUCY2D* mutations *(119)*. These data *(24,76,119)* suggest the possibility that patients with LCA with the *GUCY2D* genotype may have viable retinas and may also be amenable to gene therapy.

In conclusion, the LCA disease phenotype associated with *GUCY2D* is a severe congenital cone-rod dystrophy associated with high hyperopia, a relatively normal retinal aspect and a relatively stable course *(76,79,117)*. The obligate heterozygous parents of LCA offspring develop a recognizable subclinical cone dysfunction *(109)*. Histological and autofluorescence studies of affected patients suggest that some photoreceptors remain intact in patients with the *GUCY2D* LCA *(24,119)*. The molecular mechanism appears to be a depression of the level of cGMP (Table 1).

RPE65

The *RPE65* gene was cloned by Hamel et al. *(120)* was found to be exclusively expressed by the RPE and plays an important role in the vitamin A cycle. The gene for *RPE65* is located on 1p22 *(121)* and is composed of 14 exons, whereas the protein has 533 amino acids *(122)*. *RPE65* mutations are associated with a variety of overlapping severe retinal dystrophies ranging from LCA (most severe), AR childhood-onset severe retinal dystrophy, and to juvenile RP (least severe) *(7,123,124)*. In an attempt to discern the function of RPE65, a *RPE65*$^{-/-}$ KO mouse was constructed *(125)*, which revealed that OS discs of rod photoreceptors in *RPE65*$^{-/-}$ mice are disorganized compared to those of *RPE65*$^{+/-}$ and *RPE65*$^{+/+}$ mice. Rod function, as measured by electroretinography, was abolished in *RPE65*$^{-/-}$ mice, although some cone function remained. The remaining visual function in the RPE65$^{-/-}$ mouse was later determined to be generated by the rods, not cones as shown by Seeliger et al. *(126)*.

RPE65[−/−] mice lack rhodopsin as a result of an inability to recycle 11-*cis* retinal, but not opsin apoprotein, whereas all-*trans*-retinyl esters overaccumulate in the RPE, therefore, *RPE65* is necessary for the production of rhodopsin and prevention of the accumulation of retinyl esters *(125)*. The logical conclusion could be that in humans and animals lacking *RPE65*, the accumulation of the retinyl esters is somehow toxic to the RPE and, therefore, involved in the pathogenesis and death of the photoreceptor cells. However, there is also evidence that persistant opsin or rhodopsin signaling can cause photoreceptor pathology, and Fain et al. have named this phenomenon the *equivalent light hypothesis (127,128)*. In an attempt to distinguish these two possibilities (accumulation of retinyl esters vs light independent signaling), Woodruff et al. compared *RPE65* null mice with *RPE65* null/transducin (that have a block in the phototransduction cascade) mice and hypothesized that continuous, light independent opsin (unbound to its ligand 11-*cis* retinal) activity causes photoreceptor degeneration, without retinyl ester accumulation playing a role *(129)*. They postulated that if spontaneous opsin activity simulates the light adapted state of the rod, then there must be a diminished circulating current, a reduced light sensitivity, an accelerated light response kinetics, closure of the cGMP gated channels, and subsequent reduction of the intracellular calcium levels *(129)*. If these predictions are accurate, they argued, then by blocking the transducin-ediated signaling, the photoreceptor degeneration must be prevented. In the *RPE65* KO mouse, Woodruff et al. showed that indeed the circulating currents are much lower, the light sensitivity is diminished, and the turn off of the photoresponse was accelerated in the *RPE65* KO mouse compared to the WT. To test whether spontaneous-unliganded opsin causes photoreceptor death through transducin-mediated signaling, the authors crossbred mice null for *RPE65* with mice null for transducin and examined the retinal histology, by counting the number of photoreceptor nuclei in the ONL. Compared to WT, null/transducin mice, and *RPE65* heterozygotes, which maintain 9–11 rows of nuclei, the *RPE65* KO mice have 6–7 rows at 40 wk. The crossbred mice (KO for *RPE65* and transducin) had 8–10 rows of nuclei at 40 wk and were completely protected from the retinal degeneration. Dark rearing of the *RPE65* KO mouse did not protect the retinal cells from degeneration as expected, because the opsin activation is light independent. Retinyl esters accumulations were found to be 14 times higher in the double KO than in the control mice, showing that the retinyl esters are not responsible for the degeneration. Finally, Ca^{2+} levels were measured and found to be very low as expected in the *RPE65* KO, illustrating the consequences of closed cGMP gated channels *(129)*. How aberrant opsin signaling, closure of the rod-cGMP channels, and low Ca^{2+} leads to death is not currently known, but low Ca^{2+} is known to trigger apoptosis in neurons and photoreceptors *(128,130)*. In addition to the KO mouse, a natural KO of the *RPE65* gene exists in the Briard dog, which harbors a 4-bp deletion in *RPE65*. Because the retinal histology was essentially normal except for RPE lipid vacuoles, the retinal phenotype of the Briard dog was initially called congenital stationary night blindness *(131)*. Further analysis of the phenotype and retinal architecture revealed that the diagnosis was more a progressive retinal dystrophy *(132)*, and a good model for human *RPE65* type retinal dystrophies.

Perrault et al. suggested that, based on the divergence of the underlying molecular defects for patients with *GUCY2D-* and *RPE65*-type LCA, the resulting phenotypes may be identifiable and distinguishable clinically *(117)*. The observation of missense and frameshift *GUCY2D* mutations suggests that the cGMP production in photoreceptor cells is markedly reduced or abolished in LCA *(117)*. As a consequence, the excitation process of rod and cone photoreceptors would be markedly impaired because of constant closure of cGMP-gated cation channels, with chronic hyperpolarization of the membrane. The cGMP concentration in photoreceptor cells would not be restored to the dark level, leading to a situation equivalent to constant light exposure during photoreceptor development. Thus, in contrast to *GUCY2D* mutations, *RPE65* mutations would decrease the rhodopsin production, leading to a situation equivalent to a retina kept in a constant dark state *(117)*. Although it appears possible to separate patients LCA with *GUCY2D* defects from those with *RPE65* defects *(78,117)*, the results of Woodruff et al. *(129)* strongly suggest that both *GUCY2D* and *RPE65* defects lead to similar molecular defects, i.e., abnormal lowering of cGMP and subsequent lowering of the intracellular Ca^{2+} levels, a situation consistent with the equivalent light hypothesis suggested by Fain et al. *(127,128)*.

In the patients harboring *GUCY2D* mutations, no visual improvement was observed, the pendular nystagmus remained unchanged, and visual acuity was reduced to light perception or ability to count fingers held in the visual field. In addition, the patients complained of severe photophobia and usually preferred half light, and significant hyperopia was observed, although visual fields were not recordable because of profound loss of visual acuity *(78,117)*. However, in the patients with *RPE65* mutations, transient improvement in visual function was regularly noted by the parents *(78,117, 133)*. This improvement or adaptation of visual function was later confirmed objectively by Lorenz et al. and Paunescu et al. *(133,134)*. Young children became able to follow light or objects, especially during the daytime. They complained of night blindness and usually preferred bright light. Visual acuity reached 20/100–20/200, mild or no hyperopia was observed, and mild myopia occured occasionally. Finally, the visual field in this group was usually recordable and frequently displayed a peripheric concentric reduction. They concluded that *GUCY2D* defects lead to a functional outcome that is different and recognizable from the functional outcome of patients with *RPE65* defects. Patients with *GUCY2D* develop a congenital stationary cone-rod dystrophy, whereas patients with *RPE65* defects develop a progressive rod-cone dystrophy *(78,117,133)*. Silva et al. found a modification of the *RPE65* phenotype by a *GUCY2D* missense mutation *(118)*.

The relative burden of the *RPE65* locus in LCA may be as high as 16% *(124)*, although most large studies found a burden between 6.1% (11 out of 179 patients LCA) and 8.2% (8 out of 98 patients) *(79,135)*. We screened 275 new patients with LCA and found that 25 (9%) carry at least one *RPE65* mutation, and 7 of these were novel. The new LCA microchip constructed by the Allikmets group, which contains 81 mutant *RPE65* alleles (the largest number of any gene), only identified 2.4% of 200 new patients with LCA with *RPE65* defects *(136,137)*. Strict clinical criteria must be maintained for LCA (including poor fixation, wandering nystagmus, amaurotic pupils and nondetectable ERG) to allow exclusion of milder phenotypes, such as juvenile RP and AR childhood-onset severe retinal dystrophy *(133)*. In a large consanguinous family with homozygous Y368H mutations in *RPE65*, we found a surprising variability in visual evolution, with some patients having stable vision, others declining, whereas yet others were measured to improve, implying the action of modifier effects from genetic or environmental factors or both *(138)*.

We have thus far analyzed nine obligate *RPE65* carriers and found that most carriers have characteristic foveal drusen at a young age and all have normal rod and cone ERG functions *(62,89)*. We found that affected patients with *RPE65* mutations have a characteristic retinal aspect, with a transluscent RPE and areas of atrophy. Histological analysis of the KO mouse and Briard dog shows subtle abnormalities of the RPE, but essentially normal retinal architecture *(131,132)*, although one human fetus with *RPE65* mutations had attenuation of the photoreceptor layer *(139,140)*. Autofluorescence studies confirm the block in the vitamin A recycling process, as the autofluorescence of patients with mutated *RPE65* is very abnormal and low *(119)*. Autofluorescent patterns may provide a clinical flag for patients with the *RPE65* genotype *(119)*. Measurable acuities, visual fields, and small cone ERG are possible in patients with *RPE65* and this possibly makes their phenotype distinct *(78,117,133)*.

Recently, mechanism-based pharmacological therapy has been shown to arrest photoreceptor death in the RPE65$^{-/-}$ mouse model of LCA, based on the emerging knowledge of the genes and their functions in the retina. Van Hooser et al. *(43)*, using recent knowledge that *RPE65* mutations in mice lead to an inability to form 11-*cis* retinal (which binds to rod opsin to form light sensitive rhodopsin), supplemented the mouse diet with a vitamin A derivative and consequently showed short-term improvements in rod physiology. The long-term consequences of this intervention are not known. Theoretically, adding vitamin A to a system, which is known to have a block as a result of a mutant RPE65 protein, could lead to long-term accumulation of a toxic intermediate, making the disease worse. Acland et al. *(34)* studied the effects of *RPE65* replacement in the Briard dog, which harbors a natural, homozygous 4-bp deletion in the *RPE65* gene and is blind at birth. ERG function is nondetectable, despite the normal appearance of the retinas, including the photoreceptor layer in this dog model. Subretinal injections in one eye of three dogs containing the adeno-associated virus with cDNA of dog *RPE65*, with a CMV promotor, B-actin enhancer and internal ribosome entry sequence were performed at age 4 mo *(34)*. Rod-mediated ERGs, cone-mediated ERGs, visual-evoked potential, pupillometry, and behavioral testing all showed dramatic improvements in visual function at about 8 mo of age. Genomic PCR and RT-PCR demonstrated expression of the WT message in the retina and RPE, whereas immunoblots showed persistent RPE65 protein in RPE cells. This is the first study to demonstrate restoration of visual function in a large-eyed animal model with LCA caused by an RPE gene defect, and this persisted for more than 3 yr after one injection (Jean Bennett, personal communication). It is currently not known whether the retinas of human patients with LCA with *RPE65* defects are intact, and whether the photoreceptor layer is present. The time window before the retina undergoes cell death is also not known. Also, it is not known whether photoreceptor gene replacements have similar dramatic effects, which will now have to be evaluated. A human clinical trial involving well-characterized LCA gene defects in babies with LCA may commence in 2005.

In conclusion, the disease phenotype of patients with LCA with *RPE65* defects appears to be recognizable based on retinal appearance, relative mildness of the visual defect, a transient improvement in vision, and an absence of autofluorescence, and has been termed early-onset severe retinal dystrophy *(133)*. The molecular mechanism of disease appears to be very similar to that of patients with *GUCY2D* defects, i.e., a depression of the level of cGMP *(129)*. *RPE65* defects lead to lowered cGMP levels, not through a depressed production, but as a result of constitutively active opsin, which stimulates cGMP *PDE* to continously hydrolyze cGMP *(129)*.

RPGRIP1

The *RPGRIP1* gene was discovered by Roepman et al. *(141)*, Boyle and Wright *(142)*, and Hong et al. *(143)*. *RPGRIP1* is expressed both in rod and cone photoreceptors, but surprisingly also in amacrine cells *(144)*. RPGRIP1 is the molecular partner of RPGR and both proteins co-localize to the connecting cilium, which connects the inner to the OS of the photoreceptor cell *(141–143)*. *RPGR* mutations cause several types of retinal degenerations, including x-linked forms of RP, CRD, and macular dystrophy *(145)*. The RPGRIP1 protein may be a structural component of the ciliary axoneme *(146)*. The genes associated with NPHP, an important cause of end-stage kidney disease (sometimes associated with LCA and/or RP), are also associated with ciliary function, and there is evidence that NPHP proteins interact with the RPGRIP1 protein *(147)*. The exact function of *RPGRIP1* is still unknown, however, and investigations are complicated by the multiple *RPGRIP1* isoforms, which have distinct cellular, subcellular localizations and biochemical properties in the retina *(148,149)*. To understand *RPGR* and *RPGRIP1* function in the retina, the molecular partners of *RPGR* were sought in two-yeast hybrid experiments. By screening bovine cDNA retinas with the RDH of *RPGR* as bait, a novel *RPGRIP* was subsequently identified by three independent groups and the protein bears no homology to any thus far identified retinal proteins *(141–143)*. This protein was named *RPGRIP-1*, and the gene *RPGRIP1* was mapped to chromosome 14q11 *(141–143)*. *RPGRIP1* consists of 3861 bp (i.e., 1287 amino acids) (Gerber et al. *[150]* divided more than 25 exons).

In 2000, Roepman et al. found that *RPGRIP1* contains a C-terminal *RPGR* interacting domain and a coiled-coil domain, which the authors suggested is homologous to proteins involved in vesicular trafficking *(141)*. In vivo expression experiments with *RPGR* mutations (from X linked retinitis pigmentosa [XRRP] patients) were observed to impair RPGR–RPGRIP1 interactions. Both *RPGR* and *RPGRIP1* were found in this study to co-localize to the OS of rod and cone photoreceptors. Based on these observations, the authors concluded that the site of pathology for *RPGRIP1* associated disease is in mediating vesicular transport-mediated processes. In the same year, Boylan and Wright *(142)* also isolated *RPGRIP1* and showed its expression in retina and testis. *RPGR–RPGRIP1* interactions were confirmed by co-immunoprecipitation experiments and *RPGR* mutation co-expression studies *(142)*. Hong et al. then subsequently confirmed that *RPGRIP1* is indeed a molecular partner of *RPGR* and found that both proteins co-localize in the photoreceptor connecting cilium *(143)*. Their data suggested that *RPGRIP1* is a structural component of the ciliary axoneme of both rods and cones and functions to anchor *RPGR* within the cilium. Unlike the suggestions of Roepman et al., the subcellular localization of *RPGRIP1* to the connecting cilium by Hong et al. would suggest that *RPGRIP1* could be critically important for the directional transport of nascent proteins from the IS destined for the OS and that *RPGRIP1* associated disease would be caused by an interruption of these processes.

Disagreements about the subcellular localizations of *RPGRIP1* existed as a result of Roepman et al. finding that *RPGRIP1* was localized in the outer segments and Hong et al. found localization to the connecting cilium. These differences predict important disparities in putative pathophysiology of *RPGRIP1* associated retinal diseases. Mavlyutov et al. *(149)* then showed that *RPGR* and *RPGRIP1* co-localize in restricted foci in cone and rod photoreceptors in both bovine and human retinas but NOT in mice, where the two proteins co-localize to the connecting cilium. They also determined for the first time that *RPGRIP1* expression is not restricted to the photoreceptor cells in bovine and human retinas, but is also expressed in the inner retina, specifically in amacrine cells *(149)*. Therefore, they concluded that species–specific subcellular local-ization of *RPGRIP1* and *RPGR* exists which provides a rationale for the striking disparity of the phenotypes *(149)*.

Subsequently, Castagnet et al. *(144)* provided new evidence that *RPGRIP1* isoforms have distinct cellular, subcellular, and biochemical properties. They found that in bovine and human retinas, *RPGRIP1* can be found in four distinct locales: (1) in amacrine cells, (2) in the OS of rod and cones, (3) in the cytoskeleton of photoreceptors, and (4) around microtubules. In the amacrine cells, *RPGRIP1* specifically localizes to restricted foci at the nuclear pore complexes of the nuclei and it associates with RanBP2 *(144)*. RanBP2 is a neuronal nucleoporin, which has been implicated in mediating rate-limiting steps of the nuclear-cytoplasmic trafficking system. It has also been proposed that RanBP2 forms a flexible filamentous molecule, suggesting that it comprises a major portion of the cytoplasmic fibrils implicated in initial binding of import substrates to the nuclear pore complex *(151)*. It has been proposed that docking and release reactions of cargoes occur at the RanBP2 molecule *(152)*. Therefore, in amacrine cells, it appears that *RPGRIP1* participates in nuclear-cytoplasmic trafficking *(144)*.

The *RPGRIP1* KO mouse was developed by Zhao et al. who found that the *RPGRIP1$^{-/-}$* mouse initially develops a full set of photoreceptors, based on the normal thickness of the ONL *(146)*. The subcellular histology, however, showed abnormalities of the OS, even as early as P15. OS were disorganized, contained oversized discs, and pycnotic nuclei were identified. Because of *RPGRIP1*'s location in the mouse's connecting cilium and its puta-tive role in protein trafficking between the source, the photoreceptor nucleus, and the des-tination, the OS protein localization was tested. Surprisingly, arrestin, transducin, cGMP PDE, cGMP-gated cation channel, peripherin retinal degeneration (RDS), and rod outer membrane protein-1 (ROM1) were normally distributed in the *RPGRIP1* KO mouse OS. Rod (rhodopsin) and cone opsins, however, were abnormally mislocalized to the photore-ceptor cell bodies. After 3 mo, most photoreceptors were lost, and compared to the *RPGR* KO mouse, the *RPGRIP1* KO mouse appears much more severe, which correlates well with the human phenotypes for *RPGR* and *RPGRIP1*, as LCA is much more severe initially than XRRP. Also, the heterozygous *RPGRIP1* KO mouse was followed for 6 mo with ERG measurements and histological analysis, but did not develop a phenotype at this time. The authors conclude that RPGRIP1 in mice is a stable polymer in the photoreceptor-connect-ing cilium and tethers RPGR. RPGRIP1 is required for disc morphogenesis in the mouse, perhaps by regulating cytoskeleton dynamics.

In 2001, two independent studies found that *RPGRIP1*, when mutated, causes LCA, the most severe retinal dystrophy in the RP group of retinal degenerations. Dryja et al. *(11)* found recessive mutations in *RPGRIP1* in 3 out of 57 (6%) patients with LCA. One patient was a compound heterozygote for W65X and D1203 1-bp deletion, a second patient was homozygous for a Q893 1-bp insertion mutation, and the final patient was homozygous for a K369 1-bp deletion mutation. Each of the three frameshift mutations are predicted to cause premature termination of translation and with the nonsense mutation (W65X), all are predicted to produce null alleles. The two patients available for eye examinations were found to have nystagmus and poor vision since childhood, with light perception vision in their late teens and twenties, and essentially nondetectable ERGs. The retinal appearance was similar to RP, with vessel attenuation and marked pigment degeneration in one patient and was limited to vessel attenuation, without pigment degeneration in the other *(11)*. In the same year, Gerber et al. performed a genome wide scan of seven families with LCA, unlinked to the six known LCA loci, and found homozygosity of the *RPGRIP1* markers at chromosome 14q11 in two of the seven families *(150)*. In family 1, a homozygous missense mutation, G746E, was found, and in family 2 a homozygous frameshift mutation, Y170 1-bp deletion, was identified. Out of a group of 86 patients with sporadic LCA, another 6 patients were identified with *RPGRIP1* mutations. Therefore, the frequency of *RPGRIP1* mutations in their cohort totals 8 out of 142 patients with LCA, which is 5.6% of the patients. They also determined the complete *RPGRIP1* exon–intron structure, with 9 new exons, bringing the total number of exons to 24. As in the Dryja et al. series, Gerber et al. also found that the majority of mutations likely produce null alleles, as they found five frameshift and two nonsense mutations.

Thus far, *RPGRIP1* mutations have only been identified in patients with LCA. It would seem reasonable to expect that if null mutations cause LCA, milder mutations, which result in *RPGRIP1* proteins with some residual function, a milder phenotype such as juvenile- or adult-onset RP or cone-rod degeneration (CORD) may result, as suggested by Cremers et al. *(90)*. Hameed et al. studied two Pakistani families with CORD and found linkage to marker D14S1023 at chromosome 14q11, the *RPGRIP1* locus *(153)*. Their Zmax scores were 5.17 and 4.21 and in family 1 they found a homozygous missense mutation, Arg827Leu, whereas in family 2 they found another homozygous missense mutation, Ala547Ser. We also found linkage of juvenile RP families to the *RPGRIP1* locus, but found no mutations *(154)*, despite intensive mutation screening by denaturing high-performance liquid chromatography *(155)* and automated sequencing. However, we did find that the Ala547Ser change was a very common polymorphism in the general population, putting into question the significance of the aforementioned findings *(154)*. We also determined that the heterozygous parents of LCA offspring appear to have a measurable rod-cone dysfunction *(156,157)*.

In conclusion, *RPGRIP1* defects cause LCA, and the mechanism appears to be a disruption of outer segment disc morphogenesis *(146)*. The phenotype is severe with early significant visual loss and progressive pigmentary retinal degeneration. The heterozygous phenotype appears to be a rod-cone dysfunction. Both rod and cone opsin are mislocalized as a result of *RPGRIP1* mutations and this may lead to a increased levels of cGMP, but there is no direct evidence of this in the *RPGRIP1* KO mouse *(146)*.

RDH12

The relatively new LCA gene, *RDH12,* maps to 14q23.3, consists of seven exons, and encodes a retinol dehydrogenase expressed in photoreceptors participating in the vitamin A cycle, as does *RPE65 (6,158)*. *RDH12*, however, is expressed in the photoreceptors and encodes a retinol dehydrogenase involved in the conversion of all-*trans* retinal to all-*trans* retinal. The exact consequences of *RDH12* defects are not yet known, but are hypothesized to be similar to defects in *RPE65*, with a possible inability to produce rhodopsin, resulting in unliganded opsin, which is constitutively active and leads to a chronic lowering of cGMP levels as in *RPE65* defects.

SUMMARY

LCA is a severe congenital form of RP associated with a surprising variability in disease severity, final visual outcome, visual evolution, retinal appearance, histopathology and genotypes. Both longitudinal visual function studies and histopathological studies suggest that three functional groups of LCA exist. Visual function studies suggest that visual function of some patients with LCA remains stable, whereas others decline, and a small group improves. Histopathology studies suggest that some patients have retinal degeneration, others appear to have a normal retinal architecture, whereas yet others appear to have an aplasia. There are currently seven known LCA genes accounting for approx 45% of the cases in large series *(78,79,159)* and the LCA microchip studies *(136,137)*, leaving more than 60% of the LCA genes still to be discovered. Unlike other retinal dystrophies with genetic heterogeneity such as the Bardet-Biedl syndrome where all gene products appear to function in the same retinal function/physiology (i.e., basal body dysfunction of the cilia) *(160)*, the gene products of LCA appear to function in a wide variety of retinal pathways. LCA gene products have thus far been implicated in the phototransduction process (*GUCY2D*), the retinoid (vitamin A) cycle (*RPE65* and *RDH12*), farnesylation or chaperoning of *PDE* (*AIPL1*), OLM and zonula adherens formation (*CRB1*), OS disc morphogenesis (*RPGRIP1*) and OS formation, and transactivation of several key retinal genes (*CRX*). However, after careful review of the molecular consequences of each LCA gene defect, it may be possible to classify the LCA disease into one of three molecular categories. In Table 1, an attempt is made to provide a preliminary molecular classification of LCA. Three gene defects may lead to *increased cGMP production*, i.e., *AIPL1, CRX,* and *RPGRIP1*. Three gene defects have been postulated to lead to *decreased cGMP levels*, i.e., *GUCY2D*, *RPE65* and *RDH12*. *OLM fragmentation and abnormal zonula adherens formation* (and not cGMP metabolism) appears to be the pathology of LCA with *CRB1* defects. How both abnormally low and abnormally high cGMP can both trigger apoptotic photoreceptor cell death remains to be determined. The phenotypes of LCA heterozygotes are measurable in many cases and may be specific for the gene defect and modifier alleles appear to play a role in the LCA phenotype. Much clinical and basic laboratory work remains to be done to test the molecular classification system and to correlate LCA gene defects to a variety of clinical parameters, including the type of visual function, visual evolution and histopathology so that treatment options can be matched to the disease type.

ACKNOWLEDGMENTS

The author is much indebted to the Foundation Fighting Blindness of Canada for financial support. Support has also been received from the CIHR, FRSQ, and Le Reseau de l'axe Retine de FRSQ. All LCA patients and their parents are acknowledged for their contributions. Many colleagues contributed to this review through their thoughtful discussions, including Drs. Rando Allikmets, Frans Cremers, Gerry Fishman, James Hurley, Irma Lopez, Irene Maumenee, and Melanie Sohocki.

REFERENCES

1. Alstrom CH, Olson O. Heredo-retinopathia congenitalis monohybrida recessiva autosomalis. Hereditas 1957;43:1–178.
2. Koenekoop RK. Major Review: an overview of recent developments in Leber congenital amaurosis: a model to understand human retinal development. Invited publication. Surv Ophthalmol 2004;49(4):379–398.
3. Leber T. Uber retinitis pigmentosa und angeborene amaurose. Graefes Arch Klin Ophthalmol 1869;15:1–25.
4. Franceschetti A, Dieterlé P. Die Differentaldiagnostische Bedeutung des ERG's bei tapeto-retinalen Degenerationen: Elektroretinographie. Bibl Ophth 1956;48:161.
5. Perrault I, Rozet JM, Calvas P, et al. Retinal-specific guanylate cyclase gene mutations in Leber's congenital amaurosis. Nat Genet 1996;14:461–464.
6. Janecke AR, Thompson DA, Utermann G, et al. Mutations in RDH12. encoding a photoreceptor cell retinol dehydrogenase cause childhood-onset severe retinal dystrophy. Nat Genet 2004;36(8):850–854.
7. Marlhens F, Bareil C, Griffoin J-M, et al. Mutations in RPE65 cause Leber's congenital amaurosis. (Letter) Nature Genet 1997;17:139–141.
8. De novo mutations in the CRX homeobox gene associated with Leber congenital amaurosis Carol L. Freund, Qing-Ling Wang, Shiming Chen, Brenda L. Muskat, Carmella D. Wiles, Val C. Sheffield, Samuel G. Jacobson, Roderick R. McInnes, Donald J. Zack, & Edwin M. Stone. Nature Genetics 18, 311–312 (1998).
9. den Hollander AI, Heckenlively JR, van den Born LI, et al. Leber congenital amaurosis and retinitis pigmentosa with Coats-like exudative vasculopathy are associated with mutations in the crumbs homologue 1 (CRB1) gene. Am J Hum Genet 2001;69:198–203.
10. Sohocki MM, Bowne SJ, Sullivan LS, et al. Mutations in a new photoreceptor-pineal gene on 17p cause Leber Congenital amaurosis. Nature Genet 2000;24:79–83.
11. Dryja TP, Adams SM, Grimsby JL, et al. Null RPGRIP1 alleles in patients with Leber Congenital Amaurosis. Am J Hum Genet 2001;68:1295–1298.
12. Dharmaraj S, Li Y, Robitaille J, et al. A novel locus for Leber congenital amaurosis maps on chromosome 6q. Am J Hum Genet 2000a;66:319–326.
13. Stockton DW, Lewis RA, Abboud EB, et al. A novel locus for Leber congenital amaurosis on chromosome 14q24. Hum Genet 1998;103:328–333.
14. Keen TJ, Mohamed MD, McKibbin M, et al. Identification of a locus (LCA9) for Leber's congenital amaurosis on chromosome 1p36. Eur J Hum Genet 2003;11:420–423.
15. Aubineau M. Retinite pigmentaire congenitale familiale. Examen anatomique. Ann Oculistique 1903;129:432–439.
16. Flanders M, Lapointe ML, Brownstein S, et al. Keratoconus and Leber's congenital amaurosis: a clinicopathological correlation. Can J Ophthalmol 1984;19:310–314.
17. François J, Hanssens M. Étude histo-pathologique de deux cas de dégénérescence tapéto-rétinienne congénitale de Leber. Ann Oculist 1969;202:127–155.

18. Kroll AJ, Kuwabara T. Electron Microscopy of a Retinal Abiotrophy. Arch Ophthalmol 1964;71:683–690.
19. Ramamurthy V, Niemi GA, Reh TA, Hurley JB. Leber congenital amaurosis linked to AIPL1: A mouse model reveals destabilization of cGMP phosphodiesterase. PNAS 2004;101(38):13,897–13,902.
20. Sorsby A, Williams CE. Retinal aplasia as a clinical entity. Br Med J 1960;1:293–297.
21. Sullivan TJ, Heathcote JG, Brazel SM, Musarella MA. The ocular pathology in Leber's congenital amaurosis. Aust N Z J Ophthalmol 1994;22:25–31.
22. Babel J. Constatations histologiques dans l'amaurose infantile de Leber et dans diverses formes d'héméralopie. Ophthalmologica 1962;145:399–402.
23. Horsten GP. Development of the retina of man and animals. Arch Ophthalmol 1960; 63:232–242.
24. Milam AH, Barakat MR, Gupta N, et al. Clinicopathologic effects of mutant GUCY2D in Leber congenital amaurosis. Ophthalmology 2003;110(3):549–558.
25. Gillespie FD. Congenital Amaurosis of Leber. Am J Ophthalmol 1966;61:874–880.
26. Vrabec F. Un cas de degenerance pigmentaire congenitale de la retine examinee histolo-quement. Ophthalmologica 1951;122:65–75.
27. Fariss RN, Li ZY, Milam AH. Abnormalities in rod photoreceptors, amacrine cells, and horizontal cells in human retinas with retinitis pigmentosa. Am J Ophthalmology 2000; 129: 215–223.
28. Li Z, Kljavin I, Milam A. Rod photoreceptor sprouting in retinitis pigmentosa. J Neurosci 1995;15(8):5429–5438.
29. Milam A, Li Z, Fariss R. Histopathology of the human retina in retinitis pigmentosa. Prog Retin Eye Res 1998;17(2):175–205.
30. Brecelj J, Stirn-Kranjc B. ERG and VEP follow up study in children with Leber congenital amaurosis. Eye 1999;13:47–54.
31. Fulton AB, Hansen RM, Mayer DL. Vision in Leber congenital amaurosis. Arch Ophthalmol 1996;114:698–703.
32. Heher KL, Traboulsi EI, Maumenee IH. The natural history of Leber's Congenital Amaurosis. Ophthalmology 1992;99:241–245.
33. Lambert SR, Kriss A, Taylor D, et al. Follow-up and diagnostic reappraisal of 75 patients with Leber's congenital amaurosis. Am J Ophthalmol 1989;107:624–631.
34. Acland GM, Aguire GD, Ray J, et al. Gene therapy restores vision in a canine model of childhood blindness. Nat Genet 2001;28:92–95.
35. Ali RR, Sarra G-M, Stephens C, et al. Restoration of photoreceptor ultra structure and function in retinal degeneration slow mice by gene therapy. Nature Genet 2000; 25:306–310.
36. Bush RA, Kononen L, Machida S, et al. The effect of calcium channel blocker diltiazem on photoreceptor degeneration in the rhodopsin Pro23His rat. Invest Ophthalmol Vis Sci 2000;41:2697–2701.
37. Frasson M, Sahel JA, Fabre M, et al. Retinitis pigmentosa: rod photoreceptor rescue by a calcium-channel blocker in the rd mouse. Nature Med 1999;5:1183–1187.
38. LaVail MM, Yasumura D, Matthes MT, et al. Protection of mouse survival factors in reti-nal degenerations. Invest Ophthalmol Vis Sci 1998;39(3):592–602.
39. Lem J, Flannery JG, Li T, et al. Retinal degeneration is rescued in transgenic rd mice by expression of the cGMP phosphodiesterase beta subunit. Proc Natl Acad Sci USA 1992;89:4422–4426.
40. Lewin AS, Drenser KA, Hausworth WW, et al. Ribozyme rescue of photoreceptor cells in a transgenic rat model of autosomal dominant retinitis pigmentosa. Nat Med 1998;4(8): 967–971.
41. Lindsay RM, Wiegand SJ, Altar CA, DiStefano PS. Neurotrophic factors: from molecule to man. Trends Neurosci 1987;28:1131–1137.
42. Travis GR, Groshan KR, Lloyd MB, Bok D. Complete rescue of photoreceptor dysplasia and degeneration in transgenic retinal degeneration slow (rds) mice. Neuron 1992;9:113–119.

43. Van Hooser JP, Aleman TS, He YG, et al. Rapid restoration of visual pigment and function with oral retinoid in a mouse model of childhood blindness. Proc Natl Acad Sci USA 2000;97(15):8623–8628.

44. Weiss A, Biersdorf W. Visual sensory disorders in congenital nystagmus. Ophthalmology 1989;96(4):517–523.

45. Sundin OH, Yang JM, Li Y, et al. Genetic basis of total colourblindness among the Pingelapese islanders. Nat Genet 2000;25(3):289–293.

46. Wissinger B, Gamer D, Jagle H, et al. CNGA3 mutations in hereditary cone photoreceptor disorders. Am J Hum Genet 2001;69(4):722–737.

47. Aligianis IA, Forshew T, Johnson S, et al. Mapping of a novel locus for achromatopsia (ACHM4) to 1p and identification of a germline mutation in the alpha subunit of cone transducin (GNAT2). J Med Genet 2002;39(9):656–660.

48. Weleber RG, Tongue AC. Congenital stationary night blindness presenting as Leber's congenital amaurosis. Arch Ophthalmol 1987;105(3):360–365.

49. Dryja TP. Molecular genetics of Oguchi disease, fundus albipunctatus, and other forms of stationary night blindness: LVII Edward Jackson Memorial Lecture. Am J Ophthalmol 2000;130(5):547–563.

50. Sohocki MM, Perrault I, Leroy BP, et al. Prevalence of AIPL1 mutations in inherited retinal degenerative disease. Mol Genet Metab 2000;70(2):142–150.

51. Ramamurthy V, Roberts M, van den Akker F, Niemi G, Reh TA, Hurley JB. AIPL1, a protein implicated in Leber's congenital amaurosis, interacts with and aids in processing of farnesylated proteins. Proc Natl Acad Sci USA 2003;100(22):12,630–12,635.

52. Akey DT, Zhu X, Dyer M, et al. The inherited blindness associated protein AIPL1 interacts with the cell cycle regulator protein NUB1. Hum Mol Genet 2002;11(22):2723–2733.

53. van der Spuy J, Cheetham ME. The leber congenital amaurosis protein AIPL1 modulates the nuclear translocation of NUB1 and suppresses inclusion formation by NUB1 fragments. J Biol Chem 2004;279(46):48,038–48,047.

54. Kanaya K, Sohocki MM, Kamitani T. Abolished interaction of NUB1 with mutant AIPL1 involved in Leber congenital amaurosis. Biochem Biophys Res Commun 2004;317(3):768–773.

55. Choy E, Chiu VK, Silletti J, et al. Endomembrane trafficking of ras: the CAAX motif targets proteins to the ER and Golgi. Cell. 1999;98(1):69–80.

56. Dyer MA, Donovan SL, Zhang J, et al. Retinal degeneration in Aipl1-deficient mice: a new genetic model of Leber congenital amaurosis. Brain Res Mol Brain Res 2004;132(2):208–220.

57. Liu X, Bulgakov OV, Wen XH, et al. AIPL1, the protein that is defective in Leber congenital amaurosis, is essential for the biosynthesis of retinal rod cGMP phosphodiesterase. Proc Natl Acad Sci USA 2004;101(38):13,903–13,908.

58. Bowes C, Li T, Danciger M, et al. Retinal degeneration in the rd mouse is caused by a defect in the β-subunit of rod cGMP-phosphodiesterase. Nature 1990;347:677–680.

59. Farber DB, Lolley RN. Cyclic guanosine monophosphate: Elevation in degenerating photoreceptor cells of the C3H mouse retina. Science 1974;186:449–451.

60. Lolley RN, Farber DB, Rayborn ME, et al. Cyclic GMP accumulation causes degeneration of phototreceptor cells: simulation of an inherited disease. Science 1977;196:664–666.

61. Dharmaraj S, Leroy BP, Sohocki MM, et al. Maumenee. A distinct phenotype for Leber congenital amaurosis patients with AIPL1 mutations: a cross sectional genotype-phenotype evaluation of 26 AIPL1 patients and comparisons with other LCA phenotypes. Arch Ophthalmol 2004;122:1029–1037.

62. Galvin JA, Fishman GA, Stone EM, Koenekoop RK. Clinical phenotypes in carriers of Leber congenital amaurosis mutations. Ophthalmology 2005;112:349–356.

63. Ortiz A, Xiaoshan W, Lopez I, Koenekoop RK, Sohocki MM. Functional correlations of selected AIPL1 mutations found in Leber congenital amaurosis patients and their parents. Invest Ophthalmol Vis Sci 2004;45(4):S5109.

64. van der Spuy J, Chapple JP, Clark BJ, Luthert PJ, Sethi CS, Cheetham ME. The Leber congenital amaurosis gene product AIPL1 is localized exclusively in rod photoreceptors of the adult human retina. Hum Mol Genet 2002;11(7):823–831.

65. van der Spuy J, Kim JH, Yu YS, et al. The expression of the Leber congenital amaurosis protein AIPL1 coincides with rod and cone photoreceptor development. Invest Ophthalmol Vis Sci 44(12):5396–5403.

66. Damji KF, Sohocki MM, Khan R, et al. Leber congenital amaurosis with anterior keratoconus in Pakistani families is caused by the Trp278X mutation in the AIPL1 gene on 17p. Can J Ophthalmol 2001;36(5):252–259.

67. Silva E, Yang JM, Li Y, et al. A CRX Null Mutation Is Associated with Both Leber Congenital Amaurosis and a Normal Ocular Phenotype. Invest Ophthalmol Vis Sci 2000; 41:2076–2079.

68. Freund CL, Gregory-Evans CY, Furukawa T, et al. Cone-rod dystrophy due to mutations in a novel photoreceptor-specific homeobox gene (CRX) essential for maintenance of the photoreceptor. Cell 1997;91:543–553.

69. Furukawa T, Morrow EM, Li T, et al. Retinopathy and attenuated circadian entrainment in Crx-deficient mice. Nat Genet 1999;23:466–470.

70. Furukawa T, Morrow EM, Cepko CL. Crx, a novel otx-like homeobox gene, shows photoreceptor-specific expression and regulates photoreceptor differentiation. Cell 1997; 91:531–541.

71. Pignatelli V, Cepko CL, Strettoi E. Inner retinal abnormalities in a mouse model of Leber's congenital amaurosis. J Comp Neurol 2004;469(3):351–359.

72. Livesey FJ, Furukawa T, Steffen MA, Church GM, Cepko CL. Microarray analysis of the transcriptional network controlled by the photoreceptor homeobox gene Crx. Curr Biol 2000;10(6):301–310.

73. Tsang SH, Gouras P, Yamashita CK, et al. Retinal degeneration in mice lacking the gamma subunit of the rod cGMP phosphodiesterase. Science 1996;272(5264):1026–1029.

74. Rivolta C, Berson EL, Dryja TP. Dominant Leber congenital amaurosis, cone-rod degeneration, and retinitis pigmentosa caused by mutant versions of the transcription factor CRX. Hum Mutat 2001;18(6):488–498.

75. Swaroop A, Wang QL, Wu W, et al. Leber congenital amaurosis caused by a homozygous mutation (R90W) in the homeodomain of the retinal transcription factor CRX: direct evidence for the involvement of CRX in the development of photoreceptor function. Hum Mol Genet 1999;8:299–305.

76. Sohocki MM, Sullivan LS, Mintz-Hittner HA, et al. A range of clinical phenotypes associated with mutations in CRX, a photoreceptor transcription-factor gene. Am J Hum Genet 1998;63:1307–1315.

77. Koenekoop RK, Loyer M, Dembinska O, Beneish R. Visual improvement in Leber congenital amaurosis and the CRX genotype. Ophthalmic Genet 2002;23(1):49–59.

78. Dharmaraj S, Silva E, Pina A-L, et al. Mutational Analysis and Clinical Correlation in LCA. Ophthalmic Genetics, 2000;21(3):135–150.

79. Hanein S, Perrault I, Gerber S, et al. Leber congenital amaurosis: comprehensive survey of the genetic heterogeneity, refinement of the clinical definition, and genotype-phenotype correlations as a strategy for molecular diagnosis. Hum Mutat 2004;23(4):306–317.

80. den Hollander AI, ten Brink JB, de Kok YJM, et al. Mutations in a human homologue of Drosophila crumbs cause retinitis pigmentosa (RP12). Nature Genet 1999;23:217–221.

81. den Hollander AI, van Driel MA, de Kok YJM, et al. Isolation and mapping of novel candidate genes for retinal disorders using suppression subtractive hybridization. Genomics 1999;58:240–249.

82. den Hollander AI, Davis J, van der Velde-Visser SD, et al. CRB1 mutation spectrum in inherited retinal dystrophies. Hum Mutat 2004;24(5):355–369.

83. Pellikka M, Tanentzapf G, Pinto M, et al. Crumbs, the Drosophila homologue of human CRB1/RP12, is essential for photoreceptor morphogenesis. Nature 2002;416(6877): 143–149.

84. Mehalow AK, Kameya S, Smith RS, et al. CRB1 is essential for external limiting membrane integrity and photoreceptor morphogenesis in the mammalian retina. Hum Mol Genet 2003;12(17):2179–2189.

85. Lotery AJ, Jacobson SG, Fishman GA, et al. Mutations in the CRB1 gene cause Leber congenital amaurosis. Arch Ophthalmol 2001;119:415–420.

86. van de Pavert SA, Kantardzhieva A, Malysheva A, et al. Crumbs homologue 1 is required for maintenance of photoreceptor cell polarization and adhesion during light exposure. J Cell Sci 2004;117(Pt 18):4169–4177.

87. Jacobson SG, Cideciyan AV, Aleman TS, et al. Crumbs homolog 1 (CRB1) mutations result in a thick human retina with abnormal lamination. Hum Mol Genet 2003;12(9): 1073–1078.

88. McKay GJ, Clarke S, Davis JA, Simpson DA, Silvestri G. Pigmented paravenous chorioretinal atrophy is associated with a mutation within the Crumbs homolog 1 (CRB1) gene. Invest Ophthalmol Vis Sci 2005;46(1):322–328.

89. Galvin JA, Fishman GA, Stone EM, Lopez I, Koenekoop RK. Clinical phenotypes in patients & carriers of various genotypes in Leber congenital amaurosis (LCA). Invest Ophthalmol Vis Sci 2004;45(4):S4063.

90. Cremers FP, van den Hurk JA, den Hollander AI. Molecular genetics of Leber congenital amaurosis. Hum Mol Genet 2002;11(10):1169–1176.

91. Shyjan AW, de Sauvage FJ, Gillett NA, Goeddel DV, Lowe DG. Molecular cloning of a retina-specific membrane guanylyl cyclase. Neuron 1992;9:727–737.

92. Oliveira L, Miniou P, Viegas-Pequignot E, Rozet J-M, Dollfus H, Pittler SJ. Human retinal guanylate cyclase (GUC2D) maps to chromosome 17p13.1. Genomics 1994;22:478–481.

93. Camuzat A, Dollfus H, Rozet JM, et al. A gene for Leber's congenital amaurosis maps to chromosome 17p. Hum Mol Genet 1995;4:1447–1452.

94. Camuzat A, Rozet JM, Dollfus H, et al. Evidence of genetic heterogeneity of Leber's congenital amaurosis (LCA) and mapping of LCA1 to chromosome 17p13. Hum Genet 1996;97:798–801.

95. Kelsell RE, Gregory-Evans K, Payne AM, et al. Mutations in the retinal guanylate cyclase (RETGC-1) gene in dominant cone-rod dystrophy. Hum Mol Genet 1998;7:1179–1184.

96. Dizhoor AM, Lowe DG, Olshevskaya EV, et al. Expression patterns of RetGC-1 in rod and cone photoreceptors. Neuron 1994;12:1345–1352.

97. Liu X, Seno K, Nishizawa Y, et al. Ultrastructural localization of retinal guanylate cyclase in human and monkey retinas. Exp Eye Res 1994;59:761–768.

98. Dizhoor AM, Hurley JB. Inactivation of EF-hands makes GCAP-2 (p24) a constitutive activator of photoreceptor guanylyl cyclase by preventing a Ca2+-induced "activator-to-inhibitor" transition. J Biol Chem 1996;271:19,346–19,350.

99. Dizhoor AM, Olshevskaya EV, Henzel WJ, et al. Cloning, sequencing, and expression of a 24-kDa Ca(2+)-binding protein activating photoreceptor guanylyl cyclase. J Biol Chem 1995;270:25,200–25,206.

100. Olshevskaya EV, Hughes RE, Hurley JB, Dizhoor AM. Calcium binding, but not a calcium-myristoyl switch, controls the ability of guanylyl cyclase-activating protein GCAP-2 to regulate photoreceptor guanylyl cyclase. J Biol Chem 1997;272:14,327–14,333.

101. Palczewski K, Subbaraya I, Gorczyca WA, et al. Molecular cloning and characterization of retinal photoreceptor guanylyl cyclase-activating protein. Neuron 1994;13:395–404.

102. Dizhoor AM, Lowe DG, Olshevskaya EV, et al. The human photoreceptor membrane guanylyl cyclase, RetGC, is present in outer segments and is regulated by calcium and a soluble activator. Neuron 1994;12:1345–1352.

103. Lagnado L, Baylor D. Signal flow in visual transduction. Neuron 1992;8:995–1002.
104. Laura RP, Dizhoor AM, Hurley JB. The membrane guanylyl cyclase, retinal guanylyl cyclase-1, is activated through its intracellular domain. J Biol Chem 1996;271:11,646–11,651.
105. Duda T, Venkatarama V, Goraczniak R, Lange C, Koch K-W, Sharma RK. Functional consequences of a rod outer segment membrane guanylate cyclase (ROS-GC1) gene mutation linked with Leber's congenital amaurosis. Biochemistry 1999;38:509–515.
106. Rozet JM, Perrault I, Gerber S, et al. Complete abolition of the retinal-specific guanylyl cyclase (retGC-1) catalytic ability consistently leads to leber congenital amaurosis (LCA). Invest Ophthalmol Vis Sci 2001;42(6):1190–1192.
107. Tucker C, Ramamurthy V, Pina AL, et al. Functional analyses of mutant recessive GUCY2D alleles identified in Leber congenital amaurosis patients: protein domain comparisons and dominant negative effects. Mol Vision 2004;10:297–303.
108. Tucker CL, Hurley JH, Miller TR, Hurley JB. Two amino acid substitutions convert a guanylyl cyclase, RetGC-1, into an adenylyl cyclase. Proc Natl Acad Sci USA 1998;95: 5993–5997.
109. Koenekoop RK, Fishman GA, Iannaccone A, et al. Electroretinographic (ERG) abnormalities in parents of Leber Congenital Amaurosis children with known GUCY2D mutations. Arch Ophthalmol 2002;120(10):1325–1330.
110. Koenekoop RKV, Ramamurthy AL, Pina M, et al. Biochemical consequences of RetGC-1 mutations found in children with Leber congenital amaurosis. Invest Ophthalmol Vis Sci 2000;41(4):S200 (abstract 1050).
111. Chinkers M, Wilson EM. Ligand-independent oligomerization of natriuretic peptide receptors. Identification of heteromeric receptors and a dominant negative mutant. J Biol Chem 1992;267:18,589–18,597.
112. Thompson DK, Garbers DL. Dominant negative mutations of the guanylyl cyclase-A receptor. Extracellular domain deletion and catalytic domain point mutations. J Biol Chem 1995;270:425–430.
113. Yang RB, Robinson SW, Xiong WH, et al. Disruption of a Retinal Guanylyl Cyclase gene leads to cone-specific dystrophy and paradoxical rod behavior. J Neurosci 1999;19: 5889–5897.
114. Semple-Rowland S, Lee NR, Van Hooser JP, et al. A null mutation in the photoreceptor guanylate cyclase gene causes the retinal degeneration chicken phenotype. Proc Natl Acad Sci USA 1998;95:1271–1276.
115. Coleman JE, Semple-Rowland SL. GC1 deletion prevents light-dependent arrestin translocation in mouse cone photoreceptor cells. Invest Ophthalmol Vis Sci 2005;46(1):12–16.
116. Perrault I, Rozet J, Gerber S, et al. Spectrum of RetGC1 mutations in Leber congenital amaurosis. Eur J Hu Genet 2000;8:578–582.
117. Perrault I, Rozet JM, Ghazi I, et al. Different functional outcome of RetGC1 and RPE65 gene mutations in Leber congenital amaurosis. Am J Hum Genet 1999;64(4):1225–1228.
118. Silva E, Dharmaraj S, Li YY, et al. A missense mutation in GUCY2D acts as a genetic modifier in RPE65-related Leber congenital amaurosis. Ophthalmic Genet 2004;25(3):205–217.
119. Lorenz B, Wabbels B, Wegscheider E, Hamel CP, Drexler W, Preising MN. Lack of fundus autofluorescence to 488 nanometers from childhood on in patients with early-onset severe retinal dystrophy associated with mutations in RPE65. Ophthalmology 2004;111(8): 1585–1594.
120. Hamel CP, Tsilou E, Pfeffer BA, Hooks JJ, Detrick B, Redmond TM. Molecular cloning and expression of RPE65, a novel retinal pigment epithelium-specific microsomal protein that is post-transcriptionally regulated in vitro. J Biol Chem 1993;268:15,751–15,757.
121. Hamel CP, Jenkins NA, Gilbert DJ, Copeland NG, Redmond TM. The gene for the retinal pigment epithelium-specific protein RPE65 is localized to human 1p31 and mouse 3. Genomics 1994;20:509–512.

122. Nicoletti A, Wong DJ, Kawase K, et al. Molecular characterization of the human gene encoding an abundant 61 kDa protein specific to the retinal pigment epithelium. Hum Mol Genet 1995;4(4):641–649.

123. Gu S, Thompson DA, Srikumari CRS, et al. Mutations in RPE65 cause autosomal recessive childhood-onset severe retinal dystrophy. Nature Genet 1997;17:194–197.

124. Morimura H, Fishman GA, Grover SA, Fulton AB, Berson EL, Dryja TP. Mutations in the RPE65 gene in patients with autosomal recessive retinitis pigmentosa or Leber congenital amaurosis. Proc Nat Acad Sci USA 1998;95:3088–3093.

125. Redmond TM, Yu S, Lee E, et al. Rpe65 is necessary for production of 11-cis-vitamin A in the retinal visual cycle. Nature Genet 1998;20:344–351.

126. Seeliger MW, Grimm C, Stahlberg F, et al. New views on RPE65 deficiency: the rod system is the source of vision in a mouse model of Leber congenital amaurosis. Nat Genet 2001;29(1):70–74.

127. Fain GL, Lisman JE. Photoreceptor degeneration in vitamin A deprivation and retinitis pigmentosa: the equivalent light hypothesis. Exp Eye Res 1993;57(3):335–340.

128. Fain GL, Lisman JE. Light, Ca2+, and photoreceptor death: new evidence for the equivalent-light hypothesis from arrestin KO mice. Invest Ophthalmol Vis Sci 1999; 40(12):2770–2772.

129. Woodruff ML, Wang Z, Chung HY, Redmond TM, Fain GL, Lem J. Spontaneous activity of opsin apoprotein is a cause of Leber congenital amaurosis. Nat Genet 2003;35(2):158–164.

130. Franklin JL, Sanz-Rodriguez C, Juhasz A, Deckwerth TL, Johnson EM Jr. Chronic depolarization prevents programmed death of sympathetic neurons in vitro but does not support growth: requirement for Ca^{2+} influx but not Trk activation. J Neurosci 1995;15:643–664.

131. Aguirre GD, Baldwin V, Pearce-Kelling S, Narfstrom K, Ray K, Acland GM. Congenital stationary night blindness in the dog: common mutation in the RPE65 gene indicates founder effect. Mol Vis 1998;4:23.

132. Veske A, Nilsson SE, Narfstrom K, Gal A. Retinal dystrophy of Swedish briard/briard-beagle dogs is due to a 4-bp deletion in RPE65. Genomics 1999;57(1):57–61.

133. Lorenz B, Gyurus P, Preising M, et al. Early-onset severe rod cone dystrophy in young children with RPE 65 mutations. Invest Ophthalmol Vis Sci 2000;41(9):2735–2742.

134. Paunescu K, Wabbels B, Preising MN, Lorenz B. Longitudinal and cross-sectional study of patients with early-onset severe retinal dystrophy associated with RPE65 mutations. Graefes Arch Clin Exp Ophthalmol 2005;243(5):417–426.

135. Simovich MJ, Miller B, Ezzeldin H, et al. Four novel mutations in the RPE65 gene in patients with Leber congenital amaurosis. Hum Mutat 2001;18(2):164.

136. Allikmets R, Zernant J, Külm M, et al. The genotyping microarray (disease chip) for Leber congenital amaurosis : mutation identification and modifier alleles. Invest Ophthalmol Vis Sci 2004;45(4):S2444.

137. Zernant J, Külm M, Dharmaraj S, et al. Genotyping microarray (disease chip) for Leber congenital amaurosis: detection of modifier alleles. Invest Ophthalmol Vis Sci 2005;46(9):3052–3059.

138. Yzer S, van den Born LI, Schuil J, et al. A Tyr368His RPE65 founder mutation is associated with variable expression and progression of early onset retinal dystrophy in 10 families of a genetically isolated population. J Med Genet 2003;40(9):709–713.

139. Koenekoop RK. Abnormal retinal architecture in a 33-week-old fetus with LCA and a homozygous C330Y mutation in RPE65. Ophthalmic Genet 2003;24(2):125–126.

140. Porto FB, Perrault I, Hicks D, et al. Prenatal human ocular degeneration occurs in Leber's congenital amaurosis (LCA2). J Gene Med 2002;4(4):390–396.

141. Roepman R, Bernoud-Hubac N, Schick D, et al. The Retinitis Pigmentosa GTPase Regulator (RPGR) interacts with novel transport-like proteins in the outer segments of rod photoreceptors. Hum Mol Genet 2000;9:2095–2105.

142. Boylan JP, Wright AF. Identification of a novel protein interacting with RPGR. Hum Mol Genet 2000;9:2085–2093.
143. Hong D-H, Yue G, Adamian M, Li T. Retinitis pigmentosa GTPase regulator (RPGR)-interacting protein is stably associated with the photoreceptor ciliary axoneme and anchors RPGR to the connecting cilium. J Biol Chem 2001;276:12,091–12,099.
144. Castagnet P, Mavlyutov T, Cai Y, Zhong F, Ferreira P. RPGRIP1s with distinct neuronal localization and biochemical properties associate selectively with RanBP2 in amacrine neurons. Hum Mol Genet 2003;12(15):1847–1863.
145. Koenekoop RK, Loyer M, Hand C, et al. Novel RPGR mutations with distinct retinitis pigmentosa phenotypes in French-Canadian families. Am J Ophthalmol 2003;136(4):678–687.
146. Zhao Y, Hong DH, Pawlyk B, et al. The retinitis pigmentosa GTPase regulator (RPGR)-interacting protein: subserving RPGR function and participating in disk morphogenesis. Proc Natl Acad Sci USA 2003;100(7):3965–3970.
147. Roepman R, Letteboer S, Cremers FPM. Novel interactors link the RPGR/RPGRIP1 multi-subunit assembly complex to different key processes of the retina. Invest Ophthalmol Vis Sci 2004;45(4):S2438.
148. Lu X, Ferreira P. Identification and expression profile of a novel RPGR-independent RPGRIP1 isoform. Invest Ophthalmol Vis Sci 2004;45(4):S5088.
149. Mavlyutov TA, Zhao H, Ferreira PA. Species-specific subcellular localization of RPGR and RPGRIP isoforms: implications for the phenotypic variability of congenital retinopathies among species. Hum Mol Genet 2002;11(16):1899–1907.
150. Gerber S, Perrault I, Hanein S, et al. Complete exon-intron structure of the RPGR-interacting protein (RPGRIP1) gene allows the identification of mutations underlying Leber congenital amaurosis. Eur J Hum Genet 2001;9(8):561–571.
151. Delphin C, Guan T, Melchior F, Gerace L. RanGTP targets p97 to RanBP2, a filamentous protein localized at the cytoplasmic periphery of the nuclear pore complex. Mol Biol Cell 1997;8(12):2379–2390.
152. Yaseen NR, Blobel G. GTP hydrolysis links initiation and termination of nuclear import on the nucleoporin nup358. J Biol Chem 1999;274(37):26,493–26,502.
153. Hameed A, Abid A, Aziz A, Ismail M, Mehdi SQ, Khaliq S. Evidence of RPGRIP1 gene mutations associated with recessive cone-rod dystrophy. J Med Genet 2003;40(8):616–619.
154. Koenekoop RK, Lopez I, Fossarello M, Mansfield D, Wright A. RPGRIP1 mutations in juvenile retinitis pigmentosa: a linkage and mutation study. Invest Ophthalmol Vis Sci 2004;45(4):S4727.
155. Marsh DJ, Theodosopoulos G, Howell V, et al. Rapid mutation screening of genes associated with familial cancer syndromes using denaturing high performance liquid chromatography. Neoplasia 2001;3(3):236–244.
156. Dharmaraj S, Lopez I, Fishman G, et al. Recessive RPGRIP1 mutations can cause rod and cone dysfunction in the heterozygous parents. Invest Ophthalmol Vis Sci 2004;45(4):S4728.
157. Lopez I, Fishman GAF, Racine J, et al. Functional studies of recessive RPGRIP1 mutations from Leber congenital amaurosis patients: rod and cone ERG dysfunction in the obligate heterozygotes. Invest Ophthalmol Vis Sci 2005;46(4):S1705.
158. Perrault I, Hanein S, Gerber S, et al. Retinal dehydrogenase 12 (RDH12) mutations in leber congenital amaurosis. Am J Hum Genet 2004;75(4):639–646.
159. Lotery AJ, Namperumalsamy P, Jacobson SG, et al. Mutation analysis of three genes in patients with Leber congenital amaurosis. Arch Opthalmol 2000;118:538–543.
160. Ansley SJ, Badano JL, Blacque OE, et al. Basal body dysfunction is a likely cause of pleiotropic Bardet-Biedl syndrome. Nature 2003;425(6958):628–633.

4

Macular Degeneration

Aging Changes and Novel Therapies

Peter E. Liggett, MD and Alejandro J. Lavaque, MD

CONTENTS

INTRODUCTION

Owing to the lack of an effective prevention and appropriate treatment, age-related macular degeneration (AMD) continues being the leading cause of central vision loss in patients older than 65 yr of age in the first world, and the third cause in developing countries. Despite a relatively low prevalence of choroideal neovascularization (CNV) in AMD, approximately a quarter of these patients will develop the complication *(1,2)*. As expected, the prevalence is likely to increase as a consequence of increasing longevity.

In wet AMD, a destabilization of the retinal and choroidal microenvironment leads to the formation of new blood vessels, which ultimately results in a decrease of visual acuity. Degenerative changes of the retinal pigment epithelial (RPE) and Bruch's membrane are the primary factors responsible for the disease. The pathophysiology of the disease is still not completely understood. The putative role of specific genes in the degenerative process in AMD is less clear. Although certain genes may predispose some patients to develop AMD, the genetic linkage remains controversial and, to date, the genetic of AMD is largely unknown. Genetic susceptibility to AMD is probably multifactorial and

From: *Ophthalmology Research: Retinal Degenerations: Biology, Diagnostics, and Therapeutics*
Edited by: J. Tombran-Tink and C. J. Barnstable © Humana Press Inc., Totowa, NJ

thus will not be amenable to gene therapy directed at the germinal line. In the absence of a well-defined genetic defect which gives rise to AMD, gene therapy will likely focus on somatic therapy using growth factors and anti-apoptosis therapy to prolong the survival of the RPE and retinal photoreceptors.

GROWTH FACTOR GENE THERAPY FOR AMD

There is experimental evidence that growth factors play an important role in maintaining the health of RPE cells and in enabling them to respond to injury. RPE cells express growth factors and their receptors, demonstrating the autocrine and paracrine functions of these substances. Theoretically, it may be possible to enhance RPE cell survival by somatic modulation of growth factor gene expression in patients with AMD. For instance, age-related phagocytic dysfunction and incomplete digestion of photoreceptors membranes by the RPE result in loss of RPE cells and in geographic atrophy, perhaps because of the cytotoxicity of these deposits on the surrounding cells. Enhancing phagocytic activity in aging RPE cells using gene therapy is a potential approach to the treatment of AMD. Basic-fibroblast growth factor has been shown to stimulate phagocytic activity and prolong retinal survival in animals models *(3)*. An important number of other growth factors and secondary messengers of the intracellular-signaling pathways have demonstrated a neuroprotective effect on the retinal neurons in animal models of retinal degeneration. Gene transfer and expression of these growth factor proteins may similarly inhibit retinal degeneration by a neuroprotective effect in AMD *(4)*.

The RPE synthesizes proteins that are antiangiogenic, such as tissue inhibitors of metaloproteinases and pigment epithelium-derived factor (PEDF). Thus, potential gene therapy applications to CNV include anti-angiogenic growth factor gene therapy, antisense or ribozyme therapy directed at angiogenic factors, and suicide gene therapy directed at neovascular tissue. Moreover, recently it was demonstrated that expression of angiostatin in experimental CNV significantly reduces the size of CNV lesions *(5)*.

TRANSPLANTATION OF GENETICALLY MODIFIED IRIS PIGMENT EPITHELIAL CELLS

Submacular transplantation of autologous iris pigment epithelial (IPE) cells has been proposed to replace the damage RPE following surgical removal of the CNV *(6)*. The IPE is anatomically continuous with the RPE and has the same embryonic origin. In vitro IPE cells share functional properties with the RPE cells such as phagocytosis, degradation of rod outer segments, and synthesis of trophic factors. However, autologous transplantation of IPE cells alone has not resulted in a prolonged improvement of the vision patients with AMD, potentially because the lack of expression of one or several factors that is an important part of RPE function. Semkova et al. *(6)* have suggested a treatment for AMD based on transplantation of genetically modified autologous IPE cells. The most significant findings are summarized as follows: first, IPE cells were readily transduced with a high-capacity adenovirus (HC-ad) vector. Second, IPE cells secreted functionally active PEDF after HC-ad-mediated gene transfer. Third, subretinal transplantation of PEDF-expressing IPE cells inhibited neovascularization in models of retinal neoangiogenesis, and prevent photoreceptor degeneration.

ADENOVIRAL VECTORS GENE THERAPY

Preclinical proof, using either recombinant adenovectors to carry the genes encoding PEDF and endostatin or recombinant adeno-associated viruses (AAVs) carrying the transgene encoding for angiostatin, has recently been published and demonstrated that significant inhibition of the CNV in various animals models. Recently the intravenous administration of an adenoviral construct carrying the murine endostatin gene in a murine model of CNV was tested and found an almost complete inhibition of the neovascular activity. Similarly, subcutaneous injection of an AAV carrying a truncated angiostatin gene resulted in significant inhibition of retinal neovascularization. These encouraging results with endostatin and angiostatin suggest a potential role of anti-angiogenesis in ocular disease.

Recently, PEDF received major attention. PEDF was first described in 1989 by Tombran-Tink in conditioned-medium form-cultured, fetal RPE cells as a potent neurotrophic factor *(7)*. Subsequently, PEDF has been purified and cloned both from humans and mice *(8)*. The gene is expressed as early as 17 wk in human fetal RPE cells, suggesting that PEDF is intimately involved in early neuronal development. PEDF attracted even more attention when Dawson et al. demonstrated that PEDF is one of the most potent natural inhibitors of angiogenesis *(9)*. In addition, PEDF is an inhibitor of endothelial cell migration. The amount of inhibitory PEDF produced by retinal cells was positively correlated with oxygen concentrations, suggesting that its loss plays a permissive role in ischemia-driven retinal neovascularization. Moreover, a correlation between changes in the vascular endothelial growth factor (VEGF)/PEDF ratio and the degree of retinal neovascularization in a rat model was demonstrated.

The AdPEDF *(11)* is a replication deficient adenovirus vector designed to deliver the human PEDF gene. Intravitreous injection of AdPEDF resulted in increased expression of PEDF mRNA in the eye compared with AdNull (the same vector without the transgene) or with uninjected controls. PEDF trail was present not only in the retina, but also in other parts of the eye, including the iris, the lens, and the corneal epithelium. Afterward, the subretinal injection of AdPEDF was strongly detected in the RPE cells compared with other ocular structures *(8)*.

PHARMACOLOGICAL INHIBITION OF THE VEGF IN PATIENTS WITH WET AMD

Background

Ocular neovascularization is a key factor of the most common causes of blindness in humans in the developed word: AMD and proliferative diabetic retinopathy are two well-known examples. Prevention of ocular neovascularization by development of anti-angiogenic drugs represents a rational and appealing therapeutic approach. However, because these are chronic diseases characterized by ongoing new vessel formation, long-term inhibition of the angiogenic stimuli is likely to be needed.

The development of new blood vessels from pre-existing ones, a process known as angiogenesis, is a physiological process that is fundamental to normal healing, reproduction, and embryonic development *(10)*. Angiogenesis plays an important role in a variety of pathological processes including proliferative retinopathies, AMD, rheumatoid arthritis, psoriasis, and cancer.

Over the past decade, our understanding of the complex processes involved in new vessel development has lead to the isolation of a family of angiogenic stimulators known collectively as VEGF.

VEGF-A is a pivotal angiogenic stimulator that binds to VEGF receptors, promoting endothelial cell migration, proliferation, and increasing vascular permeability.

Recognition of the central role of VEGF-A in angiogenesis has led to the hypothesis that its inhibition may represent a novel and effective approach to the treatment of choroidal neovascular membranes in wet AMD and other conditions characterized by pathologic angiogenesis.

VEGF AND ITS RECEPTORS

There are currently six known members of the VEGF family: VEGF-A, placental growth factor, VEGF-B, VEGF-C, VEGF-D, and VEGF-E.

VEGF, also known as vascular permeability factor (VPF), is a diffusable endothelial cell-specific mitogen and a pro-angiogenic factor that regulates vascular permeability. VEGF is highly specific for vascular endothelium and its potent angiogenic action has been demonstrated in arteries, veins, and lymphatic vessels, but not in other cell lines (Fig. 1). In addition, there is also evidence that VEGF functions as an anti-apoptotic factor for endothelial cells in newly formed vessels. Several mechanisms may regulate VEGF expression, the most important of which may be hypoxia. Other factors of importance include cytokines, such as epidermal growth factor, transforming growth factor-β, keratinocytes growth factor, cell differentiation, and oncogenes *(11)*.

The biological effects of VEGF are mediated by two receptor tyrosine kinases: VEGFR-1 and VEGFR-2 *(12)*. The expression of these receptors is largely restricted to the vascular endothelium and it is assumed, but not proven, that all of the effects of VEGF on the vascular endothelium may be mediated by these receptors *(13)*.

THE ANGIOGENESIS PROCESS: HOW DO NEW VESSELS GROW?

Physiological adaptation to hypoxia is a necessity for organisms having an oxygen-based metabolism. In mammals, these adjustments include vasodilation, angiogenesis, upregulation of glucose transport, activation of glycolysis, and apoptosis (programmed cell death) *(14)*.

Angiogenesis is a multistep process that is regulated by a fine balance between pro- and anti-angiogenic growth factors released in response to hypoxia, hypoglycemia, mechanical stress, release of inflammatory proteins, and genetic alterations (Table 1). Some of these factors are highly specific for the endothelium (e.g., VEGF), whereas others have a wide range of activities (e.g., matrix metalloproteinase inhibitors [MMIs]).

A series of complex and interrelated steps are necessary for angiogenesis to take place in a tissue (Fig. 2). The initial step requires the injured cell to release pro-angiogenic growth factors (mainly VEGF) into the surrounding tissues. The released VEGF binds to healthy adjacent endothelial cells located in the walls of normal blood vessels. VEGF exerts its action on endothelial cells through the VEGFR-1 and VEGFR-2 receptors *(12)*. VEGF induces genetic modifications in the endothelial cell that results in the intracellular synthesis of lytic enzymes (e.g., matrix metalloproteinases [MMPs]) responsible for

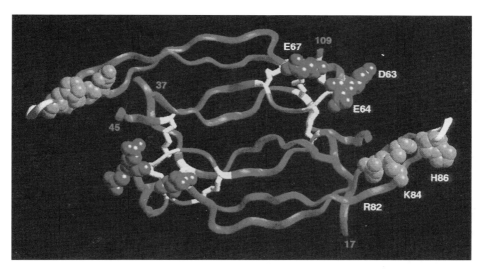

Fig. 1. Vascular endothelial growth factor molecule. Courtesy of Genentech.

Table 1
Pro-Angiogenic and Anti-Angiogenic Factors (Angiogenesis Foundation)

Pro-angiogenic factors	*Anti-angiogenic factors*
Angiogenin	Angiostatin (plasminogen fragment)
Angiopoetin-1	Antiangiogenic antithrombin III
Fibroblast growth factor (acid and basic)	Cartilage-derived inhibitor (CDI)
Granulocyte colony-stimulating factor (G-CSF)	Endostatin (collagen XVIII fragment)
Hepatocyte growth factor (HGF)	Fibronectin fragment
Interleukin-8 (IL-8)	Human chorionic gonadotropin (hCG)
Leptin	Interferon and interferon induced protein
Midkine	Interleukin-12 (IL-12)
Placental growth factor (PFG)	Tissue inhibitors of metalloproteinases (TIMPs)
Platelet-derived growth factor	2-Methoxyestradiol
Pleiotrophin (PTN)	Placental ribonuclease inhibitor
Proliferin	Plasminogen activator inhibitor
Transforming growth factor (α and β)	Platelet factor 4 (PF4)
Tumor necrosis factor-α (TNF-α)	Prolactin 16kD fragment
Vascular endothelial growth factor (VEGF)	Retinoids
	Tetrahydrocortisol-S
	Transforming growth factor-β (TGF-β)
	Vasculostatin
	Vasostatin (calreticulin fragment)

the break down of vessel walls and the extracellular matrix. Proliferating endothelial cells migrate through these holes into the extravascular space, differentiate, and organize into hollow tubes creating new blood vessel walls. The newly formed vessels anastomose and form new functional vascular loops structurally supported by pericytes. The new

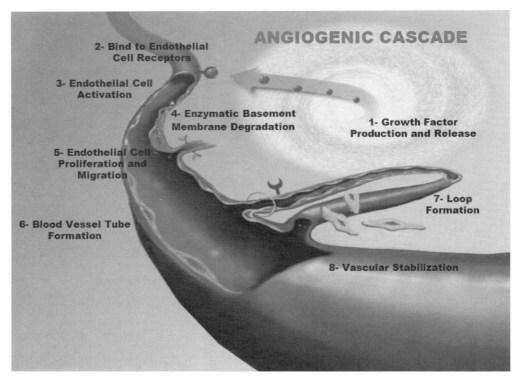

Fig. 2. Angiogenesis: the formation of new blood vessels. *Source:* The angiogenesis Foundation (www.angio.org).

vascular loops carry blood back to the initially injured tissue in an effort to reverse the initial hypoxic injury.

ROLE OF VEGF IN PHYSIOLOGICAL AND PATHOLOGICAL ANGIOGENESIS IN AMD

During embryogenesis, blood vessels are formed by two distinct processes: vasculogenesis and angiogenesis. Vasculogenesis involves the *de novo* differentiation of endothelial cells from mesodermal precursors, whereas angiogenesis involves the generation of new vessels from pre-existing ones. Vasculogenesis, which occurs only during embryonic development, leads to the formation of a primary vascular plexus.

VEGF is naturally expressed in retinal tissues with especially high levels concentrated in the RPE (Fig. 3). In the normal eye, VEGF may play a protective role in maintaining adequate blood flow to the RPE and photoreceptors *(15)*. Deficiencies in blood flow to the choriocapillaris, oxidative stress, and alterations in Bruch's membrane have all been demonstrated to trigger the initial over expression of pathologic levels of VEGF in the RPE and the retina *(16)*.

As mentioned earlier, VEGF promotes proliferation, chemotaxis of endothelial cells, and increases vascular permeability. These vascular permeability changes result in an artificial increase in interstitial fluid pressure, which produces leakage of plasma proteins. The increased oncotic pressure in the extravascular spaces results in the

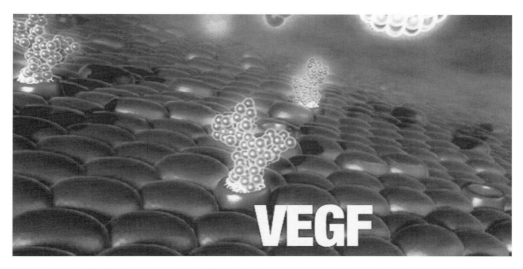

Fig. 3. The VEGF has a specific affinity for its receptors at the level of the RPE and endothelial cells. Overexpressed VEGF promotes angiogenesis, increased vascular permeability, and express proteases and other cytokines. Courtesy of Eyetech Pharmaceuticals and Pfizer, Inc.

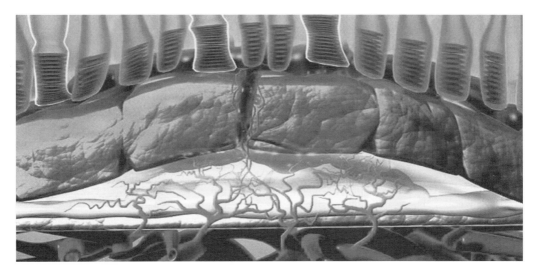

Fig. 4. The VEGF cascade contributes to CNV formation with breakdown of the blood–retinal barrier and violation of the subretinal space. Courtesy of Eyetech Pharmaceuticals and Pfizer, Inc.

formation of a fibrin gel, which provides a substrate for endothelial cell growth and migration (Fig. 4).

Localization of high levels of VEGF in the choroid in patients with wet AMD strongly suggests its direct role in the progression of this disease *(17)*.

CLASSIFICATION OF ANGIOGENESIS INHIBITORS

Several agents targeting angiogenesis have been developed and can be grouped into four categories based on their mechanisms of action (Table 2) *(18)*.

Table 2
Inhibitors of the Angiogenesis

1. Matrix metalloproteinase inhibitors
    a. Marimastat (BB2516)
    b. Prinomastat (AG3340)
    c. BMS 275291
    d. BAY 12-9566
    e. Neovastat (AE-941)
2. Drugs blocking endothelial cell signaling
    a. RhuMAb VEGF
    b. Inhibitors of the VEGF receptors
    c. SU5416
    d. SU6668
    e. ZD6474
    f. CP-547,632
    g. Other tyrosine kinase inhibitors
3. Endogenous inhibitors of angiogenesis
    a. Endostatin
    b. Interferons
4. Novel agents inhibiting endothelial cells
    a. Thalidomide
    b. Squalamine
    c. Celecoxib
    d. ZD6126
    e. Integrin antagonists
    f. TNP-470

1. *Matrix metalloproteinase inhibitors:* Degradation of the extracellular matrix and basement membrane is one of the first steps in angiogenesis. The MMPs are a family of secreted zinc-dependent enzymes that are capable of degrading the components of the extracellular matrix and basement membrane. The MMP family includes four principal classes of molecules: collagenases, gelatinases, stromelysins, and membrane-type MMPs *(19)*. Because the upregulation and activation of proteinases represents a common pathway in the process of retinal neovascularization, pharmacological intervention of this pathway may be an alternative therapeutic approach to neovascular diseases *(20)*.

2. *Drugs blocking endothelial cell signaling and migration:* Once degradation of the extracellular membrane and basement membrane occurs, the next step in angiogenesis is endothelial cell migration, proliferation, and differentiation. Clearly, VEGF and its receptors play a critical role during the stimulation of the endothelial cells and is a logical target for antiangiogenic therapies *(17,21,22)*.

3. *Endogenous inhibitors of angiogenesis:* Endostatin and inteferons are both endogenous inhibitors of angiogenesis. Endostatin, a C-terminal fragment of collagen XVIII formed by proteolysis, specifically inhibits endothelial cell migration and proliferation in vitro and potently inhibits angiogenesis and tumor growth in vivo. Reduced levels of endostatin in Bruch's membrane, RPE basal lamina, intercapillary septa, and choriocapillaris in eyes with AMD may be permissive for choroidal neovascularization *(23)*. The therapeutic implications of endostatin remain to be investigated. Inteferons were not better than the natural history of AMD when tested in a well-controlled clinical trial as an endogenous inhibitor of angiogenesis *(24)*.

4. Other agents: Thalidomide, an immunomodulator and antiangiogenic drug, has been tested as a treatment for wet AMD. Although the enrollment of patients is finished, the results are not yet known *(25)*. Squalamine, an aminosterol originally derived from the liver of the dogfish shark, has shown potent anti-angiogenesis effects both in vitro and in vivo *(26)*.

ANTI-ANGIOGENESIS THERAPY FOR EXUDATIVE AMD

Many clinical trials have been performed with the hope of finding a safe and efficacious pharmacological treatments for exudative AMD. Most of these drugs are directed at interrupting neovascularization at various points along the angiogenic pathway as illustrated in Table 3.

The recent Food and Drug Administration approval of Macugen in December of 2004 in the United States for the treatment of wet AMD represents a very important addition to the armamentarium of vitreo-retinal specialist.

Macugen®: Pegaptanib Sodium (Eyetech Pharmaceuticals)

Macugen is an aptamer with a high affinity and specificity for the extracellular pathological isoform VEGF known as VEGF-165. Macugen is injected into the vitreous cavity every 6 wk. By selectively targeting VEGF-165, Macugen prevents VEGF uptake by endothelial cells receptors, inhibits new vessel formation and prevents leakage from existing new vessels (Fig. 5).

Preliminary results from phase I/II clinical trials on Macugen were quite promising *(22,27)*. Patients were randomly assigned to 0.3, 1, or 3 mg of pegaptanib sodium or to sham treatment. Injections were given intravitreally, through the pars plana every 6 wk, in most cases, for a total of nine treatments.

The subjects were then followed for 54 wk. Macugen showed evidence of efficacy at all three dosage levels. The 0.3 mg dose was chosen as the lowest efficacious dose and, at this concentration, 87.5% of the patients treated with Macugen showed stabilization or improvement in vision during a period of 3 mo. A 60% three-line gain at 3 mo was noted in patients who received both the anti-VEGF aptamer and photodynamic therapy *(22)*.

Ranibizumab: RhuFab (Lucentis, Genentech)

RhuFab is a recombinant humanized anti-VEGF antibody fragment designed to bind and inactivate all the VEGF isoforms. Like Macugen, it is injected into the vitreous cavity in an office-based procedure. Lucentis is administered every 4 wk, whereas Macugen is administered every 6 wk. The antibody fragment consists solely of the antigen-binding portion of the entire antibody, which facilitates RhuFab's retinal penetration. Results of a phase I/II clinical trial at 6 mo have been reported. In early testing on 53 patients with subfoveal CNVM, 50 (94) had stable or improved vision at day 98 following intravitreal injection of Lucentis. The proportion of patients who had an improvement in vision equivalent to three or more lines of vision was 26%, a result almost identical to Macugen's. Two different arms of a phase III trial are under evaluation. MARINA evaluates the efficacy of Lucentis in patients with wet AMD. ANCHOR compares Lucentis vs photodynamic therapy with Verterporfin in patients with wet AMD. The only adverse effect in the treated group has been the development of a transient, mild-to-moderate inflammatory reaction following injections with Lucentis.

Table 3
Leading Angiogenesis Inhibitors

	Macugen	*Lucentis*	*Retaane*	*Squalamine*
Drug class/type	PEGylated aptamer	Humanized antibody fragment	Synthetic angiostatic steroids	Small-molecule aminosterol
Location of action	Extracellular	Extracellular	Intracellular	Intracellular
Mechanism of action	VEGF-165 inhibition	VEGF (all isoforms) inhibition	Vascular endothelial cell proliferation and migration inhibition	Inhibition of VEGF (in addition to other growth factors), cytoskeleton formation, and integrin expression
Route of administration	Intravitreal	Intravitreal	Posterior juxtascleral depot	Intravenous
Administration frequency	Every 6 wk	Every 4 wk	Every 6 mo	Likely 4–8 wk
Comments	The attachment of polyethylene glycol to the aptamer slows its metabolism allowing for treatments to be slightly longer than Lucentis	Inhibition of all of the isoforms of VEGF	Should be effective against multiple stimuli because it acts downstream	Squalamine is administered intravenously eliminating the risks of eye infection or injury
Product company and developer	Eyetech/Pfizer	Genentech/ Novartis	Alcon	Genaera

Anecortave Acetate (Retaane, Alcon)

Anecortave acetate is an angiostatic steroid. Owing to structural modifications, it has no glucocorticoid activity and does not elevate intraocular pressure nor increase the risk of cataract formation. The drug inhibits both urokinase-like plasminogen activator and matrix metallopeptidase 3, two enzymes necessary for vascular endothelial cell migration during blood vessel growth. Anecortave acetate seems to inhibit neovascularization independent of the angiogenic stimulus and it, therefore, has the potential of nonspecific angiogenesis inhibition. The drug is administered by means of a posterior juxtascleral injection with a specially designed cannula. Preclinical and clinical trials have shown that Retaane is far superior to placebo in maintaining positive visual outcomes, preventing severe visual loss, and inhibiting lesion growth *(21)*.

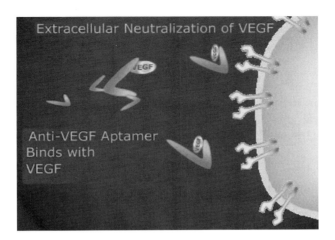

Fig. 5. VEGF is blocked from binding with its natural receptor after Macugen® (pegaptanib sodium injection) binds with VEGF. Courtesy of Eyetech Pharmaceuticals and Pfizer, Inc.

Squalamine (Genaera)

Squalamine is a naturally occurring, pharmacologically active, small molecule that belongs to the aminosterols family. It is the first compound to be tested as a clinical drug candidate in this group of agents. Squalamine is a potent molecule with a unique multifaceted mechanism of action that blocks the formation of the cytoskeleton, integrin expression, and the action of a number of angiogenic growth factors, including VEGF *(26,28)*. The drug is administered intravenously (25–50 mg/m^2). In a phase I/II study, patients were treated weekly with an iv dose of Squalamine. After 4 mo of follow-up, 10 out of 40 subjects had three or more lines of visual improvement, and 29 (72.5%) maintained their initial visual acuity or did not lose less than three lines of vision.

Other Agents Under Investigation

1. Combretastatin A4 Prodrug (CA4P From Oxigene Inc.)
   Combretastatin represents a new class of therapeutic compounds known as vascular targeting agents. It was originally derived from the root bark of the *Combretum caffrum* tree, also known as the Cape Bushwillow. Zulu warriors utilized a substance made from this tree to poison the tips of their arrows and spears as a charm to ward off their enemies. Combretastatin works by microtubule inhibition present in the cytoskeleton of endothelial cells lining the abnormal blood vessels. When this tubulin structure is disrupted, endothelial cell morphology changes from flat to round, stopping blood flow through the capillary. The drug was tested in two different models of ocular neovascularization. Combretastatin suppresses the development of VEGF-induced retinal neovascularization and also blocks and promotes regression of choroidal neovascular membranes *(29)*.
2. Small interfering RNA (siRNA) targeting VEGF (Cand5 from Acuity Pharmaceuticals)
   RNA interference (RNAi) mediated by siRNAs is a technology that allows the silencing of mammalian genes with great specificity and potency. The highly potent RNAi mechanism of action of Cand5 stops production of VEGF at the source, whereas other compounds inhibit VEGF after it is produced, with VEGF remaining present at significant and potentially pathogenic levels. Cand5's potency reflects the ability of a single RNAi drug molecule to stop the

production of VEGF protein molecules. RNAi mechanism also has the potential to translate into a longer duration of action resulting in a lower required dosing frequency compare to other compounds, which bind directly to VEGF *(30)*.

CONCLUSION

Several different classes of agents that target angiogenesis have been developed recently for the treatment of wet AMD, in which angiogenesis is thought to be the primary mechamism. It is well known that VEGF plays a critical role in the genesis of the abnormal vessels. Antiangiogenic therapy targeting the VEGF signaling pathway is a promising new strategy for the management of wet AMD. The development of dosing strategies and combination therapies will become an important clinical challenge for the vitreo-retinal specialist.

ACKNOWLEDGMENTS

The authors have no proprietary interest in any aspect of this study.

REFERENCES

1. Leibowitz HM, Krueger DE, Maunder LR, et al. The Framingham Eye Study Monograph: VI. Macular Degeneration. Surv Ophthalmol 1980;24(suppl):428–427.
2. Klein R, Klein BE, Linton KL. Prevalence of age-related maculopathy: the Beaver Dam Eye Study. Ophthalmology 1992;99:933–944.
3. Chaum E, Hatton MP. Gene Therapy for Genetic and Acquired Retinal Disease. Surv Opthalmol 2002;47:449–469.
4. Garcia Valenzuela E, Sharma SC. Rescue of retinal ganglion cells from axotomy-induced apoptosis through TRK oncogene transfer. Neuroreport 1998;9:165–170.
5. Lai CC, Wu WC, Chen SL, et al. Suppression of choroidal neovascularization by adeno-associated virus vector expressing angiostatin. Invest Ophthalmol Vis Sci 2001;42:2401–2407.
6. Semkova I, Kreppel F, Welsandt G, et al. Autologous transplantation of genetically modified iris pigment epithelial cells: A promising concept for the treatment of age-related macular degeneration and other disorders of the eye. Proc Natl Acad Sci USA 2002;99: 13,090–13,095.
7. Tombran-Tink J, Johnson L. Neurontal differentiation of retinoblastoma cells induced by medium conditioned by human RPE cells. Invest Ophthalmol Vis Sci 1989;30:1700–1709.
8. Rasmussen HS, Rasmussen CS, Durham RG, et al. Looking into anti-angiogenic gene therapies for disorders of the eye. Drug Discov Today 2001;22:1171–1175.
9. Dawson DW, Volpert OV, Gillis P, et al. Pigment epithelium-derived factor: a potent inhibitor of angiogenesis. Science 1999;285:245–248.
10. Hyder SM, Stancel GM. Regulation of angiogenic growth factors in the female reproductive tract by estrogens and progestins. Mol Endocrinol 1999;13:806–811.
11. Ferrara N, Davis-Smyth T. The biology of vascular endothelial growth factor. Endocr Rev 1997;18:4–25.
12. Klagsbrun M, D'Amore PA. Vascular endothelial growth factor and its receptors. Cytokine Growth Factor Rev 1996;7:259–270.
13. Rosen LS. Clinical Experience With Angiogenesis Signaling Inhibitors: Focus on Vascular Endothelial Growth Factor (VEGF) Blockers. Cancer Control 2002;9:36–44.
14. Jain RK. Tumor Angiogenesis and Accessibility: role of Vascular Endothelial Growth Factor. Semin Oncol 2002;29:3–9.

15. Ambati J, Ambati BK, Yoo SH, et al. Age-related macular degeneration: etiology, pathogenesis, and therapeutic strategies. Surv Ophthalmol 2003;48:257–293.
16. Witmer AN, Vrensen GF, Van Noorden CJ, et al. Vascular endothelial growth factors and angiogenesis in eye disease. Prog Retin Eye Res 2003;22:1–29.
17. Ferrara N. Role of Vascular Endothelial Growth Factor in Physiologic and Pathologic Angiogenesis: Therapeutic Implications. Semin Oncol 2002;29:10–14.
18. Sridhar SS, Shepherd FA. Targeting angiogenesis: a review of angiogenesis inhibitors in the treatment of lung cancer. Lung Cancer 2003;42:81–91.
19. Ray JM, Stetler-Stevenson WG. The role of matrix metalloproteases and their inhibitions in tumor invasion, metastasis and angiogenesis. Eur Respir J 1994;7:2062–2072.
20. Holz FG, Miller DW. Pharmacological therapy for age-related macular degeneration. Current developments and perspectives. Ophthalmologe 2003;100:97–103.
21. The Anecortave Acetate Clinical Study Group. Anecortave Acetate as Monotherapy for the Treatment of Subfoveal Lesions in Patients with Exudative Age-related Macular Degeneration (AMD). Interim (6 months) Analysis of Clinical Safety and Efficacy. Retina 2003;23:14–23.
22. The Eyetech Study Group. Anti-vascular Endothelial Growth Factor Therapy for Subfoveal Choroidal Neovascularization Secondary to Age-related Macular Degneration. Phase II Study Results. Ophthalmology 2003;110:979–986.
23. Bhutto IA, Kim SY, McLeod DS, et al. Localization of collagen XVIII and the endostatin portion of collagen XVIII in aged human control eyes and eyes with age-related macular degeneration. Invest Ophthalmol Vis Sci 2004;45:1544–1552.
24. Pharmacological Therapy for Macular Degeneration Study Group. Interferon alfa-2a is ineffective for patients with choroidal neovascularization secondary to age-related macular degeneration. Results of a prospective randomized placebo-controlled clinical trial. Arch Ophthalmol 1997;115:865–872.
25. Ciardella AP, Donsoff IM, Guyer DR, et al. Antiangiogenesis agents. Opthalmol Clin North Am 2002;15:453–458.
26. Ciulla TA, Criswell MH, Danis RP, et al. Squalamine Lactate Reduces Choroidal Neovascularization in a Laser-injury Model in the Rat. Retina 2003;23:808–814.
27. Eyetech Study Group. Preclinical and phase 1A clinical evaluation of an anti-VEGF pegylated aptamer (EYE001) for the treatment of exudative age-related macular degeneration. Retina 2002;22:143–152.
28. Higgins RD, Sanders RJ, Tan Y, et al. Squalamine Improves Retinal Neovascularization. Invest Ophthalmol Vis Sci 2000;41:1507–1512.
29. Nambu H, Nambu R, Melia M, Campochiaro PA. Combretastatin A-4 Phosphate Suppresses Development and Induces Regression of Choroidal Neovascularization. Invest Ophthalmol Vis Sci 2003;44:3650–3655.
30. Reich SJ, Fosnot J, Kuroki A, et al. Small interfering RNA (siRNA) targeting VEGF effectively inhibits ocular neovascularization in a mouse model. Mol Vis 2003;9:210–216.

5

Stargardt Disease

From Gene Discovery to Therapy

Rando Allikmets, PhD

INTRODUCTION

When the adenosine triphosphate (ATP)-binding cassette (ABC) transporter gene, *ABCA4* (originally named *ABCR*), was cloned and characterized in 1997 as the causal gene for autosomal recessive Stargardt disease (arSTGD or STGD1) *(1)* it seemed as if just another missing link was added to the extensive table of genetic determinants of rare monogenic retinal dystrophies. Now, 9 yr later, the *ABCA4* gene continues to emerge as the predominant determinant of a wide variety of retinal degeneration phenotypes. *ABCA4* has caused exciting and sometimes intense discussions among ophthalmologists and geneticists, resulting in more than 150 publications during this time.

In this chapter, I will summarize our current knowledge of the role of *ABCA4* in STGD. Substantial progress via extensive genetic and functional studies has allowed for major advances in diagnostic and therapeutic applications for STGD, which most recently seemed impossible. Although *ACBA4* has proven to be a complex and difficult research target, I hope to convince the reader that treatment of all *ABCA4*-associated disorders, and especially STGD, should be possible in the near future.

STARGARDT DISEASE

STGD1 (MIM 248200) is arguably the most common hereditary recessive macular dystrophy (estimated frequency of 1 out of 8,000 to 10,000 in the United States *[2]*) and

From: *Ophthalmology Research: Retinal Degenerations: Biology, Diagnostics, and Therapeutics*
Edited by: J. Tombran-Tink and C. J. Barnstable © Humana Press Inc., Totowa, NJ

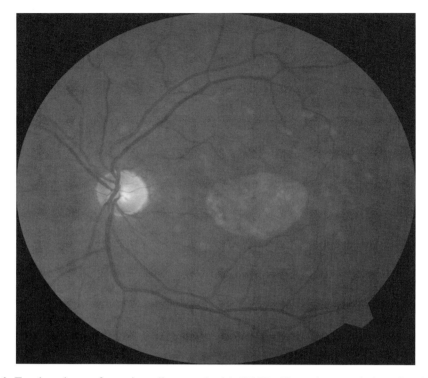

Fig. 1. Fundus photo of a patient diagnosed with STGD. Note characteristic yellowish flecks around the macula and a defined area of central macular degeneration.

is characterized by a highly variable age of onset and clinical course. Most cases present with juvenile to young-adult onset, evanescent to rapid central visual impairment, progressive bilateral atrophy of the foveal retinal pigment epithelium (RPE) and photoreceptors, and the frequent appearance of yellow-orange flecks distributed around the macula and/or the midretinal periphery *(3,4)* (Fig.1). In a large fraction of patients with STGD, a "dark" or "silent" choroid is seen on fluorescein angiography, which reflects the accumulation of lipofuscin. Electroretinographic (ERG) findings vary and are not usually considered diagnostic for the disease. A clinically similar retinal disorder, fundus flavimaculatus (FFM), often manifests later onset and slower progression *(5–7)*. Despite historical separation *(8,9)*, results of linkage and mutational analysis confirmed that STGD and FFM are allelic autosomal recessive disorders with slightly different clinical manifestations, caused by mutations of a single gene located within an approx 2-cM interval between markers *D1S406* and *D1S236,* at chromosome 1p13-p21 *(3,10–13)*.

The wide variation in clinical expression of the disease in patients with STGD has been classified into three major clinical phenotypes according to Fishman *(14)*. Phenotype I is characterized by an atrophic-appearing macular lesion, localized perifoveal yellowish-white flecks, the absence of a dark choroid, and normal ERG amplitudes. Patients in the phenotype II group present with a dark choroid, more diffuse yellowish-white flecks in the fundus, and inconsistent ERG amplitudes. Phenotype III group includes patients with extensive atrophic-appearing changes of the RPE and reduced ERG amplitudes of both cones and rods.

GENETIC PREDISPOSITION: THE *ABCA4* (*ABCR*) GENE

Several laboratories independently described *ABCA4* in 1997 as the causal gene for arSTGD *(1,15,16)*. There is no definitive evidence of genetic heterogeneity of arSTGD because, as described here previously, all families segregating the disorder have been linked to the *ABCA4* locus on human chromosome 1p13-p22. Hence, the role of the *ABCA4* gene as the only causal gene for arSTGD has not been disputed. Subsequently, several cases were reported where *ABCA4* mutations segregated with retinal dystrophies of substantially different phenotype, such as autosomal recessive cone-rod dystrophy (arCRD) *(17,18)* and autosomal recessive retinitis pigmentosa (arRP) *(17,19,20)*.

Disease-associated *ABCA4* alleles have shown an extraordinary heterogeneity *(1,14,21–24)*. Currently, about 500 disease-associated *ABCA4* variants have been identified, allowing comparison of this gene to *CFTR*, one of the best-known members of the ABC superfamily, encoding the cystic fibrosis transmembrane conductance regulator (CFTR) *(25)*. What makes *ABCA4* a more difficult diagnostic target than *CFTR* is that the most frequent disease-associated *ABCA4* alleles, e.g., G1961E, G863A/delG863, and A1038V, have each been described in only about 10% of patients with STGD in a distinct population, whereas the delF508 allele of *CFTR* accounts for close to 70% of all cystic fibrosis alleles *(26)*.

Several studies have identified frequent "ethnic group-specific" *ABCA4* alleles, such as the 2588G > C variant resulting in a dual effect, G863A/delG863, as a founder mutation in Northern European patients with STGD *(22)*, and a complex allele (two mutations on the same chromosome), L541P/A1038V, in both STGD and CRD patients of German origin *(24,27)*. Complex *ABCA4* alleles are not uncommon in STGD *(21)*, in fact, they are detected in about 10% of all patients with STGD *(28)*.

Because the *ABCA4* gene has been screened mainly in European patients with STGD, the estimates of allele frequencies and pathogenicity have been made based on these data. However, several studies have suggested substantial differences in frequencies of specific *ABCA4* alleles between the Caucasian general population and those of other ethnic/racial groups. For example, an *ABCA4* allele T1428M, which is very rare in populations of European descent, is apparently frequent (~8%) in the Japanese general population *(29)*.

An even more intriguing case involves one of the most frequent *ABCA4* mutations, G1961E. Although its frequency in the Caucasian general population is approx 0.2%, and in patients with STGD of the same ethnic origin it reaches greater than 10% *(30)*, this allele has been detected at substantially higher rate in East African countries (tribes from Somalia, Kenya, Ethiopia, etc.). It is likely that approx 10% of the general population from Somalia carries the allele in a heterozygous form *(31)*. At the same time, it is almost absent in individuals of West African descent, including African Americans (Rando Allikmets, unpublished data). Although the exact cause and consequences of this phenomenon are still to be determined, it is likely that the prevalence of STGD in East Africa is much higher than expected and/or observed.

The summarized data presented here establish allelic variation in *ABCA4* as the most prominent cause of retinal dystrophies with Mendelian inheritance patterns. The latest estimates suggest that the carrier frequency of *ABCA4* alleles in general population is in the range of 5 to 10% *(22,32,33)*. This finding, that at least 1 out of 20 people carry a disease-associated *ABCA4* allele, has enormous implications for the amount of retinal pathology attributable to *ABCA4* variation and suggests re-evaluation of current prevalence estimates.

Soon after the discovery of *ABCA4*, a disease model was proposed that suggested a direct correlation between the continuum of disease phenotypes and residual *ABCA4* activity/function *(34,35,36)*. According to the predicted effect on the *ABCA4* transport function, Maugeri et al. classified *ABCA4* mutant alleles as "mild," "moderate," and "severe" *(22)*. Different combinations of these were predicted to result in distinct phenotypes in a continuum of disease manifestations, the severity of disease manifestation being inversely proportional to the residual *ABCA4* activity. This model, although widely accepted, was recently disputed by Cideciyan and co-workers *(37)*, who did not find the suggested inverse relationship between residual *ABCA4* activity and clinical disease severity. The latter was defined by intensity of autofluorescence, rod and cone sensitivity, and rod adaptation delay. However, their study cohort included only 15 patients with known *ABCA4* variants on both alleles and only a few of these were severe mutations, compared to mild mutations (3 vs 28). Furthermore, classification of mutations by severity through circumstantial evidence, e.g., by predicting the effect of an amino acid change on the protein function, would often lead to erroneous conclusions, as clearly demonstrated by the available in vitro assay, which currently provides the best estimate of the functional effect of *ABCA4* variants *(28,38)*. In summary, all experimental, clinical, and genetic variables have to be carefully assessed when making conclusions either on the proposed model of *ABCA4*, or on genotype/phenotype correlations in patients with STGD.

MOLECULAR DIAGNOSIS

Allelic heterogeneity has substantially complicated genetic analyses of *ABCA4*-associated retinal disease. In the case of arSTGD, the mutation detection rate has ranged from approx 25% *(14,39)* to approx 55–60% *(21,22,24,40)*. In each of these studies, conventional mutation detection techniques, such as single-stranded conformational polymorphism (SSCP), heteroduplex analysis, and denaturing gradient gel electrophoresis (DGGE), were applied. Direct sequencing, which is still considered the "gold standard" of all mutation detection techniques, enabled a somewhat higher percentage of disease-associated alleles to be identified, from 66 to 80% *(28,32)*.

Other important parameters influencing mutation analysis are the effort and the cost of screening the entire gene. *ABCA4* presents a special challenge because, similar to most full-sized ABC transporters, it is comprised of 50 exons and has an open reading frame exceeding 6800 bp. Therefore, all mutation detection techniques that remain exclusively polymerase chain reaction-based, are relatively inefficient, expensive, and labor intensive. Efforts related to mutation detection and genotyping become especially crucial in allelic association or genotype/phenotype correlation studies, in which screening of thousands of samples is needed to achieve enough statistical power, as is the case with *ABCA4*, in which multiple rare variants must be studied *(41)*.

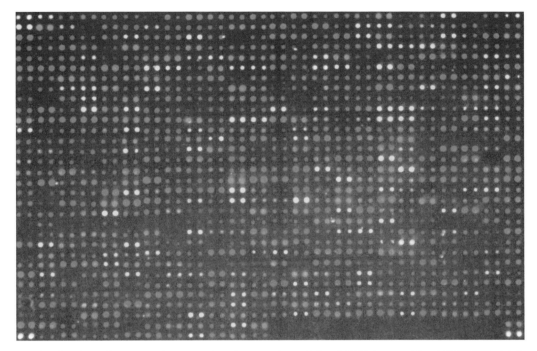

Fig. 2. The microarray for mutational screening of the *ABCA4* gene (the ABCR400 chip). Each allele (mutation) is queried by duplicate oligonucleotides from both strands, i.e., represented on the array by four dots. Four colors discriminate four dideoxynucleotide triphosphates labeled with four different fluorescent dyes.

To overcome these challenges and to generate a high-throughput and a cost-effective screening tool, the *ABCA4* genotyping microarray was developed *(33)*. The ABCR400 microarray contains all currently known disease-associated genetic variants (>450) and many common polymorphisms of the *ABCA4* gene (Fig. 2). The chip is more than 99% effective in screening for mutations and it has been used for highly efficient, systematic screening of patients with *ABCA4*-associated pathologies, especially those affected by STGD. When a cohort of patients with STGD is analyzed by both the array and one of the conventional mutation detection methods, e.g., SSCP, the mutation detection rate reaches 70–75% of all disease-associated alleles, thus being comparable to direct sequencing *(33)*. The chip alone will detect approx 55–65% of these alleles, an efficiency similar to that in the best studies with conventional (SSCP, DGGE, etc.) mutation detection methods *(33,42,43)*. However, the effort and the expense of the array-based screening are each at least one order of magnitude lower than any other method, whereas the speed and reliability are much higher. A complete screening of one DNA sample takes only a few hours and the current all-inclusive cost is estimated at $0.20/genotype. The reliability of the chip, i.e., the efficiency of detection of any mutation on the chip, is more than 99% in any given experiment. Another feature of the ABCR400 array allows for detection of any sequence variation at each and all positions included on the chip, as it directly sequences all queried positions. Several new, previously not described, mutations have been found by the ABCR400 array in several studies *(33,43)*. In conclusion, although the assessment of disease-associated variation of the *ABCA4* gene still represents a difficult task for ophthalmologists and geneticists, the ABCR400 microarray offers a diagnostic tool that can advance our knowledge substantially, specifically in genotype/phenotype correlation studies of STGD. The latter will facilitate the counseling of patients on their visual prognosis. This information will also enhance future therapeutic trials in patients with STGD (*see* below in Emerging Therapeutic Options).

FUNCTIONAL STUDIES OF ABCA4

The ABCA4 protein was first described in mid 1970s as an abundant component of photoreceptor outer segment disk rims *(44,45)*. Hence, it was called a Rim protein for the next 20 yr. After the encoding gene, *ABCA4*, was cloned in 1997 and characterized as a member of the ABC transporter superfamily, it suggested a transport function of some substrate in photoreceptor outer segments *(1,16)*. All-*trans* retinal, the isoform of rhodopsin chromophore, was identified as a potential substrate of ABCA4 by its ability to stimulate ATP hydrolysis by reconstituted ABCA4 protein in vitro, suggesting that retinal could also be the in vivo substrate for ABCA4 *(46)*. Studies of *Abca4* knockout mice fully supported this hypothesis, proposing ABCA4 as a "flippase" of the complex of all-*trans* retinal and phosphatidylethanolamine (*N*-retinylidene-phosphatidylethanolamine) *(47)*. The most recent experimental data from Dr. Molday's laboratory fully confirmed this hypothesis *(48)*.

Mice lacking the functional *Abca4* gene demonstrated delayed dark adaptation, increased all-trans retinal following light exposure, elevated phosphatidylethanolamine (PE) in rod outer segments, accumulation of the *N*-retinylidene-PE, and striking deposition of a major lipofuscin fluorophore (A2E) in RPE. Based on these findings, it was suggested that the *ABCA4*-mediated retinal degeneration may result from "poisoning" of the RPE owing to A2E accumulation, with secondary photoreceptor degeneration due to loss of the RPE support role *(47)*. A2E, a pyridinium *bis*-retinoid, which is derived from two molecules of vitamin A aldehyde and one molecule of ethanolamine, has been characterized as one of the major components of retinal pigment epithelial lipofuscin *(49)*. Accumulation of lipofuscin in the macular region of RPE is characteristic to aging eyes and is the hallmark of both STGD1 and age-related macular degeneration (AMD) (Fig. 1).

Together, these data define ABCA4 as the "rate keeper" of the retinal transport in the visual cycle. ABCA4 is apparently not absolutely essential for this process, as individuals completely lacking the functional protein (e.g., some patients with RP-like phenotype) maintain some eyesight for several years. Over time, however, even mild dysfunction of ABCA4 affects the vision irreparably.

EMERGING THERAPEUTIC OPTIONS

The scientific progress in determining the role of the *ABCA4* gene in STGD and other retinal pathology has been remarkable. We have significantly expanded our knowledge of the extensive range of phenotypes caused by various combinations of *ABCA4* mutations. *ABCA4* research has lead to the screening of thousands of patients with STGD, encompassing large cohorts of ethnically diverse samples. We also know *ABCA4* function as the transporter of *N*-retinylidene-PE, and have a mouse model that reproduces some features of the human disease. Considering all the above, our efforts should now be directed towards finding therapeutic solutions to benefit patients with STGD. I will discuss the available options and most recent advances in this area below (Fig. 2).

Modifying Exposure to Environmental Factors

Avoiding Sunlight to Delay the Disease

Experimental and clinical data have documented the damaging effects of light exposure on photoreceptor cells and several lines of evidence point to retinoids or retinoid derivatives as chromophores that can mediate light damage. Studies from Jeremy Nathans' laboratory have concluded that ABCA4 is unusually sensitive to photooxidation damage mediated by all-*trans*-retinal in vitro (50). Specifically, it was demonstrated that photodamage to ABCA4 in vitro causes it to aggregate in sodium dodecyl sulfate gels and results in the loss of retinal-stimulated ATPase activity. These observations suggested ABCA4 and several other photorecepter (PR) outer segment proteins as targets of photodamage. This is especially relevant to assessing the risk of light exposure in those individuals who already have diminished ABCA4 activity owing to mutation in the ABCA4 gene, such as patients with STGD.

Moreover, studies on *Abca4-/-* mice (47) suggested that A2E did not accumulate in RPE of animals kept in the dark. Therefore, patients with STGD can be advised to avoid light exposure by wearing (ultraviolet-blocking) sunglasses and limiting the exposure of eyes to direct light. However, the effectiveness of this approach has not been demonstrated in epidemiological studies, nor is it likely to be conductive to a productive and satisfying lifestyle.

Limiting Dietary Intake of Vitamin A

Based on our knowledge of the ABCA4 function in the visual cycle, a sound recommendation to all patients with *ABCA4*-associated pathology would be to limit the dietary intake of vitamin A. Vitamin A has been recommended to patients with RP51 and also as a dietary supplement for patients with AMD (52). However, in a substantial fraction of patients with both disorders, the disease phenotype is caused, at least in part, by *ABCA4* mutations. Elevated intake of vitamin A by these patients would only worsen the disease prognosis by stimulating the visual cycle and, consequently, increasing the accumulation of A2E in RPE with predicted grave consequences. This observation further justifies thorough diagnostic screening of patients for underlying genetic variation to pinpoint the molecular cause and/or mechanism of the disease before recommending therapeutic intervention.

Drugs Modifying Functional (ATP-Binding and Hydrolysis) Activity of ABCA4

The in vitro assay developed by Hui Sun and coworkers (46) suggested several compounds that either enhanced or diminished the (ATPase) activity of the ABCA4 protein. Small-molecule drugs, synergistically enhancing the ATPase activity of *ABCA4*, may represent potentially beneficial compounds for patients with STGD and/or for a subset of individuals at risk for AMD. Although clear predictions regarding the in vivo effect on ABCA4 function of compounds like amiodarone and digitonin are currently impossible, the existence of such compounds suggests that environmental and/or drug effects may be relevant to *ABCA4*-associated degenerative retinal diseases. In this context, Sun et al. suggested to study visual function in patients receiving amiodarone, an Food and Drug Administration-approved drug that has been widely used in the treatment of cardiac arrhythmias at tissue concentrations at 40 µ*M* or greater *(46)*.

Drugs Slowing Down the Visual Cycle

The vertebrate visual cycle consists of enzymatic reactions that contribute to the generation of 11-*cis* retinal, the visual chromophore. One consequence of the constant recycling of vitamin A, as described here previously, is the formation of the fluorophores, such as the bi-retinal conjugate A2E, its photoisomer iso-A2E, that constitute the lipofuscin of RPE cells *(53–58)*. Because of its unusual pyridinium *bis*-retinoid structure *(59)*, A2E cannot be enzymatically degraded and, thus, accumulates in RPE cells of patients with STGD and *Abca–/–* mice *(47)*. At sufficient concentrations, A2E perturbs cell membranes *(49,60)*, confers a susceptibility to blue light-induced apoptosis *(61–63)*, and alters lysosomal function *(64)*. In light of this adverse behavior, there is considerable interest in retarding A2E formation as a means to prevent vision loss in STGD and perhaps in AMD. Therefore, it is significant that studies have shown that RPE lipofuscin is substantially reduced when the 11-*cis* and all-*trans*-retinal chromophores are absent either because of dietary deficiency or gene knockout *(65,66)*. Light exposure, an obvious determinant of the rate of flux of all-*trans*-retinal through the visual cycle, can also moderate the rate of A2E synthesis *(57,58,67)*. In addition, Radu et al. *(68)* reported that isotretinoin (13-*cis*-retinoic acid), an acne medication Accutane known previously to delay dark adaptation *(69)*, dampens the deposition of A2E in RPE cells. Although heralded as a potential therapeutic agent for STGD, Accutane (i.e., retinoic acids) cannot be used for chronic treatment of patients as a result of the numerous systemic side effects associated with the use of these drugs, including teratogenic effects, depression, birth defects, dryness of mucosal membranes, and skin flaking *(70)*. Therefore, identification of other, better-targeted compounds, which will slow down the visual cycle with minimal and/or acceptable side effects (such as mild night blindness), could lead to the availability of treatment options for patients with juvenile-onset macular dystrophies and possibly to preventive patient care.

Possible protein targets of these compounds, similar to Accutane, could include 11-*cis*-retinol dehydrogenase *(71)*, or RPE65 *(72)*. Recent studies have demonstrated that slowing the kinetics of the visual cycle curtails the accumulation of A2E *(73)*. These results also point to RPE65 as a rate-limiting step in A2E formation that can be therapeutically targeted. Therefore, a class of compounds that could provide a specific, controlled inhibition of the visual cycle includes antagonists of the physiological substrate of RPE65, all-*trans*-retinyl esters. *(74)* Furthermore, the fact that RPE65 is essentially unique to the visual system strongly suggests that its inhibitors are unlikely to be generally toxic.

Gene Therapy

Although the therapeutic applications discussed here previously would only delay or modify the disease progress, the gene therapy approach should provide "the cure" for

STGD and other *ABCA4*-associated diseases. Because all *ABCA4*-associated diseases, including STGD, are recessive, introducing a normal, functioning copy of the gene to photoreceptors would restore visual function. Another advantage of the gene therapy approach in the specific case of STGD relies on the fact that degeneration of the retinal cells in this disease is relatively delayed, allowing a reasonable time window for therapeutic intervention.

The *Abca4* knockout mice described earlier allows testing various gene therapy approaches in vivo. The extent of the restoration of the retinal function in the mouse model can be judged by a combination of biochemical and functional tests, such as non-invasive electroretinogram and quantitation of A2E accumulation either by high-performance liquid chromatography or autofluorescence. The functioning human *ABCA4* gene can be introduced into mouse photoreceptors via several techniques, the most proven and robust of which are those utilizing viral vectors, which efficiently transduce photoreceptors. The two main groups of viral vectors used in gene therapy to date are adeno-associated viruses (AAV) and lentiviruses. Lentiviral vectors offer several key advantages in the specific case of *ABCA4 (75,76)*. First and foremost, lentiviruses are capable of delivering genes stably and permanently into the genome of infected cells in vivo. Second, they can transduce nondividing cells, a crucial requirement for terminally differentiated cells such as photoreceptors. Third, they can carry large inserts, a distinct advantage over AAV-based vectors because the coding region of the human *ABCA4* gene is extraordinarily large (>6800 bp), exceeding the capacity of AAV.

Although the efficiency of transducing photoreceptors by lentiviruses has been a certain concern, several studies, e.g., those from the laboratory of Inder Verma *(77)* utilizing HIV-based vectors, allow substantial optimism. Moreover, the photoreceptor transduction efficiency could be further increased by using vectors on the equine infectious anemia virus backbone *(78)*, especially those pseudotyped with the rabies virus envelope protein, which has showed greatly improved neurotropism *(79)*.

Alternatively, AAV-based vectors, which have demonstrated high PR-tropism, could be utilized. In this case, however, the *ABCA4* gene has to be delivered to the same photoreceptor cell in two clones because of the capacity constraints. Subsequently, the functional *ABCA4* gene can be reassembled by either *trans*-splicing technique *(80)* or the functional ABCA4 protein can be reconstituted from the two, structurally almost identical, *ABCA4* domains *(81)*. The well-publicized success in restoring vision in the dog model of Leber congenital amaurosis by introducing the functional *RPE65* gene via AAV vectors *(82)* allows for optimism in other recessive retinal diseases, such as STGD.

OUTLOOK

The scientific progress in determining the role of the ABCA4 gene in STGD and in overall retinal pathology has been remarkable. We have significantly expanded our knowledge of the extensive range of phenotypes caused by various combinations of *ABCA4* mutations, and have efficient diagnostic tools, such as genotyping arrays. *ABCA4* research has lead to the formation of multicenter studies, encompassing large cohorts of ethnically diverse samples. Knowledge of the ABCA4 function as the transporter of *N*-retinylidene-PE, and availability of the mouse model that reproduces several features of the human disorders have allowed rapid advancement to the next stage of research directed towards finding therapeutic solutions for *ABCA4*-mediated retinal disease. One, or a combination, of the therapeutic options discussed in this chapter should be available relatively soon for application toward the therapy of the entire spectrum of *ABCA4*-associated retinal disorders, which were most recently considered incurable.

ACKNOWLEDGMENTS

The author sincerely appreciates helpful discussions and useful suggestions from many colleagues and collaborators over the years of STGD and *ABCA4* research. Support by National Eye Institute/National Institutes of Health, Foundation Fighting Blindness, and Research to Prevent Blindness is gratefully acknowledged.

REFERENCES

1. Allikmets R, Singh N, Sun H, et al. A photoreceptor cell-specific ATP-binding transporter gene (ABCR) is mutated in recessive Stargardt macular dystrophy. Nat Genet 1997a; 15:236–246.
2. Blacharski PA: Fundus flavimaculatus. In: Newsome DA, (ed.): Retinal Dystrophies and Degenerations. New York: Raven Press, 1988:135–159.
3. Anderson KL, Baird L, Lewis RA, et al. A YAC contig encompassing the recessive Stargardt disease gene (STGD) on chromosome 1p. Am J Hum Genet 1995;57:1351–1363.
4. Stargardt K. Über familiäre, progressive Degeneration in der Maculagegend des Auges. Albrecht von Graefes Arch Ophthalmol 1909;71:534–550.
5. Franceschetti A. Über tapeto-retinale Degenerationen im Kindesalter. In: von Sautter, H., ed. Entwicklung und Fortschritt in der Augenheiklunde. Stuttgart: Ferdinand Enke Verlag, 1963:107–120.
6. Hadden OB, Gass JD. Fundus flavimaculatus and Stargardt's disease. Am J Ophthalmol 1976;82:527–539.
7. Noble KG, Carr RE. Stargardt's disease and fundus flavimaculatus. Arch Ophthalmol 1979;97:1281–1285.
8. McKusick VA. Mendelian inheritance in man: catalogs of autosomal dominant, autosomal recessive, and X-linked phenotypes. 10th ed., Baltimore, MD: Johns Hopkins University Press, 1992.
9. Mendelian inheritance in man: a catalog of human genes and genetic diseases. 12th ed. Johns Hopkins University Press, Baltimore, MD: Johns Hopkins University Press, 1998.
10. Gerber S, Rozet JM, Bonneau D, et al. A gene for late-onset fundus flavimaculatus with macular dystrophy maps to chromosome 1p13. Am J Hum Genet 1995;56:396–399.
11. Hoyng CB, Poppelaars F, van de Pol TJ, et al. Genetic fine mapping of the gene for recessive Stargardt disease. Hum Genet 1996;98:500–504.
12. Kaplan J, Gerber S, Larget-Piet D, et al. A gene for Stargardt's disease (fundus flavimaculatus) maps to the short arm of chromosome 1. Nat Genet 1993;5:308–311.
13. Weber BH, Sander S, Kopp C, et al. Analysis of 21 Stargardt's disease families confirms a major locus on chromosome 1p with evidence for non-allelic heterogeneity in a minority of cases. Br J Ophthalmol 1996;80:745–749.
14. Fishman GA, Stone EM, Grover S, Derlacki DJ, Haines HL, Hockey RR. Variation of clinical expression in patients with Stargardt dystrophy and sequence variations in the ABCR gene. Arch Ophthalmol 1999;117:504–510.
15. Azarian SM, Travis GH. The photoreceptor rim protein is an ABC transporter encoded by the gene for recessive Stargardt's disease (ABCR). FEBS Lett 1997;409:247–252.

16. Illing M, Molday LL, Molday RS. The 220-kDa rim protein of retinal rod outer segments is a member of the ABC transporter superfamily. J Biol Chem 1997;272:10,303–10,310.

17. Cremers FP, van de Pol DJ, van Driel M, et al. Autosomal recessive retinitis pigmentosa and cone-rod dystrophy caused by splice site mutations in the Stargardt's disease gene ABCR. Hum Mol Genet 1998;7:355–362.

18. Rozet JM, Gerber S, Souied E, et al. Spectrum of ABCR gene mutations in autosomal recessive macular dystrophies. Eur J Hum Genet 1998;6:291–295.

19. Martinez-Mir A, Paloma E, Allikmets R, et al. Retinitis pigmentosa caused by a homozygous mutation in the Stargardt disease gene ABCR. Nat Genet 1998;18:11–12.

20. Rozet JM, Gerber S, Ghazi I, et al. Mutations of the retinal specific ATP binding transporter gene (ABCR) in a single family segregating both autosomal recessive retinitis pigmentosa RP19 and Stargardt disease: evidence of clinical heterogeneity at this locus. J Med Genet 1999;36:447–451.

21. Lewis RA, Shroyer NF, Singh N, et al. Genotype/Phenotype analysis of a photoreceptor-specific ATP-binding cassette transporter gene, ABCR, in Stargardt disease. Am J Hum Genet 1999;64:422–434.

22. Maugeri A, van Driel MA, van de Pol DJ, et al. The 2588G—>C Mutation in the ABCR Gene Is a Mild Frequent Founder Mutation in the Western European Population and Allows the Classification of ABCR Mutations in Patients with Stargardt Disease. Am J Hum Genet 1999;64:1024–1035.

23. Fumagalli A, Ferrari M, Soriani N, et al. Mutational scanning of the ABCR gene with double-gradient denaturing-gradient gel electrophoresis (DG-DGGE) in Italian Stargardt disease patients. Hum Genet 2001;109:326–338.

24. Rivera A, White K, Stohr H, et al. A comprehensive survey of sequence variation in the ABCA4 (ABCR) gene in Stargardt disease and age-related macular degeneration. Am J Hum Genet 2000;67:800–813.

25. Riordan JR, Rommens JM, Kerem B, et al. Identification of the cystic fibrosis gene: cloning and characterization of complementary DNA. Science 1989;245:1066–1073.

26. Zielenski J, Tsui LC. Cystic fibrosis: genotypic and phenotypic variations. Annu Rev Genet 1995;29:777–807.

27. Maugeri A, Klevering BJ, Rohrdchneider K, et al. Mutations in the ABCA4 (ABCR) gene are the major cause of autosomal recessive cone-rod dystrophy. Am J Hum Genet 2000;67:960–966.

28. Shroyer NF, Lewis RA, Yatsenko AN, Wensel TG, Lupski JR. Cosegregation and functional analysis of mutant ABCR (ABCA4) alleles in families that manifest both Stargardt disease and age-related macular degeneration. Hum Mol Genet 2001;10:2671–2678.

29. Kuroiwa S, Kojima H, Kikuchi T, Yoshimura N. ATP binding cassette transporter retina genotypes and age related macular degeneration: an analysis on exudative non-familial Japanese patients. Br J Ophthalmol 1999;83:613–615.

30. Allikmets R. Further evidence for an association of ABCR alleles with age-related macular degeneration. The International ABCR Screening Consortium. Am J Hum Genet 2000;67:487–491.

31. Guymer RH, Heon E, Lotery AJ, et al. Variation of codons 1961 and 2177 of the Stargardt disease gene is not associated with age-related macular degeneration. Arch Ophthalmol 2001;119:745–751.

32. Yatsenko AN, Shroyer NF, Lewis RA, Lupski JR. Late-onset Stargardt disease is associated with missense mutations that map outside known functional regions of ABCR (ABCA4). Hum Genet 2001;108:346–355.

33. Jaakson K, Zernant J, Kulm M, et al. Genotyping microarray (gene chip) for the ABCR (ABCA4) gene. Hum Mutat 2003;22:395–403.

34. van Driel MA, Maugeri A, Klevering BJ, Hoyng CB, Cremers FP. ABCR unites what ophthalmologists divide(s). Ophthalmic Genet 1998;19:117–122.

35. Shroyer NF, Lewis RA, Allikmets R, et al. The rod photoreceptor ATP-binding cassette transporter gene, ABCR, and retinal disease: from monogenic to multifactorial. Vision Res 1999;39:2537–2544.

36. Allikmets R. Molecular genetics of age-related macular degeneration: current status. Eur J Ophthalmol 1999;9:255–65.

37. Cideciyan AV, Aleman TS, Swider M, et al. Mutations in ABCA4 result in accumulation of lipofuscin before slowing of the retinoid cycle: a reappraisal of the human disease sequence. Hum Mol Genet 2004;13:525–534.

38. Sun H, Smallwood PM, Nathans J. Biochemical defects in ABCR protein variants associated with human retinopathies. Nat Genet 2000;26:242–246.

39. Webster AR, Heon E, Lotery AJ, et al. An analysis of allelic variation in the ABCA4 gene. Investigative Ophthalmology Visual Science 2001;42:1179–1189.

40. Simonelli F, Testa F, de Crecchio G, et al. New ABCR mutations and clinical phenotype in Italian patients with Stargardt disease. Invest Ophthalmol Vis Sci 2000;41:892–897.

41. Allikmets R. Simple and complex ABCR: genetic predisposition to retinal disease. Am J Hum Genet 2000;67:793–799.

42. Klevering BJ, Yzer S, Rohrschneider K, et al. Microarray-based mutation analysis of the ABCA4 (ABCR) gene in autosomal recessive cone-rod dystrophy and retinitis pigmentosa. Eur J Hum Genet 2004;12:1024–1032.

43. Simonelli F, Testa F, Zernant J, et al. Association of a homozygous nonsense mutation in the ABCA4 (ABCR) gene with cone-rod dystrophy phenotype in an Italian family. Ophthalmic Res 2004;36:82–88.

44. Papermaster DS, Converse CA, Zorn M. Biosynthetic and immunochemical characterization of large protein in frog and cattle rod outer segment membranes. Exp Eye Res 1976;23:105–115.

45. Papermaster DS, Schneider BG, Zorn MA, Kraehenbuhl JP. Immunocytochemical localization of a large intrinsic membrane protein to the incisures and margins of frog rod outer segment disks. J Cell Biol 1978;78:415–425.

46. Sun H, Molday RS, Nathans J. Retinal stimulates ATP hydrolysis by purified and reconstituted ABCR, the photoreceptor-specific ATP-binding cassette transporter responsible for Stargardt disease. J Biol Chem 1999;274:8269–8281.

47. Weng J, Mata NL, Azarian SM, Tzekov RT, Birch DG, Travis GH. Insights into the function of Rim protein in photoreceptors and etiology of Stargardt's disease from the phenotype in abcr knockout mice. Cell 1999;98:13–23.

48. Beharry S, Zhong M, Molday RS. N-retinylidene-phosphatidylethanolamine is the preferred retinoid substrate for the photoreceptor-specific ABC transporter ABCA4 (ABCR). J Biol Chem 2004;279:53,972–53,979.

49. Sparrow JR, Parish CA, Hashimoto M, Nakanishi K. A2E, a lipofuscin fluorophore, in human retinal pigmented epithelial cells in culture. Invest Ophthalmol Vis Sci 1999;40:2988–2995.

50. Sun H, Nathans J. ABCR, the ATP-binding cassette transporter responsible for Stargardt macular dystrophy, is an efficient target of all-trans-retinal-mediated photooxidative damage in vitro. Implications for retinal disease. J Biol Chem 2001;276:11,766–11,774.

51. Berson EL, Rosner B, Sandberg MA, et al. Vitamin A supplementation for retinitis pigmentosa. Arch Ophthalmol 1993;111:1456–1459.

52. Age-Related Eye Disease Study Research Group. A randomized, placebo-controlled, clinical trial of high-dose supplementation with vitamins C and E, beta carotene, and zinc for age-related macular degeneration and vision loss: AREDS report no. 8. Arch Ophthalmol 2001;119:1417–1436.

53. Sparrow JR, Vollmer-Snarr HR, Zhou J, et al. A2E-epoxides damage DNA in retinal pigment epithelial cells. Vitamin E and other antioxidants inhibit A2E-epoxide formation. J Biol Chem 2003;278:18,207–18,213.

54. Eldred GE, Lasky MR. Retinal age pigments generated by self-assembling lysosomotropic detergents. Nature 1993;361:724–726.

55. Parish CA, Hashimoto M, Nakanishi K, Dillon, Sparrow J. Isolation and one-step preparation of A2E and iso-A2E, fluorophores from human retinal pigment epithelium. Proc Natl Acad Sci USA 1998;95:14,609–14,613.

56. Liu J, Itagaki Y, Ben-Shabat S, Nakanishi K, Sparrow JR. The biosynthesis of A2E, a fluorophore of aging retina, involves the formation of the precursor, A2-PE, in the photoreceptor outer segment membrane. J Biol Chem 2000;275:29,354–29,360.

57. Mata NL, Weng J, Travis GH. Biosynthesis of a major lipofuscin fluorophore in mice and humans with ABCR-mediated retinal and macular degeneration. Proc Natl Acad Sci USA 2000;97:154–159.

58. Ben-Shabat S, Parish CA, Vollmer HR, et al. Biosynthetic studies of A2E, a major fluorophore of retinal pigment epithelial lipofuscin. J Biol Chem 2002;277:7183–7190.

59. Sakai N, Decatur J, Nakanishi K, Eldred GE. Ocular age pigment "A2E": An unprecedented pyridinium bisretinoid. J Am Chem Soc 1996;118:1559–1560.

60. Suter M, Reme C, Grimm C, et al. Age-related macular degeneration. The lipofusion component N-retinyl-N-retinylidene ethanolamine detaches proapoptotic proteins from mitochondria and induces apoptosis in mammalian retinal pigment epithelial cells. J Biol Chem 2000;275:39,625–39,630.

61. Sparrow JR, Nakanishi K, Parish CA. The lipofuscin fluorophore A2E mediates blue light-induced damage to retinal pigmented epithelial cells. Invest Ophthalmol Vis Sci 2000;41: 1981–1989.

62. Schutt F, Davies S, Kopitz J, Holz FG, Boulton ME. Photodamage to human RPE cells by A2-E, a retinoid component of lipofuscin. Invest Ophthalmol Vis Sci 2000;41:2303–2308.

63. Sparrow JR, Zhou J, Cai B. DNA is a target of the photodynamic effects elicited in A2E-laden RPE by blue-light illumination. Invest Ophthalmol Vis Sci 2003;44:2245–2251.

64. Holz FG, Schutt F, Kopitz J, et al. Inhibition of lysosomal degradative functions in RPE cells by a retinoid component of lipofuscin. Invest Ophthalmol Vis Sci 1999;40:737–743.

65. Katz ML, Norberg M. Influence of dietary vitamin A on autofluorescence of leupeptin-induced inclusions in the retinal pigment epithelium. Exp Eye Res 1992;54:239–246.

66. Katz ML, Redmond TM. Effect of Rpe65 knockout on accumulation of lipofuscin fluorophores in the retinal pigment epithelium. Invest Ophthalmol Vis Sci 2001;42: 3023–3030.

67. Radu RA, Mata NL, Bagla A, Travis GH. Light exposure stimulates formation of A2E oxiranes in a mouse model of Stargardt's macular degeneration. Proc Natl Acad Sci USA 2004;101:5928–3593.

68. Radu RA, Mata NL, Nusinowitz S, Liu X, Sieving PA, Travis GH. Treatment with isotretinoin inhibits lipofuscin accumulation in a mouse model of recessive Stargardt's macular degeneration. Proc Natl Acad Sci USA 2003;100:4742–4747.

69. Sieving PA, Chaudhry P, Kondo M, et al. Inhibition of the visual cycle in vivo by 13-cis retinoic acid protects from light damage and provides a mechanism for night blindness in isotretinoin therapy. Proc Natl Acad Sci USA 2001;98:1835–1840.

70. Guzzo CA, Lazarus GS, Werth VP. Dermatological pharmacology. In: Hardman JG, Limbird LE, eds. Goodman & Gilman's The Pharmacological Basis of Therapeutics. 9th ed. New York, NY: McGraw-Hill, Health Professions Division; 1996;1593–1616.

71. Law WC, Rando RR. The molecular basis of retinoic acid induced night blindness. Biochem Biophys Res Commun 1989;161:825–829.

72. Gollapalli DR, Rando RR. The specific binding of retinoic acid to RPE65 and approaches to the treatment of macular degeneration. Proc Natl Acad Sci USA 2004;101:10,030–10,035.

73. Kim SR, Fishkin N, Kong J, Nakanishi K, Allikmets R, Sparrow JR. Rpe65 Leu450Met variant is associated with reduced levels of the retinal pigment epithelium lipofuscin fluorophores A2E and iso-A2E. Proc Natl Acad Sci USA 2004;101:11,668–11,672.\

74. Maiti P, Kong J, Kim SR, Sparrow JR, Allikmets R, Rando RR. Small molecule RPE65 antagonists limit the visual cycle and prevent lipofuscin formation. Biochemistry 2006;45: 852–860.

75. Gouze E, Pawlyk B, Pilapil C, et al. In vivo gene delivery to synovium by lentiviral vectors. Mol Ther 2002;5:397–404.

76. Kostic C, Chiodini F, Salmon P, et al. Activity analysis of housekeeping promoters using self-inactivating lentiviral vector delivery into the mouse retina. Gene Ther 2003;10:818–821.

77. Miyoshi H, Takahashi M, Gage FH, Verma IM. Stable and efficient gene transfer into the retina using an HIV-based lentiviral vector. Proc Natl Acad Sci USA 1997;94:10,319–10,323.

78. Kingsman SM. Lentivirus: a vector for nervous system applications. Ernst Schering Res Found Workshop, Oxford, UK, 2003;179–207.

79. Mazarakis ND, Azzouz M, Rohll JB, et al. Rabies virus glycoprotein pseudotyping of lentiviral vectors enables retrograde axonal transport and access to the nervous system after peripheral delivery. Hum Mol Genet 2001;10:2109–2121.

80. Yan Z, Zhang Y, Duan D, Engelhardt JF. Trans-splicing vectors expand the utility of adeno-associated virus for gene therapy. Proc Natl Acad Sci USA 2000;97:6716–6721.

81. Bungert S, Molday LL, Molday RS. Membrane topology of the ATP binding cassette transporter ABCR and its relationship to ABC1 and related ABCA transporters: identification of N-linked glycosylation sites. J Biol Chem 2001;276:23,539–23,546.

82. Acland GM, Aguirre GD, Ray J, et al. Gene therapy restores vision in a canine model of childhood blindness. Nat Genet 2001;28:92–95.

6

X-Linked Juvenile Retinoschisis

Bernhard H. F. Weber, PhD and Ulrich Kellner, MD

CONTENTS

INTRODUCTION

First described in 1898 in two affected brothers by the Austrian ophthalmologist Haas *(1)*, X-linked juvenile retinoschisis (RS) (OMIM #312700) is one of the more frequently inherited retinal disorders affecting macular function in males. The prevalence of RS has been estimated to range between 1 in 5000 to 1 in 20,000 *(2)*. The name derives from an internal splitting of the retina mostly affecting the temporal periphery of the fundus. This trait is present in less than 50% of affected individuals. The major diagnostic feature is a limited splitting of the central retina presenting as a spoke-wheel pattern in the macular area (Fig. 1A). Although present in nearly all affected males less than 30 yr of age, the alterations of the fovea may be very discrete and especially in young children can sometimes be overlooked even by experienced clinicians *(3)*. Extraocular manifestations have not been reported in RS.

Frequently, the terms "congenital" and "juvenile" are used in conjunction with the RS condition. Several cases of severe RS have been described in the first year of age, suggesting that RS indeed may be present at birth *(3)* (Fig. 1B). Less severe retinal alterations are generally associated with only moderate visual loss and are frequently diagnosed prior to school age, suggesting a juvenile onset. It is not uncommon that affected

From: *Ophthalmology Research: Retinal Degenerations: Biology, Diagnostics, and Therapeutics*
Edited by: J. Tombran-Tink and C. J. Barnstable © Humana Press Inc., Totowa, NJ

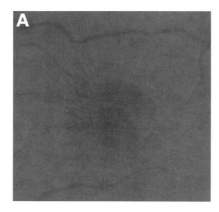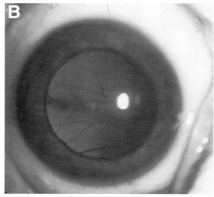

Fig. 1. Variation of retinal abnormalities in patients with RS. (**A**) Macular area with clearly visible spoke-wheel pattern surrounding the fovea in a 16-yr-old patient with RS. (**B**) Severe RS in a 3-mo-old infant. The RS involves nearly the complete retina forming two bullous schisis cavities with the retina visible behind the lens and only a single small central horizontal area with nonschitic retina.

males are misdiagnosed as amblyopic. Usually, the underlying cause is not detected until a thorough retinal investigation including electroretinography is conducted.

CLINICAL MANIFESTATION

The penetrance of RS is almost complete with the vast majority of male mutation carriers presenting with at least one sign of RS pathology, e.g., foveal changes *(3)*. In contrast, expressivity is highly variable. In our series of 86 patients with RS, manifestations ranged from nearly complete RS in both eyes at the age of 3 mo (Fig. 1B) to normal visual acuity with mild pigmentary macular abnormalities and a negative electroretinogram (ERG) in a 57-year-old male from a single RS family *(4)*. In the majority of cases, the expression of the disease is symmetrical in both eyes; however, a marked asymmetry of visual function can be present, especially in cases where additional complications occur *(5)*. Visual acuity is reduced to 20/100 in most patients, although it may vary greatly. Although macular abnormalities, such as the spoke-wheel pattern in younger patients and pigmentary changes in older patients, are present in nearly all affected males, peripheral retinal abnormalities are less common (in about 40–50% of patients). They are most frequent in the lower temporal quadrant of the retina *(4)*. In these cases, a sharply delineated RS usually is limited to the periphery or mid-periphery, but may extend from the periphery to the macula, including the fovea in some cases. If the inner sheet of the RS degenerates, the retinal vessels may remain running free through the vitreous cavity (presenting as so called vitreous veils). If additional breaks occur in the outer sheet of the RS, a retinal detachment may occur.

In the majority of patients, the disease either shows no or minimal progression *(6)*. Around the age of 30 yr, the macular alterations may change from the characteristic spoke-wheel pattern to unspecific mild retinal pigment abnormalities. In some cases, severe visual loss with increased age has been described *(7)*, although incidence data are not available as a result of the lack of long-term follow-up studies. In contrast to other X-linked disorders, female carriers have rarely been reported with retinal abnormalities or visual loss. In two cases of female RS, unilateral retinal cysts *(8,9)* and an abnormal

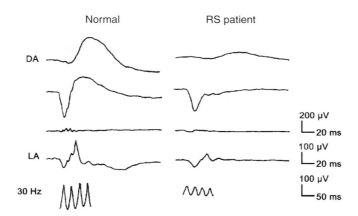

Fig. 2. Full-field ERG. Three responses at dark adaptation (DA) at low-stimulus intensity (first row), high-stimulus intensity (second row), and with special filtering for oscillatory potentials are shown for a normal individual (left column) and a patient with RS (right column). In addition, a single-flash response at light adaptation (LA) and a response to 30 Hz flicker is depicted. Note, the response in the second row reveals a severe reduction in the second part of the response (i.e., the b-wave) in the patient with RS (the so-called "negative" ERG). The cone-dependent responses at LA and 30 Hz flicker are reduced as well.

electroretinogram were noted *(9)*. In addition, two affected woman were homozygous carriers of disease mutations *(10,11)*.

CLINICAL DIAGNOSIS OF RS

Diagnosis of RS is unambiguous when a typical spoke-wheel pattern or even peripheral RS are present. In boys with bilateral reduced visual acuity, a thorough examination of the fovea with specific high-magnification lenses is necessary to detect retinal abnormalities. If a foveal RS cannot be detected or excluded, other techniques can be used to demonstrate the macular alterations. With optical coherent tomography (OCT), reflectance of retinal structures can be measured facilitating the detection of even small RS cavities *(12)*. Measurement of the fundus autofluorescence will reveal an increased foveal autofluorescence in most cases of RS, most likely because of the altered light transmission in the area of RS. Fluorescein angiography may reveal RPE alterations in older males; however, this is of limited value in children.

The major functional test for the diagnosis of RS is the full-field ERG. Typically, a bright flash of light will elicit a "negative" response from the diseased retina, in which the a-wave is larger than the b-wave in contrast to the normal findings (Fig. 2). Usually, the light adapted responses show an amplitude reduction as well. The origin of the retinal dysfunction is an abnormality in the ON- and OFF-pathways on the level of the bipolar cells *(13)*. A "negative" ERG can be associated with various retinal disorders; however, in young males, the only major differential diagnosis is congenital stationary night blindness (CSNB). The combination of macular alteration and a "negative" ERG indicates RS. Recently, detailed evaluation of macular dysfunction with the multifocal ERG has demonstrated a widespread cone dysfunction *(14)*. Other functional tests are of limited value. As in most disorders affecting the macula, color vision may be abnormal to a variable degree without any value for differential diagnosis. If peripheral RS is present, visual field testing will reveal an absolute scotoma in the corresponding area, but otherwise the visual fields will be normal.

COMPLICATIONS IN RS

Complications include vitreous hemorrhages or retinal detachment. Vitreous hemorrhage mostly clears spontaneously and only rarely requires vitreous surgery. Results of retinal detachment surgery are of limited benefit even with advanced surgical techniques *(15)*. Prophylactic treatment of RS either by laser or vitreoretinal surgery cannot be recommended owing to possible severe complications *(4,16)*. Overall, the frequency of secondary problems is approx 5% in all affected males, with complications most frequently developing in the first decade of life.

Differential Diagnosis in RS

Differential diagnosis of RS includes disorders with congenital or juvenile retinal detachment, other forms of inherited RS, and inherited disorders affecting macular function in the first decade of life.

RS is more frequently compared to disorders with retinal detachment. These include the X-linked Norrie syndrome (NS), in which complete retinal detachment is present at birth and visual function is nearly absent. NS is associated with mutations in the Norrie disease gene on Xp11.4 *(17,18)*. A less severe variant of NS is the X-linked familial exudative vitreoretinopathy (FEVR), also associated with mutations in the Norrie disease gene. Peripheral vascular retinal abnormalities are present in a variable degree, which may lead to retinal detachment. Similar alterations can be observed in autosomal dominant FEVR (Criswick-Schepens syndrome) and can be associated with mutations in the frizzled-4 *(FZD4)* gene on 11q14.2 *(19)* or the *LRP5* gene on 11q13.2 *(20)*. The ophthalmoscopic features in both forms of FEVR are distinct from RS. Incontinentia pigmentii (Bloch-Sulzberger syndrome) can present with early-onset retinal detachment; however, as the condition is lethal in males, it is not a differential diagnosis for RS. Other forms of congenital or juvenile retinal detachment, e.g., following trauma or retinopathy of prematurity, can usually be excluded by the patient's history.

Foveal RS has rarely been reported as an apparent autosomal recessive trait in families with predominantly affected females *(21–23)*. In Goldmann-Favre syndrome (alias enhanced S-cone syndrome), which is associated with mutations in the *NR2E3* gene on 15q23 *(24)*, a foveal RS may be present. The ERG is quite different from RS and distinguishes the two disorders.

Other inherited conditions affecting macular function are the macular dystrophies and CSNB. The macular dystrophies, e.g., Stargardt disease, usually show a markedly progressive course within 1 or 2 yr, which is not typical for RS. However, as expression in both disorders varies, an ERG can easily distinguish RS from other macular dystrophies because of the presence of the "negative" ERG response (Fig. 2). CSNB, most frequently inherited as an X-linked trait, presents with a similar reduction of visual function and shows a "negative" ERG similar to RS *(25)*. In contrast to RS, the retina is normal on ophthalmoscopy.

TREATMENT OPTIONS

In most cases, treatment of RS is limited to the prescription of low-vision aids. Surgical interventions are indicated in vitreous hemorrhage without spontaneous resolution or when retinal detachment occurs.

ISOLATION OF THE RS1 GENE AND STRUCTURAL FEATURES OF ITS GENE PRODUCT

In 1969, a first tentative association of RS with Xg blood group markers was reported suggesting a localization of the genetic defect to the distal short arm of the X chromosome *(26)*. Later, a number of polymorphic DNA markers derived from Xp became available and mapped the RS locus more precisely to chromosomal region Xp22.1-p22.2 *(27–29)*. In the following years, the DNA marker map of the X chromosome grew increasingly dense, eventually facilitating the refinement of the *RS1* gene locus to an approx 1000-kb interval on Xp22.13 flanked by markers DXS418 on the distal side and DXS999 on the telomeric side *(30,31)*. Following the initial localization, the minimal RS region on Xp22.13 was subsequently cloned and searched for disease gene candidates. From a total of 14 genes positioned within the DXS418-DXS999 interval *(32)*, one transcript, later termed *RS1* (alias *XLRS1*), was found to be abundantly expressed exclusively in the retina *(33)*. Mutational analysis of *RS1* in a number of multigeneration RS families revealed distinct mutations segregating with the disease, thus providing strong and convincing evidence for a causal role of the gene in the etiology of RS *(33)*.

The *RS1* gene locus spans 32.4 kb of genomic DNA and is organized in six exons coding for retinoschisin, a 224-amino acid protein of which the N-terminal 23 amino acids reveal a signature characteristic for proteins destined for cellular secretion *(33)*. Significant homology exists between a large portion of the predicted retinoschisin sequence (157 amino acid residues including codon 63 to codon 219) and the so-called discoidin domain, which was first identified in the discoidin I protein of the slime mold *Dictyostelium discoidium (34)*. This motif is highly conserved in a number of secreted and transmembrane proteins from many eukaryotic species and is known to be involved in functions such as neuronal development and cellular adhesion *(35–37)*. Two minor portions of the RS1 protein are unique with no homologies to other known proteins. N terminal to the discoidin domain is a 38-amino acid RS1-specific sequence, whereas 5 unique amino acid residues flank the discoidin domain on the C-terminal side. Functional properties of the two RS1-specific protein domains have not been identified so far, but may be confined to structural aspects of subunit assembly in protein complex formation *(38,39)*.

SPECTRUM AND MOLECULAR PATHOLOGY OF RS-ASSOCIATED MUTATIONS

As of January 2005, 127 distinct disease-associated mutations have been reported from patients with RS of various genealogical ancestry, such as British *(40)*, Chinese *(41)*, Colombian *(42)*, Danish *(43)*, Finnish *(44)*, French *(45)*, German *(33,45)*, Greek *(40)*, Icelandic *(46)*, Italian *(47)*, Japanese *(48)*, North American *(33,49)*, Swedish *(50)*, and Taiwanese *(51)*. These findings further confirm the initial disease-association of the

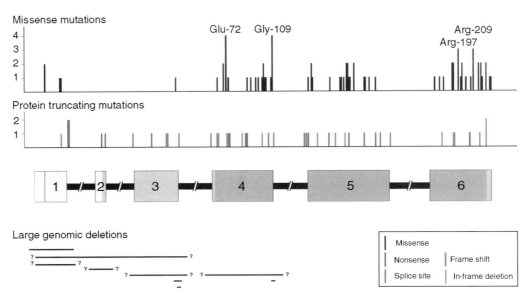

Fig. 3. RS-associated mutations in the RS1 gene. The RS1 locus is drawn schematically with six coding exons represented by bars and intronic sequences by lines. Bar colors indicate the functionally distinct domains of the RS1 protein encoded by the respective exonic sequences (yellow = N-terminal 23-amino acid signal peptide; gray = RS1-specific domain; orange = discoidin domain). Distinct mutations reported in the RS mutation database (http://www.dmd.nl/) are divided in three categories including missense mutations, mutations resulting in protein truncation (nonsense, splice site, frame shift, in-frame deletion), and larger genomic rearrangement causing the loss of partial or complete exons.

RS1 gene and demonstrate locus homogeneity within the studied ethnic backgrounds (although with considerable allelic heterogeneity). Of the identified mutations, 75 (59.1%) are missense, 18 (14.2%) cause a shift in the reading frame consequently resulting in a truncation of the protein, 13 (10.2%) represent a nonsense mutation, 11 (8.7%) affect the correct splicing of the pre-messenger RNA (mRNA), and 1 (0.8%) results in an in-frame deletion of a single amino acid (Asn85del) (Fig. 3). Nine mutations involve genomic deletions of entire or partial exons and range in size from 8 bp (321_326+1del) to at least 15 kb (1-?-184+?del). Three of the smaller genomic deletions (173-184+21del, 181_184+10del, 321_326+1del) affect splice donor sequences and are likely to interfere with correct splicing of the pre-mRNA sequence. With the exception of the missense mutations and the single in-frame deletion, the other sequence changes represent true null alleles and thus should produce no protein or truncated, non-functional, versions of the RS1 protein.

Although some families segregate unique mutations (e.g. Gly70Arg or Asn104Lys), a number of sequence alterations are recurrent (for a comprehensive listing of mutations, *see* RS1 mutation database at http://www.dmd.nl/). Most notably, the latter type of sequence changes are exclusively missense mutations with the most frequent disease alleles Glu72Lys and Arg102Trp reported 50 and 25 times, respectively. The missense mutations Gly70Ser, Trp96Arg, Arg102Gln, Gly109Arg, Arg141Cys, Pro192Ser,

Arg200Cys were found in eight or more apparently unrelated families. In addition, the Glu-72 and the Gly-109 residues have been affected by four different amino acid substitutions, codons Arg-197 and Arg-209 were affected three times each (Fig. 3). These commonly altered amino acid residues could be indicative of mutational hotspots and/or functionally important moieties of the RS1 protein.

The distribution of missense mutations along the coding sequence of the RS1 gene is nonrandom with significant clustering of this type of alterations in exons 4 to 6 (Fig. 3). This uneven distribution coincides with the extent of the discoidin domain encoded by codons 63 through 219. Although the 157 amino acid residues of the discoidin domain are affected by 68 missense mutations (0.43 mutations per codon), there are only 7 missense mutations reported in the remaining 67 amino acids of RS1 (0.11 mutations per codon). This amounts to an approximately fourfold excess of missense mutations within the discoidin domain compared to the rest of the protein. In contrast, other types of mutations combined (i.e., frame shift, nonsense, splice site, and in-frame deletion mutations) are randomly distributed within the coding region. Compared to 23 such mutations within the discoidin domain (0.15 mutations per codon), 8 mutations are localized outside of the conserved motif (0.12 mutations per codon). Together, these data suggest that absence of RS1 or loss of protein function may be the main molecular mechanism underlying RS pathology. In particular, the discoidin domain appears to be most crucial for RS1 function with strong constraints on the proper amino acid sequence.

To assess the pathological mechanisms of RS1 missense mutations, Wang and associates *(52)* have conducted in vitro studies expressing mutant RS1 protein (Leu12His, Cys59Ser, Gly70Ser, Arg102Trp, Gly109Arg, Arg141Gly, Arg213Trp). Their findings have led to the suggestion that in the majority of cases failure of cellular secretion (either partial or complete) may underlie disease pathology. Based on an extended series of elegant biochemical experiments, Wu and Molday *(38)* and Wu and associates *(39)* have developed a structural model of the RS1 discoidin domain that identifies particular cysteine residues as crucial components for intra- and intermolecular disulfide bond formation and thus proper protein folding and subunit assembly (Fig. 4A,B). Essentially, intramolecular disulfide bonds are formed between residues Cys-63 and Cys-219 and between Cys-110 and Cys-142 thus playing a central role in proper subunit folding (Fig. 4A). Eight such subunits are then joined together by disulfide bonds between the discoidin-flanking cysteine residues Cys-59 and Cys-223, which are crucial for subunit oligomerization. Within this complex, the homomeric subunits are further organized into dimers by Cys-40-Cys40 disulfide bond formation with dimerization and octamerization processes evidently independent of each other *(39)* (Fig. 4B).

Based on this model, Wu and Molday *(38)* have pointed out three specific disease mechanisms underlying RS. First, pathological missense mutations in the discoidin domain (e.g., Glu72Lys, Gly109Glu, Cys110Tyr, Arg141Cys, Cys142Trp, Asp143Val, Arg182Cys, Pro203Leu, Cys219Arg) cause protein misfolding and consequently retention of the mutant protein in the endoplasmatic reticulum (ER). Second, missense mutations affecting the disulfide-linked subunit assembly (e.g., Cys59Ser, Cys223Arg) do not significantly interfere with cellular secretion but cause a failure to assemble a functional oligomeric RS1 protein complex. Third, missense mutations in the 23-amino acid leader sequence of RS1 (e.g., Leu13Pro) prevent proper insertion of the protein into the

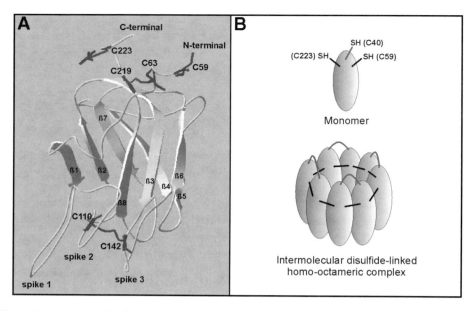

Fig. 4. Topology and subunit organization of the retinoschisin discoidin domain. **(A)** Ribbon diagram of the retinoschisin-discoidin domain modeled after the F5/8-type C-domain fold. The eight core β strands are labeled β1–β8. The three loops ("spikes") demonstrating amino acid sequence-dependent ligand affinities are marked. The intramolecular disulfide bonds C63-C219 and C110-C142 as well as the C-terminal (C223) and N-terminal (C59) cysteine residues participating in intermolecular disulfide bond formation are shown *(38)*. **(B)** Schematic illustrating the formation of the homo-octameric RS1 complex from intermolecular disulfide-linked monomers. Subunits within the octamer are further organized into dimers mediated by C40–C40 disulfide bonds *(39)*.

ER membrane resulting in cellular mislocalization and thus defective secretion. Together, these studies suggest that the known disease-associated RS1 mutations (i.e., missense mutations and protein-truncating mutations) represent loss-of-function mutations. This is also consistent with the observation that disease severity in RS does not appear to be correlated with the specific type of mutational change.

FUNCTIONAL PROPERTIES OF RETINOSCHISIN

The Discoidin Domain and Its Putative Role in Protein Function

Retinoschisin belongs to a family of proteins that contain one or two phylogenetically highly conserved discoidin domains *(53)*. This motif, also known as the F5/8 type C-domain fold, has been found in a variety of proteins such as coagulation factors V and VIII (F5, F8), milk fat globule-epidermal growth factor (EGF) factor 8 (MFGE8), EGF-like repeats- and discoidin I-like domains-containing protein-3 (EDIL3), neurexin IV (NRXN4), neuropilin-1 and -2 (NRP1 and NRP2, respectively), discoidin domain receptors-1 and -2 (DDR1 and DDR2, respectively), or aortic carboxypeptidase-like protein. Many of these proteins are expressed only transiently, in response to specific stimuli or during development and are associated with cellular adhesion, migration, or

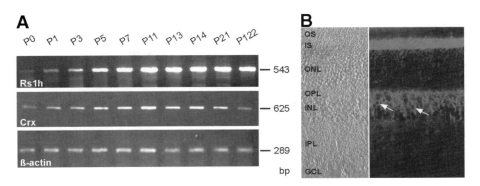

Fig. 5. Expression of murine Rs1h. **(A)** Reverse-transcriptase polymerase chain reaction analysis of Rs1h, the murine ortholog of human RS1, in comparison to the cone-rod homeobox-containing gene (Crx) and β-actin. Rs1h mRNA expression is detectable starting at P0 with increasing expression until postnatal day P7. This level of expression is then maintained throughout life. This is in contrast to the Crx transcript level where peak expression around P11 decreases in development and only low levels of expression remain during adulthood. β-actin serves as a control for cDNA integrity. **(B)** RS1 immunolabelling of outer and inner murine retina. Arrows point to the labeling of bipolar cell surfaces.

aggregation via membrane surfaces. High-resolution crystal structures of human F5 and F8 C2 domains have been reported *(54,55)* and provide a framework for the homology-based modeling of the three-dimensional structure of the RS1 discoidin domain (Fig. 4A). Its central feature is an eight-stranded antiparallel β-barrel consisting of tightly packed five-stranded (β1, β2, β4, β5, β7) and three-stranded (β3, β6, β8) β-sheets. This conserved β-barrel scaffold exhibits three protruding loops ("spikes") (Fig. 4A) that have been shown to display specific amino acid-dependent affinities *(56)*. A number of ligands have been identified for discoidin-containing proteins and include negatively charged membrane surfaces with defined phospholipids (F5 *[57]*, F8 *[58]*, MFGE8 *[59]*), integrin receptors that bind to the classic integrin-binding RGD sequence (discoidin I protein of *D. discoideum [60]*; EDIL3 *[61]*), as well as collagen (DDR1, DDR2 *[62,63]*). Modelling of the RS1 loops indicate that several hydrophobic amino acid residues may be exposed at their spike apexes. This could suggest that similar to coagulation factors F5 or F8, the discoidin domain of RS1 could also interact with phospholipid membranes.

Expression of RS1 and Localization of the Protein in the Mammalian Retina

Expression of *RS1* is restricted to the retina as shown by Northern blot hybridization to a number of human *(33)* and mouse *(64)* tissues. Furthermore, *in situ* hybridization experiments revealed *RS1* mRNA transcripts in rod and cone photoreceptor inner segments *(65,66)* and also in other cell bodies of the retinal layers namely in bipolar cells, amacrine cells, and retinal ganglion cells *(67)*. In postnatal eye development, measurable levels of *RS1* expression resume around P1 and reach a maximum between P5 and P7 (Fig. 5A). This level of expression is then maintained throughout adult live indicating that continued *de novo* synthesis of RS1 is required and is essential for the maintenance of retinal integrity.

As suggested by the presence of an N-terminal signal peptide sequence, retinoschisin was shown experimentally to be secreted from the cell after removal of the first 23-amino acid residues *(38)*. To localize the protein within the mammalian retina, a number of mono- and polyclonal antibodies were raised against the RS1-specific N-terminus of the mature protein *(66–69)*. In the adult retina, prominent immunolabeling is consistently observed at the extracellular surfaces of the inner segments of both the rod and cone photoreceptors, most bipolar cells, as well as the two plexiform layers *(66–68,70)* (Fig. 5B). There is still some controversy with regard to the immunoreactivity of ganglion cells *(66,67)* and Müller cells *(70)*.

Developmental Expression of Retinoschisin

Immunostaining of the developing rat retina revealed weak RS1 labeling at P6 within the neuroblastic zone *(68)*. The staining intensity of the outer retina increased over time with intense staining of the newly formed inner segment layer at P10. Adult pattern labeling was reached around P12. Developmental expression of RS1 was also investigated in the mouse retina *(67)*. Similar to the rat, inner segment labeling of the photoreceptors became evident by P7 and more prominent by P10 to P14 when finally an adult pattern of immunostaining developed. Interestingly, Takada and associates *(67)* observed a transient pattern of expression of RS1 in retinal ganglion cells in murine stages E16.5, P1, and P3, but not at later time points of development. With subsequent layer formation, this expression was found to move posteriorily through the retinal layers as additional types of neurons became differentiated. If confirmed, this would suggest that RS1 is produced locally at several defined neuronal cell populations *(67)*, arguing against the need for a *trans*-retinal transport system of photoreceptor- or bipolar-secreted retinoschisin as suggested by Reid and associates *(70)*.

Mouse Model for X-Linked Juvenile Retinoschisis

The ability to develop mouse strains with targeted mutations in a defined protein has tremendously expanded our spectrum of experimental tools allowing us to gain insight into physiological mechanisms controlled by these proteins. In particular, because photoreceptor cells cannot be maintained in vitro, the study of animal models with a mutant photoreceptor protein is instrumental in understanding pathological processes and in testing therapeutic approaches. With these objectives in mind, a mouse line deficient in *Rs1h*, the murine orthologe of the human *RS1* gene *(64)*, was generated and the clinical phenotype characterized by electroretinography, histology, and immunohistochemistry *(69)*. The Rs1h$^{-/Y}$ mouse shares several diagnostic features with human RS, including the typical "negative" ERG response and the development of cystic structures In addition, the diseased murine retina shows a general disorganization and a disruption of the synapses between the photoreceptors and bipolar cells accompanied by a dramatic and progressive loss of photoreceptor cells (Fig. 6). Starting at postnatal day 14, minor and major gaps within the inner nuclear layer become evident. Some of the schisis cavities are filled with fragmented nerve cell terminals containing synaptic vesicles. In accordance with the observed lack of immunoreactivity of Müller cells *(67,68)*, glial cells are relatively unaffected in the Rs1h$^{-/Y}$ retina *(69)*.

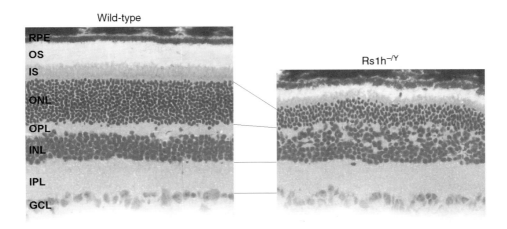

Fig. 6. Retinal cryosections of 12-mo-old wild-type and retinoschisin-deficient mice (Rs1h$^{-/Y}$) stained with hematoxylin and eosin. In the Rs1h$^{-/Y}$ retina, note the striking reduction of photoreceptor cell bodies of the outer nuclear layer (ONL) and the layer disorganization/cell displacement greatly affecting the outer plexiform layer (OPL) and the inner nuclear layer (INL). The outer (OS) and inner segments (IS) of the photoreceptor are markedly shortened in the knockout mouse, whereas histology of the inner plexiform layer (IPL) and ganglion cell layer (GCL) appears normal.

Evidently, the site of schisis in the retinoschisin-deficient mouse retina within the inner nuclear layer is not consistent with the classic view of human RS pathology formerly described as a splitting between the inner limiting membrane and the nerve fiber layer *(2,71)*. However, recent histopathological data from a 19-year-old patient with RS, have emphasized a more general disorganization of the retinal layers with splitting readily apparent in the inner and outer plexiform layers (IPL and OPL, respectively) *(72)*. Similarly, cross sectional views of younger patients with RS by OCT revealed schisis cavities in multiple layers, both superficial and deeper in the retina including the outer and inner nuclear layers (ONL and INL, respectively) *(5,12,73)*.

Gene Therapy in RS

Because RS pathology arises as a consequence of a missing functional RS1 protein *(38,66)*, delivery of the normal *RS1* gene product to the retina may be an amenable treatment option for RS. To test this, we have delivered an adeno-associated virus (AAV) serotype 5 vector containing the human RS1 cDNA under the control of the mouse opsin promoter (AAV5-mOPs-RS1) into the subretinal space of the right eyes of *Rs1h*-deficient 15-day-old mice *(74,75)*. At 2 and 3 mo after treatment, ERG recordings revealed a significant improvement of a- and b-waveform characteristics in injected over noninjected left control eyes. At 5-mo, five treated animals were sacrificed and retinal cryosections analyzed by immunohistochemistry (Fig. 7). Retinal tissues from treated eyes showed an RS1 immunostaining pattern similar to that of wild-type with intense RS1 labeling present in the inner segment layer while more moderate staining was seen in the ONL and

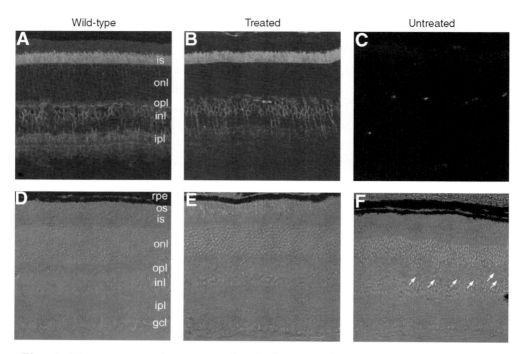

Fig. 7. Fluorescence microscopy of retinal cryosections from wild-type **(A,D)** and retinoschisin-deficient mice 5 mo after subretinal delivery of a serotype 5 AAV vector containing the wild-type human RS1 cDNA driven by the mouse opsin promoter (AAV5-mOPs-RS1) to the right eye of 15-d-old mice **(B,E)**. The left eyes of the Rs1h$^{-/Y}$ mice were not injected and served as internal controls **(C,F)**. **(A–C)** Sections were labeled with RS1 3R10 anti-RS1 monoclonal antibody *(69)*. **(D–F)** sections were stained with DAPI (blue) and imaged with DIC microscopy. Note the wild-type like expression pattern of RS1 and the recovery of the retinal structures in the treated animals. The arrows point to schisis cavities in the untreated animals.

OPL (Fig. 7A,B). The INL and to a lesser degree the IPL was also labeled. As expected, no immunostaining was observed in the uninjected left control retina (Fig. 7C). It is interesting to note that in the treated eyes retinoschisin can be found at physical distances as far away from the site of secretion (i.e., photoreceptors) as the IPL. This may indicate that secreted retinoschisin is able to spread through the various retinal layers. In agreement with this, we also observed an impressive lateral movement of secreted retinoschisin away from the site of injection in all animals analyzed.

The expression of RS1 coincided with a marked improvement in the structural organization of the retinal layers as visualized in differential interference contrast images merged with DAPI nuclear stain (Fig. 7D–F). The treated retina was organized into characteristic layers with a distinct separation of the INL and ONL and an absence of gaps between bipolar cells. An increased thickness of the ONL, indicative of enhanced photoreceptor survival, was also seen in the treated retina (Fig. 7E). In contrast, the untreated eye showed the known manifestations of advanced disease (Fig. 7F).

Recently, Zeng and associates *(73)* reported on a similar approach to treat retinoschisin-deficient mice, although with some minor variations. Instead of the human gene, they delivered an AAV construct containing the orthologous mouse *Rs1h* cDNA under the control of a cytomegalovirus (CMV) promoter. By intraocular instead of subretinal injections, a serotype 2 AAV vector was administered to adult knockout mice of 13 wk of age. Similar to our results, their preliminary data indicate retinoschisin expression in all retinal layers. So far, no data are available documenting the effects of the treatment on the morphology of the retinal layers or on photoreceptor cell survival. Despite the late delivery of retinoschisin at an advanced stage of disease progression, ERG recordings nevertheless showed a reversal of the electronegative a-wave and restoration of the normal positive b-wave *(73)*. The latter findings may be encouraging for the future design of gene therapy protocols aimed at adult RS patients with advanced pathology.

CONCLUSIONS AND FUTURE DIRECTIONS

Although a relatively rare Mendelian condition, RS is an important disease to study, not only to achieve a better understanding of retinal physiology in the normal and diseased eye but also more importantly to define targeted treatment options for the patient. From the early descriptions of the clinical features of RS more than 100 yr ago, advances in molecular research over the last 10 yr have given us tremendous insight into RS pathology not without the promise for novel therapeutic avenues. The gene mutated in RS codes for retinoschisin, a protein secreted as a disulfide-linked oligomeric complex from the photoreceptors, to a minor degree from bipolar cells and possibly other neuronal cells of the retina. Retinoschisin is firmly associated with membrane surfaces owing to a highly conserved discoidin motif, which is known in other proteins to mediate cell adhesion/aggregation properties. Particularly, high concentrations of RS1 are found along the entire length of the photoreceptor inner segment membranes. Most bipolar cell types also markedly bind retinoschisin. Müller cells, however, long thought to play a crucial role in RS pathology, appear devoid of retinoschisin strongly arguing against a pivotal part of this cell type in disease etiology. Rather, the characteristic retinal expression pattern of retinoschisin brings primary pathology of photoreceptor and bipolar cells more into focus.

Analysis of the molecular pathology of disease-associated *RS1* mutations has suggested that the mutant gene product is either absent or nonfunctional in patients with RS, leading to a complete absence of retinoschisin in males. Consequently, retinoschisin-deficient mice closely mimic human RS pathology by forming cystic structures within the inner retina and revealing a characteristic electronegative ERG waveform pattern. In addition, a striking reduction of rod and, to a greater extent, cone photoreceptor cells is seen in several week old mutant animals. In postnatal retinal development, retinoschisin deficiency results in destabilization of the retinal organisation and decreased cohesion of cell structures but appears also critical for the formation and maintenance of retinal synapses in the outer and inner plexiform layers.

The disease mechanism in RS suggests that replacement of the normal gene product in the retina may provide an adequate therapeutic approach to improve the outcome for patients with RS. Toward this end, gene delivery to the retinoschisin-deficient mouse retina via AAV particles has successfully been attempted. The findings are most promising and provide proof-of-concept for the feasibility of protein replacement in RS. AAV-mediated *RS1* gene therapy lies ahead of us as an optional treatment for patients with RS.

ACKNOWLEDGMENTS

The authors wish to thank Andreas Janssen and Andrea Gehrig (Institute of Human Genetics, University of Würzburg, Germany) for their help with the figures and Robert S. Molday for his support and his critical comments to the manuscript.

REFERENCES

1. Haas J. Ueber das Zusammenvorkommen von Veraenderungen der Retina und Choroidea. Arch Augenheilkd 1898;37:343–348.
2. George ND, Yates JR, Moore AT. X linked retinoschisis. Br J Ophthalmol 1995;79:697–702.
3. Sieving P. Juvenile retinoschisis, in Traboulsi EI, ed. Genetic diseases of the eye. Oxford: Oxford University Press, 1998:347–355.
4. Kellner U, Brümmer S, Foerster MH, Wessing A. X-linked congenital retinoschisis. Graefe`s Arch Clin Exp Ophthalmol 1990;228:432–437.
5. Tantri A, Vrabec TR, Cu-Unjieng A, Frost A, Annesley WH, Donoso LA. X-linked retinoschisis: a clinical and molecular genetic review. Surv Ophthalmol 2004;49:214–230.
6. Roesch MT, Ewing CC, Gibson AE, Weber BH. The natural history of X-linked retinoschisis. Can. J Ophthalmol 1998;33:149–158.
7. George ND, Yates JR, Moore AT. Clinical features in affected males with x-linked retinoschisis. Arch Ophthalmol 1996;114:274–280.
8. Gieser EP, Falls HF. Hereditary retinoschisis. Am J Ophthalmol 1961;51:1193–1200.
9. Wu G, Cotlier E, Brodie S. A carrier state of X-linked juvenile retinoschisis. Ophthalmic Paediatr Genet 1985;5:13–17.
10. Forsius H, Ericksson A, Nuutila A, Vainio-Mattila B. Geschlechtsgebundene, erbliche retinoschisis in zwei Familien in Finnland. Klein Monatsbl Augenheilkd 1963;143: 806–816.
11. Ali A, Feroze AH, Rizvi ZH, Rehman TU. Consanguineous marriage resulting in homozygous occurrence of X-linked retinoschisis in girls. Am J Ophthalmol 2003;136:767–769.
12. Eriksson U, Larson E, Holmstrom G. Optical coherence tomography in the diagnosis of juvenile X-linked retinoschisis. Acta Ophthalmol Scand 2004;82:218–223.
13. Khan NW, Janison JA, Kemp JA, Sieving PA. Analysis of photoreceptor function and inner retinal activity in juvenile X-linked retinoschisis. Vision Res 2001;41:3931–3942.
14. Piao CH, Kondo M, Nakamura M, Terasaki H, Miyake Y. Multifocal electroretinograms in X-linked retinoschisis. Invest Ophthalmol Vis Sci 2003;44:4920–4930.
15. Rosenfeld PJ, Flynn HW, McDonald HR, et al. Outcomes of vitreoretinal surgery in patients with X-linked retinoschisis. Ophthalmic Surg Lasers Imaging 1998;29:190–197.
16. Sobrin L, Berrocal AM, Murray TG. Retinal detachment 7 years after prophylactic schisis cavity excision in juvenile X-linked retinoschisis. Ophthalmic Surg Lasers Imaging 2003; 34:401–402.
17. Berger W, Meindl A, van de Pol TJR, et al. Isolation of a candidate gene for Norrie disease by positional cloning. Nat Genet 1992;1:199–203.
18. Chen ZY, Hendriks RW, Jobling MA, et al. Isolation and characterization of a candidate gene for Norrie disease. Nat Genet 1992;1:204–208.
19. Robitaille J, MacDonald MLE, Kaykas A, et al. Mutant frizzled-4 disrupts retinal angiogenesis in familial exudative vitreoretinopathy. Nat Genet 2002;32:326–330.
20. Toomes C, Bottomley HM, Jackson RM, et al. Mutations in LRP5 or FZD4 underlie the common familial exudative vitreoretinopathy locus on chromosome 11q. Am J Hum Genet 2004;74:721–730.
21. Cibis PA. Retinoschisis-retinal cysts. Trans Am Ophthalmol Soc 1965;53:417–453.

22. Lewis RA, Lee GB, Martonyi CL, Barnett JM, Falls HF. Familial foveal retinoschisis. Arch Ophthalmol 1977;95:1190–1196.
23. Perez Alvarez MJ, Clement Fenandez F. No X-chromsosme linked juvenile foveal retinoschisis. Arch Soc Esp Oftalmol 2002;77:443–448.
24. Haider NB, Jacobson SG, Cideciyan AV, et al. Mutation of a nuclear receptor gene, NR2E3, causes enhanced S cone syndrome, a disorder of retinal cell fate. Nat Genet 2000;24:127–131.
25. Bradshaw K, Allen T, Trump D, Hardcastle A, George N, Moore A. A comparison of ERG abnormalities in XLRS and XLCSNB. Doc Ophthalmol 2004;108:135–145.
26. Vainio-Mattila B, Eriksson AW, Forsius H. X-chromosomal recessive retinoschisis in the Region of Pori. An ophthalmo-genetical analysis of 103 cases. Acta Ophthalmol (Copenh) 1969;47:1135–1148.
27. Wieacker P, Wienker TF, Dallapiccola B, Bender K, Davies KE, Ropers HH. Linkage relationships between Retinoschisis, Xg, and a cloned DNA sequence from the distal short arm of the X chromosome. Hum Genet 1983;64:143–145.
28. Gal A, Wienker TF, Davies K, et al. Further linkage studies between retinoschisis and cloned DNA sequences from distal Xp. Human gene mapping 8. CytoGenet Cell Genet 1985;40:634.
29. Alitalo T, Karna J, Forsius H, de la Chapelle A. X-linked retinoschisis is closely linked to DXS41 and DXS16 but not DXS85. Clin Genet 1987;32:192–195.
30. Van de Vosse E, Bergen AA, Meershoek EJ, et al. An Xp22.1-p22.2 YAC contig encompassing the disease loci for RS, KFSD, CLS, HYP and RP15: refined localization of RS. Eur J Hum Genet 1996;4:101–104.
31. Huopaniemi L, Rantala A, Tahvanainen E, de la Chapelle A, Alitalo T. Linkage disequilibrium and physical mapping of X-linked juvenile retinoschisis. Am J Hum Genet 1997;60:1139–1149.
32. Warneke-Wittstock R, Marquardt A, Gehrig A, Sauer CG, Gessler M, Weber BH. Transcript map of a 900-kb genomic region in Xp22.1-p22.2: identification of 12 novel genes. Genomics 1998;51:59–67.
33. Sauer CG, Gehrig A, Warneke-Wittstock R, et al. Positional cloning of the gene associated with X-linked juvenile retinoschisis. Nat Genet 1997;17:164–170.
34. Poole S, Firtel RA, Lamar E, Rowekamp W. Sequence and expression of the discoidin I gene family in Dictyostelium discoideum. J Mol Biol 1981;153:273–289.
35. Wood WI, Capon DJ, Simonsen CC, et al. Expression of active human factor VIII from recombinant DNA clones. Nature 1984;312:330–337.
36. Kawakami A, Kitsukawa T, Takagi S, Fujisawa H. Developmentally regulated expression of a cell surface protein, neuropilin, in the mouse nervous system. J Neurobiol 1995;29:1–17.
37. Baumgartner S, Littleton JT, Broadie K, et al. A Drosophila neurexin is required for septate junction and blood-nerve barrier formation and function. Cell 1996;87:1059–1068.
38. Wu WW, Molday RS. Defective discoidin domain structure, subunit assembly, and endoplasmic reticulum processing of retinoschisin are primary mechanisms responsible for X-linked retinoschisis. J Biol Chem 278:28,139–28,146.
39. Wu WW, Wong JP, Kast J, Molday RS. RS1, a discoidin domain containing retinal cell adhesion protein associated with X-linked retinoschisis, exists as a novel disulfide-linked octamer. J Biol Chem 2005;280:10,721–10,730.
40. Gehrig A, White K, Lorenz B, Andrassi M, Clemens S, Weber BH. Assessment of RS1 in X-linked juvenile retinoschisis and sporadic senile retinoschisis. Clin Genet 1999;55:461–465.
41. Yu P, Li J, Li R, Zhang W. Identification of mutation of the X-linked juvenile retinoschisis gene. Zhonghua Yi Xue Yi Chuan Xue Za Zhi (translation) (Chinese Journal of Medical Genetics) 2001;18:88–91.

42. Mendoza-Londono R, Hiriyanna KT, Bingham EL, et al. A Colombian family with X-linked juvenile retinoschisis with three affected females finding of a frameshift mutation. Ophthalmic Genet 1999;20:37–43.

43. Huopaniemi L, Tyynismaa H, Rantala A, Rosenberg T, Alitalo T. Characterization of two unusual RS1 gene deletions segregating in Danish retinoschisis families. Hum Mutat 2000;16:307–314.

44. Huopaniemi L, Rantala A, Forsius H, Somer M, de la Chapelle A, Alitalo T. Three widespread founder mutations contribute to high incidence of X-linked juvenile retinoschisis in Finland. Eur J Hum Genet 1999;7:368–376.

45. The Retinoschisis Consortium. Functional implications of the spectrum of mutations found in 234 cases with X-linked juvenile retinoschisis. Hum Mol Genet 1998;7:1185–1192.

46. Tantri A, Vrabec TR, Cu-Unjieng A, Frost A, Annesley WH, Donoso LA. X-linked retinoschisis: report of a family with a rare deletion in the XLRS1 gene. Am J Ophthalmol 2003;136:547–549.

47. Simonelli F, Cennamo G, Ziviello C, et al. Clinical features of X-linked juvenile retinoschisis associated with new mutations in the XLRS1 gene in Italian families. Br J Ophthalmol 2003;87:1130–1134.

48. Shinoda K, Ishida S, Oguchi Y, Mashima Y. Clinical characteristics of 14 japanese patients with X-linked juvenile retinoschisis associated with XLRS1 mutation. Ophthalmic Genet 2000;21:171–180.

49. Hiriyanna KT, Bingham EL, Yashar BM, et al. Novel mutations in XLRS1 causing retinoschisis, including first evidence of putative leader sequence change. Hum Mutat 1999;14:423–427.

50. Eksandh LC, Ponjavic V, Ayyagari R, et al. Phenotypic expression of juvenile X-linked retinoschisis in Swedish families with different mutations in the XLRS1 gene. Arch Ophthalmol 2000;118:1098–1104.

51. Tsai FJ, Yang C, Wu JY, Lin HJ, Lee CC, Tsai CH. A novel mutation K167X of the XLRS1 gene (RS1) in a Taiwanese family with X-linked juvenile retinoschisis. Hum Mut 2000;16:449.

52. Wang T, Waters CT, Rothman AM, Jakins TJ, Romisch K, Trump D. Intracellular retention of mutant retinoschisin is the pathological mechanism underlying X-linked retinoschisis. Hum Mol Genet 2002;11:3097–3105.

53. Baumgartner S, Hofmann K, Chiquet-Ehrismann R, Bucher P. The discoidin domain family revisited: new members from prokaryotes and a homology-based fold prediction. Protein Sci 1998;7:1626–1631.

54. Macedo-Ribeiro S, Bode W, Huber R, et al. Crystal structures of the membrane-binding C2 domain of human coagulation factor V. Nature 1999;402:434–439.

55. Pratt KP, Shen BW, Takeshima K, Davie EW, Fujikawa K, Stoddard BL. Structure of the C2 domain of human factor VIII at 1.5 A resolution. Nature 1999;402:439–442.

56. Fuentes-Prior P, Fujikawa K, Pratt KP. New insights into binding interfaces of coagulation factors V and VIII and their homologues lessons from high resolution crystal structures. Curr Protein Pept Sci 2002;3:313–339.

57. Srivastava A, Quinn-Allen MA, Kim SW, Kane WH, Lentz BR. Soluble phosphatidylserine binds to a single identified site in the C2 domain of human factor Va. Biochemistry 2001;40:8246–8255.

58. Gilbert GE, Arena AA. Unsaturated phospholipid acyl chains are required to constitute membrane binding sites for factor VIII. Biochemistry 1998;37:13,526–13,535.

59. Andersen MH, Graversen H, Fedosov SN, Petersen TE, Rasmussen JT. Functional analyses of two cellular binding domains of bovine lactadherin. Biochemistry 2000;39:6200–6206.

60. Gabius HJ, Springer WR, Barondes SH. Receptor for the cell binding site of discoidin I. Cell 1985;42:449–456.

61. Hidai C, Zupancic T, Penta K, et al. Cloning and characterization of developmental endothelial locus-1: an embryonic endothelial cell protein that binds the alphavbeta3 integrin receptor. Genes Dev 1998;12:21–33.
62. Shrivastava A, Radziejewski C, Campbell E, et al. An orphan receptor tyrosine kinase family whose members serve as nonintegrin collagen receptors. Mol Cell 1997;1:25–34.
63. Vogel W, Gish GD, Alves F, Pawson T. The discoidin domain receptor tyrosine kinases are activated by collagen. Mol Cell 1997;1:13–23.
64. Gehrig AE, Warneke-Wittstock R, Sauer CG, Weber BH. Isolation and characterization of the murine X-linked juvenile retinoschisis (Rs1h) gene. Mamm Genome 1999;10:303–307.
65. Reid SN, Akhmedov NB, Piriev NI, Kozak CA, Danciger M, Farber DB. The mouse X-linked juvenile retinoschisis cDNA: expression in photoreceptors. Gene 1999;227:257–266.
66. Grayson C, Reid SN, Ellis JA, et al. Retinoschisin, the X-linked retinoschisis protein, is a secreted photoreceptor protein, and is expressed and released by Weri-Rb1 cells. Hum Mol Genet 2000;9:1873–1879.
67. Takada Y, Fariss RN, Tanikawa A, et al. A retinal neuronal developmental wave of retinoschisin expression begins in ganglion cells during layer formation. Invest Ophthalmol Vis Sci 2004;45:3302–3312.
68. Molday LL, Hicks D, Sauer CG, Weber BH, Molday RS. Expression of X-linked retinoschisis protein RS1 in photoreceptor and bipolar cells. Invest Ophthalmol Vis Sci 2001; 42:816–825.
69. Weber BH, Schrewe H, Molday LL, et al. Inactivation of the murine X-linked juvenile retinoschisis gene, Rs1h, suggests a role of retinoschisin in retinal cell layer organization and synaptic structure. Proc Natl Acad Sci USA 2002;99:6222–6227.
70. Reid SN, Yamashita C, Farber DB. Retinoschisin, a photoreceptor-secreted protein, and its interaction with bipolar and muller cells. J Neuro Sci 2003;23:6030–6040.
71. Condon GP, Brownstein S, Wang NS, Kearns JA, Ewing CC. Congenital hereditary (juvenile X-linked) retinoschisis. Histopathologic and ultrastructural findings in three eyes. Arch Ophthalmol 1986;104:576–583.
72. Mooy CM, Van Den Born LI, Baarsma S, et al. Hereditary X-linked juvenile retinoschisis: a review of the role of Müller cells. Arch Ophthalmol 2002;120:979–984.
73. Zeng Y, Takada Y, Kjellstrom S, et al. RS-1 gene delivery to an adult Rs1h knockout mouse model restores ERG b-wave with reversal of the electronegative waveform of X-linked retinoschisis. Invest Ophthalmol Vis Sci 2004;45:3279–3285.
74. Molday LL, Min SH, Weber BH, Molday RS, Hauswirth WW. Structural and functional rescue of RS1 knockout mice by AAV-mediated delivery of retinoschisin – a gene therapy model for X-linked juvenile retinoschisis. 11th International Symposium on Retinal Degeneration. August 23–28, 2004, Perth, Western Australia.
75. Min SH, Molday LL, Seeliger MW, et al. Recovery of retinal structure and function after gene therapy in a Rs1h-deficient mouse model of human X-linked juvenile retinoschisis. Mol Ther 2005;12:644–651.

7

Retinal Degeneration in Usher Syndrome

David S. Williams, PhD

CONTENTS

INTRODUCTION

Inherited retinitis pigmentosa (RP) in combination with deafness was reported in the 19th century (1,2), but became known as Usher syndrome from a report by Charles Usher in 1914 (3). Usher syndrome is autosomal recessive (3) and responsible for more than half of the cases involving deafness and blindness (4). It affects about 1 in 23,000 within the United States (5). Estimates are slightly lower from Scandinavia at 1 in 29,000, and as high as 1 in 12,500 from Germany (6). Because the frequency of RP is 1 per 4000 persons (7), Usher syndrome accounts for about 17% of all cases of RP in the United States.

CLINICAL SUBTYPES AND GENETICS

Usher syndrome is clinically and genetically heterogeneous. It includes three general subtypes, types 1, 2, and 3, which are distinguished from each other primarily by the extent and onset of the deafness, which results from defective hair cells of the inner ear. Patients with Usher 1 are profoundly deaf from birth. They also have vestibular dysfunction, which results in retarded motor development. The deafness in Usher 2 is less severe, and vestibular function is normal. Patients with Usher 3 also have milder deafness, but, unlike in Usher 2, the hearing loss is progressive, and about half have vestibular dysfunction (8). RP, which is clinically similar to nonsyndromic RP, develops in all types of Usher syndrome (9,10). RP has been reported to have a slightly earlier onset in Usher 1 (9).

Usher 1 can be caused by mutations in any one of seven different genes, and mutations in any one of three different genes can result in Usher 2. Only one reported locus

From: *Ophthalmology Research: Retinal Degenerations: Biology, Diagnostics, and Therapeutics*
Edited by: J. Tombran-Tink and C. J. Barnstable © Humana Press Inc., Totowa, NJ

Table 1
Usher Syndrome Genes

Subtype	Locus	Gene (protein)	References
Usher 1A	14q32	Unknown	69
Usher 1B	11q13.5	MYO7A (myosin VIIa)	70,71
Usher 1C	11p15.1	USH1C (harmonin)	30,72
Usher 1D	10q21-22	CDH23(cadherin23)	15,31,73
Usher 1E	21q21	Unknown	74
Usher 1F	10q21-22	PCDH15 (protocadherin15)	32,33
Usher 1G	17q24-25	SANS (sans)	34,75
Usher 2A	1q41	USH2A (usherin)	76,77
Usher 2B	3p23-24	Unknown	78
Usher 2C	5q14-21	VLGR1	37,79
Usher 3	3q21-25	USH3A (clarin)	38,80

has been linked to Usher 3. Thus, so far, Usher syndrome can be divided further into eleven different genetic loci, for which eight of the genes have been identified (Table 1). It should be noted that mutations in these genes do not always result in Usher syndrome. Cases of nonsyndromic deafness have been linked to mutations in the Usher 1B, 1C and 1D genes (11–15). Conversely, mutations in USH2A and USH3A can cause autosomal recessive RP without reported hearing loss (16–18).

Usher 1 and 2 are the most common forms of Usher syndrome, with Usher 3 contributing to a large proportion of cases only in isolated areas, such as Finland (19) and Birmingham, UK (20).

VISUAL IMPAIRMENT IN USHER SYNDROME

The visual loss in Usher syndrome begins with deterioration of peripheral and night vision, as in other forms of RP. Measurements of the kinetics of visual field loss in patients with Usher 2 were determined to be comparable to those of nonsyndromic RP (21). Among the different types of Usher syndrome, some differences in visual impairment have been identified, although such differences are much less evident than differences in vestibular and auditory dysfunction. Fishman's group has observed that the visual loss with respect to age is significantly greater in Usher 1 than in Usher 2, on the grounds of visual acuity and visual field (22,23), and according to the probability of developing foveal lesions (24). In a multifocal electroretinogram (ERG) analysis (25), it was found that the ERG amplitude of Usher 1, Usher 2, and nonsyndromic patients with RP was reduced to a similar extent and in the same pattern across the retina (i.e., more reduced in the periphery). However, the waveform of the ERG differed between patients with Usher 1 and Usher 2. The time to the peak of the response (latency or implicit time) was unaffected in patients with Usher 1, but this time was significantly longer in patients with Usher 2 or nonsyndromic RP.

The finding of distinguishing characteristics between Usher 1 and 2 has not extended to differences among different genotypes of the same type of Usher syndrome. No significant difference in visual acuity, visual field, ERG amplitude, and ERG implicit time was detected between Usher 1B and other Usher 1 genotypes *(25,26)*. Similarly, no distinguishing characteristics between Usher 2A and 2C were found, using psychophysical, electrophysiological, and retinal imaging analyses *(27)*.

RETINAL FUNCTION OF USHER PROTEINS

The function of proteins, involved in phototransduction and related events, is best understood in the photoreceptor and retinal pigment epithelial (RPE) cells. Although the cell biology of the photoreceptor and RPE cells—these cells are among the most specialized cells in our bodies—is clearly important, considerably less is known about the proteins involved in their structural organization. Yet, hair cells of the inner ear and the photoreceptor and RPE cells of the retina are more similar structurally than they are functionally. In particular, all possess regions of amplified plasma membrane: the stereocilia in the hair cells, the apical microvilli of the RPE cells, and the disk membranes that make up the photoreceptor outer segments. Not surprisingly, then, the Usher proteins appear to be more related to cell structure than function.

The proteins encoded by the known Usher genes are listed in Table 1, their distributions in the retina are shown in Fig. 1, and their structural organization is depicted in Figs. 2 and 3. *MYO7A* was predicted to encode an unconventional myosin, i.e., a molecular motor that uses energy from adenosine triphosphate (ATP) hydrolysis to move along actin filaments. Direct experiments have now shown this to be the case; myosin VIIa is a bona fide actin-based motor *(28,29)*. The *USH1C* gene generates a number of different isoforms, belonging to three different classes of harmonin. The isoforms each contain two or three PDZ domains (modular protein interaction domains), so that harmonin is predicted to be a scaffolding protein *(30)*. The Usher 1D and 1F genes are both predicted to encode cadherins *(15,31–33)*. The Usher 1G protein, sans, has no indicated function *(34)*. Usherin, encoded by *USH2A*, is an extracellular matrix protein that binds type IV collagen *(35)*. The Usher 2B gene is not known, although a likely candidate is that encoding the sodium bicarbonate cotransporter, NBC3. *NBC3* is at the Usher 2B locus, and mice lacking NBC3 undergo degeneration of the retina and inner ear *(36)*. VLGR1 appears to be a G protein-coupled receptor with a large N-terminal region *(37)*. Lastly, the only reported Usher 3 gene encodes clarin1, which has been speculated to function in synaptic shaping and maintenance, based on loose homology with a protein known to function in this manner in the cerebellum *(38)*.

The precise retinal functions of these proteins are largely unknown. It has been proposed that many of the proteins might function together in a common cellular mechanism. Such a unifying hypothesis is attractive and, certainly, the similarities of clinical phenotype within the different types of Usher syndrome suggest that the mutated genes of each type might affect a common cellular mechanism. Experimental evidence to support this notion has come from studies indicating that some of the Usher 1 proteins can interact with each other.

In 1956, it was reported that a genetic interaction in ear function was evident from crossing shaker1 mice with waltzer mice *(39)*. We now know that shaker1 mice carry mutations in *Myo7a (40,41)*, and waltzer mice carry mutations in *Cdh23 (42)*, so that this result indicates that these two Usher 1 proteins might interact. More recently, in studying the retinas of shaker1 and waltzer mutant crosses, no interaction was found by ERG analysis *(43)*. However, after approx 12 mo, double homozygous mutants were

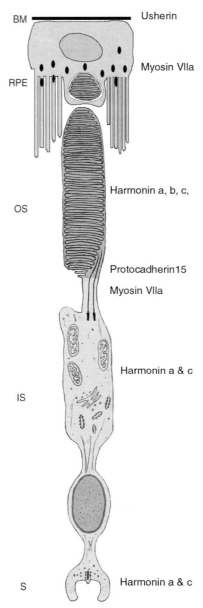

Fig. 1. Distribution of Usher proteins in the retina. The Usher 2A protein, usherin is present in Bruch's membrane, a basement membrane that supports the retinal pigment epithelial (RPE) *(81)*. The Usher 1B protein, myosin VIIa, is found mainly in the apical RPE *(53)*, but also in the connecting cilium between the photoreceptor inner and outer segment (IS and OS, respectively) *(59)*. All three classes of isoforms of the Usher 1C protein, harmonin, are represented in the OS, whereas isoforms of the smaller-sized classes, a and c, are present in the IS and synaptic terminal *(82)*. The Usher 1F protein, protocadherin15 is present at the base of the OS *(83)*. Localization data that remain controversial have not been considered for this figure. There are no reports on the distribution of the Usher 1G protein, sans, and the Usher 2C protein, VLGR1. The Usher 3 protein, clarin1, is absent from the RPE and photoreceptor cells; it is present in the inner retina *(62)*. Modified from ref. *84*, with permission from Elsevier.

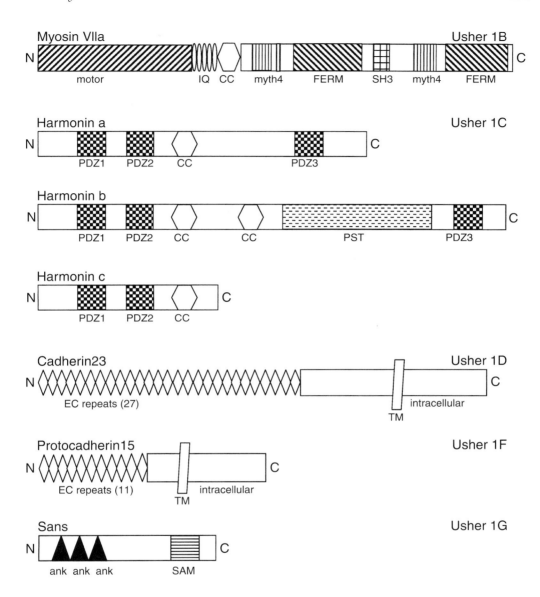

Fig. 2. Structural organization of the known Usher 1 proteins. The Usher 1B protein, myosin VIIa, has an N-terminal myosin motor domain, which contains binding motifs for ATP and actin filaments. The rest of the "head" of the myosin contains five IQ motifs, to which light chains, mostly calmodulin, bind, and a short coiled-coil (CC) domain, which may effect dimerization. The "tail" contains a duplicated tandem repeat of myth4 (myosin tail homology 4) and FERM (protein 4.1, ezrin, radixin, and moesin-like) domains, with a weak SH3 (src homology 3) domain between the repeats *(84,85)*. The Usher 1C protein, harmonin, has three different classes of isoforms, a, b, and c *(30)*. All have the first two PDZ domains and at least one CC region. The Usher 1D protein, cadherin23, is a very large molecule with 27 extracellular cadherin (EC) repeats 15, 31. The Usher 1F protein, protocadherin15, is also a cadherin, but with fewer (eleven) EC repeats *(32,33)*. The Usher 1G protein, sans, contains three ankyrin-like domains (ank) in tandem, and a sterile α motif (SAM) 34. PST, proline, serine, threonine-rich region.

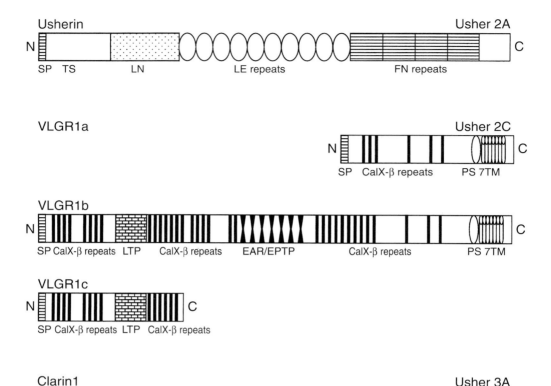

Fig. 3. Structural organization of the known Usher 2 and Usher 3 proteins. The Usher 2A protein, usherin, has four main structural elements. Following a signal peptide sequence (SP), is a domain with homology to the thromospondin family of extracellular matrix proteins (TS), and a laminins (LN) module, which is a globular domain found in many laminins. Like laminins, this domain is followed by 10 consecutive repeats of laminin-epithelial growth factor-like modules (LE). Lastly, it contains three repeats of fibronectin type III (FN) 86. Three different isoforms of the Usher 2C protein, VLGR1, are predicted, with isoforms a and c being abbreviated versions of isoform b. VLGR1b is a G protein coupled receptor (GPCR) with a very large extracellular component, containing 35 CalX-β domains, a LamG/TspN/PTX homology domain (LTP), 7 EAR/EPTP repeats (epilepsy-associated repeat/Epitempin repeat, forming a protein interaction domain), and a GPCR proteolysis site (PS). 7TM, seven transmembrane domains. *See* references in ref. *37*. The Usher 3 protein, clarin1, has four predicted transmembrane regions 38.

observed to undergo a small amount of retinal degeneration, which was not observed in mice that were homozygous mutant for only one of the genes *(44)*.

In protein-binding studies with ear isoforms, two groups found binding of the cytoplasmic region of cadherin23 to the second PDZ domain of harmonin *(45,46)*. One of the groups also reported binding between the first PDZ domain (PDZ1) of harmonin and myosin VIIa *(45)*, whereas the other group reported that the PDZ1 domain of harmonin binds the isoform of cadherin23 that is found in tissues other than the ear, such as kidney, brain, and retina *(46)*. The binding of sans to the harmonin PDZ1 domain has also been reported *(34)*.

Rather controversially, Wolfrum and colleagues have proposed that all the Usher 1 and Usher 2 proteins, including the putative Usher 2B protein, NBC3, form a supramolecular complex in the photoreceptor synapse *(47)*. This suggestion is based on protein-binding data and immunolocalization studies. However, it has been argued that the reported immunolocalization of some of these proteins to the synapse is artefactual *(48)*. Moreover, none of the mouse models for Usher syndrome have been shown to have defective synaptic transmission from the photoreceptor cells. Detailed ERG analyses have been made on mice carrying mutations in *Myo7a*, *Cdh23*, and *Pcdh15* (orthologs of the Usher 1B, 1D, and 1F genes) *(43,49,50)*, and although the *Myo7a* and *Cdh23* mutants have a reduced a-wave amplitude (indicating a reduced photoreceptor response), the ratio of the b-wave amplitude to a-wave amplitude appeared normal in all three mutants. This ratio is a good indicator of signal transmission from the photoreceptor cells to the second-order neurons; it is reduced in a variety of mice with photoreceptor synaptic defects *(51,52)*.

Although it is likely that some Usher proteins function together, on balance, most evidence points to independent functions of Usher 1 proteins in the retina. There are several subcellular regions in which only one protein is found (Fig. 1). The clearest example of an independent function is the RPE function of myosin VIIa, the most extensively-studied Usher protein.

Most of the retinal myosin VIIa is present in the RPE *(53)*, where it performs more than one function. No other Usher protein has been detected in the RPE. Studies on mutant *Myo7a* mice, including a line that has a null allele, have shown that myosin VIIa is required for the correct localization of the RPE melanosomes *(54)*, by transporting and tethering melanosomes in the apical RPE *(55–57)*. In addition, myosin VIIa functions in the delivery of phagosomes to the basal RPE and, in this role, is important for the efficient degradation of phagosomes *(58)*.

Myosin VIIa is also present in the connecting cilium of photoreceptor cells *(59)*, where it is required for the normal transport of opsin *(60)*. Cadherin23 also appears to be associated with the connecting cilium *(46,61)* (which may explain the synergistic effect of the combined loss of MYO7A and CDH23 on photoreceptor death *[44]*). So far, there have been no reports of any other Usher protein associated with this structure, however, protocadherin15 appears to be "nearby," at the base of the outer segment *(83)*.

Little is known about the Usher 2 and Usher 3 proteins that have been identified. As noted previously, usherin is an extracellular matrix protein. In the retina, it has been described as a component of Bruch's membrane, which supports the RPE *(88)*. A different group has found it to be localized also in the photoreceptor cells *(89)*.

The Usher 3 protein, clarin1, is unusual among RP proteins in that it is present in the second-order neurons of the retina and, thus, apparently functions in one or more post-synaptic processes *(62)*.

RETINAL PATHOGENESIS

Our understanding of the cellular bases of retinal degeneration in Usher syndrome has been limited by the availability of animal models that mimic Usher syndrome. Curiously, none of the Usher 1 mouse models has been found to undergo retinal degeneration. Such models include different alleles for *Myo7a, Ush1c, Cdh23*, and *Pcdh15*. Animal models for Usher 2 and Usher 3 have yet to be published.

The most thoroughly studied model is the shaker1 mouse, which has mutant *Myo7a* *(40)*, with the 4626SB allele having a null mutation *(60,63)*. As noted here previously, studies of these mice have shown that myosin VIIa is required for normal photorecep-tor electrophysiology *(49)*, and several normal cellular functions in the RPE and photoreceptor cells *(54,56–58,60)*, although there is no photoreceptor cell death. Of par-ticular interest to pathogenesis is its role in the turnover of phototransductive mem-brane. Myosin VIIa is required for both the efficient phagosome delivery to the basal RPE, and its ensuing degradation, as well as the normal flow of opsin along the con-necting cilium. Either one or both of these roles results in a retarded turnover of the pho-totransductive disk membranes *(60)*. Although in the mouse retina, these defects do not result in photoreceptor cell death (at least within the life span of a mouse), this turnover process is clearly critical. Faults at any one of several stages of disk membrane turnover underlie various forms of retinal degeneration, whether it is in the transport of proteins to the outer segment, the formation of new disk membranes (as in *Rds* mutations) *(64–66)*, or in the phagocytosis of outer segment disks (as in *Mertk* mutations) *(67,68)*. The retinal degeneration found in patients with Usher 1B might therefore be linked to defects in the turnover of phototransductive disk membrane.

SUMMARY

Most of the genes responsible for Usher syndrome have been identified. Clinical stud-ies of retinal defects have identified minor differences between Usher 1 and Usher 2, but no differences have yet been detected among the different genotypes of a given Usher type. Most of the Usher proteins are present in the photoreceptor cells, although the Usher 1B protein also functions in the RPE cells, the Usher 2A protein is a component of Bruch's membrane, and the Usher 3 protein is found only in the inner retinal cells. The proteins are more likely to function in cellular organization rather than in processes more directly related to phototransduction.

The previously reported locus for Usher syndrome type 1A *(69)* has now been shown to be false *(90)*. Hence, there are presently six, not seven, known loci for Usher 1.

REFERENCES

1. von Graefe A. Vereinzelte Beobachtungen und Bemerkungen Exceptionelle Verhalten des Gesichtsfeldes bei Pigmentenartung des Netzhaut. Arch Klin Ophthalmol 1858;4: 250–253.
2. Leibreich R. Abkunft und Eben unter Blutsverwandten als Grund von Retinitis Pigmentosa. Deutsch Klin 1861;13:53–55.
3. Usher C. On the inheritance of retinitis pigmentosa with notes of cases. R Lond Ophthalmol Hosp Rep 1914;19:130:236.
4. Vernon M. Usher's syndrome—deafness and progressive blindness. Clinical cases, preven-tion, theory and literature survey. J Chronic Dis 1969;22:133–151.
5. Boughman J, Vernon M, Shaver K. Usher syndrome: definition and estimate of prevalence from two high-risk populations. J Chronic Dis 1983;36:595–603.
6. Otterstedde CR, Spandau U, Blankenagel A, Kimberling WJ, Reisser C. A new clinical clas-sification for Usher's syndrome based on a new subtype of Usher's syndrome type I. Laryngoscope 2001;111:84–86.
7. Berson EL. Retinitis pigmentosa. Invest Ophthalmol Vis Sci 1993;34:1659–1676.

8. Sadeghi M, Cohn ES, Kimberling WJ, Tranebjaerg L, Moller C. Audiological and vestibular features in affected subjects with USH3: a genotype/phenotype correlation. Int J Audiol 2005;44:307–316.

9. Fishman GA, Kumar A, Joseph ME, Torok N, Anderson RJ. Usher's syndrome. Ophthalmic and neuro-otologic findings suggesting genetic heterogeneity. Arch Ophthalmol 1983; 101:1367–1374.

10. Smith RJ, Berlin CI, Hejtmancik JF, et al. Clinical diagnosis of the Usher syndromes. Usher Syndrome Consortium. Am J Med Genet 1994;50:32–38.

11. Liu XZ, Walsh J, Mburu P, et al. Mutations in the myosin VIIA gene cause non-syndromic recessive deafness. Nat Genet 1997;16:188–190.

12. Liu XZ, Walsh J, Tamagawa Y, et al. Autosomal dominant non-syndromic deafness caused by a mutation in the myosin VIIA gene. Nat Genet 1997;17:268–289.

13. Weil D, Kussel P, Blanchard S, et al. The autosomal recessive isolated deafness, DFNB2, and the Usher 1B syndrome are allelic defects of the myosin-VIIA gene. Nat Genet 1997;16:191–193.

14. Ahmed ZM, Smith TN, Riazuddin S, et al. Nonsyndromic recessive deafness DFNB18 and Usher syndrome type IC are allelic mutations of USHIC. Hum Genet 2002;110:527–531.

15. Bork JM, Peters LM, Riazuddin S, et al. Usher syndrome 1D and nonsyndromic autosomal recessive deafness DFNB12 are caused by allelic mutations of the novel cadherin-like gene CDH23. Am J Hum Genet 2001;68:26–37.

16. Rivolta C, Sweklo EA, Berson EL, Dryja TP. Missense mutation in the USH2A gene: Association with recessive retinitis pigmentosa without hearing loss. Am J Hum Genet 2000;66:1975–1978.

17. Seyedahmadi BJ, Berson EL, Dryja TP. USH3A mutations in patients with a prior diagnosis of Usher syndrome type I, Usher syndrome type II, and nonsyndromic recessive retinitis pigmentosa. Invest Ophthalmol Vis Sci 2004;45:E-Abstract 4726.

18. Seyedahmadi BJ, Rivolta C, Keene JA, Berson EL, Dryja TP. Comprehensive screening of the USH2A gene in Usher syndrome type II and non-syndromic recessive retinitis pigmentosa. Exp Eye Res 2004;79:167–173.

19. Pakarinen L, Tuppurainen K, Laippala P, Mantyjarvi M, Puhakka H. The ophthalmological course of Usher syndrome type III. Int Ophthalmol 1996;19:307–311.

20. Hope CI, Bundey S, Proops D, Fielder AR. Usher syndrome in the city of Birmingham—prevalence and clinical classification. Br J Ophthalmol 1997;81:46–53.

21. Iannaccone A. Usher syndrome: correlation between visual field size and maximal ERG response b-wave amplitude. Adv Exp Med Biol 2003;533:123–131.

22. Piazza L, Fishman GA, Farber M, Derlacki D, Anderson RJ. Visual acuity loss in patients with Usher's syndrome. Arch Ophthalmol 1986;104:1336–1339.

23. Edwards A, Fishman GA, Anderson RJ, Grover S, Derlacki DJ. Visual acuity and visual field impairment in Usher syndrome. Arch Ophthalmol 1998;116:165–168.

24. Fishman GA, Anderson RJ, Lam BL, Derlacki J. Prevalence of foveal lesions in type 1 and type 2 Usher's syndrome. Arch Ophthalmol 1995;113:770–773.

25. Seeliger MW, Zrenner E, Apfelstedt-Sylla E, Jaissle GB. Identification of Usher syndrome subtypes by ERG implicit time. Invest Ophthalmol Vis Sci 2001;42:3066–3071.

26. Bharadwaj AK, Kasztejna JP, Huq S, Berson EL, Dryja TP. Evaluation of the myosin VIIA gene and visual function in patients with Usher syndrome type I. Exp Eye Res 2000; 71:173–181.

27. Schwartz SB, Aleman TS, Cideciyan, et al. Disease expression in Usher syndrome caused by VLGR1 gene mutation (USH2C) and comparison with USH2A phenotype. Invest Ophthalmol Vis Sci 2005;46:734–743.

28. Udovichenko IP, Gibbs D, Williams DS. Actin-based motor properties of native myosin VIIa. J Cell Sci 2002;115:445–450.

29. Inoue A, Ikebe M. Characterization of the motor activity of mammalian myosin VIIA. J Biol Chem 2003;278:5478–5487.

30. Verpy E, Leibovici M, Zwaenepoel I, et al. A defect in harmonin, a PDZ domain-containing protein expressed in the inner ear sensory hair cells, underlies Usher syndrome type 1C. Nat Genet 2000;26:51–55.

31. Bolz H, von Brederlow B, Ramirez A, et al. Mutation of CDH23, encoding a new member of the cadherin gene family, causes Usher syndrome type 1D. Nat Genet 2001;27:108–112.

32. Ahmed ZM, Riazuddin S, Bernstein SL, et al. Mutations of the protocadherin gene PCDH15 cause Usher syndrome type 1F. Am J Hum Genet 2001;69:25–34.

33. Alagramam KN, Yuan H, Kuehn MH, et al. Mutations in the novel protocadherin PCDH15 cause Usher syndrome type 1F. Hum Mol Genet 2001;10:1709–1718.

34. Weil D, El-Amraoui A, Masmoudi S, et al. Usher syndrome type I G (USH1G) is caused by mutations in the gene encoding SANS, a protein that associates with the USH1C protein, harmonin. Human Mol Genet 2003;12:463–471.

35. Bhattacharya G, Kalluri R, Orten DJ, Kimberling WJ, Cosgrove D. A domain-specific usherin/collagen IV interaction may be required for stable integration into the basement membrane superstructure. J Cell Sci 2004;117:233–242.

36. Bok D, Galbraith G, Lopez I, et al. Blindness and auditory impairment caused by loss of the sodium bicarbonate cotransporter NBC3. Nat Genet 2003;34:313–319.

37. Weston MD, Luijendijk MW, Humphrey KD, Moller C, Kimberling WJ. Mutations in the VLGR1 gene implicate G-protein signaling in the pathogenesis of Usher syndrome type II. Am J Hum Genet 2004;74:357–366.

38. Adato A, Vreugde S, Joensuu T, et al. USH3A transcripts encode clarin-1, a four-transmembrane-domain protein with a possible role in sensory synapses. Eur J Hum Genet 2002; 10:339–350.

39. Deol MS. The anatomy and development of the mutants pirouette, shaker-1 and waltzer in the mouse. Proc Roy Soc (London) 1956;145:206–213.

40. Gibson F, Walsh J, Mburu P, et al. A type VII myosin encoded by mouse deafness gene shaker-1. Nature 1995;374:62–64.

41. Mburu P, Liu XZ, Walsh J, et al. Mutation analysis of the mouse myosin VIIA deafness gene. Genes Funct 1997;1:191–203.

42. Di Palma F, Holme RH, Bryda EC, et al. Mutations in Cdh23, encoding a new type of cadherin, cause stereocilia disorganization in waltzer, the mouse model for Usher syndrome type 1D. Nat Genet 2001;27:103–107.

43. Libby RT, Kitamoto J, Holme RH, Williams DS, Steel KP. Cdh23 mutations in the mouse are associated with retinal dysfunction but not retinal degeneration. Exp Eye Res 2003; 77:731–739.

44. Lillo C, Kitamoto J, Liu X, Quint E, Steel KP, Williams DS. Mouse models for Usher syndrome 1B. Adv Exp Med Biol 2003;533:143–150.

45. Boeda B, El-Amraoui A, Bahloul A. Myosin VIIa, harmonin and cadherin 23, three Usher I gene products that cooperate to shape the sensory hair cell bundle. Embo J 2002; 21:6689–6699.

46. Siemens J, Kazmierczak P, Reynolds A, Sticker M, Littlewood-Evans A, Muller U. The Usher syndrome proteins cadherin 23 and harmonin form a complex by means of PDZ-domain interactions. Proc Natl Acad Sci USA 2002;99:14,946–14,951.

47. Wolfrum U, Marcker T, Van Wijk E, et al. Molecular linkage between Usher syndrome 1 and 2 by interacting within supramolecular Usher protein complexes. Invest Ophthalmol Vis Sci 2005;46:E-Abstract 5173.

48. Gibbs D, Williams DS. Usher 1 protein complexes in the retina. Invest Ophthalmol Vis Sci 2004;45:e-letter (May 26).

49. Libby RT, Steel KP. Electroretinographic anomalies in mice with mutations in Myo7a, the gene involved in human Usher syndrome type 1B. Invest Ophthalmol Vis Sci 2001;42:770–778.

50. Ball SL, Bardenstein D, Alagramam KN. Assessment of retinal structure and function in Ames waltzer mice. Invest Ophthalmol Vis Sci 2003;44:3986–3992.

51. Peachey NS, Ball SL. Electrophysiological analysis of visual function in mutant mice. Doc Ophthalmol 2003;107:13–36.

52. Libby RT, Lillo C, Kitamoto J, Williams DS, Steel KP. Myosin Va is required for normal photoreceptor synaptic activity. J Cell Sci 2004;117:4509–4515.

53. Hasson T, Heintzelman MB, Santos-Sacchi J, Corey DP, Mooseker MS. Expression in cochlea and retina of myosin VIIa, the gene product defective in Usher syndrome type 1B. Proc Natl Acad Sci USA 1995;92:9815–9819.

54. Liu X, Ondek B, Williams DS. Mutant myosin VIIa causes defective melanosome distribution in the RPE of shaker-1 mice. Nat Genet 1998;19:117–118.

55. El-Amraoui A, Schonn JS, Kussel-Andermann P, et al. MyRIP, a novel Rab effector, enables myosin VIIa recruitment to retinal melanosomes. EMBO Rep 2002;3:463–470.

56. Futter CE, Ramalho JS, Jaissle GB, Seeliger MW, Seabra MC. The role of Rab27a in the regulation of melanosome distribution within retinal pigment epithelial cells. Mol Biol Cell 2004;15:2264–2275.

57. Gibbs D, Azarian SM, Lillo C, et al. Role of myosin VIIa and Rab27a in the motility and localization of RPE melanosomes. J Cell Sci 2004;117:6473–6483.

58. Gibbs D, Kitamoto J, Williams DS. Abnormal phagocytosis by retinal pigmented epithelium that lacks myosin VIIa, the Usher syndrome 1B protein. Proc Natl Acad Sci USA 2003;100:6481–6486.

59. Liu X, Vansant G, Udovichenko IP, Wolfrum U, Williams DS. Myosin VIIa, the product of the Usher 1B syndrome gene, is concentrated in the connecting cilia of photoreceptor cells. Cell Motil Cytoskel 1997;37:240–252.

60. Liu X, Udovichenko IP, Brown SDM, Steel KP, Williams DS. Myosin VIIa participates in opsin transport through the photoreceptor cilium. J Neurosci 1999;19:6267–6274.

61. Lillo C, Siemens J, Kazmierczak P, Mueller U, Williams DS. Roles and interactions of three USH1 proteins in the retina and inner ear. Invest Ophthalmol Vis Sci 2005;46:E-Abstract 5176.

62. Geller SF, Isosomppi J, Makela H, Sankila E, Johnson PT, Flannery JG. Vision loss in Usher syndrome type III is caused by mutations in clarin-1, an inner retinal protein. Invest Ophthalmol Vis Sci 2004;45:E-Abstract 5123.

63. Hasson T, Walsh J, Cable J, Mooseker MS, Brown SDM, Steel KP. Effects of shaker-1 mutations on myosin-VIIa protein and mRNA expression. Cell Motil Cytoskeleton 1997;37:127–138.

64. Sanyal S, Zeilmaker GH. Development and degeneration of retina in rds mutant mice: light and electron microscopic observations in experimental chimaeras. Exp Eye Res 1984; 39:231–246.

65. Kajiwara K, Hahn LB, Mukai S, Travis GH, Berson EL, Dryja TP. Mutations in the human retinal degeneration slow gene in autosomal dominant retinitis pigmentosa. Nature 1991; 354:480–483.

66. Farrar GJ, Kenna P, Jordan SA, et al. A three-base-pair deletion in the peripherin-RDS gene in one form of retinitis pigmentosa. Nature 1991;354:478–480.

67. Bok D, Hall MO. The role of the pigment epithelium in the etiology of inherited retinal dystrophy in the rat. J Cell Biol 1971;49:664–682.

68. Gal A, Li Y, Thompson DA, et al. Mutations in MERTK, the human orthologue of the RCS rat retinal dystrophy gene, cause retinitis pigmentosa. Nat Genet 2000;26:270–271.

69. Kaplan J, Gerber S, Bonneau D, et al. A gene for Usher syndrome type I (USH1A) maps to chromosome 14q. Genomics 1992;14:979–987.

70. Kimberling WJ, Moller CG, Davenport S, et al. Linkage of Usher syndrome type I gene (USH1B) to the long arm of chromosome 11. Genomics 1992;14:988–994.

71. Weil D, Blanchard S, Kaplan J, et al. Defective myosin VIIA gene responsible for Usher syndrome type 1B. Nature 1995;374:60–61.

72. Smith RJ, Lee EC, Kimberling WJ, et al. Localization of two genes for Usher syndrome type I to chromosome 11. Genomics 1992;14:995–1002.

73. Wayne S, Derkaloustian VM, Schloss M, et al. Localization of the usher syndrome type id gene (ush1d) to chromosome 10. Hum Mol Genet 1996;5:1689–1692.

74. Chaib H, Kaplan J, Gerber S, et al. A newly identified locus for Usher syndrome type I, USH1E, maps to chromosome 21q21. Hum Mol Genet 1997;6:27–31.

75. Mustapha M, Chouery E, Torchard-Pagnez D, et al. A novel locus for Usher syndrome type I, USH1G, maps to chromosome 17q24-25. Hum Genet 2002;110:348–350.

76. Eudy JD, Weston MD, Yao S, et al. Mutation of a gene encoding a protein with extracellular matrix motifs in Usher syndrome type IIa. Science 1998;280:1753–1757.

77. Eudy JD, Yao S, Weston MD, et al. Isolation of a gene encoding a novel member of the nuclear receptor superfamily from the critical region of Usher syndrome type IIa at 1q41. Genomics 1998;5:382–384.

78. Hmani M, Ghorbel A, Boulila-Elgaied A, et al. A novel locus for Usher syndrome type II, USH2B, maps to chromosome 3 at p23-24.2. Eur J Hum Genet 1999;7:363–367.

79. Pieke-Dahl S, Moller CG, Kelley PM, et al. Genetic heterogeneity of Usher syndrome type II: localisation to chromosome 5q. J Med Genet 2000;37:256–262.

80. Sankila EM, Pakarinen L, Kaariainen H, et al. Assignment of an Usher syndrome type III (USH3) gene to chromosome 3q. Hum Mol Genet 1995;4:93–98.

81. Pearsall N, Bhattacharya G, Wisecarver J, Adams J, Cosgrove D, Kimberling W. Usherin expression is highly conserved in mouse and human tissues. Hear Res 2002;174:55–63.

82. Reiners J, Reidel B, El-Amraoui A, et al. Differential distribution of harmonin isoforms and their possible role in Usher-1 protein complexes in mammalian photoreceptor cells. Invest Ophthalmol Vis Sci 2003;44:5006–5015.

83. Reiners J, Marker T, Jurgens K, Reidel B, Wolfrum U. Photoreceptor expression of the Usher syndrome type 1 protein protocadherin 15 (USH1F) and its interaction with the scaffold protein harmonin (USH1C). Mol Vis 2005;11:347–355.

84. Williams DS. Transport to the photoreceptor cell outer segment by moysin VIIa and kinesin II. Vision Res 2002;42:392–403.

85. Chen ZY, Hasson T, Kelley PM, et al. Molecular cloning and domain structure of human myosin-VIIa, the gene product defective in Usher syndrome 1B. Genomics 1996;36:440–448.

86. Weil D, Levy G, Sahly I, et al. Human myosin VIIA responsible for the Usher 1B syndrome: a predicted membrane-associated motor protein expressed in developing sensory epithelia. Proc Natl Acad Sci USA 1996;93:3232–3237.

87. Weston MD, Eudy JD, Fujita S, et al. Genomic structure and identification of novel mutations in Usherin, the gene responsible for Usher syndrome type IIa. Am J Hum Genet 2000;66:1199–1210.

88. Bhattacharya G, Miller C, Kimberling WJ, et al. Localization and expression of usherin: a novel basement membrane protein defective in people with Usher's syndrome type IIa. Hear Res 2002;163:1–11.

89. Reiners J, van Wijk E, Marker T, et al. Scaffold protein harmonin (USH1C) provides molecular links between Usher syndrome type 1 and type 2. Hum Mol Genet 2005;14:3933–3943.

90. Gerber S, Bonneau D, Gilbert B, et al. USH1A: Chronicle of a Slow Death. Am J Hum Genet 2006;78:357–359.

8

Mouse Models of RP

**Bo Chang, MD, Norman L. Hawes, BS, Muriel T. Davisson, PhD,
and J. R. Heckenlively, MD**

Contents

INTRODUCTION

Retinitis pigmentosa (RP) is the name given to a group of eye diseases often characterized by night blindness and the gradual loss of peripheral vision RP causes the degeneration of photoreceptor cells in the retina. As these cells degenerate and die, patients experience progressive vision loss to eventual blindness. The most common feature of all forms of RP is a gradual degeneration of photoreceptor cells: rod cells and cone cells. The rods and cones are the cells responsible for converting light into electrical impulses that are transmitted to the brain where "seeing" actually occurs. RP is an inherited, genetically heterogeneous condition, i.e., RP can result from mutations in many different genes. It is caused by mutations in genes that are active in retinal cells. RP is a common form of retinal degeneration (RD). Mouse models of RD have been investigated for many years to understand the causes of photoreceptor cell death. Knowledge about these mutations is essential for proper selection of mouse models for use in research. In this chapter, we review these naturally occurring mouse mutants that manifest degeneration of photoreceptors in the retina with preservation of all other retinal cell types. The mutations are described in chronological order and provide a list of the mouse strains that carry each mutation.

TYPES OF RETINAL DEGENERATION

Retinal Degeneration 1 (Pde6b^{rd1})

The first RD, discovered by Keeler more than 80 yr ago, is *Pde6b^{rd1}* (formerly *rd1*, *rd*, identical with Keeler rodless retina, *r*) *(1–3)*. Mice homozygous for the *Pde6b^{rd1}* mutation

From: *Ophthalmology Research: Retinal Degenerations: Biology, Diagnostics, and Therapeutics*
Edited by: J. Tombran-Tink and C. J. Barnstable © Humana Press Inc., Totowa, NJ

Table 1
Mouse Strains With Different Retinal Degenerations

Gene	Chromosome mouse	Chromosome human	Retinal ONL loss by mo	Strains
Pde6b[rd1]	5	4p16	1	ABJ/Le; BDP/J; BUB/BnJ; C3H, all substrains; CBA/J; CBA/NJ; FVB/NJ; JGBF/Le; MOLD/Rk; MOLF/Ei; NFS/N; NON/LtJ; P/J; PL/J; RSV/Le; SF/CamEi; SF/CamRk; SK/CamEi; ST/bJ; SJL/J; SWR/J; WB/ReJ; WC/ReJ
Agtpbp1[pcd]	13	9q21.33	13	B6.BR-pcd; B6C3Fe-a/a-pcd/+; BALB/cByJ-pcd[3J]
nr	8	8p21-p11.2	10	BALB/cByJ-nr; C3Fe.CGr-nr
Rds[Rd2]	17	6p21.2-p12.3	12	O20/A; C3.BliA-Rds[Rd2]
rd3	1	1q32	4	RBF/DnJ; Stock Rb4Bnr; RBJ/Dn; Stock In30Rk
Cln8[mnd]	8	8p23	6	B6.KB2-mnd/MsrJ
Rd4	4	1p36	2	Stock In56Rk Rd4
Tub[tub]	7	11p15.5	8	C57BL/6J-tub/+
Mitf[mi-vit]	6	3p14.2-p14.1	10	C57BL/6J-Mitf[mi-vit]
Mfrp[rd6]	9	11q23	24	B6.C3Ga-rd6/rd6;
Nr2e3[rd7]	9	15q22.32	30[a]	77-2C2a-special; Stock Nr2e3[rd7]
Cln6[nclf]	9	15q23	13	B6.Cg-nclf; Stock a/a nclf/nclf
Crb1[rd8]	1	1q31-q32.1	30[b]	B6-Crb1[rd8]
Rd9	X	Xp12.1	30	C57BL/6J-Rd9
Pde6b[rd10]	5	4p16	2	Stock Pde6b[rd10]
Rd11	13	5p15	2	Stock rd11
Rpe65[rd12]	3	1p31	20	B6-Rpe65[rd12]
rd13	15	12q13	4	Stock rd13
rd14	18	18q21	20	Stock rd14
rd15	7	19q13.3	9	Stock rd15
rd16	10	12q22	2	Stock rd16

[a]Nr2e3[rd7] mice have a five-cell layer in ONL at 16 mo of age and still have a five-cell layer when mice reach 30 mo of age.

[b]Crb1[rd8] mice lose ONL only in disease areas.

ONL, outer nuclear layer.

have an early-onset severe RD because of a lack of activity of rod cGMP-phosphodiesterase (PDE) because of a murine viral insert and a second nonsense mutation in exon 7 of the Pde6b gene encoding the β-subunit of cGMP-PDE (4–7). This mutation has been found in several common laboratory inbred strains, as well as wild-derived inbred strains including MOLD/Rk, MOLF/Ei, derived from *Mus musculus molossinus* trapped in Japan; SF/CamRk, SF/CamEi, derived from *Mus musculus domesticus* trapped in California; and SK/CamEi, derived from *Mus musculus domesticus* trapped on Skokholm Island off Pembrokeshire (8). It is important for investigators evaluating eyes to be aware of Pde6b[rd1] and its associated morphological findings, as it is a frequent strain background characteristic (*see* Table 1). Because the *Pde6b^rd1* mutation is common in mice, it is important to avoid mouse strains or stocks carrying the *Pde6b^rd1* allele or exclude the *Pde6b^rd1* allele contamination in studying new retinal disorders. Mice with the *Pde6b^rd1* mutation can be easily typed by phenotype based on vessel attenuation and pigment patches in the fundus *(9,10)* and by genotype with polymerase chain reaction for *Dde I*, which reveals a restriction fragment length polymorphism *(7)*. Mutations in the gene encoding the β-subunit of cGMP-PDE have been found in human patients suffering from autosomal recessive (AR) RP (OMIM 180072), a disorder bearing phenotypic resemblance to that caused by the mouse *Pde6b^rd1* mutation *(11)*. An array of mutations in the catalytic domain of the human homolog of *Pde6b* occurs in patients with RP *(12)*. Using this model, the photoreceptor degeneration has been rescued by gene therapy *(13)* and cone photoreceptor degeneration has been prevented by incorporating autologous bone marrow-derived lineage-negative hematopoietic stem cells into the degenerating blood vessels *(14)*.

Purkinje Cell Degeneration (Agtpbp1^pcd)

A second, slower RD is associated with Purkinje cell degeneration (*pcd*) as a result of a sequence alteration in the axotomy-induced gene *Nna1 (15)*. *Nna1* encodes an adenosine triphosphate/guanosine triphosphate binding protein 1 and the gene symbol has been updated to *Agtpbp1^pcd*. *Agtpbp1^pcd* is an AR mutation that arose in the C57BR/cdJ strain. *Agtpbp1^pcd* homozygotes show moderate ataxia beginning at 3 to 4 wk of age because of the degeneration of Purkinje cells of the cerebellum (beginning at 15 to 18 d of age). A slower degeneration of the photoreceptor cells of the retina, mitral cells of the olfactory bulb, and some thalamic neurons also occurs in *Agtpbp1^pcd* homozygotes *(16)*. Pycnotic nuclei begin to appear in the photoreceptor cells between 18 and 25 d of age, and the outer rod segments become disorganized. Degeneration of the photoreceptor cells proceeds slowly to completeness over the course of 1 yr *(17,18)*. Male *Agtpbp1^pcd* homozygotes have abnormal sperm and are sterile. *Agtpbp1^pcd* was mapped to mouse chromosome 13; its human homolog should be on 9q21.33.

Nervous (nr)

A third RD was discovered in nervous (*nr*) mice. The *nr* mutant arose spontaneously in a BALB/cGr subline carrying *tk* (tail-kinks). Homozygous *nr* mice may be recognized at 2 to 3 wk by their smaller size and hyperactive ataxic behavior. There is a 90% loss of cerebellar Purkinje cells between 23 and 50 d of age. Purkinje cell abnormalities are preceded by partial degeneration of cells of the external granular layer at 6 to 8 d of age. In the retina, degeneration of the photoreceptors is already present at 13 d. Whorls of outer segment (OS) membranes form and the OS eventually disappear completely *(19–21)*. *nr* was mapped to mouse chromosome 8 *(22,23)*, and the human homolog should be on chromosome 8 (8p21-p11.2).

Retinal Degeneration 2 (Rds^Rd2)

The fourth RD is RD-slow, formerly *rds*. Later it was found to be partially dominant and so renamed *Rds*. It has since been renamed *Rd2* in the RD series because its slow progression is no longer unique and the gene symbol has been updated to *Rds^Rd2*. A mutation causing slow RD was identified in the 020/A inbred strain *(24)*; it is caused by an insertion of foreign DNA into an exon of *Rds*, causing transcription of an abnormally large messenger RNA *(25)*. Mutations in the human gene for peripherin have been shown to cause slow RD (OMIM 179605) similar to that caused by the mouse Rds^{Rd2} mutation *(26)*. The primary site of the Rds^{Rd2} defect is the OS of the photoreceptor cell, and the expression of Rds is specific to these cells *(27)*. Thus, although the apical retinal pigment epithelium (RPE) of retinas of Rds^{Rd2} mice is reduced compared to wild-type (WT) *(28)*, transplants of WT RPE cells into mutant homozygotes do not rescue the photoreceptor cells *(29)*. Introduction of a transgene construct including the WT peripherin-coding region causes complete reversion to WT morphology of the retina *(30)*. Rds^{Rd2} was mapped to mouse chromosome 17; its human homolog should be on chromosome 6 (p21.2-p12.3), which is the location of the human peripherin gene.

Retinal Degeneration 3 (rd3)

A fifth RD is *rd3*. This spontaneous AR mutation that arose in the RBF/DnJ mouse strain at the Jackson Laboratory causes early-onset RD. The *rd3* mutation was mapped to mouse chromosome 1 in a region homologous to the region of human Chr 1q to which Usher syndrome type IIA has been located *(31)*. *rd3* has been proposed as a candidate gene for orthology to human *USH2A* (OMIM 276901), which causes a mild form of Usher syndrome deafness. High-frequency progressive hearing loss was identified in RBF/DnJ mice, but did not segregate with *rd3* *(32)*. Genetic and physical maps of the mouse *rd3* locus also have excluded the mouse ortholog of *USH2A* from the *rd3* locus *(33)*. *rd3* was mapped to mouse Chr 1; its human homolog should be on chromosome 1q32.

Motor Neuron Degeneration (Cln8^mnd)

The sixth RD was found in motor neuron degeneration (*mnd*). The *mnd* mutation arose in the C57BL/6.KB2/Rn (now B6.KB2-*mnd*/MsrJ) congenic inbred strain. The RD and motor neuron degeneration are caused by a single nucleotide insertion (267-268C, codon 90, predicts a frameshift and a truncated protein) in the ceroid-lipofuscinosis neuronal 8 (*Cln8*) gene *(34)* and the gene symbol has been updated to *Cln8^mnd*. It is a good model for human Batten disease and its mode of inheritance is clearly recessive *(35)*. In homozygous *Cln8^mnd* mice, the RD occurs before neuromuscular dysfunction. The RD is nearly complete by 6 mo of age, when the motor neuron abnormalities are just beginning *(36)*. Genetic analysis showed that the RD in *Cln8^mnd* mice is inherited as a single autosomal gene with recessive expression and is mapped to the proximal end of mouse chromosome 8 *(37)* and a human homolog should be on 8p23.

Retinal Degeneration 4 (Rd4)

A seventh RD is *Rd4*. This autosomal dominant RD mutation was found in a stock carrying the chromosomal inversion In(4)56Rk, which was induced in a DBA/2J male mouse. The inversion is homozygous lethal and in heterozygotes is always associated with RD. *Rd4* has not recombined with the inversion in an outcross, suggesting that the *Rd4* locus is located very close to or is disrupted by one of the breakpoints of the inversion on Chr 4 *(38)*. Analysis of the backcross data placed the inversion breakpoints within 1 cM of the telomere and 7 cM of the centromere. Fluorescent *in situ* hybridization analysis narrowed the region in which the breakpoint lies at the distal end of the chromosome and showed that the proximal breakpoint is in the centromere itself *(39)*. Therefore, the *Rd4* gene must be disrupted by the telomeric breakpoint of the inversion. The telomeric region of mouse chromosome 4 is homologous to human 1p36. No inherited retinal diseases have been identified previously in either of these chromosomal regions. Therefore, the *Rd4* gene is novel and necessary for the normal integrity and function of the mouse retina. Its identification will provide another candidate gene for the site of mutations responsible for inherited human RDs.

Retinal Degeneration 5 (Tub^tub)

An eighth RD was found in mice of the C57BL/6J-*tub* strain and named *rd5*, now a subline of C57BL/6J wherein a mutation to tubby (*tub*) occurred in 1977. Based on the tight linkage of the two phenotypes and the molecular defect underlying the *tub* mutation, it seemed likely that *tub* and *rd5* were the same gene *(40,41)*. This was further supported by a knockout of the tub gene that recapitulated the RD phenotype *(42)* and the gene symbol has been updated to *Tub^tub*. The involvement of *Tub^tub* in obesity, RD, and hearing loss suggests it as a model for the Alstrom (*ALMS1*, OMIM 203800) or Bardet-Biedl (*BBS2*, OMIM 209900) *(41)* or Usher (*USH1A*, OMIM 276900) syndromes *(40)*. *Tub^tub* was mapped to mouse chromosome 7; its human homolog should be on chromosome 11p15.5.

Vitiligo (Mitf^mi-vit)

The ninth RD was found in vitiligo (*vit*) mice. This spontaneous mutation in C57BL/6J mice causes lighter initial coat color than normal, with extensive white spotting. After 8 wk of age, the mice produce increasing numbers of white hairs with each molt, thus phenotypically resembling human vitiligo *(43)*. The vitiligo mutation was mapped to Chr 6 near *Mitf (44)*, and lack of complementation between vitiligo and *Mitf^{Mi-wh}* was shown *(45)*, establishing vitiligo as a mutation in the microphthalmia-associated transcription factor (*Mitf*) gene and the gene symbol has been updated to *Mitf^{mi-vit}*. An amino acid substitution in the first helix is the molecular lesion *(46)*. *Mitf^{mi-vit}* homozygotes show a slow progressive loss of photoreceptor cells, cosegregating with the gradual depigmentation *(47)*. The outer plexiform layer (OPL) is significantly thinner by 4 mo of age and the rows of photoreceptor cells are lost at a rate of about 1/mo beginning at 2 mo of age. By 8 mo, the photoreceptor cell nuclei have diminished to only two to three rows *(48)*. Furthermore, retinal detachment from the RPE, the displacement of darkly staining cells into the subretinal space, and the influx of macrophage-like cells in the area of the retina OS (ROS) are present in *Mitf^{mi-vit}/Mitf^{mi-vit}* mice *(49)*. *Mitf^{mi-vit}* was mapped to mouse chromosome 6; its human homolog should be on chromosome 3p14.2-p14.1.

Retinal Degeneration 6 (Mfrp^rd6)

The tenth RD is *rd6*. The distinctive small, white retinal spots and progressive photoreceptor degeneration *(50)* is the result of a spontaneous mutation resulting in a splice donor mutation leading to loss of one exon *(51)* in a gene encoding a membrane-type frizzled-related protein (*Mfrp*) and the gene symbol has been updated to *Mfrp^rd6*. The inheritance pattern of *Mfrp^rd6* is AR and it maps to mouse chromosome 9 about 24 cM from the centromere, suggesting the human homolog may be on chromosome 11q23. Ophthalmoscopic examination of mice homozygous for *Mfrp*[rd6] reveals discrete subretinal spots oriented in a regular pattern across the retina. The retinal spots appear clinically by 8 to 10 wk of age and persist through advanced stages of RD. Histological examination reveals large cells in the subretinal space, typically juxtaposed to the RPE. The white dots seen on fundus examination correspond both in distribution and size to these large cells. By 3 mo of age, the cells are filled with membranous profiles, lipofuscin-like material, and pigment. These cells react strongly with an antibody directed against a mouse macrophage-associated antigen. Photoreceptor cells progressively degenerate with age and an abnormal electroretinogram (ERG) is initially detected between 1 and 2 mo of age *(52)*.

Retinal Degeneration 7 (Nr2e3[rd7])

An eleventh RD is *rd7*. The *rd7* mouse is a model for hereditary RD characterized clinically by retinal spotting throughout the fundus, late-onset RD, and histologically by retinal dysplasia manifesting as folds and whorls in the photoreceptor layer *(53)*. The retinal abnormalities in *rd7* mice are caused by a deletion of exons 4 and 5, resulting in the absence of 380 bp from a nuclear receptor subfamily 2, group E, member 3 gene (*Nr2e3*) and the gene symbol has been updated to *Nr2e3*[rd7]. The predicted protein expressed from this allele would lack 127 amino acids, including sequences corresponding to the DNA-binding domain. The deletion also introduces a frameshift and creates a premature stop codon. Our results demonstrate that NR2E3 expression is critical for the integrity and function of mouse photoreceptor cells *(54)*. The function of *Nr2e3* may be involved in cone cell proliferation, and mutations in this gene lead to retinal dysplasia and degeneration by disrupting normal photoreceptor cell topography as well as cell–cell interactions *(55)*. Another study demonstrates that NR2E3 is involved in regulating the expression of rod photoreceptor-specific genes and support its proposed role in transcriptional regulatory network(s) during rod differentiation *(56)*. The inheritance pattern of *Nr2e3*[rd7] is AR, it maps to mouse chromosome 9 and the human homolog may be on chromosome 15q22.32.

Neuronal Ceroid Lipofuscinosis (Cln6[nclf])

The twelfth RD was found in the neuronal ceroid lipofuscinosis (*nclf*) mutant mouse. The phenotypes of progressive ataxia, myelination defects, neurodegeneration, neuromuscular defects, and RD in *nclf* homozygotes are the result of a single nucleotide insertion of a cysteine, located within a run of cysteines in exon four in ceroid-lipofuscinosis, neuronal 6 gene and the gene symbol has been updated to *Cln6*[nclf]. The insertion produces a frameshift at amino acid 103, followed by a premature stop codon *(57,58)*. Numerous mutations were identified in the human ortholog in families with variant late infantile ceroid lipofuscinosis *(57)*. In the mouse mutant, paralysis and death occur by around 9 mo of age *(59)*. *nclf* maps to chromosome 9. This suggests homology to the human *CLN6* (OMIM 601780) gene on chromosome 15q23, encoding a late infantile variant of ceroid lipofuscinosis *(59)*. In homozygous *nclf* mice, the outer nuclear layer of the retina shows cell loss by 4 mo after birth. By 6 mo, the peripheral outer retina is severely affected, and by 9 mo, the entire retina is atrophied.

Retinal Degeneration 8 (Crb1^rd8)

A thirteenth RD is *rd8*. Mice homozygous for *rd8* show slow RD with large white retinal deposits covering the inferior quadrant of the retina. Histology at 5 wk of age shows areas of retinal folding that correspond to the white retinal spots *(60)*. The mutation in the *rd8* mouse has been identified as a single base deletion in the *Crb1* gene and the gene symbol has been updated to *Crb1^rd8*. This deletion causes a frame shift and a premature stop codon that truncates the transmembrane and cytoplasmic domain of the protein *(61)*. Mutations in a human homolog of *CRB1* cause RP *(RP12)* *(62)*. The inheritance pattern of *Crb1^rd8* is AR and it maps to mouse chromosome 1 and the human homolog may be on Chr 1q31-q32.1.

Retinal Degeneration 9 (Rd9)

A fourteenth RD is *Rd9*. *Rd9* is an X-linked semidominant RD model. In *Rd9* mutants, the retina is covered with diffuse white spots (mottled). The fundus appears normal at weaning age and the retina becomes mottled starting at 6 wk of age in heterozygous females. The fundus of homozygous females and hemizygous males has a blond appearance. Retinal pigment loss and decreasing ERG response progress with age *(60)*. *Rd9* was mapped to mouse chromosome X between *DXMit124* and *DXMit224*; its human homolog should be on chromosome Xp12.1.

Retinal Degeneration 10 (Pde6b^rd10)

A fifteenth RD is *rd10*. Mice homozygous for the *rd10* mutation show RD with sclerotic retinal vessels at 4 wk of age. Histology at 3 wk of age shows RD. ERGs of *rd10/rd10* mice are never normal. The maximal response occurs at 3 wk of age and is nondetectable at 2 mo of age. Genetic analysis shows that this disorder is caused by an AR mutation that maps to mouse Chr 5. Sequence analysis shows that the RD is caused by a missense mutation in exon 13 of the beta subunit of the rod phosphodiesterase gene. Therefore, the gene symbol for the *rd10* mutation is now *Pde6b^rd10*. The exon 13 missense mutation is the first known occurrence of a remutation in the *Pde6b* gene in mice and may provide a good model for studying the pathogenesis of ARRP in humans. It also may provide a better model than *Pde6b^rd1* for experimental pharmaceutical-based therapy for RP because of its later-onset and milder RD *(63)*. The *Pde6b^rd10* model has been used in molecular therapy *(64)* and cone photoreceptor degeneration has been prevented by incorporating autologous bone marrow-derived lineage-negative hematopoietic stem cells into the degenerating blood vessels *(14)*.

Retinal Degeneration 11 (rd11)

The sixteenth RD is *rd11*. Mice homozygous for the *rd11* mutation show a clinical RD with white retinal vessels at 4 wk of age. Histology at 3 wk of age shows RD. ERGs of *rd11/rd11* mice are never normal. The maximal response occurs at 3 wk of age and is nondetectable at 2 mo of age. The *rd11* mutation is a new AR RD in mice and may provide a good model for studying the pathogenesis of ARRP in humans. *rd11* maps to mouse chromosome 13 approx 40 cM from the centromere, suggesting that the human homolog may be on chromosome 5p15 *(65)*.

Retinal Degeneration 12 (Rpe65^{rd12})

The seventeenth RD is *rd12*. Mice homozygous for the *rd12* mutation show retinal spots at 7 mo of age with RD starting at 12 mo of age. Despite the relatively late onset of RD in the *rd12* homozygous mutant mice, their eyes show a poor rod and good cone ERG response at 3 wk of age. Genetic analysis showed that *rd12* is an AR mutation and maps to mouse chromosome 3 closely linked to *D3Mit19*, suggesting that the human homolog may be on chromosome 1p31 where the human *RPE65* gene is located. Sequence analysis showed that the RD is caused by a nonsense mutation in exon 3 of the *Rpe65* gene and the gene symbol for the *rd12* mutation has been changed to *Rpe65^{rd12} (66)*. Mutations in the RPE gene encoding RPE65 are a cause of an early-onset AR form of human RP, known as Leber congenital amaurosis, which results in blindness or severely impaired vision in children. The natural arising *rd12* mutation will provide a good model for studying the pathogenesis of ARRP in humans. In adeno-associated virus (AAV)5-CBA-hRPE65 treated *Rpe65^{rd12}* eyes, dark-adapted ERG waveforms were restored as soon as 1 wk after treatment. Treated eyes also maintained normal cone ERGs, although they were delayed and exhibited progressive loss in amplitude in partner untreated eyes. AAV-mediated gene therapy can restore a significant fraction of the visual function in this naturally occurring model lacking RPE65 *(67)*.

Retinal Degeneration 13 (rd13)

The eighteenth RD is *rd13*. In the course of routine histological screening, a retinal abnormality was detected in *nmf5* mice affected with a neurological disorder. Genetic analysis shows that the retinal abnormality is caused by mutation of a distinct locus, designated *rd13 (68)*. The *rd13* mutation is a new AR RD in mice and may provide a good model for studying the pathogenesis of ARRP in humans. *rd13* was mapped to the distal 4 Mb of mouse chromosome 15, suggesting that the human homolog may be on chromosome 12q13.

Retinal Degeneration 14 (rd14)

The nineteenth RD is *rd14*. Ophthalmoscopic examination of mice homozygous for *rd14* mutation revealed white retinal spots. Histological examination revealed large cells in the subretinal space, juxtaposed to the RPE. White dots seen on fundus examination corresponded both in distribution and size to these large cells. Blood was observed on some retinas, which decreased cone and rod function but otherwise *rd14* has a normal ERG. Homozygous mutants also display a hopping gait when young, but appear neurologically normal after 2 to 3 wk of age. The inheritance pattern of the *rd14* mutant allele is AR. Linkage studies mapped this new mutation to mouse chromosome 18, in a region between markers *D18Mit103* and *D18Mit186*, suggesting the human homolog may be on chromosome 18q21. RD and the neurological phenotype combined with our genetic data suggest that this is a new mutation not previously described in mouse or human. This provides a novel mouse model for a RD associated with neurological defects *(69)*.

Retinal Degeneration 15 (rd15)

The twentieth RD is *rd15*. Mice homozygous for the *rd15* mutation show a normal fundus, but no rod ERG b-wave and a poor cone ERG by 4 wk of age. Histology shows poor retinal OPL at 5 mo of age and RD at 9 mo of age. The inheritance pattern of *rd15* mutant allele is AR. Linkage studies mapped this new mutant to mouse chromosome 7, in a region between markers *D7Mit230* and *D7Mit82*, suggesting the human homolog may be on chromosome 19q13.3. The early onset of rod ERG b-wave loss and impaired retinal cone function combined with our genetic data suggest that this is a new mutation not previously described in mouse or human. RD 15 (*rd15*) may provide a novel mouse model for a RD associated retinal OPL dystrophy *(70)*.

Retinal Degeneration 16 (rd16)

The twenty-first RD is *rd16*. It was submitted as an Association for Research in Vision and Ophthalmology (ARVO) abstract in 2005 as *rd13*. Mice homozygous for *rd16* show RD with white retinal vessels at 1 mo of age. Histology at 3 wk of age shows RD. ERGs of *rd16/rd16* mice are never normal. The maximal response occurs at 3 wk of age and is nondetectable at 2 mo of age. The inheritance pattern of the *rd16* mutant allele is AR. Linkage studies mapped this new mutation to mouse chromosome 10, in a region between markers *D10Mit96* and *D10Nds2*, suggesting the human homolog may be on chromosome 12q22. The early onset of RD combined with our genetic data suggest that this is a new mutation not previously described in mouse or human. RD 16 (*rd16*) may provide a novel mouse model for pathogenesis of ARRP in humans *(71)*.

SUMMARY

Table 1 summarizes the progression of RD for all twenty-one mutations and lists the available mouse strains carrying each mutation. Although the end point in RD is very similar in appearance among mice with the different mutations, there is richness in variation of onset and progression, offering opportunities for study and experiments to ameliorate the conditions.

REFERENCES

1. Keeler C. The inheritance of a retinal abnormality in white mice. Proc Natl Acad Sci USA 1924;10:329.
2. Keeler C. Retinal degeneration in the mouse is rodless retina. J Hered 1966;57(2):47–50.
3. Pittler SJ, Keeler CE, Sidman RL, Baehr W. PCR analysis of DNA from 70-year-old sections of rodless retina demonstrates identity with the mouse rd defect. Proc Natl Acad Sci USA 1993;90(20):9616–9619.
4. Drager UC, Hubel DH. Studies of visual function and its decay in mice with hereditary retinal degeneration. J Comp Neurol 1978;180(1):85–114.
5. Bowes C, Li T, Danciger M, Baxter LC, Applebury ML, Farber DB. Retinal degeneration in the rd mouse is caused by a defect in the beta subunit of rod cGMP-phosphodiesterase. Nature 1990;347(6294):677–680.
6. Bowes C, Li T, Frankel WN, et al. Localization of a retroviral element within the rd gene coding for the beta subunit of cGMP phosphodiesterase. Proc Natl Acad Sci USA 1993;90(7):2955–2959.
7. Pittler SJ, Baehr W. Identification of a nonsense mutation in the rod photoreceptor cGMP phosphodiesterase beta-subunit gene of the rd mouse. Proc Natl Acad Sci USA 1991;88(19): 8322–8326.
8. Bonhomme F, Guenet J. The laboratory mouse and its wild relatives. In: Lyon MF, Rastan S, and Brown SDM, eds. Genetic variants and strains of the laboratory mouse, 3rd ed., Oxford: Oxford University Press, 1996:1577–1659.

9. Hawes NL, Smith RS, Chang B, Davisson M, Heckenlively JR, John SWM. Mouse fundus photography and angiography: a catalogue of normal and mutant phenotypes. Molecular Vision 1999;5:22.

10. Chang B, Hawes NL, Hurd RE, Davisson MT, Nusinowitz S, Heckenlively JR. Retinal degeneration mutants in the mouse, Vision Res 2002;42(4):517–525.

11. McLaughlin ME, Sandberg MA, Berson EL, Dryja TP. Recessive mutations in the gene encoding the beta-subunit of rod phosphodiesterase in patients with retinitis pigmentosa. Nat Genet 1993;4(2):130–134.

12. McLaughlin ME, Ehrhart TL, Berson EL, Dryja TP. Mutation spectrum of the gene encoding the beta subunit of rod phosphodiesterase among patients with autosomal recessive retinitis pigmentosa. Proc Natl Acad Sci USA 1995;92(8):3249–3253.

13. Takahashi M, Miyoshi H, Verma IM, Gage FH. Rescue from photoreceptor degeneration in the rd mouse by human immunodeficiency virus vector-mediated gene transfer. J Virol 1999;73:7812–7816.

14. Otani A, Dorrell MI, Kinder K, et al. Rescue of retinal degeneration by intravitreally injected adult bone marrow-derived lineage-negative hematopoietic stem cells. J Clin Invest 2004;114(6):765–774.

15. Fernandez-Gonzalez A, La Spada AR, Treadaway J, et al. Purkinje cell degeneration (pcd) phenotypes caused by mutations in the axotomy-induced gene, Nna1. Science 2002;295(5561):1904–1906.

16. Mullen RJ, Eicher EM, Sidman RL. Purkinje cell degeneration, a new neurological mutation in the mouse. Proc Natl Acad Sci USA 1976;73(1):208–212.

17. Blanks JC, Mullen RJ, LaVail MM. Retinal degeneration in the pcd cerebellar mutant mouse II. Electron microscopic analysis. J Comp Neurol 1982;212(3):231–246.

18. LaVail MM, Blanks JC, Mullen RJ. Retinal degeneration in the pcd cerebellar mutant mouse I. Light microscopic and autoradiographic analysis. J Comp Neurol 1982;212(3):217–230.

19. Sidman RL, Green MC. "Nervous," a new mutant mouse with cerebellar disease. In: Sabourdy M, ed. Les Mutants Pathologiques Chez l'Animal. (Paris: Centre National de la Recherche Scientifique, 1970:69–79.

20. Mullen RJ, LaVail MM. Two new types of retinal degeneration in cerebellar mutant mice. Nature 1974;258:528–530.

21. LaVail MM, White MP, Gorrin GM, Yasumura D, Porrello KV, Mullen RJ. Retinal degeneration in the nervous mutant mouse. I. Light microscopic cytopathology and changes in the interphotoreceptor matrix. J Comp Neurol 1993;333(2):168–181.

22. Campbell DB, Hess EJ. Chromosomal localization of the neurological mouse mutations tottering (tg): Purkinje cell degeneration (pcd): and nervous (nr). Brain Res Mol Brain Res 1996;37(1–2):79–84.

23. De Jager PL, Harvey D, Polydorides AD, Zuo J, Heintz N. A high-resolution genetic map of the nervous locus on mouse chromosome 8. Genomics 1998;48(3):346–353.

24. van Nie R, Ivanyi D, Demant P. A new H-2-linked mutation, rds, causing retinal degeneration in the mouse. Tissue Antigens 1978;12(2):106–108.

25. Ma J, Norton JC, Allen AC, et al. Retinal degeneration slow (rds) in mouse results from simple insertion of a haplotype-specific element into protein-coding exon II. Genomics 2,1995; 8(2):212–219.

26. Travis GH, Hepler JE. A medley of retinal dystrophies. Nat Genet 1993;3(3):191–192.

27. Travis GH, Brennan MB, Danielson PE, Kozak CA, Sutcliffe JG. Identification of a photoreceptor-specific mRNA encoded by the gene responsible for retinal degeneration slow (rds). Nature 1989;338(6210):70–73.

28. Carter-Dawson L, Burroughs M. Interphotoreceptor retinoid-binding protein (IRBP) in the postnatal developing rds mutant mouse retina: EM immunocytochemical localization. Exp Eye Res 1989;49(5):829–841.

29. Li L, Sheedlo HJ, Turner JE. Retinal pigment epithelial cell transplants in retinal degeneration slow mice do not rescue photoreceptor cells. Invest Ophthalmol Vis Sci 1993;34(6): 2141–2145.

30. Travis GH, Groshan KR, Lloyd M, Bok D. Complete rescue of photoreceptor dysplasia and degeneration in transgenic retinal degeneration slow (rds) mice. Neuron 1992;9(1): 113–119.

31. Chang B, Heckenlively JR, Hawes NL, Roderick TH. New mouse primary retinal degeneration (rd-3). Genomics 1993;16(1):45–49.

32. Pieke-Dahl S, Ohlemiller KK, McGee J, Walsh EJ, Kimberling WJ. Hearing loss in the RBF/DnJ mouse, a proposed animal model of Usher syndrome type IIa. Hear Res 1997;112(1–2):1–12.

33. Danciger JS, Danciger M, Nusinowitz S, Rickabaugh T, Farber DB. Genetic and physical maps of the mouse rd3 locus, exclusion of the ortholog of USH2A. Mamm Genome 1999;10(7):657–661.

34. Ranta S, Zhang Y, Ross B, et al. The neuronal ceroid lipofuscinoses in human EPMR and mnd mutant mice are associated with mutations in CLN8. Nat Genet 1999;23(2):233–236.

35. Bronson RT, Lake BD, Cook S, Taylor S, Davisson MT. Motor neuron degeneration of mice is a model of neuronal ceroid lipofuscinosis (Batten's disease). Ann Neurol 1993;33(4):381–385.

36. Chang B, Bronson RT, Hawes NL, et al. A retinal degeneration in motor neuron degeneration: a mouse model of ceroid lipofuscinosis. Invest Ophthal Vis Sci 1994;35:1071–1076.

37. Chang B, Bronson RT, Hawes NL, et al. Improved genetic map for the mnd gene, using the retinal degeneration aspect of the phenotype. MGI Direct Data Submission 1998. MGI:86 557.

38. Roderick TH, Chang B, Hawes NL, Heckenlively JR. A new dominant retinal degeneration (Rd4) associated with a chromosomal inversion in the mouse. Genomics 1997;42(3):393–396.

39. Danciger M, Hendrickson J, Rao N, et al. Positional cloning studies of the Rd4 retinal degeneration. Invest Ophthalmol Vis Sci 2000;41(Suppl.):S203 (ARVO-Abstract).

40. Heckenlively JR, Chang B, Erway LC, et al. Mouse model for Usher syndrome: linkage mapping suggests homology to Usher type I reported at human chromosome 11p15. Proc Natl Acad Sci USA 1995;92:11,100–11,104.

41. Noben-Trauth K, Naggert JK, North MA, Nishina PM. A candidate gene for the mouse mutation tubby. Nature 1996;380(6574):534–538.

42. Stubdal H, Lynch CA, Moriarty A, et al. Targeted deletion of the tub mouse obesity gene reveals that tubby is a loss-of-function mutation. Mol Cell Biol 2000;20(3):878–882.

43. Lerner AB. Vitiligo (vit). Mouse News Lett 1986;74:125.

44. Tang M, Neumann PE, Kosaras B, Taylor BA, Sidman RL. Vitiligo maps to mouse chromosome 6 within or close to the mi locus. Mouse Genome 1992;90(3):441–443.

45. Lamoreux ML, Boissy RE, Womack JE, Nordlund JJ. The vit gene maps to the mi (microphthalmia) locus of the laboratory mouse. J Hered 1992;83(6):435–439.

46. Steingrimsson E, Moore KJ, Lamoreux ML, et al. Molecular basis of mouse microphthalmia (mi) mutations helps explain their developmental and phenotypic consequences. Nat Genet 1994;8(3):256–263.

47. Smith SB, McCool DJ. Slow progressive loss of photoreceptor cells in vitiligo mice cosegregates with gradual depigmentation of the pelage. Mouse Genome 1995;93(3):871–873.

48. Smith SB. Evidence of a difference in photoreceptor cell loss in the peripheral versus posterior regions of the vitiligo (C57BL/6J-mi(vit)/mi(vit)) mouse retina. Exp Eye Res 1995;60(3):333–336.

49. Smith SB, Cope BK, McCoy JR. Effects of dark-rearing on the retinal degeneration of the C57BL/6-mivit/mivit mouse. Exp Eye Res 1994;5;8(1):77–84.

50. Chang B, Hageman GS, Heckenlively JR, et al. A mouse model for retinitis punctata albescens: a new pathologic finding for retinal white dots. Invest Ophthalmol Vis Sci 1996;37(Suppl.):S505 (ARVO-Abstract).

51. Kameya S, Hawes NL, Chang B, Heckenlively JR, Naggert JK, Nishina PM. Mfrp, a gene encoding a frizzled related protein, is mutated in the mouse retinal degeneration 6. Hum Mol Genet 2002;11(16):1879–1886.

52. Hawes NL, Chang B, Hageman GS, et al. Retinal degeneration 6 (rd 6): a new mouse model for human retinitis punctata albescens. Invest Ophthalmol Vis Sci 2000;41(10):3149–3157.

53. Chang B, Heckenlively JR, Hawes NL, Davisson MT. A new mouse model of retinal dysplasia and degeneration (rd7). Invest Ophthalmol Vis Sci 1998;39 (Suppl.):S880 (ARVO-Abstract).

54. Akhmedov NB, Pirie NI, Chang B, et al. A deletion in a photoreceptor-specific nuclear receptor mRNA causes retinal degeneration in the rd7 mouse. Proc Natl Acad Sci USA 2000;97:5551–5556.

55. Haider NB, Naggert JK, Nishina PM. Excess cone cell proliferation due to lack of a functional NR2E3 causes retinal dysplasia and degeneration in rd7/rd7 mice. Hum Mol Genet 2001;10(16):1619–1626.

56. Cheng H, Khanna H, Oh EC, Hicks D, Mitton KP, Swaroop A. Photoreceptor-specific nuclear receptor NR2E3 functions as a transcriptional activator in rod photoreceptors. Hum Mol Genet 2004;13(15):1563–1575.

57. Gao H, Boustany RM, Espinola JA, et al. Mutations in a novel CLN6-encoded transmembrane protein cause variant neuronal ceroid lipofuscinosis in man and mouse. Am J Hum Genet 2002;70(2):324–335.

58. Wheeler RB, Sharp JD, Schultz RA, Joslin JM, Williams RE, Mole SE. The gene mutated in variant late-infantile neuronal ceroid lipofuscinosis (CLN6) and in nclf mutant mice encodes a novel predicted transmembrane protein. Am J Hum Genet 2002;70(2):537–542.

59. Bronson RT, Donahue LR, Johnson KR, Tanner A, Lane PW, Faust JR. Neuronal ceroid lipofuscinosis (nclf):a new disorder of the mouse linked to chromosome 9. Am J Med Genet 1998;77(4):289–297.

60. Chang B, Hawes NL, Nishina PM, Smith RS, Davisson MT, Heckenlively JR. Two new mouse models of retinal degeneration (rd8 and Rd9). Invest Ophthalmol Vis Sci 1999;40 (Suppl.):S976 (ARVO-Abstract).

61. Mehalow AK, Kameya S, Smith RS, et al. CRB1 is essential for external limiting membrane integrity and photoreceptor morphogenesis in the mammalian retina. Hum Mol Genet 2003;12(17):2179–2189.

62. den Hollander AI, ten Brink JB, de Kok YJ, et al. Mutations in a human homologue of Drosophila crumbs cause retinitis pigmentosa (RP12). Nat Genet 1999;23(2):217–221.

63. Chang B, Hawes NL, Hurd RE, Davisson MT, Nusinowitz S, Heckenlively JR. A new mouse retinal degeneration (rd10) caused by a missense mutation in exon 13 of the beta-subunit of rod phosphodiesterase gene. Invest Ophthalmol Vis Sci 2000;41 (Suppl):S533 (ARVO-Abstract).

64. Rex TS, Allocca M, Domenici L, et al. Systemic but not intraocular Epo gene transfer protects the retina from light-and genetic-induced degeneration. Mol Ther 2004;10(5):855–861.

65. Hawes NL, Chang B, Hurd RE, Nusinowitz S, Heckenlively JR, Davisson MT. A new mouse model of retinal degeneration (rd11). Invest Ophthalmol Vis Sci 2002;43:ARVO E-Abstract 3669.

66. Chang B, Hawes NL, Hurd RE, Davisson MT, Nusinowitz S, Heckenlively JR. A point mutation in the RPE65 gene causes retinal degeneration (rd12) in mice. Invest Ophthalmol Vis Sci 2002;43:ARVO E-Abstract 3670.

67. Pang J, Chang B, Heckenlively J, et al. Gene therapy restores vision in a natural model of rpe65 leber congenital amaurosis: the rd12 mouse. Invest Ophthalmol Vis Sci 2004;45: ARVO E-Abstract 3486.

68. Buchner DA, Seburn KL, Frankel WN, Meisler MH. Three ENU-induced neurological mutations in the pore loop of sodium channel Scn8a (Na(v)1.6) and a genetically linked retinal mutation, rd13. Mamm Genome 2004;15(5):344–351.

69. Zhang J, Hawes NL, Wang J, et al. A new mouse model of retinal degeneration (rd14). Invest Ophthalmol Vis Sci 2005;46:ARVO E-Abstract 3170.
70. Hawes NL, Hurd RE, Wang J, et al. A new mouse model of retinal degeneration (rd15) with retinal outer plexiform dystrophy. Invest Ophthalmol Vis Sci 2005;46:ARVO E-Abstract 3175.
71. Chang B, Hawes NL, Hurd RE, et al. A new mouse model of retinal degeneration (rd13). Invest Ophthalmol Vis Sci 2005;46:ARVO E-Abstract 3173.

III
MECHANISMS UNDERLYING RETINAL DEGENERATIONS

9

The Impact of Diabetes on Neuronal, Glial, and Vascular Cells of the Retina

Implications for the Pathogenesis of Diabetic Retinopathy

Sylvia B. Smith, PhD

INTRODUCTION

Diabetic retinopathy is the leading cause of blindness in working-aged Americans *(1)*; the seriousness of the disease is underscored by the burgeoning literature in this field. Reviews of pathogenesis and mechanisms of the disease abound. Indeed, a PubMed search of the topic yields more than 14,000 papers dating back to the late 1940s. It is beyond the scope of this chapter to review such a staggering volume of literature and largely unnecessary given the plethora of outstanding reviews on the topic of mechanisms of diabetic retinopathy that have been published in the last decade.

Consequently, the focus of this chapter is on cell types affected in the retina during diabetes and newer models that have shed light on various cell types affected in this disease. In 2002, Gardner and colleagues *(2)* published a review article in which they pointed out that there is a prevalent assumption that diabetic retinopathy is solely a microvascular abnormality. They emphasized that this perception may be the result of the ease with which vascular-associated changes can be observed ophthalmoscopically. They stressed that, in addition to changes in the vascular component of the retina, other classes of cells in the retina (which they divided into neurons, glial cells, and microglia)

From: *Ophthalmology Research: Retinal Degenerations: Biology, Diagnostics, and Therapeutics*
Edited by: J. Tombran-Tink and C. J. Barnstable © Humana Press Inc., Totowa, NJ

are also affected by diabetes. The present chapter expands on this and reviews recent literature about involvement of various retinal cell types in diabetic retinopathy with an emphasis on work published in the last 3–5 yr.

GLIAL CELL INVOLVEMENT IN DIABETIC RETINOPATHY

Müller cells are the chief retinal glial cell *(3)* and play numerous roles in maintaining normal retinal function, including modulating levels of excitatory neurotransmitters in the retina, transporting nutrients and ions, mediating glycogen metabolism, and facilitating aerobic and anaerobic glycolysis. These cells play a crucial role in neuronal survival by providing trophic substances and precursors of neurotransmitters to neurons *(4)*. Müller cells form the primary scaffolding of the retina, spanning the entire thickness of the retina, contacting and ensheathing every type of neuronal cell body and process in the retina. Virtually every disease of the retina is associated with a reactive Müller cell gliosis. Müller cells are the only glial cells in the outer half of the retina, but in the inner portions accessory glial cells are present. These include astrocytes, microglia, and perivascular glia. Müller cells, astrocytes, and perivascular glia are of glioblast origin deriving from the primitive neural tube, whereas microglia are derived from mesoderm. Frequently, studies that examine Müller cells will analyze astrocytes and microglia as well.

Evidence that Müller cells may be involved very early in the pathogenesis of diabetic retinopathy comes from studies by Li and co-workers *(5)*, in which diabetic rats were monitored for changes in retinal function using electroretinograms (ERGs). The b-wave activity that originates from Müller cells is altered as early as 2 wk postonset of diabetes. These observations preceded changes in expression of glial fibrillary acid protein (GFAP), which is associated with severe Müller cell stress. Others have reported structural gliosis as early as 4 wk postonset in the diabetic rat model *(6)*. The expression of GFAP in the diabetic rat model has been reported by several laboratories to be increased by about 3 mo of diabetes. Interesting metabolic studies from Lieth and colleagues reported the increased glial reactivity as well as impaired glutamate metabolism in diabetic rats *(7)*. They observed a marked reduction in the capacity of Müller cells to convert glutamate to glutamine in diabetes and hypothesized that altered glutamate metabolism could lead to elevated glutamate during diabetes. Elevated glutamate levels have been reported in retinas of diabetic rat models *(8)* and in the vitreous of patients with diabetes *(9)*. Significantly, changes in the b-wave have been reported in studies of human patients with diabetic retinopathy as well as elevation of GFAP *(10)*.

The altered levels of glutamate observed in retinas of diabetic humans and rats caused investigators from the Puro laboratory to analyze the function of the sodium-dependent glutamate transporter in Müller cells of diabetic rats *(11)*. Müller cells were freshly isolated from normal and diabetic rat retinas and Müller cell sodium-dependent glutamate transporter activity was monitored using a perforated-patch-clamp technique. As early as 4 wk postonset of diabetes, significant dysfunction of the Müller cell glutamate transporter was observed and by 3 mo its activity was reduced by nearly 70%. Our laboratory has used an in vitro system to study the effects of hyperglycemia on the uptake of radiolabeled glutamate in primary Müller cells isolated from mice. Interestingly, short periods of hyperglycemia alone do not seem sufficient to alter glutamate uptake *(12)*; however, experiments are underway to dissect the myriad factors associated with diabetes that can compromise the function of the EAAT1 (GLAST) transporter in these cells as well as the function of system x_c^-, the sodium-independent glutamate/cystine exchanger. Altered function of glutamate transporters has implications for neuronal toxicity in the retina because of the possible accumulation of glutamate in the extracellular milieu. The intracellular glutamate concentration in neurons, which lack glutamine synthetase (the enzyme that converts glutamate to glutamine), is as high as 10 mM, whereas the extracellular glutamate concentration is in the micromolar range *(13)*. Clearance of extracellular glutamate is a key function of Müller cells and alterations of this function could have significant implications on the function of neuronal cells in the retina.

There have been a number of studies that have evaluated oxidative stress, thought to be a key player in the pathogenesis of diabetic retinopathy *(14)* on Müller cell function *(15–17)*. The development of a rat Müller cell line (rMC-1) by Sarthy and colleagues *(18)* has benefited the field of retinal research considerably, particularly studies of diabetic retinopathy. Culturing the rMC-1 under hyperglycemic conditions led to increased production of nitric oxide (NO), prostaglandin E(2) (PGE[2]), inducible nitric oxide synthase (iNOS), and cyclooxygenase-2 *(15)*. Human subjects with diabetes have also demonstrated increased expression of iNOS *(16)*. Interestingly, the immunoreactivity was associated with Müller cells in retinas of these subjects, suggesting that high levels of NO produced by neural NOS could contribute to neurotoxicity and angiogenesis that occur in diabetic retinopathy. Studies from the Kern laboratory have shown that culturing Müller cells in hyperglycemia leads to increased production of superoxide as well *(17)*.

In addition to elevated levels of glutamate and NO in retinas of patients with diabetes and subsequent effects on Müller cells, Inokuchi and colleagues reported that insulin-like growth factor (IGF)-1 levels were elevated significantly in vitreous obtained from patients with diabetes and suggested that at least one source of IGF-1 was the Müller cell *(19)*. Similar observations have been reported independently from other laboratories *(20)*. The observation is intriguing in light of a 2004 report from Ruberte and co-workers showing that increasing the ocular levels of IGF-1 in a transgenic mouse model leads to diabetes-like alteration in the eye including thickening of the basement membrane of capillaries, microvascular abnormalities, neovascularization, increased GFAP expression, and cataract *(21)*. IGF has been shown to generate tractional forces by Müller cells. Recent studies from the Guidry laboratory *(22,23)* have focused on the role of Müller cells in proliferative diabetic retinopathy (PDR), which is an end-stage complication of diabetes. In PDR, fibrovascular tissues grow into the vitreous and tractional forces originating within these tissues threaten retinal anatomy. Guidry and co-workers *(24)* have data that suggest that Müller cells play a key role in fibrocontractile retinal disorders and that they function as an effector cell type in traction retinal detachment associated with PDR. The group has exploited the ability to isolate pure cultures of Müller cells permitting characterization of altered cell function and changes in proteins. They show that Müller cells have remarkable capacity to alter their shape becoming polygonal and fibroblast-like over a period of weeks and that the myofibroblastic Müller cell phenotype observed in culture is in fact present in human fibrocontractile disorders. The relevance of their findings to human patients with diabetes was demonstrated recently when vitreous samples of patients with diabetes had considerably greater activity in a Müller cell tractional force generation bioassay than patients without diabetes *(24)*.

Given the abundance of data implicating Müller cells in the pathogenesis of diabetic retinopathy, Gerhardinger and co-workers have initiated a systematic assessment of alterations in gene expression in Müller cells as a consequence of diabetes *(25)*. Their work has used the streptozotocin-induced diabetic rat model and focused on changes that occurred 6 mo postonset of diabetes. Their gene expression profile studies identified 78 genes that were differentially expressed in Müller cells isolated from diabetic rats. Notable among these were acute-phase response proteins including α2-macroglobulin, ceruloplasmin, complement components, lipocalin-2, metallothionein, serine protease inhibitor-2, transferrin, tissue inhibitor of metalloproteases-1, transthyretin, and the transcription factor C/EBPδ. The acute-phase response of Müller cells in diabetes was associated with upregulation of interleukin (IL)-1β in the retina suggesting this cytokine as a mediator of the acute-phase response.

Regarding the alterations of microglia in diabetic retinopathy, there have been a few reports in the literature. Rungger-Brandle et al. *(6)* studied diabetic rats from 2 wk postonset of diabetes through 20 wk. They reported not only a significant increase in Müller cells with 4 wk of diabetes, but also an increase in microglia. Interestingly, the number of astrocytes was significantly reduced. Thus, microglial activation with astrocytes regression was an early event in diabetes, which the authors believe may contribute to the onset and development of neuropathy in the diabetic retina. A study by Kuiper and colleagues *(26)* described differential expression of connective tissue grown factor (CTGF) in microglia and pericytes in humans with diabetic retinopathy. CTGF stimulates extracellular matrix formation, fibrosis, and angiogenesis. Immunohistochemical analysis of CTGF expression patterns in human control and diabetic retinas revealed distinct and specific staining of CTGF in microglia of control retinas and a shift to microvascular pericytes in retinas from human subjects with diabetes.

In summary, Müller cells show changes in function early in experimental diabetes as evidenced by alterations in the ERG, altered capacity to transport glutamate, increased oxidative stressors including NO and superoxide, production of IGF-1, and involvement in the tractional forces involved in PDR. A few additional reports have noted changes in microglial and astrocytic cells during diabetes as well.

NEURONAL CELL LOSS IN DIABETIC RETINOPATHY

The involvement of retinal neurons in diabetic retinopathy has gained considerable attention over the past several years, although the first studies documenting the histological loss of neurons in retinas of patients with diabetic retinopathy were actually published in the early 1960s *(27,28)*. The histological observations of neuronal loss have been corroborated by more recent electrophysiological studies of human patients with diabetes in which loss of color and contrast sensitivity was observed within 2 yr of diabetes onset *(29,30)*. Studies using focal ERGs, which detect electrical responses from ganglion and amacrine cells, have revealed dysfunction of these cells early in diabetes *(31–33)*. In the late 1990s Barber and colleagues published important observations of apoptotic neurons in retinas of patients with diabetes compared to controls *(43)*. Similar results were observed by Bek et al. *(35)* and by Kerrigan et al. *(36)*.

GANGLION CELLS

Ganglion cells are among the retinal neurons most studied in diabetic retinopathy. Recently, a comprehensive immunohistochemical analysis of markers of apoptosis was undertaken in retinas of human subjects with diabetes mellitus and compared to retinas of nondiabetic subjects *(37)*. Antibodies against GFAP, caspase-3, Fas, Fas ligand (FasL), Bax, Bcl-2, survivin, p53, extracellular signal-regulated kinases (ERK1/2), and p38 were used in the study. All diabetic retinas showed cytoplasmic immunoreactivity for caspase-3, Fas, and Bax in ganglion cells. The authors concluded that ganglion cells in diabetic retinas express several proapoptosis molecules, suggesting that these cells are the most vulnerable population of cell types in the diabetic retina for apoptosis. Interestingly, they found that glial cells in diabetic retinas are activated and express several antiapoptosis molecules, but also the cytotoxic effector molecule FasL, suggesting a possible role of glial cells in induction of apoptosis in ganglion cells. Given that ganglion cell axons form the nerve fiber layer of the retina, it is perhaps not surprising that a reduction in the retinal nerve fiber layer has been reported in human patients with poorly controlled diabetes *(38,39)*.

The studies of Barber and colleagues *(34)* reported not only ganglion cell apoptosis in retinas of humans with diabetes, but also a 10% reduction in cells of the ganglion cell layer in streptozotocin-induced diabetic rats. They also reported marked reductions in the thickness of the inner plexiform layers (IPLs) and inner nuclear layers (INLs) and a 10-fold increase in the numbers of apoptotic nonvascular cells. The data lend strong support to the notion that diabetic retinopathy has a significant neurodegenerative component. Interestingly, treating the rats with insulin largely prevented this neuronal cell death *(34)*. Others using the streptozotocin rat model of diabetes have confirmed the loss of ganglion cells *(40–47)*. Some of these studies have demonstrated a loss of the ganglion cell axonal fibers in diabetic rat retinas *(41,43,44)*. The studies by Zhang were particularly interesting because they showed impairment of retrograde axonal transport especially in type 1 diabetes, with less impairment in type 2 diabetes *(41)*. The authors speculated that impaired retrograde transport may precede or be a consequence of metabolic dysfunction of the large and medium-sized ganglion cells eventually leading to optic nerve atrophy. Additional electrophysiological experiments in diabetic rats detected reduced ERG responses as early as 2 wk postonset of diabetes *(48)*.

The observations that ganglion cells die in diabetic retinopathy are not restricted to humans and to rats. Mice have been used also in studies of diabetic retinopathy and it is clear that they develop features of diabetic retinopathy. Kern and co-workers have described vascular changes in galactose-induced and streptozotocin-induced mouse models of diabetes *(49)*. Hammes and colleagues have confirmed vascular changes in diabetic mice *(50)*. Mohr et al. reported increased levels of caspase activation, a marker of apoptosis in retinas of diabetic mice *(51)*. Recently, we performed a comprehensive analysis of apoptosis in the retina of streptozotocin-induced diabetic mice *(52)*. The mice were made diabetic at 3 wk and studied over the subsequent 14 wk of diabetes. They were not maintained on insulin. The eyes were subjected to morphometric analysis and detection of apoptotic cells by TUNEL analysis, activated caspase-3 and electron microscopic analysis of the ultrastructural features of apoptosis. The morphometric analysis of retinal cross sections of mice that had been diabetic 14 wk showed approx

20 to 25% fewer cells in the ganglion cell layer compared with age-matched control mice. There was a modest, but significant, decrease in the thickness of the whole retina and the INLs and outer nuclear layers (ONLs) in mice that had been diabetic for 10 wk. TUNEL analysis and detection of active caspase-3 revealed that cells of the ganglion cell layer were dying by apoptosis. Electron microscopic analysis detected morphologic features characteristic of apoptosis, including margination of chromatin and crenated nuclei of cells in the ganglion cell layer. These data suggest that as with the diabetic rat, neurons in the ganglion cell layer of diabetic mouse retinas die and this death occurs through an apoptotic pathway. The observation is an important one given the abundance of transgenic mice available that can permit analysis of various genetic contributions to diabetic retinopathy. It is probably not surprising that ganglion cells are affected in diabetic mice given the findings of Kowluru *(53)* that metabolic changes in retinas of diabetic mice were quite similar to those of diabetic rats. Both species showed increased oxidative stress, protein kinase C (PKC) activity, and NO levels in the retina. To determine whether other factors associated with diabetes can exaggerate the neuronal cell death, we have exploited a mouse model with a defect in the cystathionine-β-synthase gene, which has an elevation of plasma homocysteine. Elevated levels of homocysteine have been reported in patients with diabetic retinopathy *(54,55)*. Preliminary studies suggest that inducing diabetes in these mice leads to increased death of cells in the ganglion cell layer and alterations of the nerve fiber layer *(56)*.

Ganglion cell loss has been reported in other mouse models of diabetes. One of the most promising models to be characterized recently is the spontaneously diabetic Ins2^{Akita} mouse *(57)*. This mouse carries a spontaneous mutation of one insulin gene and heterozygous mice develop diabetes within 4 to 5 wk after birth. The mice have been studied through 28 wk morphometrically and by TUNEL analysis. The morphometric analysis revealed approx 23% fewer cells in the ganglion cell layer of affected mice, in addition to significant decreases in the thickness of the IPL and the INL. There is increased apoptosis and accelerated loss of neurons in the retina of Ins2^{Akita} mice. The investigators contend that these changes parallel and extend observations of neuronal degeneration in streptozotocin-diabetic rats and postmortem tissue from humans with diabetes and may explain the loss of vision in diabetic retinopathy.

Apoptosis of cells in the ganglion cell layer has been reported also in a mouse model for type 2 diabetes. Ning and colleagues *(58)* investigated neuroretinal apoptosis in the KKAY mouse and reported significantly more TUNEL positive cells in the ganglion cell layer than in control mice. Many apoptotic cells were observed in the inner portion of the INL as well. The investigators also reported microangiopathy in early stage of diabetes.

A number of proteins, thought to be involved in the pathogenesis of diabetic retinopathy, especially the angiogenic component, have been investigated and several of these proteins have been localized to the retinal ganglion cell layer. Included among this group is IGF-1, a growth-promoting polypeptide that can act as an angiogenic agent in the eye. IGF-1 is thought to play a central role in the pathogenesis of proliferative diabetic retinopathy. Interestingly messenger RNA (mRNA) encoding IGF-1 and its receptor is expressed throughout the neural retina, including retinal ganglion cells, as well as vascular, Müller and retinal pigment epithelial (RPE) cells *(59)*. The angiopoietin (Ang)/Tie-2 system, also thought to play a role in vascular integrity and angiogenesis, has been analyzed in diabetic retinas *(60)*. Molecular methods were used to study the expression of Ang-1, Ang-2, and Tie-2 in the retinas of streptozotocin-induced diabetic rats. Results of *in situ* hybridization analysis revealed Ang-1, Ang-2, and Tie-2 mRNA expression in the ganglion cell layer and the INL. In particular, Ang-2 mRNA expression increased in the ganglion cell layer and was accompanied by a reduction of α smooth-muscle actin-positive perivascular cells. These changes may suggest a role for Ang-2 in the mechanism of pericyte loss in diabetic retinopathy. CTGF has been postulated to have prosclerotic and angiogenic properties and has been characterized also in the diabetic rat model *(61)*. In comprehensive molecular studies, diabetes was associated with a greater than twofold increase in CTGF mRNA levels and the major site of CTGF gene and protein expression in the retina of diabetic rats was the ganglion cell layer. The authors theorized that ciliary neurotrophic factor (CNTF) plays a key role in mediating diabetes-associated retinal pathology. Platelet-derived growth factor (PDGF) has been studied in diabetic retina and found to be expressed in the ganglion cell layer *(62)*. Until recently, the source of PDGF was not known, but studies from the Gardiner group have confirmed ganglion cells as the principal source of PDGF. *In situ* hybridization studies using streptozotocin diabetic mice demonstrated that PDGF-A and -B were predominantly expressed by the retinal ganglion cells/nerve fiber layer in both normal and diabetic mice, and this localization pattern did not alter in diabetes. Interestingly, there was a significant decrease of PDGF-B mRNA levels in diabetic retina when compared to nondiabetic controls, which may have significant implications for the vascular complications of diabetic retinopathy. Expression of vascular endothelial growth factor (VEGF), implicated in retinal neovascularization, is upregulated in ganglion cells, the IPL and cells of the INL in studies of retinas of human patients with diabetic retinopathy *(63)*. Others have demonstrated that advanced glycation end products (AGEs) can increase VEGF mRNA levels in the ganglion, INLs and RPE cell layers of the rat retina *(64)*. The data lend support to a role for AGEs in the pathogenesis of diabetic retinopathy through their ability to increase retinal VEGF gene expression.

In addition to proteins associated with angiogenesis, other groups have studied various proteins that might be linked with diabetes. Among these studies is work showing that glutamate receptors and calcium binding proteins are increased in diabetes *(65)*. The vulnerability of neuronal cells to excessive levels of glutamate, coupled with the observation by several groups that glutamate levels may be increased in diabetes *(7–9)* prompted the study of expression of glutamate receptors and calcium-binding proteins in streptozotocin-induced diabetic rats. Upregulation of glutamate receptors (*N*-methyl-D-aspartate receptor [NMDAR]1 and GluR2/3) was observed in the ganglion, amacrine, and bipolar cells as well as in the IPL and outer plexiform layer at 1 mo of diabetes and was further enhanced at 4 mo. Immunoreactivity of calcium-binding proteins (calbindin and parvalbumin) was also increased. The data suggest changes on glutamate and calcium metabolism in the diabetic retina.

Owing to observations of ganglion cell death in diabetes, our group has been interested in neuroprotective proteins that are expressed in diabetes. Among these we have analyzed the expression of the type 1 sigma receptor (σR1) in retinas of diabetic mice *(66)*. σR1 is a nonopiate and nonphencyclidine binding site that has numerous pharmacological and physiological functions. In particular, agonists for σR1 have been shown to afford neuroprotection against overstimulation of the NMDAR. We analyzed whether ganglion cells cultured under hyperglycemic conditions and ganglion cells of diabetic mice continue to express σR1. Reverse-transcriptase (RT)-polymerase chain reaction (PCR), *in situ* hybridization, Western blot analysis, and immunolocalization studies confirmed that ganglion cells cultured under hyperglycemic conditions continue to express σR1 as do ganglion cells of diabetic mice. Studies are currently underway to determine whether ligands for σR1 may prove useful in inhibiting the apoptosis observed in diabetic retinopathy.

AMACRINE, BIPOLAR, AND HORIZONTAL CELLS

The cell type most studied within the INL of the retina during diabetes is the Müller cell. The three neuronal cell types of this layer, amacrine, bipolar, and horizontal have not been investigated extensively in diabetic retinopathy. This may be in part because of the fact that Müller cells have distinctive radial fibers and are also positive for GFAP and other markers allowing them to be analyzed more easily. There are known markers for bipolar, amacrine, and horizontal cells now available and as investigators use these it will be possible to determine how extensively these neuronal cell types are involved in diabetes. There certainly have been studies reporting that cells of the INL undergo apoptosis *(34)* and reporting that the thickness of this layer is reduced in diabetic retinopathy *(34,52,57,67)*. Further, reports that the IPL, which is the synaptic layer between the neurons of the INL and the ganglion cell layer, is reduced in diabetic retinopathy *(34)*, are highly suggestive of involvement of neurons in the INL during the disease.

That said, the literature in this area is sparse compared to that for Müller or ganglion cells. Studies of the diabetic rat models have revealed a 22 and 14% reduction in the thickness of the IPL and INL, respectively *(34)*. This study did not examine which cell types of the INL might have been lost during the disease. Studies with the Ins2[Akita] mice revealed a reduction of the IPL of 17 and 27% in the central vs peripheral retinas, respectively *(57)*. Additionally, the INL was reduced by 16% in these diabetic mice. In this study, a marker of starburst amacrine cells, choline acetyltransferase, revealed significantly fewer subtypes of this amacrine cell in the diabetic mice than controls. Studies from streptozotocin-induced diabetic mice showed a reduced thickness of the INL, though specific cell types that were lost were not reported *(52)*. Seki and colleagues reported that dopaminergic amacrine cells degenerate in diabetic rat retinas, as revealed by reduction in tyrosine hydroxylase immunoreactivity *(68)*. The levels of this protein in diabetic retinas were decreased to one-half of controls. Sophisticated electrophysiological studies of diabetic rat retinas by Kaneko and coworkers evaluated the scotopic threshold response and the oscillatory potential of the ERG *(69)*. Their data showed a significantly prolonged mean latency of the oscillatory potential, which reflects altered function of dopaminergic amacrine cells. These data suggest disordered neurotransmission of the amacrine cells in the inner retinal layers of diabetic rats. In studies by Park and colleagues *(67)*, postsynaptic processes of horizontal cells in the deep invaginations of the photoreceptors of streptozotocin diabetic rats showed degenerative changes within 1 wk of diabetes onset. Some amacrine cells and a few horizontal cells showed necrotic features at 12 wk postonset of diabetes. Agardh and co-workers have observed a decrease in the numbers of horizontal cells and decreased branching of their terminals in diabetic rats *(70)*. Studies of proteins whose expression might be altered in neurons of the INL in diabetic retinopathy have been quite limited, although Ng and colleagues showed that amacrine and bipolar cells alter their expression of glutamate receptors and calcium-binding proteins *(71)*.

PHOTORECEPTOR CELLS

Alterations in photoreceptor cells have been reported in the diabetic retina. In a 2003 study using streptozotocin-induced diabetic rats, which were not maintained on insulin, Park et al. *(67)* reported a remarkable reduction in the thickness of the ONL 24 wk after the onset of diabetes. TUNEL analysis of these retinas showed some apoptotic nuclei at 4 wk postonset of diabetes and this number increased as diabetes progressed. The authors suggest that visual loss associated with diabetic retinopathy could be attributed to an early phase of substantial photoreceptor loss, in addition to later microangiopathy. Nork used histological methods to distinguish L/M cones from S-cones in humans with diabetic retinopathy and found "selective and widespread" loss of the S-cones was found in diabetic retinopathy *(72)*. The investigator concluded that in diabetic retinopathy the acquired tritan-like color vision loss could be caused, or contributed to, by selective loss of the S-cones. Landenranta et al. *(73)* point out that PDR is rarely associated clinically with retinitis pigmentosa (RP). They note than humans with diabetic retinopathy demonstrate a spontaneous regression of retinal neovascularization associated with long-standing diabetes mellitus when RP becomes clinically evident. They test this theory using the rd mouse (rd1/rd1), which has a marked reduction of photoreceptor cells within the first 3 wk of postnatal life. When the rd mice are subjected to conditions of oxygen-induced proliferative retinopathy they fail to mount reactive retinal neovascularization. Their data would suggest that marked photoreceptor cell loss, characteristic of RP is incompatible with PDR.

RETINAL PIGMENT EPITHELIUM

The RPE is a remarkable cell type with numerous critical functions that maintain the health and integrity of the retina, particularly the adjacent photoreceptor cells. It not only plays a key role in retinoid metabolism and the regeneration of rhodopsin, it also phagocytoses shed outer segment photoreceptor disks, contains pigment granules thought to reduce light scattering, and mediates the transport of numerous nutrients. Owing to tight junctions between cells, the RPE contributes to the blood retinal barrier. Glucose transport across this barrier is mediated by the sodium-independent glucose transporter GLUT1. Alterations in the blood–retinal barrier may affect glucose delivery to the retina and may have major implications in the development of two major diabetic complications, namely insulin-induced hypoglycemia and diabetic retinopathy *(74)*. Thus, it is not surprising that studies of the role of RPE, especially during proliferative stages have been reported. In addition to the glucose transporter, another transporter has been examined in cultured RPE and in retina/RPE of diabetic mice. This work, from our laboratory, explored the regulation of a transporter for the essential vitamin folate under diabetic conditions. Folate is essential for synthesis of RNA, DNA, and proteins and is thus critical for cell survival. Our lab characterized the polarized distribution of two folate transport proteins in RPE *(75)* and then studied the regulation of reduced-folate transporter (RFT-1), under hyperglycemic conditions *(76)*. Exposure of RPE cells to 45 m*M* glucose for as short an incubation time as 6 h resulted in a 35% decrease in methyltetrahydrofolate (MTF) uptake. Kinetic analysis showed that the hyperglycemia-induced attenuation was associated with a decrease in the maximal velocity of the transporter with no significant change in the substrate affinity. Semiquantitative RT-PCR demonstrated that the mRNA encoding RFT-1 was significantly decreased in cells exposed to high glucose, and Western blot analysis showed a significant decrease in protein levels. The uptake of (^{3}H)-MTF in RPE of diabetic mice was reduced by approx 20%, compared with that in nondiabetic, age-matched control animals. Semiquantitative RT-PCR demonstrated that the mRNA encoding RFT-1 was decreased significantly in RPE of diabetic mice. These findings demonstrated for the first time that hyperglycemic conditions reduce the expression and activity of RFT-1 and may have profound implications for the transport of folate by RPE in diabetes. Stevens and colleagues reported a similar finding of downregulation of the human taurine transporter by glucose in cultured RPE cells *(77)*. The implications for decreased taurine are potentially serious owing to the apparent role of this amino acid as an osmolyte and as an antioxidant.

Other laboratories have examined various proteins in RPE during diabetes to attempt to understand the role of RPE in the pathogenesis of diabetic retinopathy. Rollin et al. *(78)* examined atrial natriuretic peptide (ANP), an endogenous inhibitor of the synthesis and angiogenic action of VEGF, in plasma and vitreous humor of human subjects with PDR. They found that vitreous ANP concentrations were significantly higher in patients with active PDR compared to patients with quiescent PDR, diabetes without PDR, or controls. ANP was detected in the fibrovascular epiretinal tissue of patients with PDR. They concluded that patients with diabetes with active neovascularization have significantly higher levels of ANP in the vitreous humor than those without active PDR. Patients with diabetes without PDR were also found to have significantly higher vitreous ANP levels than nondiabetic patients. In the fibrovascular epiretinal tissue of these patients, ANP was localized to vascular, glial, fibroblast-like, and RPE cells.

One of the treatments used commonly in PDR is thermal laser photocoagulation. Framme and Roider *(79)* used fundus autofluorescence imaging to study the RPE of nearly 190 human subjects who had undergone the laser photocoagulation. Initially, autofluorescence decreases in the area of laser lesions 1 h after laser treatment, but more than 1 mo posttreatment, it increases. The authors point out the usefulness of this method in evaluating RPE destruction and subsequent proliferation after continuous wave laser photocoagulation. A zone of atrophy noted as dark spots can signify denaturation of neurosensory retinal tissue. Further studies *(80)* suggest that autofluorescence imaging may replace invasive fluorescein angiography in many cases to verify therapeutic laser success and reveal extent of RPE damage.

Bensaoula and Ottlecz have evaluated the RPE of the diabetic rats using electron microscopic methods *(81)*. Their ultrastructural studies revealed a significant deepening of basal infoldings in the RPE and a noticeable increase in the size of the extracellular space between the basal infoldings of 5-mo streptozotocin-induced diabetic rats. They suggest that this increase extracellular space may contribute to increased alteration of the RPE barrier function in progressive diabetes.

ENDOTHELIAL CELLS

The retina receives blood from two sources: the choroid, which supplies the outer retina, and the central retinal artery, which ramifies on the inner region of the retina. Within the vasculature of the inner retina, the capillary network is distributed in a superficial network within the nerve fiber layer and an intraretinal network at the level of the INL. The capillaries of the INL tend to be more involved in diabetes. Two cell types are found in the capillaries. The endothelial cells, the cytoplasm of which makes up the lining or wall of the capillaries, and the pericytes, which are on the surface of the capillaries. The effects of diabetes on protein expression and function of retinal endothelial cells have been studied extensively. This review highlights some of the more recent reports (within the last 3 yr) of the effects of diabetes on retinal endothelial cells.

In an effort to understand the pathogenesis of diabetic retinopathy, especially the proliferative phase, a number of investigations have been performed to determine whether hyperglycemia induces potentially damaging molecules in retinal endothelial cells. Members of the Kern laboratory reported that hyperglycemia increases NO and PGE(2) in endothelial cells. They believe that this increase in these molecules contributes to the pathogenesis of diabetic retinopathy *(15,82)*. Additional studies from this group examined the role of poly(ADP-ribose) polymerase (PARP) in the development of diabetic retinopathy *(83)*. Activity of PARP increased in whole retina and in endothelial cells and pericytes of diabetic rats. They found that inhibition of PARP or nuclear factor (NF)-κB inhibited the hyperglycemia-induced cell death in retinal endothelial cells. Thus, PARP activation plays an important role in the diabetes-induced death of retinal capillary cells, at least in part via its regulation of NF-κB. Similar results were obtained by Kowluru and co workers *(84)*, who demonstrated that high glucose activates NF-κB and elevates NO and lipid peroxides in both retinal endothelial cells and pericytes. They further reported that the effects could be inhibited by antioxidants. Additional studies from the Kowluru laboratory examined the role of advanced AGEs in accelerating retinal capillary cell death under in vitro conditions using bovine retinal endothelial cells and pericytes *(85)*. Her studies incubating these cells with AGE-bovine serum albumin showed a 60% increase in oxidative stress and NO. She proposed that inhibiting AGEs in the retinal capillary cells could prevent their apoptosis, and perhaps ultimately the development of retinopathy in diabetes. Other studies from this laboratory have examined the cytokine IL-1β and have determined that it too accelerates apoptosis of retinal capillary cells via activation of NF-κB *(86)*, and showed that the process is exacerbated in high-glucose conditions.

Interesting studies published in 2005 by members of the Grant and Scott laboratories *(87)* have shown that stromal cell-derived factor (SDF)-1 increases as PDR progresses and it induces human retinal endothelial cells to increase expression of vascular cell adhesion molecule-1, a receptor for very late antigen-4 found on many hematopoietic progenitors. They found also that SDF-1 reduced tight cellular junctions by reducing occludin expression. The investigators used a mouse model of proliferative retinopathy and found that the new vessels that developed originated from hematopoietic stem cell-derived endothelial progenitor cells. SDF-1, when injected intravitreally in mice can induce retinopathy. The authors theorized that SDF-1 plays a major role in proliferative retinopathy and may be an ideal target for the prevention of proliferative retinopathy.

Kondo and colleagues have also exploited mice to study the pathogenesis of neovascularization. They used mice in which the insulin receptor or the IGF-1 receptor was eliminated from endothelial cells and found decreased neovascularization in these experimental animals *(88,89)*. They conclude that insulin and IGF-1 signaling in endothelium play a role in retinal neovascularization through the expression of vascular mediators, with the effect of insulin being most important in this process.

Other studies have focused on the expression and function of the sodium dependent glucose transporter, GLUT1 in endothelials cells. Fernandes et al. *(90)* studied GLUT1 in endothelial cells cultured under conditions of experimental diabetes and reported that levels were decreased in the in vivo models and in the retinal endothelial cells exposed to elevated glucose concentrations. The decreased abundance of GLUT1 may be associated with its increased degradation by an ubiquitin-dependent mechanism. Electron microscopic immunohistological methods have also been used to study GLUT1 transporter in retinas of diabetic rats *(91)*, but no difference in immunoreactivity for GLUT1 was observed between control and diabetic retinas. In contrast, Zhang et al. *(92)* reported that high glucose actually increased GLUT1-mediated glucose transport in cultured retinal endothelial cells.

Interesting studies have been reported recently about the effects that leptin, an adipocyte-derived hormone, might have on endothelial proliferation and angiogenesis *(93)*. Suganami and colleagues used a mouse model of ischemia and demonstrated more pronounced retinal neovascularization in 17-d-old transgenic mice overexpressing leptin than in age-matched wild-type littermates. Interestingly, the ischemia-induced retinal neovascularization was markedly suppressed in 17-d-old leptin-deficient ob/ob mice. This suggests that leptin can stimulate ischemia-induced retinal neovasucularization possibly through the upregulation of endothelial VEGF, thereby suggesting that leptin antagonism may offer a novel therapeutic strategy to prevent or treat diabetic retinopathy. Chibber and co-workers have demonstrated that PKC β2-dependent phosphorylation of core 2 GlcNAc-T promotes leukocyte-endothelial cell adhesion *(94)*. They suggest that this may represent a novel regulatory mechanism in mediating increased leukocyte-endothelial cell adhesion and capillary occlusion in diabetic retinopathy.

Recent studies have reported that oncofetal fibronectin may play a role in the proliferation of endothelial cells during diabetic retinopathy. Khan et al. studied the expression of oncofetal fibronectin variants in human vitreous samples obtained from patients undergoing vitrectomy for PDR *(95)*. They also used an animal model of diabetes and cultured endothelial cells in their work and showed that diabetes-induced upregulation of oncofetal fibronectin is, in part, dependent on hyperglycemia-induced transforming growth factor-β1 and endothelin-1. Their data may suggest that oncofetal fibronectin is involved in endothelial cell proliferation.

An interesting comparative study by Grammas and Riden *(96)* showed that retinal endothelial cells are more susceptible to oxidative stress and increased permeability than brain-derived endothelial cells. They reported that retinal endothelial cells release higher levels of superoxide, have less glutathione peroxidase activity, and lower levels of superoxide dismutase and ZO-1 than brain-derived endothelial cells. Unlike brain-derived endothelial cells in which ZO-1 levels increase in response to glucose, in retinal endothelial cells, ZO-1 levels are unaffected by glucose. These findings suggest that greater oxidative stress and lower-junctional protein levels in retinal endothelial cells may contribute to blood/retinal barrier dysfunction in diabetic retinopathy.

Hammes and colleagues *(97)* have studied the role of pericyte function in control of endothelial cell proliferation. They used PDGF-B (PDGF-B[+/−] mice) and studied pericyte numbers. They showed that retinal capillary coverage with pericytes is crucial for the survival of endothelial cells, particularly under stress conditions such as diabetes. When VEGF levels are high, as in diabetic retinopathy, pericyte deficiency may contribute to endothelial proliferation. Others have reported that the adhesive characteristics of pericytes are impaired when they are cultured with endothelial cells that have been exposed to high-hexose concentrations *(98)*. These data suggest considerable interactions of endothelial cells and adjacent pericytes in maintaining capillary integrity during diabetes.

Thus, the studies of endothelial cells and their associated pericytes have focused primarily on various factors that induce their death or their proliferation during diabetes. The availability of the bovine retinal endothelial cell line has facilitated these studies in much the same way that availability of the Müller cell line and isolated primary Müller cells has made the dissection of the role of that cell type possible in diabetes.

SUMMARY

It is clear from the literature reviewed above that every cell type in the retina has been reported to be affected to some degree in diabetic retinopathy. Certain cell types, such as endothelial cells, are obvious cell types to study owing to the dramatic vascular changes that appear particularly later in diabetes. The availability of endothelial cells for in vitro studies has made the analysis of these cells very feasible. Similarly, the availability of a Müller cell line *(18)*, a ganglion cell line (RGC-5) *(99)*, and a RPE cell line (ARPE-19) *(100)* make studies of the effects of hyperglycemia on these cell types not only very doable, but likely to yield informative results. Other cell types such as amacrine, horizontal, bipolar have not been cultured to purity and no cell lines have yet been developed to ask similar questions about them. The recently available antibody markers for these cell types should yield more data about the consequences of diabetes on their function. A few studies of photoreceptor cells in diabetes have provided intriguing results that deserve repetition. The availability of the 661W photoreceptor cell line *(101)* will permit more thorough investigation of the susceptibility of these cells to hyperglycemia.

The importance of the availability of these cells for study and of various animal models for analysis is the information they may be able to provide regarding the pathogenesis of diabetes. Which cell type or types is/are the first to be affected by diabetes is not known. Many studies have focused on the long-term consequences of diabetes, but fewer studies have probed in models of diabetes precisely which cells are affected at the outset of diabetes. An alternative approach could be to determine which genes encoding which proteins are the first to be affected in diabetes. The field of diabetes research is already benefiting from data generated heretofore about dysfunction of various cells and it is predicted that sophisticated molecular and cell biological methods will lead to important discoveries in the next few years.

ACKNOWLEDGMENT

I thank those present and former members of my laboratory, H. Naggar, P. Martin, B. Mysona, T. Van Ells, S. Ola, A. El-Sherbeny, and P. Roon, who worked diligently in trying to understand the consequences of diabetes on various functions of several retinal cell types. I also thank my colleague and collaborator, Dr. Vadivel Ganapathy, for the many useful discussions over the past several years concerning diabetic retinopathy. Studies from our laboratory described in this review were supported by National Institutes of Health, National Eye Institute: EY014560, EY12830, and EY13089.

REFERENCES

1. Klein R, Klein BEK. Vision Disorders in Diabetes. National Diabetes Data Group Diabetes in America. NIH, Bethesda, MD pp. 293–337. NIH publication no 95–1468, 1995.
2. Gardner TW, Antonetti DA, Barber AJ, LaNoue KF, Levison SW. Diabetic retinopathy. More than meets the eye. Surv Ophthalmol 2002;47:S253–S262.
3. Bringmann A, Reichenbach A. Role of Müller cells in retinal degenerations. Front Bio Sci 2001;6:E72–E92.
4. Poitry S, Poitry-Yamate C, Ueberfeld J, MacLeish PR, Tsacopoulos M. Mechanisms of glutamate metabolic signaling in retinal glial (Müller) cells. J Neurosci 2000;20:1809–1821.
5. Li Q, Zemel E, Miller B, Perlman I. Early retinal damage in experimental diabetes: electroretinographical and morphological observations. Exp Eye Res 2002;74:615–625.
6. Rungger-Brandle E, Dosso AA, Leuenberger PM. Glial reactivity, an early feature of diabetic retinopathy. Invest. Ophthalmol Vis Sci 2000;41:1971–1980.
7. Lieth E, Barber AJ, Xu B, et al. Glial reactivity and impaired glutamate metabolism in short-term experimental diabetic retinopathy. Diabetes 1998;47:815–820.
8. Kowluru RA, Engerman RL, Case GL, Kern TS. Retinal glutamate in diabetes and effect of antioxidants. Neurochem Int 2001;38:385–390.
9. Ambati J, Chalam KV, Chawla DK, et al. Elevated gamma-aminobutyric acid, glutamate, and vascular endothelial growth factor levels in the vitreous of patients with proliferative diabetic retinopathy. Arch Ophthalmol 1997;115:1161–1166.
10. Mizutani M, Gerhardinger C, Lorenzi M. Müller cell changes in human diabetic retinopathy. Diabetes 1998;47:445–449.
11. Li Q, Puro DG. Diabetes induced dysfunction of the glutamate transporter in retinal Muller cells. Invest Ophthalmol Vis Sci 2002;43:3109–3116.
12. Mysona BA, Rankin D, Van Ells TK, Ganapathy V, Smith SB. Effects of glucose and insulin on glast and x_c^- transporter function in primary mouse Müller cells. Invest Ophthalmol Vis Sci 2005;46 Abstract #2967.
13. Trotti D, Rossi D, Gjesdal O, et al. Peroxynitrite inhibits glutamate transporter subtypes. J Biol Chem 1996;271:5976–5979.
14. van Reyk DM, Gillies MC, Davies MJ. The retina: oxidative stress and diabetes. Redox Rep 2003;8:187–192.
15. Du Y, Sarthy VP, Kern TS. Interaction between NO and COX pathways in retinal cells exposed to elevated glucose and retina of diabetic rats. Amer J Physiol Reg Integ Comp Phys 2004;287:R735–R741.

16. Abu El-Asrar AM, Desmet S, Meersschaert A, Dralands L, Missotten L, Geboes K. Expression of the inducible isoform of nitric oxide synthase in the retinas of human subjects with diabetes mellitus. Amer J Ophthalmol 2001;132:551–556.

17. Du Y, Miller CM, Kern TS. Hyperglycemia increases mitochondrial superoxide in retina and retinal cells. Free Radical Biol Medicine 2003;35:1491–1499.

18. Sarthy VP, Brodjian SJ, Dutt K, Kennedy BN, French RP, Crabb JW. Establishment and characterization of a retinal Müller cell line. Invest Ophthalmol Vis Sci 1998;39:212–216.

19. Inokuchi N, Ikeda T, Imamura Y, et al. Vitreous levels of insulin-like growth factor-I in patients with proliferative diabetic retinopathy. Curr Eye Res 2001;23:368–371.

20. Guidry C, Feist R, Morris R, Hardwick CW. Changes in IGF activities in human diabetic vitreous. Diabetes 2004;53:2428–2435.

21. Ruberte J, Ayuso E, Navarro M, et al. Increased ocular levels of IGF-1 in transgenic mice lead to diabetes-like eye disease. J Clin Invest 2004;13:1149–1157.

22. King JL, Guidry C. Müller cell production of insulin-like growth factor-binding proteins in vitro: modulation with phenotype and growth factor stimulation. Invest Ophthalmol Vis Sci 2004;45:4535–4542.

23. Guidry C. The role of Müller cells in fibrocontractive retinal disorders. Prog Retin Eye Res 2005;24:75–86.

24. Guidry C, Bradley KM, King JL. Tractional force generation by human Müller cells growth factor responsiveness and integrin receptor involvement. Invest Ophthalmol Vis Sci 2003;44:1355–1363.

25. Gerhardinger C, Costa MB, Coulombe MC, Toth I, Hoehn T, Grosu P. Expression of acute-phase response proteins in retinal Müller cells in Diabetes Invest Ophthalmol Vis Sci 2005;46:349–357.

26. Kuiper EJ, Witmer AN, Klaassen I, Oliver N, Goldschmeding R, Schlingemann RO. Differential expression of connective tissue growth factor in microglia and pericytes in the human diabetic retina. Br J Ophthalmol 2004;88:1082–1087.

27. Wolter JR. Diabetic retinopathy. Am J Ophthalmol 1961;51:1123–1139.

28. Bloodworth JMB. Diabetic retinopathy. Diabetes 1962;2:1–22.

29. Roy M, Gunkel R, Podgor M. Color vision defects in early diabetic retinopathy. Arch Ophthalmol 1986;104:225–228.

30. Hirsh J, Puklin J. Reduced contrast sensitivity may precede clinically observable retinopathy in type 1 diabetes. In: Henkind P, ed. Acta XXIV International Congress of Ophthalmology. New York: Lippincott, 1982:719–724.

31. Greco AV, Di Leo MA, Caputo S, et al. Early selective neuroretinal disorder in prepubertal type 1 (insulin-dependent) diabetic children without microvascular abnormalities. Acta Diabetol 1994;31:98–102.

32. Ghirlanda G, Di Leo MA, Caputo S, et al. Detection of inner retina dysfunction by steady-state focal electroretinogram pattern and flicker in early IDDM. Diabetes 1991;40:1122–1127.

33. Shimada Y, Li Y, Bearse MA Jr, Sutter EE, Fung W. Assessment of early retinal changes in diabetes using a new multifocal ERG protocol. Br J Ophthalmol 2001;85:414–419.

34. Barber AJ, Lieth E, Khin SA, Antonetti DA, Buchanan AG, Gardner TW. Neural apoptosis in the retina during experimental and human diabetes: early onset and effect of insulin. J Clin Invest 1998;102:783–791.

35. Bek T. Transretinal histopathological changes in capillary-free areas of diabetic retinopathy. Acta Ophthalmol 1994;72:409–415.

36. Kerrigan LA, Zack DJ, Quigley HA, Smith SD, Pease ME. TUNEL-positive ganglion cells in human primary open-angle glaucoma. Arch Ophthalmol 1997;115:1031–1035.

37. Abu-El-Asrar AM, Dralands L, Missotten L, Al-Jadaan IA, Geboes K. Expression of apoptosis markers in the retinas of human subjects with diabetes. Invest Ophthalmol Vis Sci 2004;45:2760–2766.

38. Lonneville YH, Ozdek SC, Onol M, Yetkin I, Gurelik G, Hasanreisoglu B. The effect of blood glucose regulation on retinal nerve fiber layer thickness in diabetic patients. Ophthalmologica 2003;217:347–350.
39. Ozdek S, Lonneville YH, Onol M, Yetkin I, Hasanreisoglu BB. Assessment of nerve fiber layer in diabetic patients with scanning laser polarimetry. Eye 2002;16:761–765.
40. Asnaghi V, Gerhardinger C, Hoehn T, Adeboje A, Lorenzi M. A role for the polyol pathway in the early neuroretinal apoptosis and glial changes induced by diabetes in the rat. Diabetes 2003;52:506–511.
41. Zhang L, Inoue M, Dong K, Yamamoto M. Retrograde axonal transport impairment of large- and medium-sized retinal ganglion cells in diabetic rat. Curr Eye Res 2000;20:131–136.
42. Lieth E, Gardner TW, Barber AJ, Antonetti DA. Retinal neurodegeneration: early pathology in diabetes. Clin Experiment Ophthalmol 2000;28:3–8.
43. Scott TM, Foote J, Peat B, Galway G. Vascular and neural changes in the rat optic nerve following induction of diabetes with streptozotocin. J Anat 1986;144:145–152.
44. Chihara E, Matsuoka T, Ogura Y, Matsumura M. Retinal nerve fiber layer defect as an early manifestation of diabetic retinopathy. Ophthalmology 1993;100:1147–1151.
45. Hammes HP, Federoff HJ, Brownlee M. Nerve growth factor prevents both neuroretinal programmed cell death and capillary pathology in experimental diabetes. Mol Med 1997;1:527–534.
46. Sima AA, Zhang WX, Cherian PV, Chakrabarti S. Impaired visual evoked potential and primary axonopathy of the optic nerve in the diabetic BB/W-rat. Diabetologia 1992;35:602–607.
47. Kim YS, Kim YH, Cheon EW, et al. Retinal expression of clusterin in the streptozotocin-induced diabetic rat. Brain Res 2003;976:53–59.
48. Li Q, Zemel E, Miller B, Perlman I. Early retinal damage in experimental diabetes: electroretinographical and morphological observations. Exp Eye Res 2002;74:615–625.
49. Kern TS, Engerman RL. A mouse model of diabetic retinopathy. Arch Ophthalmol 1996;114:986–990.
50. Hammes HP, Lin J, Renner O, et al. Pericytes and the pathogenesis of diabetic retinopathy. Diabetes 2002;51:3107–3112.
51. Mohr S, Xi X, Tang J, Kern TS. Caspase activation in retinas of diabetic and galactosemic mice and diabetic patients. Diabetes 2002;51:1172–1179.
52. Martin PM, Roon P, Van Ells TK, Ganapathy V, Smith SB. Death of retinal neurons in streptozotocin-induced diabetic mice. Invest Ophthalmol Vis Sci 2004;45:3330–3336.
53. Kowluru RA. Retinal metabolic abnormalities in diabetic mouse: comparison with diabetic rat. Curr Eye Res 2002;24:123–128.
54. Goldstein M, Leibovitch I, Yeffimov I, Gavendo S, Sela BA, Loewenstein A. Hyperhomocysteinemia in patients with diabetes mellitus with and without diabetic retinopathy. Eye 2004;18:460–465.
55. Neugebauer S, Baba T, Kurokawa K, Watanabe T. Defective homocysteine metabolism as a risk factor for diabetic retinopathy. Lancet 1997;349:473–474.
56. Smith SB, Van Ells TK, Mysona B, Martin PM, Roon P, Ganapathy V. Assessment of retinas of diabetic mice with mild elevation of plasma homocysteine. Invest Ophthalmol Vis Sci 2004;45 Abstract #3231.
57. Barber AJ, Antonetti DA, Kern TS, et al. The Ins2Akita mouse as a model of early retinal complications in Diabetes Invest Ophthalmol Vis Sci 2005;46:2210–2218.
58. Ning X, Baoyu Q, Yuzhen L, Shuli S, Reed E, Li QQ. Neuro-optic cell apoptosis and microangiopathy in KKAY mouse retina. Int J Mol Med 2004;13:87–92.

59. Lambooij AC, van Wely KH, Lindenbergh-Kortleve DJ, Kuijpers RW, Kliffen M, Mooy CM. Insulin-like growth factor-I and its receptor in neovascular age-related macular degeneration. Invest Ophthalmol Vis Sci 2003;44:2192–2198.

60. Ohashi H, Takagi H, Koyama S, et al. Alterations in expression of angiopoietins and the Tie-2 receptor in the retina of streptozotocin induced diabetic rats. Mol Vis 2004; 26:10:608–617.

61. Tikellis C, Cooper ME, Twigg SM, Burns WC, Tolcos M. Connective tissue growth factor is up-regulated in the diabetic retina: amelioration by angiotensin-converting enzyme inhibition. Endocrinology 2004;145:860–866.

62. Cox OT, Simpson DA, Stitt AW, Gardiner TA. Sources of PDGF expression in murine retina and the effect of short-term diabetes. Mol Vis 2003;9:665–672.

63. Vinores SA, Youssri AI, Luna JD, et al. Upregulation of vascular endothelial growth factor in ischemic and non-ischemic human and experimental retinal disease. Histol Histopath 1997;12:99–109.

64. Lu M, Kuroki M, Amano S, et al. Advanced glycation end products increase retinal vascular endothelial growth factor expression. J Clin Invest 1998;101:1219–1224.

65. Ng YK, Zeng XX, Ling EA. Expression of glutamate receptors and calcium-binding proteins in the retina of streptozotocin-induced diabetic rats. Brain Res 2004;1018:66–72.

66. Ola MS, Moore P, Maddox D, et al. Analysis of sigma receptor (σR1) expression in retinal ganglion cells cultured under hyperglycemic conditions and in diabetic mice. Brain Res Mol Brain Res 2002;107:97–107.

67. Park SH, Park JW, Park SJ, et al. Apoptotic death of photoreceptors in the streptozotocin-induced diabetic rat retina. Diabetologia 2003;46:1260–1268.

68. Seki M, Tanaka T, Nawa H, et al. Involvement of brain-derived neurotrophic factor in early retinal neuropathy of streptozotocin-induced diabetes in rats: therapeutic potential of brain-derived neurotrophic factor for dopaminergic amacrine cells. Diabetes 2004;53:2412–2419.

69. Kaneko M, Sugawara T, Tazawa Y. Electrical responses from the inner retina of rats with streptozotocin-induced early diabetes mellitus. Acta Societ Ophthalmol Jap 2000;104: 775–778.

70. Agardh E, Bruun A, Agardh CD. Retinal glial cell immunoreactivity and neuronal cell changes in rats with STZ-induced diabetes. Curr Eye Res 2001;23:276–284.

71. Ng YK, Zeng XX, Ling EA. Expression of glutamate receptors and calcium-binding proteins in the retina of streptozotocin-induced diabetic rats. Brain Res 2004;1018:66–72.

72. Nork TM. Acquired color vision loss and a possible mechanism of ganglion cell death in glaucoma. Trans Amer Ophthalmol Soc 2000;98:331–363.

73. Lahdenranta J, Pasqualini R, Schlingemann RO, et al. An anti-angiogenic state in mice and humans with retinal photoreceptor cell degeneration. Proc Natl Acad Sci USA 2001;98: 10,368–10,373.

74. Kumagai AK. Glucose transport in brain and retina: implications in the management and complications of diabetes. Diabetes/Metab Res Rev 1999;15:261–273.

75. Chancy CD, Kekuda R, Huang W, et al. Expression and Differential Polarization of the Reduced-Folate Transporter-1 and the Folate Receptor α in Mammalian Retinal Pigment Epithelium. J Biol Chem 2000;275:20,676–20,684.

76. Naggar H, Ola MS, Moore P, et al. Downregulation of reduced-folate transporter by glucose in cultured RPE cells and in RPE of diabetic mice. Invest Ophthalmol Vis Sci 2002;43:556–563.

77. Stevens MJ, Hosaka Y, Masterson JA, Jones SM, Thomas TP, Larkin DD. Downregulation of the human taurine transporter by glucose in cultured retinal pigment epithelial cells. Am J Physiol 1999;277:E760–E771.

78. Rollin R, Mediero A, Martinez-Montero JC, et al. Atrial natriuretic peptide in the vitreous humor and epiretinal membranes of patients with proliferative diabetic retinopathy. Mol. Vis 2004;10:450–457.

79. Framme C, Roider J. Immediate and long-term changes of fundus autofluorescence in continuous wave laser lesions of the retina. Ophthal Surg Lasers Imag 2004;35:131–138.

80. Framme C, Brinkmann R, Birngruber R, Roider J. Autofluorescence imaging after selective RPE laser treatment in macular diseases and clinical outcome: a pilot study. Br J Ophthalmol 2002;86:1099–1106.

81. Bensaoula T, Ottlecz A. Biochemical and ultrastructural studies in the neural retina and retinal pigment epithelium of STZ-diabetic rats: effect of captopril. J Ocul Pharmacol Therapeut 2001;17:573–586.

82. Du Y, Smith MA, Miller CM, Kern TS. Diabetes-induced nitrative stress in the retina, and correction by aminoguanidine. J Neuro Chem 2002;80:771–779.

83. Zheng L, Szabo C, Kern TS. Poly(ADP-ribose) polymerase is involved in the development of diabetic retinopathy via regulation of nuclear factor-kappaB. Diabetes 2004;53:2960–2967.

84. Kowluru RA, Koppolu P, Chakrabarti S, Chen S. Diabetes-induced activation of nuclear transcriptional factor in the retina, and its inhibition by antioxidants. Free Rad Res 2003;37:1169–1180.

85. Kowluru RA. Effect of advanced glycation end products on accelerated apoptosis of retinal capillary cells under in vitro conditions. Life Sciences 2005;76:1051–1060.

86. Kowluru RA, Odenbach S. Role of interleukin-1beta in the pathogenesis of diabetic retinopathy. Br J Ophthalmol 2004;88:1343–1347.

87. Butler JM, Guthrie SM, Koc M, et al. SDF-1 is both necessary and sufficient to promote proliferative retinopathy. J Clin Invest 2005;115:86–93.

88. Kondo T, Vicent D, Suzuma K, et al. Knockout of insulin and IGF-1 receptors on vascular endothelial cells protects against retinal neovascularization. J Clin Invest 2003;111:1835–1842.

89. Kondo T, Kahn CR. Altered insulin signaling in retinal tissue in diabetic states. J Biol Chem 2004;279:37,997–38,006.

90. Fernandes R, Carvalho AL, Kumagai A, et al. Downregulation of retinal GLUT1 in diabetes by ubiquitinylation. Mol Vis 2004;10:618–628.

91. Fernandes R, Suzuki K, Kumagai AK. Inner blood-retinal barrier GLUT1 in long-term diabetic rats: an immunogold electron microscopic study. Invest Ophthalmol Vis Sci 2003;44:3150–3154.

92. Zhang JZ, Gao L, Widness M, Xi X, Kern TS. Captopril inhibits glucose accumulation in retinal cells in Diabetes Invest Ophthalmol Vis Sci 2003;44:4001–4005.

93. Suganami E, Takagi H, Ohashi H, et al. Leptin stimulates ischemia-induced retinal neovascularization: possible role of vascular endothelial growth factor expressed in retinal endothelial cells. Diabetes 2004;53:2443–2448.

94. Chibber R, Ben-Mahmud BM, Mann GE, Zhang JJ, Kohner EM. Protein kinase C beta2-dependent phosphorylation of core 2 GlcNAc-T promotes leukocyte-endothelial cell adhesion: a mechanism underlying capillary occlusion in diabetic retinopathy. Diabetes 2003;52:1519–1527.

95. Khan ZA, Cukiernik M, Gonder JR, Chakrabarti S. Oncofetal fibronectin in diabetic retinopathy. Invest Ophthalmol Vis Sci 2044;45:287–295.

96. Grammas P, Riden M. Retinal endothelial cells are more susceptible to oxidative stress and increased permeability than brain-derived endothelial cells. Microvasc Res 2003;65:18–23.

97. Hammes HP, Lin J, Renner O, et al. Deutsch Pericytes and the pathogenesis of diabetic retinopathy. Diabetes 2002;51:3107–3112.

98. Beltramo E, Pomero F, Allione A, D'Alu F, Ponte E, Porta M. Pericyte adhesion is impaired on extracellular matrix produced by endothelial cells in high hexose concentrations. Diabetologia 2002;45:416–419.
99. Krishnamoorthy RR, Agarwal P, Prasanna G, et al. Characterization of a transformed rat retinal ganglion cell line. Brain Res Mol Brain Res 2001;86:1–12.
100. Dunn KC, Aotaki-Keen AE, Putkey FR, Hjelmeland LM. ARPE-19, a human retinal pigment epithelial cell line with differentiated properties. Exp Eye Res 1996;62:155–169.
101. Tan E, Ding XQ, Saadi A, Agarwal N, Naash MI, Al-Ubaidi MR. Expression of cone-photoreceptor-specific antigens in a cell line derived from retinal tumors in transgenic mice. Invest Ophthalmol Vis Sci 2004;45:764–768.

10
Statins and Age-Related Maculopathy

Gerald McGwin, Jr, MS, PhD and Cynthia Owsley, MSPH, PhD

CONTENTS

AGE-RELATED MACULOPATHY

Age-related maculopathy (ARM) is the leading cause of irreversible vision loss among older adults in the industrialized world *(1–4)*. ARM is a progressive, degenerative disorder of the retina that results in a degradation of the sharp, central vision necessary for most everyday activities, including reading and driving. The prominent histopathological and clinical lesions in ARM involve Bruch's membrane, a specialized vascular intima separating the photoreceptors and their support cells, the retinal pigment epithelium (RPE), from their blood supply. ARM is characterized according to the presence and distinctiveness of drusen that lie at the border between the RPE and Bruch's membrane. Currently, the gold standard for the definition of ARM is based on the use of retinal grading systems that provide standardized means for the grading of visible lesions and pigmentary changes associated with the disease *(4–7)*. The number and size of these drusen are used to classify ARM into various stages—early, intermediate, or advanced—and linked to the progression of the disease *(8)*.

Even in the early and intermediate stages of ARM, patients experience difficulty in performing everyday activities *(9–14)*, emotional distress *(15,16)*, driving cessation *(17,18)*, and decrements in health-related quality of life *(9,19,20)*. Treatments for ARM that have some proven effectiveness are photodynamic therapy *(12)*, anti-oxidant therapy *(13)* and intravitreal pegaptanib *(14)*. However, although these breakthroughs are scientifically and clinically important, it is also important to emphasize that they are limited in that they apply to only select groups of patients who have

From: *Ophthalmology Research: Retinal Degenerations: Biology, Diagnostics, and Therapeutics*
Edited by: J. Tombran-Tink and C. J. Barnstable © Humana Press Inc., Totowa, NJ

ARM in the later phases. In addition, although results from studies indicate that these treatments slow progression in the later stages of the disease, they have not been shown to reverse vision loss.

The public health impact of ARM is currently substantial and will grow rapidly with the aging of the population *(5)*. Currently, there are no effective treatments for the earliest stages of ARM, although some treatments slow the loss of visual function in later stages of the disease *(6)*. Although the search for effective treatments to reverse or slow the progression of ARM is of importance, identifying characteristics associated with the development of the disease is equally important. The identification of modifiable risk factors could lead to disease prevention initiatives. The impact on health care resources and quality of life is likely to be significantly greater by preventing ARM rather than by treating it. For nonmodifiable risk factors, disease prevention is precluded owing to their nature. However, should effective treatments become available, particularly treatments for early ARM, they allow for high-risk individuals to be identified and targeted for early intervention.

ARM is the leading cause of blindness among older adults in the United States and other industrialized nations *(21)*. In the United States alone, more than 7 million people have drusen of sufficient size and number so that they are at substantial risk for severe visual loss *(22)*. The prevalence of ARM is estimated to be approx 10% among those in their early 40s and 50s and nearly 40% among those 80 yr of age and older. In addition to age, the incidence and prevalence of ARM also differs according to gender and ethnicity. Males and whites appear to be at increased risk, although these findings are not consistent across all studies *(22–24)*.

Epidemiological research points toward a variety of risk factors for ARM, including demographic, medical, nutritional, and genetic characteristics *(24–27)*. Research on potentially modifiable risk factors has been extensive, focusing predominantly on lifestyle characteristics such as smoking, alcohol consumption, and diet as well as cardiovascular disease (CVD) characteristics including hypertension and atherosclerosis. To date, however, much of this research has been equivocal with the exception of smoking wherein most studies have reported an increased risk among smokers *(28,29)*. One of the reasons for the varying results found in the current body of literature may relate to several study design issues. First, the definition of ARM varies from study to study with some relying on self report and others on fundus appearance as well as different studies using differing definitions of fundus appearance. Second, many studies have focused on late ARM or an unspecified combination of patients with early and late disease. If particular risk factors are differentially associated with certain disease subtypes and different levels of disease severity, then such studies may not provide a clear characterization of the etiology of ARM throughout the natural history of the disease, including its earliest phases. Obviously then, additional work is necessary to properly evaluate currently hypothesized etiologic risk factors.

HMG-CoA REDUCTASE INHIBITORS (STATINS)

Recent investigations have identified 3-hydroxy-3-methylgluatryl coenzyme A (HMG-CoA) reductase inhibitors or "statins" as being associated with a reduced risk of

ARM *(30–32)*. Statins are prescribed to help reduce low-density lipoprotein (LDL)-cholesterol levels by inhibiting cholesterol production and increasing LDL-cholesterol removal from plasma. HMG-CoA reductase is a key enzyme not only for cholesterol biosynthesis but also for the biosynthesis of numerous nonsteroidal isoprenoid compounds *(33)*. Numerous biological processes associated with atherosclerotic progression are modulated by HMG-CoA reductase inhibition (e.g., endothelial cell health, thrombosis, angiogenesis) *(34,35)*. Statins were developed to lower plasma cholesterol levels in patients with atherosclerotic CVD. Recent recommendations for the aggressive use of antihyperlipidemic medications to lower LDL-cholesterol levels has contributed to a rising trend in the prescription of statins to reduce risk of CVD and CVD-related outcomes such as myocardial infarction and stroke *(36)*. Statins have few side effects and those that occur tend to be mild (e.g., fatigue, nausea). Reports have also emerged linking statins with lowered risk of fracture, dementia, and depression, although findings have been inconsistent *(37–43)*. However, the benefits of statin use may extend beyond that which can be explained by the direct effect of lowering plasma lipids concentrations *(35)*.

HMG-CoAreductase is a key enzyme not only for cholesterol biosynthesis, but also for the biosynthesis of numerous nonsteroidal isoprenoid compounds *(44)*. Mevalonate, the end product of the HMG-CoA reductase reaction, is a precursor to many other pathways requiring isoprenoids, including posttranslational modification of proteins involved in diverse cellular functions. Among the biological processes associated with atherosclerotic progression that are modulated by HMG-CoA reductase inhibition are endothelial cell health, inflammation, myocyte proliferation and migration, plaque stability, thrombosis, angiogenesis, and platelet activation *(34,35)*. Statins were originally developed to lower plasma cholesterol levels in patients with atherosclerotic CVD, a condition for which elevated plasma cholesterol is a well-established risk factor. It is now recognized that the benefits of statin usage extend far beyond that which can be explained by direct effect of lowering plasma lipids concentrations to the so-called pleiotrophic effects *(34,35)*. The association between statins and nonatherosclerotic conditions, such as Alzheimer's disease *(37–40)*, may be caused by an improvement of vascular health via plasma lipid lowering or by pleiotrophic effects, including a direct effect on neurons *(45,46)*.

BIOLOGICAL PLAUSIBILITY FOR STATINS AS BENEFICIAL FOR ARM

The eye tissues affected by ARM are the photoreceptors, the RPE (a cellular layer dedicated to sustaining photoreceptor health), the choriocapillaris (the blood supply to the photoreceptors and the RPE), and Bruch's membrane (a thin vascular intima between the RPE and the choriocapillaris) *(47,48)*. ARM is a multifactorial process, involving a complex interplay of genetic and environmental factors. Early ARM is characterized by minor-to-moderate vision loss associated with characteristic extracellular lesions and changes in RPE pigmentation and morphology. Lesions between the RPE basal lamina and Bruch's membrane can be either focal (drusen) or diffuse (basal linear deposits). Late ARM is characterized by severe vision loss associated with extensive RPE atrophy with or without the sequelae of choroidal neovascularization, that is, in-growth of choroidal vessels through Bruch's membrane and under the RPE in the plane of drusen and basal linear deposits.

There are potentially multiple biological bases for a potential association between statins and ARM, as statins could reduce the incidence of ARM through direct plasma-lipid lowering or through pleiotrophic effects as described above. With regard to the potential for a lipid-lowering effect, it is notable that Bruch's membrane accumulates lipids including cholesterol with normal aging, and cholesterol is a ubiquitous component of drusen in normal and ARM eyes (49–56).

The relative contributions of plasma lipoproteins and local cells to this cholesterol are still under investigation although recent evidence points to the fact these particles are intra-ocular in origin (51). Studies of normal human Bruch's membrane cholesterol composition have implicated both local cells and plasma lipoproteins as sources (50,54), but cholesterol has not been detected in Bruch's membrane following diet-induced hypercholesterolemia in mice (57–59). However, apolipoprotein B, the principal protein of the atherogenic plasma lipoproteins (60), is detectable in drusen, basal deposits, and in inner Bruch's membrane in association with basal deposits (49), where it could undergo modifications with deleterious impact on surrounding cells (61).

With regard to the potential for pleiotrophic effects, it is notable that many of the same processes that occur in the atherosclerotic intima likely also occur in ARM. Neovascularization is a major complication in both conditions (62,63). Therefore, angiogenesis and associated processes such as metalloproteinase activity (64) are potential points of statin modulation. Choroidal neovascular membranes associated with ARM include macrophages (65,66) and smooth muscle actin-positive cells (67), which may respond to statin. Drusen contain many proteins associated with inflammation and complement activation (68), and multiple lines of evidence point to a role for inflammatory processes in ARM progression (69). Statins affect RPE cell survival and morphology in vitro (70). The challenge for future laboratory research will be to determine which processes are modulated by statins in vivo and therefore are may be primarily responsible for the purported beneficial effects observed in some epidemiological studies.

EXISTING STUDIES ON CHOLESTEROL-LOWERING MEDICATIONS AND ARM

Observational studies evaluating a potential association between plasma cholesterol and ARM risk have yielded conflicting results. Although the mechanism of putative action for lipid levels in the development of ARM is unknown, laboratory evidence suggests local rather than systemic sources of cholesterol may contribute to ARM pathogenesis. First, there is a ubiquitous presence of cholesterol in drusen and the accumulation of lipid deposits in Bruch's membrane of older eyes, and second, there is evidence that the RPE is capable of producing cholesterol (49–55). In addition to the potential role of a statin in lowering plasma and potentially ocular lipid concentrations, statins have been shown to have a number of pleiotropic effects, such as helping to restore endothelial cell function, enhance atherosclerotic plaque stability, and decrease oxidative stress and vascular inflammation (35). These pleiotropic effects of statins, though well characterized for cardiovascular outcomes, have yet to be definitively linked to a reduction in ARM risk.

Table 1
Summary of Studies Investigating the Association Between Statins and Cholesterol-Lowering Medications and ARM

Reference	Reference no.	Study design	Study population	Primary finding
McCarty et al., 2001	*31*	Cohort	Population-based sample, aged 44 and older, Victoria, Australia (*n* = 580)	OR 0.2 95% CI 0.0–1.9
McCarty et al., 2001	*76*	Cross-sectional	Population-based sample, aged 40 and older, Victoria, Australia (*n* = 4345)	OR 1.7[a] 95% CI 1.1–2.5
Klein et al., 2001	*72*	Cohort	Population-based cohort, aged 43 and older, Wisconsin, US (*n* = 3,684)	OR 0.9[a] 95% CI 0.5–1.8
Hall et al., 2001	*32*	Cross-sectional	Convenience sample, aged 65 to 75, Sheffield, UK (*n* = 379)	OR 0.1 95% CI 0.0–0.8
van Leeuwen et al., 2003	*74*	Cohort	Population-based cohort, aged 55 and older, Rotterdam, the Netherlands (*n* = 5241)	HR 1.0 95% CI 0.7–1.5
Klein et al., 2003	*73*	Cohort	Population-based cohort, aged 48 and older, Wisconsin, US (*n* = 2780)	OR 1.1 95% CI 0.7–1.7
McGwin et al., 2003	*30*	Case-control	Males, 50 and older, Alabama, US (*n* = 550 ARM cases; *n* = 5500 non-ARM controls)	OR 0.3 95% CI 0.2–0.5
van Leeuwen et al., 2004	*75*	Cohort	Population-based cohort, 43 and older, pooled from the Netherlands, US, and Australia (*n* = 8649)	OR 1.0[a] 95% CI 0.6–1.6
McGwin et al., 2005	*71*	Case-control	Population-based cohort, aged 45 to 65 at enrollment, selected communities in NC, MS, MN, and MD, United States (*n* = 15,792)	OR 0.8[a] 95% CI 0.6–1.0
McGwin et al., in press	*77*	Case-control	Population-based cohort, aged 65 and older at enrollment, selected communities in NC, CA, PA, and MD, United States (*n* = 2755)	OR 1.4 95% CI 1.0–2.0

[a]Cholesterol-lowering medication rather than statin use.
OR, odds ratio; HR, hazard ratio; CI, confidence interval.

To date, there have been 10 published studies on the relationship between cholesterol-lowering medications, including statins and ARM. Table 1 summarizes the results of each of these studies. Four of these studies *(30–32,71)* provide support for the potential protective role of statins, three of which reported statistically significant results studies *(30,31,71),* whereas the fourth study was borderline significant ($p = 0.11$) *(31)*. Four studies have reported null associations *(72–75)* and two studies have reported that statins increase the risk of ARM *(76,77)*. There was geographical diversity across all of these studies though the demographic characteristics of the study populations were generally similar.

This heterogeneity in previous study results is somewhat expected given that some studies have determined the presence of ARM using standard techniques (i.e., fundus photography), whereas others have relied upon administrative data sources. Another problem, probably more serious, is that all of the studies to date have been secondary data analyses. That is, the data used in these analyses was not primarily collected for the purpose of testing the hypothesis in question. Such analyses are commonplace in clinical science and can yield valid results; however, because the primary variables of interest are not collected with an explicit hypothesis in mind, limitations often arise.

In the context of the studies in Table 1, perhaps one of the greatest limitations in this regard is the lack of explicit information pertaining to statins vs other cholesterol-lowering medications. For example, the studies by McCarthy et al. *(76)*, Klein et al. *(73)*, van Leeuwen et al. *(74)*, and McGwin et al. *(71)* presented data with respect to cholesterol-lowering drugs (e.g., statin, cholestyramine, clofibrate, colestipol, gemfibrozil, and others related to bile sequestrants, antihyperlipidemic, and vitamin B3) and not statins in particular. The consequence of aggregating statins with other non-statin cholesterol-lowering drugs depends on the hypothesized mechanism of action behind the relationship between statins and ARM, an issue discussed in greater detail in the subsequent section. Briefly, if the hypothesized mechanism is via a lowering of cholesterol levels, then one might expect a similar association for statins as for nonstatin drugs. However, if the mechanism of action is not solely via a lowering of cholesterol but rather attributable, at least in part, to a characteristic specific to statins, then studies combining statins and nonstatins would produce estimates that are biased towards the null. To date, only one study has simultaneously compared the relationship between statins and nonstatins and ARM *(30)*. McGwin et al. *(30)* reported no reduced risk for ARM associated with nonstatin cholesterol-lowering drugs, but a significantly reduced risk associated with statin use. This evidence, albeit limited, supports the feasibility of a statin-specific effect and suggests that the null associations observed in studies that did not separate nonstatin cholesterol lowering drugs from statins might be biased towards the null. Further evidence for a non-cholesterol lowering mechanism is the lack of a consistent association between elevated cholesterol levels and the risk of ARM *(78–83)*.

Because all of the studies were secondary analyses using existing cohorts, despite the large sample sizes, the ultimate number of subjects who developed ARM was often small, as in the case of the McCarthy et al. *(31)* and Klein et al. *(72,73)* studies. This, in turn, equates to lower statistical power to detect associations, particularly those that are modest in size.

Yet another limitation of many of the existing studies is failure to account for the temporal relationship between statin use and the onset of ARM. For example, the cross-sectional studies by Hall et al. *(32)* and McCarthy et al. *(76)* documented the presence of ARM and statin use simultaneously. Our own recent work also suffered from the same limitation *(71,77)*. Thus, we cannot be certain the extent to which the reported statin use preceded, and by how much, the development of ARM. Despite the use of the case-control design, McGwin et al. were able to address this issue by using a nested case-control design *(30)*. Thus, they only considered statin use that occurred prior to ARM diagnosis. As previously noted, much of the research on statins and ARM comes from secondary analysis of existing cohort studies. The cohort design will generally ensure that the primary exposure of interest is present prior to the occurrence of the primary outcome of interest. It is important to note that the follow-up time in many of these studies was relatively short, for example 5 yr in one study *(73)*. However, given that statins did not become widespread until 1993, the follow-up time for any cohort study conducted presently would be no greater than 11 yr. Further, many of these studies did not quantify duration of statin use. Thus, the null associations may be the result of a heterogeneous mixture of short- and long-term statin users; for the former group, the short duration of time between use and potential onset of ARM may be too short to have had any potential impact on the risk of ARM. Two studies have addressed the duration issue and reported differing results. van Leeuwen et al. *(74)* found no association between statin use (or use of any cholesterol-lowering drugs) and ARM, regardless of the duration of use. However, there were only 25 subjects who developed ARM during the follow-up period of this study, resulting in low statistical power to detect an association. McGwin et al. also evaluated duration of statin use and documented a stronger association for past vs current users and for those who had used statins for a longer period of time *(30)*.

Among the studies in Table 1, there are a variety of other lesser limitations that, although unlikely to account for the heterogeneity of results, hamper the ability to draw any firm conclusions from the body of research. For example, some studies relied upon self report of statin use, whereas others had independent sources of such information (e.g., pharmacy records); several studies had low response rates therefore introducing opportunity for selection bias; and finally, several studies failed to account for the impact of potentially confounding characteristics, some of which are potentially strong confounders (e.g., smoking, nutritional characteristics).

The aforementioned research described here is limited to those studies evaluating the potential role of statins in the *prevention* of ARM; there is an emerging body of research suggesting that statins may be an effective *treatment* for patients with ARM. Recently, Wilson et al. *(84)* evaluated the hypothesis that treatment with statins is associated with a reduced risk of choroidal neovascularization (CNV) in patients with ARM. The results of this study indicated that, in fact, statin use appeared to reduce the risk of CNV independent of other potentially confounding characteristics (e.g., smoking). This result is consistent with that from one other study wherein cholesterol-lowering medications were found to reduce the progression of ARM *(31)*. Although indirect, this body of research provides support for a potential effect of statins on ARM.

SUMMARY

ARM is an important cause of vision impairment. The lack of effective treatments, particularly in the earliest stages of the disease, underscores the need for continued research on risk factors for preventing ARM from occurring in the first place. The evidence from observational studies for a relationship between statins and ARM is equivocal yet there are a sufficient number of studies suggesting a protective association as well as significant limitations across all studies to leave the question open to further research. There are several mechanisms by which statins might reduce the risk of ARM. The cholesterol-lowering and anti-inflammatory properties of statins are among the most frequently mentioned mechanisms. However, additional research is needed to support or refute these mechanisms or reveal additional ones. Additional epidemiological studies are also needed. Despite the rapidly growing body of literature on the relationship between statins and ARM, all studies to date have been secondary analyses and, therefore, possessed specific limitations that likely contribute to the heterogeneity of observed results and prevent firm conclusions from being drawn. Although the evidence to date is sufficient to support the conduct of a clinical trial, there is debate regarding this issue *(85,86).* Thus, in the absence of support for such an endeavor, an observational study specifically design to test the association between statins and ARM is needed. Such a study would address the limitations of published studies to date, therefore providing further, perhaps more valid, results but also preliminary data required for a randomized clinical trial.

ACKNOWLEDGMENTS

Preparation of this chapter was made possible by National Institutes of Health grants R21-14071 and R01-AG04212, Research to Prevent Blindness, Inc, and the EyeSight Foundation of Alabama.

REFERENCES

1. Klein R, Klein BEK, Linton, KLP. Prevalence of age-related maculopathy: The Beaver Dam eye study. Ophthalmology 1992;99:933–943.
2. Mitchell P, Smith W, Attebo K, Wang JJ. Prevalence of age-related maculopathy in Australia: The Blue Mountains Eye Study. Ophthalmology 1995;102:1450–1460.
3. Vingerling JR, Dielemans I, Hofman A, et al. The prevalence of age-related maculopathy in the Rotterdam study. Ophthalmology 1995;102:205–210.
4. Bressler NM, Bressler SB, West SK, Fine SL, Taylor HR. The grading and prevalence of macular degeneration in Chesapeake Bay watermen. Archives of Ophthalmology 1989; 107:847–852.
5. Bird AC, Bressler NM, Bressler SB, et al. An international classification and grading system for age-related maculopathy and age-related macular degeneration. Survey of Ophthalmology 1995;39:367–374.
6. Klein R, Davis MD, Magli YL, Segal P, Klein BEK, Hubbard L. The Wisconsin age-related maculopathy grading system. Ophthalmology 1991;98:1128–1134.
7. Group, "Age-Related Eye Disease Study Research Group"(A.-r. E. D. S. R.) Risk factors associated with age-related macular degeneration. A case-control study in the age-related eye disease study: Age-related eye disease study report number 3. Ophthalmology 2000;107:2224–2232.
8. Klein R, Klein BE, Tomany SC, Meuer SM, Huang GH. Ten-year incidence and progression of age-related maculopathy: The Beaver Dam eye study. Ophthalmology 2002; 109:1767–1779.
9. Mangione CM, Gutierrez PR, Lowe G, Orav EJ, Seddon JM. Influence of age-related maculopathy on visual functioning and health-related quality of life. American Journal of Ophthalmology 1999;128:45–53.
10. Scott IU, Smiddy WE, Schiffman J, Feuer WJ, Pappas CJ. Quality of life of low-vision patients and the impact of low-vision services. American Journal of Ophthalmology 1999;128:54–62.

11. Scilley K, Jackson GR, Cideciyan AV, et al. Early age-related maculopathy and self-reported visual difficulty in daily life. Ophthalmology 2002;109:1235–1242.
12. Bressler NM, Bressler SB. Photodynamic therapy with verteprofin (Visudyne): impact on ophthalmology and visual science. Invest Ophthalmol Vis Sci 2000;41:624–628.
13. Age-related Eye Disease Study Research Group. A randomized, placebo-controlled, clinical trial of high-dose supplementation with vitamins C and E, beta carotene, and zinc for age-related macular degeneration and vision loss. Arch Ophthalmol 2001;119:1417–1436.
14. Gragoudas ES, Adamis AP, Cunningham ET, Feinsod M, Guyer DR, Group, VEGF Inhibition Study in Ocular Neovascularization Clinical Trial Study Group. Pegaptanib for neovascular age-related macular degeneration. N Engl J Med 2004;351:2805–2816.
15. Williams RA, Brody BL, Thomas RG, Kaplan RM, Brown SI. The psychosocial impact of macular degeneration. Arch Ophthalmol 1998;116:514–520.
16. Rovner BW, Casten RJ, Tasman WS. Effect of depression on vision function in age-related macular degeneration. Arch Ophthalmol 2002;120:1041–1044.
17. Marottoli RA, Ostfeld AM, Merrill SS, Perlman GD, Foley DJ, Cooney LM, Jr. Driving cessation and changes in mileage driven among elderly individuals. J Gerontol: Soc Sci 1993;48:S255–S260.
18. DeCarlo DK, Scilley K, Wells J, Owsley C. Driving habits and health-related quality of life in patients with age-related maculopathy. Optom Vis Sc 2003;80:207–213.
19. Scilley K, DeCarlo DK, Wells J, Owsley C. Vision-specific health-related quality of life in age-related maculopathy patients presenting for low vision services. Ophthalmic Epidemiol 2004;11:131–146.
20. Mangione CM, Lee PP, Gutierrez PR, et al. Development of the 25-item National Eye Institute Visual Function Questionnaire. Arch Ophthalmol 2001;119:1050–1058.
21. National Advisory Eye Council National Eye Institute, National Institutes of Health. US Department of Health and Human Services. Vision research. A national plan: 1999–2003. 1999; Number 99–4120.
22. Friedman DS, O'Colman BJ, Munoz B, et al. Prevalence of age-related macular degeneration in the United States. Arch Ophthalmol 2004;122:564–572.
23. Friedman DS, Katz J, Bressler NM, Rahmani B, Tielsch JM. Racial differences in the prevalence of age-related macular degeneration: The Baltimore Eye Survey. Ophthalmology 1999;106:1049–1055.
24. Klein R, Peto T, Bird AC, Vannewkirk MR. The epidemiology of age-related macular degeneration. Am J Ophthalmol 2004;137:486–495.
25. Snow KK, Seddon JM. Do age-related macular degeneration and cardiovascular disease share common antecedents? Ophthalmic Epidemiol 1999;6:125–143.
26. Gottlieb JL. Age-related macular degeneration. JAMA 2002;288:2233–2236.
27. van Leeuwen R, Klaver C, Vingerling J, Hofman A, de Jong PTVM. Epidemiology of age-related maculopathy: a review. Eur J Epidemiol 2003;18:845–854.
28. Chan D. Cigarette smoking and age-related macular degeneration. Optom Vis Sci 1998;75:476–484.
29. Solberg Y, Rosenr M, Belkin M. The association between cigarette smoking and ocular diseases. Surv Ophthalmol 1998;42:535–547.
30. McGwin G, Owsley C, Curcio CA, Crain RJ. The association between statin use and age related maculopathy. Br J Ophthalmol 2003;87:1121–1125.
31. McCarty CA, Mukesh BN, Guymer RH, Baird PN, Taylor HR. Cholesterol-lowering medications reduce the risk of age-related maculopathy progression. Med J Aust 2001;175:340.
32. Hall NF, Gale CR, Syddall H, Phillips DIW, Martyn CN. Risk of macular degeneration in users of statins: cross sectional study. BMJ 2001;323:375–376.
33. Brown MS, Goldstein JL. Multivalent feedback regulation of HMG CoA reductase a control mechanism coordinating isoprenoid synthesis and cell growth. J Lipid Res 1980;21:505–517.

34. Comparato C, Altana C, Bellosta S, Baetta R, Paoletti R, Corsini A. Clinically relevant pleiotropic effects of statins: Drug properties or effects of profound cholesterol reduction? Nutri Metab Cardiovasc Dis 2001;11:328–343.

35. Takemoto M, Liao JK. Pleiotropic effects of 3-hydroxy-3-methylglutaryl coenzyme a reductase inhibitors. Arterioscler Thromb Vasc Biol 2001;21:1712–1719.

36. Hennekens CH. Current perspectives on lipid lowering with statins to decrease risk of cardiovascular disease. Clin Cardiol 2001;24:2–5.

37. Miller LJ, Chacko R. The role of choesterol and stains in Alzheimer's disease. Ann Pharmacother 2004;38:91–98.

38. Jick H, Zornberg GL, Jick SS, Seshadri S, Drachman DA. Statins and the risk of dementia. Lancet 2000;356:1627–1631.

39. Wolozin B, Kellman W, Ruosseau P, Celesia GG, Siegel G. Decreased prevalence of Alzheimer disease associated with 3-hydroxy-3-methyglutaryl coenzyme a reductase inhibitors. Arch Neurol 2000;57:1439–1443.

40. Vaughan CJ. Prevention of stroke and dementia with statin: Effects of lipid lowering. Am J Cardiol 2003;91:23B–29B.

41. Bauer DC, Mundy GR, Jamal SA, et al. Use of statins and fracture: Results of 4 prospective studies and cumulative meta-analysis of observational studies and controlled trials. Arch Intern Med 2004;164:146–152.

42. Yaturu S. Skeletal effects of statins. Endicr Pract 2003;9:315–320.

43. Young-Xu Y, Chan KA, Liao JK, Ravid S, Blatt CM. Long-term statin use and psychological well-being. J Am Coll Cardiol 2003;42:690–697.

44. Edwards PA, Ericsson J. Sterols and isoprenoids: Signalling molecules derived from the cholesterol biosynthetic pathway. Annu Rev Biochem 1999;68:157–185.

45. Cucchiara B, Kasner SE. Use of statins in CNS disorders. J Neurol Sci 2001;187:81–89.

46. Fassbender K, Simons M, Bergmann C, et al. Simvastatin strongly reduces levels of Alzheimer's disease beta-amyloid peptides A-beta-42 and A-beta-40 in vitro and in vivo. Proc Natl Acad Sci USA 2001;98:5856–5861.

47. Sarks SH. Aging and degeneration in the macular region: A clinico-pathological study. Br J Ophthalmol 1976;60:324–341.

48. Green WR, Enger C. Age-related macular degeneration histopathologic studies: The 1992 Lorenz E. Zimmerman Lecture. Ophthalmology 1993;100:1519–1535.

49. Malek G, Li CM, Guidry C, Medeiros NE, Curcio CA. Apolipoprotein B in cholesterol-containing drusen and basal deposits of human eyes with age-related maculopathy. Am J Pathol 2003;162:413–425.

50. Curcio CA, Millican CL, Bailey T, Kruth H. Accumulation of cholesterol with age in human Bruch's membrane. Invest Ophthalmol Vis Sci 2001;42:265–274.

51. Curcio C, Presley JB, Millican CL, Medeiros N. Basal deposits and drusen in eyes with age-related maculopathy: evidence for solid lipid particles. Exp Eye Res 2005;80:761–775.

52. Farkas TG, Sylvester V, Archer D, et al. The histochemistry of drusen. Am J Ophthalmol 1971;71:1206–1215.

53. Pauleikhoff D, Harper C, Marshall J, Bird A. Aging changes in Bruch's membrane. A histochemical and morphologic study. Ophthalmology 1990;97:171–178.

54. Holz FG, Sheraidah G, Pauleikhoff D, Bird AC. Analysis of lipid deposits extracted from human macular and peripheral Bruch's membrane. Arch Ophthalmol 1994;112:402–406.

55. Haimovici R, Gantz DL, Rumelt S, Freddo TF, Small DM. The lipid composition of drusen, bruch's membrane, and sclera by hot stage polarizing light microscopy. Invest Ophthalmol Vis Sci 2001;42:1592–1599.

56. Chuan-Ming L, Chung H, Presley JB, et al. Lipoprotein-like particles and cholesteryl esters in human Bruch's membrane: Initial characterization. Invest Ophthalmol Vis Sci 2005; 46:2576–2586.

57. Dithmar S, Sharara NA, Curcio CA, et al. Murine high-fat diet and laswer photochemical model of basal deposits in Bruch membrane. Arch Ophthalmol 2001;119:1643–1649.

58. Miceli MV, Newsome DA, Tate DJ, Jr, Sarphie TG. Pathologic changes in the retinal pigment epithelium and Bruch's membrane of fat-fed atherogenic mice. Curr Eye Res 2000;20:8–16.

59. Dithmar S, Curcio C, Le N-A, Brown S, Grossniklaus HE. Ultrastructural changes in Bruch's membrane of apolipoprotein E-deficient mice. Invest Ophthalmol Vis Sci 2000; 41:2035–2042.

60. Havel RJ, Kane JP. Introduction: structure and metabolism of plasma lipoproteins. In: Scriver CR, Beaudet AL, Sly WS, Valle D, eds. The Metabolic and Molecular Basis of Inherited Disease, Vol. 2. New York: McGraw-Hill, 2001.

61. Tabas I. Nonoxidative modifications of lipoproteins in atherogenesis. Annu Rev Nutr 1999;19:123–129.

62. Moulton KS. Plaque angiogenesis and atherosclerosis. Curr Atheroscler Rep 2001;3:225–233.

63. Campachiaro PA. Retinal and choroidal neovascularization. J Cell Physiol 2000;184: 301–310.

64. Qi JH, Ebrahem Q, Yeow K, Edwards DR, Fox PL, Anand-Apte B. Expression of Sorsby's fundus dystrophy mutations in human reitnal pigment epithelial cells reduces matrix metallaoproteinase inhibition and may promote angiogenesis. J Biol Chem 2002; 277:13,394–13,400.

65. Grossniklaus HE, Cingle KA, Yoon YD, Ketkar N, L'hernault N, Brown S. Correlation of histologic 2-dimenstional reconstructional and confocal scanning laser microscopic imagng of choroidal neovascularization in eyes with age-related maculopathy. Arch Ophthalmol 2000;118:625–629.

66. Killlingsworth MC, Sarks JP, Sarks SH. Macrophages related to Bruch'sm embrane in age-related macular degeneration. Eye 1990;4:613–621.

67. Lopez PF, Sippy BD, Lambert HM, Tahck AB, Hinton DR. Transdifferentiated retinal pigment epithelial cells are immunoreactive for vacular endothelial growth factor in surgically excised age-related macular degeneration-related choroidal neovascular membranes. Invest Ophthalmol Vis Sci 1996;37:855–868.

68. Hageman GS, Luthert PJ, Chong NHV, Johnson LV, Anderson DH, Mullins RF. An integrated hypothesis that considers drusen as biomarkers of immune-mediated processes at the RPE-Bruch's membrane interface in aging and age-related macular degeneration. Prog Retin Eye Res 2001;20:705–732.

69. Penfold PL, Madigan MC, Gillies MC, Provis JM. Immunological and aetiological aspects of macular degeneration. Prog Retin Eye Res 2001;20:385–414.

70. Capeans C, Pineiro A, Pardo M, et al. Role of inhibitors of isoprenylation in proliferation, phenotype and apoptosis of human retinal pigment epithelium. Graefes Arch Clin Exp Ophthalmol 2001;239:188–198.

71. McGwin G, Xie A, Owsley C. The use of cholesterol lowering medications and age-related macular degeneration. Ophthalmology 2005;112:488–494.

72. Klein R, Klein BEK, Jensen SC, et al. Medication use and the 5-year incidence of early age-related maculopathy: the Beaver Dam Eye Study. Arch Ophthalmol 2001;119:1354–1359.

73. Klein R, Klein BEK, Tomany SC, Danforth LG, Cruickshanks KJ. Relation of statin use to the 5-year incidence and progression of age-related maculopathy. Arch Ophthalmol 2003;121:1151–1155.

74. van Leeuwen R, Vingerling JR, Hofman A, de Jong PTVM, Stricker BHC. Cholesterol lowering drugs and risk of age related maculopathy: Prospective cohort study with cumulative exposure measurement. BMJ 2003;326:255–256.

75. van Leeuwen R, Tomany SC, Wang JJ, et al. Is medication use associated with incidence of early age-related maculopathy? Ophthalmology 2004;111:1169–1175.

76. McCarty CA, Mukesh BN, Fu CL, Mitchell P, Wang JJ, Taylor HR. Risk factors for age-related maculopathy: The visual impairment project. Arch Ophthalmol 2001; 119:1455–1462.

77. McGwin G, Modjarrad K, Hall TA, Xie A, Owsley C. 3-hydroxy-3-methylglutaryl coenzyme a reductase inhibitors and the presence of age-related macular degeneration in the Cardiovascular Health Study. Arch Ophthalmol 2006;124:33–37.

78. van Leeuwen R, Klaver CCW, Vingerling JR, et al. Cholesterol and age-related macular degeneration: Is there a link? Am J Ophthalmol 2004;137:750–752.

79. Tomany SC, Wang JJ, van Leeuwen R, et al. Risk factors for incident age-related macular degeneration: Pooled findings from 3 continents. Ophthalmology 2004;111:1280–1287.

80. Eye Disease Case Control Study Group. Risk factors for neovascular age-related macular degeneration. Arch Ophthamol 1992;110:1701–1708.

81. Klein R, Klein BEK, Franke T. The relationship of cardiovascular disease and its risk factors to age-related maculopathy: The Beaver Dam Eye Study. Ophthalmology 1993; 10:406–414.

82. Delcourt C, Michel F, Colvez A, et al. Associations of cardiovascular disease and its risks factors with age-related macular degeneration: The POLA study. Ophthalmic Epidemiol 2001;8:237–249.

83. Smith W, Assink J, Klein R, et al. Risk factors for age-related macular degeneration: Pooled findings from three continents. Ophthalmology 2001;108:697–704.

84. Wilson HL, Schwartz DM, Bhatt HRF, McCulloch CE, Duncan JL. Statin and aspirin therapy are associated with decreased rates of choroidal neovascularization among patients with age-related macular degeneration. Am J Ophthalmol 2004;137:615–624.

85. Klein R, Klein BE. Do statins prevent age-related macular degeneration? Am J Ophthalmol 2004;137:747–749.

86. McGwin G, Owsley C. Statins and age-related macular degeneration. Am J Ophthalmol 2004;138:688.

11

The Role of Drusen in Macular Degeneration and New Methods of Quantification

R. Theodore Smith, MD, PhD and Umer F. Ahmad, MD

CONTENTS

INTRODUCTION

Drusen are yellowish-white subretinal deposits that vary in size and composition. Often easily visible during slit lamp biomicroscopy, they tend to present in multiple morphologies that may vary throughout the lifetime of an eye. The mystery surrounding their relevance has been at the center of attempts to understand age-related macular degeneration (AMD) over the past several decades *(1)*.

DRUSEN CHARACTERISTICS AND SUBTYPES

Drusen subtypes should be understood in the context that drusen formation and regression is an extremely dynamic process, and drusen manifestation in the same patient evolves. Subtypes have been elucidated well by Sarks and colleagues *(2)*. Most drusen may be divided into either soft or hard drusen. Hard drusen have well-defined borders and are no larger than 63 μm in diameter *(3)*. They have not been found to be associated with advanced AMD (geographic atrophy [GA] or choroidal neovascularization [CNV])

From: *Ophthalmology Research: Retinal Degenerations: Biology, Diagnostics, and Therapeutics*
Edited by: J. Tombran-Tink and C. J. Barnstable © Humana Press Inc., Totowa, NJ

and are generally considered benign when present alone. Soft drusen, sometimes postulated to form from hard drusen or from a membranous component of the basement membrane of the retinal pigment epithelium (RPE), tend to be larger in size and may be coalescent *(2)*. They can be divided into soft distinct drusen, which have uniform reflectance throughout, and soft indistinct drusen, the reflectance of which tapers off toward the boundary and often blends with the background *(2)*. Other types of drusen have also been noted. Reticular pseudodrusen are seen in the peripheral macula as a lacy, lobular pattern whose histopathological correlate is uncertain and are probably not drusen at all *(2)*. There are also cuticular drusen, a descriptive term referring to small nodules in large numbers emanating from the inner portion of Bruch's membrane *(4)*. However, Russell et al. have differed concerning the uniqueness of cuticular drusen, and suggest that they are indistinguishable from drusen found in AMD *(5)*.

DRUSEN AS A RISK FACTOR

Soft drusen are a clear risk factor for the development of advanced AMD. The Blue Mountain Eye Study demonstrated that the presence of drusen 125 µm or larger, soft indistinct or reticular drusen, were amongst the chief risk factors for the progression to CNV *(6)*. The Beaver Dam Eye study found there was a 100 times greater risk of developing CNV in eyes of patients with established early AMD with drusen formation or retinal pigmentary abnormalities when compared to eyes with no evidence of drusen or pigment change *(7)*. The macular photocoagulation study (MPS) also found that large drusen, and five or more soft drusen, were risk factors for progression to advanced AMD *(8)*. For further details on the epidemiology of drusen in AMD, please *see* the companion chapter by Klein.

DRUSEN COMPOSITION

Drusen composition has also been studied in greater detail recently. Studies by Hageman et al., as well as Crabb et al., have resulted in a more precise elucidation of the composition of macular drusen *(9,10)*. Drusen are aggregates of lipids, glycolipids, proteins and other cellular elements. Following isolation of drusen from Bruch's membrane, Crabb et al. conducted proteome analysis that isolated many common proteins in drusen including TIMP3, clusterin, vitronectin, and serum albumin *(10)*. An intriguing group of compounds isolated were carboxyethyl pyrrole protein adducts. These adducts, a result of oxidative damage to proteins, suggest that age-related oxidative damage to tissues (vascular elements or elements intrinsic to Bruch's membrane and the RPE) contributes to drusen formation.

DRUSEN IN OTHER DISEASES

Diseases other than AMD may present with drusen. For example, Doyne's honeycomb macular dystrophy presents with drusen that appear as white or brownish-white deposits occupying the posterior pole in a honeycomb pattern *(11)*. Doyne's honeycomb and a related phenotype, Mallatia Leventinese, are caused by mutations in the EFEMP1 protein, an extracellular matrix protein, providing yet another genetic link between drusen and its etiology *(12)*. Unlike AMD, these present in the third to fourth decade of life. Drusen can also be seen in systemic diseases such as membrano-proliferative glomerulonephritis *(13)*. Sometimes, choroidal melanomas will have overlying drusen *(14)*. Nevertheless, the most common cause of drusen remains by far AMD, a disease afflicting millions of elderly.

PATHOPHYSIOLOGY OF DRUSEN

Historical Theories

Drusen were first noted by Muller *(15)*. Muller theorized that drusen were congregate deposits of secretions from RPE membranes onto Bruch's membrane (the "deposition theory"). Donders proposed that rather than deposits, drusen were actually just degenerated RPE cells (the "transformation theory") *(16)*. Both theories held paramount the role of RPE dysfunction in the generation of drusen.

Although RPE dysfunction remains a central tenet in the current understanding of drusen, inflammation, inflammatory mediators, and choriocapillaris dysfunction have recently been implicated in drusen formation.

Inflammation

Hageman et al. *(9)* proposed a unifying theory of drusen biogenesis that places the role of the dendritic cell, an antigen-presenting cell, as the primary instigator of drusen formation. Hageman finds that dendritic cells accumulate in areas of drusen formation at a very early stage, and after nucleation of the drusen, the subsequent inflammatory processes direct further aggregation of compounds. These include complement mediators, complement inhibitors such as vitronectin, amyloid compounds, HLA-DR, as well as coagulation factors such as Factor X, and prothrombin. Hageman argued that these inflammatory and immune compounds play a central role in drusen biogenesis, especially if combined with genetic alterations that might predispose to decreased clearance of debris. Ambati et al. described just such an alteration *(17)*.

Inflammatory Factors and Drusen

Ambati et al. demonstrated that mice that lack a chemoattractant for macrophages develop accumulations of drusen-like material *(17)*. These ccl/r-2 deficient mice developed an AMD-like disease in the absence of functional scavenging macrophages. These data were therefore suggestive that macrophages play a critical role in maintaining homeostasis in the environment of Bruch's membrane and the RPE, and thereby prevent buildup of compounds leading to drusen formation.

The Vascular Theory

Further evidence recently has indicated that the anticholesterol drug family known as the statins decreases the rate of AMD progression. Wilson et al. *(18)* demonstrated that patients taking statins demonstrated a lower rate than controls of progression from early to advanced AMD.

This supports the notion that the choriocapillaris and vascular factors play an important role in the progression of AMD. Friedman proposed that decreased compliance of ocular tissues and progressive narrowing of the macular choriocapillaris due to infiltration with lipid result in increased intraocular vascular resistance and increased choriocapillaris pressures *(19)*. These high pressures prohibit clearance of lipoprotein debris secreted by the RPE, thereby resulting in drusen deposition. Further, calcification and fracture of Bruch's membrane may predispose to choroidal neovascularization. Here, we see that drusen may be symptomatic of vascular disease, suggesting that AMD might be slowed were hypertension and atherosclerosis better controlled.

Although theories of drusen formation, composition, and relevance abound, it is clear that drusen play a central role in the physician's ability to diagnose AMD early, and perhaps prevent its progression. Further, these theories are not inconsistent. Thus, the modern model of atherosclerosis *(20)* as immune-mediated suggests a relationship between the vascular and inflammatory concepts

Therapeutic Laser

Therapeutic attempts to treat early AMD have included laser photocoagulation for drusen *(21)*. Gass first reported that laser photocoagulation resulted in drusen resolution, and suggested such prophylactic treatment *(22,23)*. Interestingly, the MPS investigated the role of photocoagulation in treating existing CNV, but not soft drusen in early AMD *(24)*. Wetzig studied drusen resorption following photocoagulation (in a scatter pattern) on 77 eyes, without using controls. There was an impressive result of decreased overall drusen but the 12% progression to CNV raised concerns *(25)*. In a small trial in 1994, Figueroa et al. established that laser photocoagulation to the peripheral macula could result in resolution of drusen without complications, not only near the treatment site, but also throughout the rest of the macula *(26)*. In 1998, the Choroidal Neovascularization Prevention Treatment Group reported that treated eyes with early AMD had a significant rate of conversion to CNV in patients with CNV in the fellow eye *(27)*. Most of the CNV developed in the region of laser treatment. As a result, the trial has since enrolled only patients with no preexisting CNV. Now known as the Complications of AMD Prevention Trial, it is an ongoing trial of more than 1000 patients with bilateral large drusen only. Results are expected in late 2006. The European drusen laser study also found laser contra-indicated in fellow eyes of eyes with CNV *(28)*. Olk et al. are also evaluating subthreshold diode laser in the The Prophylactic Treatment of Non exudative AMD Trial, with early equivocal results (Friberg TR, Results of the Bilateral Arm of the PTAMD, Macula Society Meeting, February, 2006). Clearly, in the eyes of investigators, there remains hope that laser treatment may become both safe and effective in the treatment of patients with early AMD and no evidence of advanced disease.

IMAGING

Introduction to Retinal Image Analysis

Clinical medical retinal research, in particular, and visual science in humans, in general, is based on minimally invasive testing with imaging serving as the surrogate for biopsy. Given the transparency of ocular tissue, retinal images are able to provide large amounts of valuable information. Image analysis of the retina can be performed in a variety of settings ranging from the standard digital fundus photograph, to autofluorescence imaging to infrared imaging, all of which provide unique information to the viewer. Combined analysis of imaging data from multiple methods can reveal heretofore-unexpected relationships.

The power of engineering technology to image these lesions in a variety of ways has grown exponentially, as has the raw digital power to store and process these images.

Image analysis to interpret this wealth of data unfortunately lagged far behind. Current systems for analyzing fundus photographs are manual, subjective, and expensive. Efficient and quantitative analysis of these images in clinical research is therefore needed to provide high throughput, cost savings, and accurate conclusions in large-scale studies. As an estimate, image grading for a study on the scale of Age-Related Eye Disease Study (AREDS) *(29)*, with a study population of 3640 examined every 6 mo for 6.3 yr, a total of 46,000 exams, with film photographs graded manually at $150/exam, would cost $6.9 million. The same study with digital imaging and automatic image analysis would consume a fraction of these resources.

Color Fundus Photography

As previously mentioned, extensive drusen area as seen on the fundus photograph is the greatest risk factor for the progression of AMD *(1,8,30–35)*. Historically, the accepted standard for drusen grading in AMD was manual grading of stereo photographic pairs at the light box, for example as refined by Klein et al. in the Wisconsin grading system *(3,36)*. However, there was always difficulty in obtaining inter-observer agreement in drusen identification. Inter-observer agreement in the presence of soft drusen only was 89% and on the total number of drusen was 76% in one study by Bressler *(37)*. Further, examiners were asked to aggregate mentally the amount of drusen occupying a given macular subfield *(3)*, as in the International System where drusen areas were estimated to within 10 to 25% or 25 to 50%, and so on *(36)*. Even these semiquantitative estimates proved difficult for human observers. Clearly, there was a pressing need for techniques that allowed more precise and confident measurements of macular drusen loads, e.g., to within 5%, to improve the quality of data being gathered in clinical trials and epidemiological studies.

Autofluorescence, Drusen, Lipofuscin

In addition to standard fundus photography, autofluorescence (AF) imaging with the scanning laser ophthalmoscope (SLO) has played a greater role in understanding drusen and AMD. It is already clear that the autofluorescence of RPE lipofuscin, which contains known fluorophores including A2E, is related to AMD *(38,39)*. In a landmark study, von Ruckmann et al. demonstrated focally increased AF (FIAF) in a broad range of AMD patients *(40,41)*. For further details on the relationship of RPE lipofuscin to the cell biology of AMD, *see* the companion chapter by Sparrow. Lipofuscin is also imaged by AF as it accumulates in the flecks of juvenile macular degeneration or Stargardt disease (STGD). For greater details, *see* Chapter 5 (the companion chapter by Allikmets).

Recent work has dealt with the relationship between the distribution of drusen and increased or decreased AF. The subjective study of Lois et al. *(40)* reported that FIAF and drusen are mostly *independent* markers for Stage 3 AMD. However, evidence using image registration techniques subsequently pointed to a possible co-localization between drusen and autofluorescence when only large soft drusen are present, and changes in this relationship when pigment abnormalities or geographic atrophy are also present *(42)*. *See* Figs. 1 and 2. Image registration allows precise comparison between fundus photographs and AF images. These techniques may prove consequential in future studies of the dynamic characteristics and life history of drusen as they correspond to lipofuscin distribution in a patient with AMD.

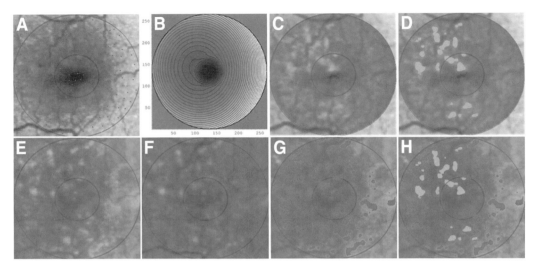

Fig. 1. (A) Original AF image. Image has been registered with the color photo (E). Note exact vascular correspondence, verifying good registration. The red dots show the selection of AF background, avoiding vessels and FIAF. (B) Geometric model constructed from the background points of AF image. (C) AF image leveled by the model constructed from the background dots. FIAF centrally, which was dim in the original, is now distinct. (D) FIAF segmentation by an Otsu threshold. (E) Original fundus image with drusen (3000-μm region). (F) Green channel of original image. (G) Drusen segmentation by leveling/thresholding (F), superimposed in green on (E). (H) FIAF superimposed on the drusen in the color photo, showing 78% of FIAF co-localized with drusen.

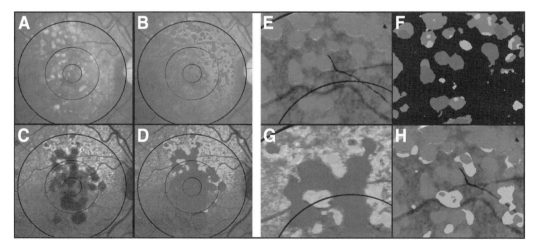

Fig. 2. (A) The original fundus photograph with drusen and GA. (B) Drusen segmented by the automated method and overlaid on (A). (C) The AF image has been registered with the fundus photograph and segmented in (D) into atrophic regions with decreased fluorescence, purple, and FIAF, pink, regions. The FIAF lesions, with excess lipofuscin, may precede atrophy. (E) Drusen (green) overlaid on the contrast-enhanced image. (G) Same detail of AF image (D). (F) Hypofluorescent (28% of this section) and hyperfluorescent (13%) regions are overlaid as fluorescent "stains" on the isolated drusen. Total drusen (28% of this section) are thus

Infrared Imaging

Near infrared (IR) reflectance imaging can also be performed with the confocal SLO (cSLO). IR imaging of the macula by cSLO was first aimed at the detection of drusen and subretinal structures *(43–47)*. Kirkpatrick *(43)* found IR less reliable than photography for drusen identification. Elsner *(44)* demonstrated the presence of small (<25 μ) hyperreflectant subretinal deposits in most normal subjects over age 20 and in all AMD subjects. These deposits were not visible in photographs, but larger deposits in AMD subjects correlated with photographically visible drusen. IR images of patients with early STGD show more abnormalities than do photographs *(48)*, owing to the sensitivity of IR imaging to pigmentation of the RPE.

A precise accounting of all AMD lesions in a given macula also provides the best possible phenotypic data to correlate with environmental and genetic risk factors. In a complex disorder like AMD, with widely varying phenotypes and multiple risk factors, these relationships can be uncovered only with very large data sets such as in the AREDS study. From this will flow a wealth of quantitative relationships, thereby improving the likelihood of discovering significant phenotype–genotype correlations. The discovery of specific DNA mutations associated with AMD lesions could form the basis for early diagnosis of individuals at risk, as well as the development of therapies based on specific molecular defects. These advances would extend profound health and social benefits to our aging population.

As the parallel tasks of segmenting drusen and RPE abnormalities in color fundus photographs and analyzing AF and IR abnormalities in SLO images proceed, image registration directly compares the features in these three imaging modalities. Segmentation and registration of macular images permits construction of disease metrics for understanding the relationships between disease components. Early work by Goldbaum et al. and Hart et al. *(49,50)* pointed this out. Many general registration techniques are available *(51)*. Montaging and mosaicing *(52–55)* have also been explored for rapid image alignment. Unfortunately, systematic research using ophthalmic image registration to study disease processes has just begun. For example, Lois et al. studied the relationship of AF abnormalities with drusen in a completely subjective manner *(40)* and concluded that they were essentially independent. However, image registration and quantitative lesion segmentation demonstrated significant co-localization of drusen and hyper AF in Stage 3 AMD *(42)*. In view of the importance of AF, registration of photographs, and SLO images for data fusion, an established technique in computer vision *(56–58)* would seem essential for understanding disease mechanisms. Digital registration of serial fundus photographs for analysis of AMD lesions such as drusen or GA, carries obvious advantages *(58)*, but manual methods continue to predominate *(59,60)*.

Fig. 2. *(Continued)* subdivided into normo- (15%), hyper- (3%), and hypofluorescent (10%) drusen. In particular, the coincidence of drusen and FIAF (3%) is a small portion of this section. **(H)** Reversing the overlay (placing the drusen on top of the FIAF) demonstrates the more extensive FIAF (pink) *adjacent* to, but not within, the drusen and GA (10% of this section). The remainder (3%) of the hyper AF that co-localized with drusen hence comprises only 3 out of 13 (23%)of the total FIAF.

Digital Fundus Photography and Segmentation of Drusen

Digital image analysis techniques faced three significant obstacles in drusen identification *(43,61–66)*. First, the inherent nature of the reflectance of the normal macula is nonuniform. There is less reflectance centrally and increasing reflectance moving out towards the arcades. Local threshold approaches to drusen segmentation met with only partial success because the background variability limited the extent to which purely histogram-based methods could succeed. This increased the need for operator intervention and has been the main obstacle to automating drusen segmentation.

The second major obstacle to drusen identification has been that of object recognition. A computer must ultimately learn to differentiate drusen from areas of retinal pigment epithelial hypopigmentation, exudates and scars. Goldbaum suggested subtleties of coloration and shape as modes of automated recognition *(67)*. However, this subject has not been developed further and, at present, the complete attention of the operator during the preprocessing phase is required to exclude such confounders in approx 20% of images *(68)*.

The third major obstacle to drusen identification and equally challenging is that of boundary definition: soft, indistinct drusen have no precise boundary and, thereby, the solution to their segmentation, by definition, cannot be precise. The central color fades into the background peripherally, and on stereo viewing there is no well-defined edge. Soft drusen, that are confluent or coalesce as part of the natural course of AMD are particularly "boundaryless." Practical segmentation of drusen then requires that areas of drusen determined by a digital method agree, in aggregate, with the judgments of a qualified grader. This approach was adopted by Shin et al. for validation of their method *(61)*. However, expert manual drawings themselves are necessarily variable. In some cases, expert manual drawing can vary as much as digital segmentation methods. Indeed, specificity and sensitivity calculations for expert manual drawings of two retinal experts can demonstrate significant inter-observer differences *(69)*. Therefore, achieving comparable accuracy in automated drusen segmentation relative to an acceptable stereo viewing expert grader remains a laudable goal during refinement of the digital segmentation methods.

An objection to digital techniques generally might be loss of information compared to color slides, but recent work by Lee et al. and Scholl et al. has demonstrated that even compressed images perform well for drusen identification *(70,71)*.

Complex images combining drusen and RPE hypopigmentation are more difficult, because these lesions may not be separable on green channel reflectance alone *(63)*. The use of red-green color space discriminants has been suggested *(67)* with very limited results. Modern neural net techniques and matrix factorization techniques merit investigation *(72–74)*. Now, we will take a closer look at digital segmentation methods as they have been developed thus far.

AUTOMATED DRUSEN MEASUREMENT BY THE MATHEMATICAL BACKGROUND MODEL

The Concept

The key concept in this method, which is called background leveling, uses a mathematical model to reconstruct the macular background from selected subsets and then remove this background variability from the entire image. This allows global threshold selection for uniform object identification, freed from the local thresholding obstacle that had so far prevented digital methods from being effective. This concept is quite general, i.e., not restricted to ophthalmic images, and appears to be original in the medical and imaging literature. In the macular applications, the algorithm gains additional power by exploiting the specific geometry of macular reflectance, as will be described. Further, despite the past inadequacy of adaptive histogram techniques, they become much more effective for automated thresholding *after the model has leveled the background.*

The first step was to demonstrate that the mathematical model, quadratic polynomials in several zones with cubic spline interpolation in blending regions between the zones, can approximate the global macular image background of a normal photograph or AF image with sufficient accuracy to allow its reconstruction and leveling *(75–77)*. The next step was to show that the model, operating on user-defined subsets of background data in abnormal images, was still capable of accurately leveling the background for reliable segmentation of drusen *(78)*. Finally, background leveling was combined with the well-known histogram-based Otsu method *(79)* for background selection and final threshold selection to achieve a largely automated method of drusen segmentation *(69)*.

Algorithm for Drusen Segmentation

The following synopsis of the key steps in the automated algorithm illustrates these principles. Many elementary image-processing details are omitted and can be found in Smith et al. *(69)*.

In this example, the region studied was the central 3000-μm diameter circle (the combined central and middle subfields defined by the Wisconsin grading template: central subfield, the 1-mm circle of diameter; middle subfield, the 3-mm annulus of outer diameter). The green channel, in which drusen have the greatest contrast, is extracted from a high-resolution digital fundus photograph or digitized film photograph.

Operator Preprocessing

Any potential confounding lesions such as GA or marked RPE hypopigmentation with elevated green channel values are removed manually. It has been estimated this would be necessary in about 20% of cases *(68)*.

Luteal Compensation

The first step is a luteal pigment correction applied to the green channel of the standardized image. The ratio of the median values of the histograms of the green channel in the middle and central subfields was calculated. This ratio was applied to a Gaussian distribution centered on the fovea and having a half-maximum at 600-μm diameter. The green channel was multiplied by this Gaussian to produce the luteal compensated image. All further processing and segmentation was carried out on this image.

Two Zone Math Model

Zone 1 is the central subfield, Zone 2 the annulus of diameters 1000 and 3000 μm. The pixel gray levels were considered to be functions of their pixel coordinates *(x,y)* in the *x–y* plane. The general quadratic $q(x, y) = ax^2 + bxy + cy^2 + dx + ey + \text{constant}$ in two variables was fit by custom software employing least squares methods to any chosen background input of green channel gray levels to optimize the six coefficients (a, b, c, d, e, constant) *(76)*. In this case, the model consisted of a set of two quadratics, one for each zone, with cubic spline interpolations at the boundary *(76)*.

Initial Background Selection by Otsu Method

The automatic histogram based thresholding technique known as the Otsu method *(79)* in each zone provided initial input to the background model. Briefly, let the pixels in the green channel be represented in L gray levels [1, 2,...,L]. Suppose the pixels are dichotomized into two classes C_0 and C_1 by a threshold at level k. C_0 denotes pixels with levels [1,...,k] and C_1 denotes pixels with levels [$k + 1$,...,L]. Ideally, C_0 and C_1 would represent background and drusen. The Otsu method uses the criterion of between-class variance and selects the threshold k that maximizes this variance *(79)*.

The Otsu method can be generalized to the case of two thresholds k and m, where there are three classes C_0, C_1, and C_2 defined by pixels with levels [1,...,k], [$k + 1$,...,m], and [$m + 1$,...,L], respectively. In a given image, these classes might represent background, objects of interest, and other objects (e.g., retinal vessels), in some permutation. The criterion for class separability is the total between-class variance.

The Otsu method may also be performed sequentially to subdivide a given class. That is, if a given class C is already defined, then C may be treated as the initial histogram and one can apply an Otsu method to subdivide C into two (or three) classes.

Operator Choices (Supervision)

The two-threshold Otsu method was used to provide an initial segmentation by thresholds k and m in each region into three desired classes: C_0 (dark nonbackground sources, e.g., vessels and pigment), C_1 (background), and C_2 (drusen). In particular, for each region there was an initial choice of background, C_1, for input to the mathematical background model. The operator could also modify the Otsu method by choosing among two other options: (1) If multiple large, soft, ill-defined drusen were present, the upper (drusen) thresholds were each reduced by four gray levels; (2) If few drusen (5% range) were present in a region, the drusen class C_2 was subdivided again by the single threshold Otsu method, with the higher values becoming the new C_2 and the lower values included in C_1. These were the only operator decisions needed to determine C_1 (the background) for input to the model. The rest of the algorithm up to final segmentation was completely automatic.

Background Leveling and Iteration

Let Z be the luteal corrected image data and let Q be the two quadratic, two-zone model blended by cubic splines at the 1000-μm boundary, that was fit to the background data C_1 in each zone determined by the Otsu method specified previously. The first leveled image Z_1 is defined as

$$Z_1 = Z - Q + 125$$

The constant offset 125 maintains an image with mean approx 125. The process can now also be iterated if desired, with the leveled image input to this algorithm. The final drusen segmentation was then obtained by applying the specified Otsu method to the final leveled image and removing any confounding lesions identified in manual preprocessing. In practice, the final results changed little after two iterations of the leveling process.

Validation

Validation with the established standard of stereo fundus photo grading was obtained manually as follows: a retinal expert drew on a graphic tablet the boundaries of all lesions identified in a suitably contrast-enhanced image. As the user drew, the 1-pixel pencil tool in Photoshop (Photoshop 7.0, Adobe Systems Inc., San Jose, CA) outlined the lesions in a transparent digital layer. Reference was also made as needed to the stereo fundus photographs to determine the exact boundary. Drusen areas were measured, and in cases in which the experts disagreed by more than 5%, the two graders collaborated to redraw to consensus. On a total of 20 images, the drusen areas were also measured by the automated method and compared to a stereo viewing drawing of an expert grader. False-positive pixels (drusen areas found by the automated method but not selected by the retinal expert) and false-negative pixels (drusen areas selected by the retinal expert but not selected by the automated method) were also identified. The specificity and sensitivity of the automated method were 0.81 and 0.70, respectively.

APPLICATION: SEGMENTATION AND CO-LOCALIZATION OF DRUSEN AND AUTOFLUORESCENCE

The model was used to level the background and perform precise segmentation of drusen in fundus photographs and FIAF in SLO images from a patients with Stage 3 AMD (large, soft drusen and no pigment abnormalities). A close relationship was demonstrated in the registered images between the spatial distribution of drusen and FIAF (Fig. 1). In other patients with Stage 4 AMD (large, soft drusen and GA), the FIAF occurs in the border zone of the GA and adjacent to some drusen, but no longer coincident with the drusen (Fig. 2). This small series suggested that focal lipofuscin accumulation in AMD as measured by autofluorescence largely co-localizes with soft drusen in Stage 3 and then the relationship changes in the presence of pigment abnormalities. This suggested that dispersal of drusen-associated lipofuscin may be a marker for disease progression in AMD.

THE FUTURE OF MACULAR IMAGE ANALYSIS

The future lies in the further development of efficient algorithms for segmentation, classification, and quantitative analysis of retinal images in register, each carrying different information, to yield powerful tools for quantitative macular research and telemedical applications for early disease detection. The global mathematical models described herein for photographs and SLO images will be applied to image data from techniques such as fluorescence lifetime *(80)* and hyperspectral *(81,82)* imaging not yet deployed in AMD research. With these data researchers will move beyond lesion identification and be able to identify in vivo individual molecular signatures. As we exploit new information from disease metrics based on data fusion from all these images in registration, there will inevitably be process failures and sources of outcome variability such as lens effects. New algorithms will be developed to remedy them.

Background leveling may be useful as a preliminary step before other image analysis tools are employed *(83,84)*. Modern hierarchical neural networks and non-negative matrix factorizations for spectral decomposition *(72–74)* are among many powerful tools in Machine Vision and Pattern Recognition that have not yet reached retinal image analysis. These digital algorithms will yield scalable integrated systems suitable for high throughput in clinical trials and the potential for a deeper understanding of the pathological pathways and genetic basis of macular disease.

REFERENCES

1. Smiddy WE, Fine SL. Prognosis of patients with bilateral macular drusen. Ophthalmology 1984;91:271–277.
2. Sarks JP, Sarks SH, Killingsworth MC. Evolution of soft drusen in age-related macular degeneration. Eye 1994;8:269–283.
3. Klein R, Davis MD, Magli YL, Segal P, Klein BE, Hubbard L. The Wisconsin age-related maculopathy grading system. Ophthalmology 1991;98:1128–1134.
4. Gass J, Jallow S, Davis B. Adult Vitelliform macular detachment occuring in patients with basal laminar drusen. Am J Ophthalmol 1985;99:445–459.
5. Russell S, Mullins R, Schnieder B, Hageman G. Location, substructure, and composition of basal laminar drusen compared with aging and aging associated with age-related macular degeneration. Am J Ophthalmol 2000;129:205–214.
6. Wang JJ, Foran S, Smith W, Mitchell P. Risk of age-related macular degeneration in eyes with macular drusen or hyperpigmentation: the Blue Mountain Eye Study cohort. Arch Ophthalmol 2003;121:658–663.
7. Klein R, Klein BE, Linton KL. Prevalence of age-related maculopathy: The Beaver Dam Eye Study. Ophthalmology 1992;99:933–943.
8. Bressler SB, Maguire MG, Bressler NM, Fine SL. Relationship of drusen and abnormalities of the retinal pigment epithelium to the prognosis of neovascular macular degeneration. The Macular Photocoagulation Study group. Arch Ophthalmol 1990;108:1442–1447.
9. Hageman GS, Luthert PJ, Victor Chong NH, Johnson LV, Anderson DH, Mullins RF. An integrated hypothesis that considers drusen as biomarkers of immune-mediated processes at the RPE-Bruch's membrane interface in aging and age-related macular degeneration. Prog Retin Eye Res 2001;20:705–732.
10. Crabb J, Miyagi M, Gu X, et al. Drusen Proteome Analysis: an approach to the etiology of age-related macular degeneration. Proc Natl Acad Sci USA 2002;23:14,682–14,687.
11. Evans K, Gregory C, Wijesuriya S, et al. Assessment of the phenotypic range seen in Doyne honeycomb retinal dystrophy. Arch Ophthalmol 1997;115:904–910.
12. Narendran N, Guymer RH, Cain M, Baird PN. Analysis of the EFEMP1 gene in individuals and families with early onset drusen. Eye 2005;19:11–15.
13. D'Souza Y, Duvall-Young J, Mcleod D, Short C, Roberts I, Bonshek R. Ten year review of drusen like lesions in mesangiocapillary glomerulonephritis. Invest Ophthalmol Vis Sci 2000;4:S164.
14. Fishman G, Apple D, Goldberg M. Retinal and pigment epithelial alterations over choroidal malignant melanomas. Ann Ophthalmol 1975;7:487–489.
15. Muller H. Anatomische beitrage zur ophthalmologie. Graefe Arch Ophthalmol 1856;2:1–69.
16. Donders F. Beitrage, zur pathologischen Anatomie des auges. Graefe's Arch Clin Exp Ophthalmol 1854;1:106–118.
17. Ambati J, Anand A, Fernandez S, et al. An animal model of age-related macular degeneration in senescent Ccl-2- or Ccr-2-deficient mice. Nature Med 2003;9:1390–1397.
18. Wilson HL, Schwartz DM, Bhatt HRF, McCulloch CE, Duncan JL. Statin and aspirin therapy are associated with decreased rates of choroidal neovascularization among patients with age-related macular degeneration. Am J Ophthalmol 2004;137:615–624.
19. Friedman E. Update of the vascular model of AMD. Br J Ophthalmol 2004;88:161–163.
20. Ibrahim M, Chain B, Katz D. The injured cell: the role of the dendritic cell system as a sentinel receptor pathway. Immunol Today 1995;16:181–186.
21. Berger JW, Fine SL, Maguire MG. Age related macular degeneration, first ed. St. Louis: Mosby, 1999;263–264.

22. Gass JD. Drusen and disciform macular detachment and degeneration. Arch Ophthalmol 1973;90:206–217.
23. Gass J. Photocoagulation of macular lesions. Trans Am Acad Ophthalmol Otolaryngol 1971;75:580–608.
24. Group MPS. Argon laser photocoagulation for neovascular maculopathy: five year results from randomized clinical trials. Arch Ophthalmol 1991;109:1109–1114.
25. Wetzig P. Photocoagulation of drusen-related aging macular degeneration by photocoagulation: a long term outcome. Trans Am Ophthalmol Soc 1994;136:276–290.
26. Figueroa MS, Regueras A, Bertrand J. Laser photocoagulation to treat macular soft drusen in age-related macular degeneration. Retina 1994;14:391–396.
27. Choroidal Neovascularization Prevention Trial Study Group. Choroidal neovascularization in the Choroidal Neovascularization Prevention Trial. Ophthalmology 1998;105:1364–1372.
28. Owens SL, Bunce C, Brannon AJ, Xing W, et al. Prophylactic laser treatment hastens choroidal neovascularization in unilateral age-related maculopathy: Final results of the drusen laser study. Am J Ophthalmol 2006;141:276–281.
29. Age-Related Eye Disease Study Research Group. A randomized, placebo-controlled, clinical trial of high-dose supplementation with vitamins C and E, beta carotene, and zinc for age-related macular degeneration and vision loss: AREDS report no. 8. Arch Ophthalmol 2001;119:1417–1436.
30. Bressler NM, Bressler SB, Seddon JM, Gragoudas ES, Jacobson LP. Drusen characteristics in patients with exudative versus non-exudative age-related macular degeneration. Retina 1998;8:109–114.
31. Holz FG, Wolfensberger TJ, Piguet B, et al. Bilateral macular drusen in age-related macular degeneration. Prognosis and risk factors. Ophthalmology 1994;101:1522–1528.
32. Zarbin MA. Current concepts in the pathogenesis of age-related macular degeneration. Arch Ophthalmol 2004;122:598–614.
33. Little HL, Showman JM, Brown BW. A pilot randomized controlled study on the effect of laser photocoagulation of confluent soft macular drusen [*see* comments]. Ophthalmology 1997;104:623–631.
34. Frennesson IC, Nilsson SE. Effects of argon (green) laser treatment of soft drusen in early age-related maculopathy: a 6 month prospective study. Br J Ophthalmol 1995;79:905–909.
35. Bressler NM, Munoz B, Maguire MG, et al. Five-year incidence and disappearance of drusen and retinal pigment epithelial abnormalities. Waterman study. Arch Ophthalmol 1995;113:301–308.
36. Bird AC, Bressler NM, Bressler SB, et al. An international classification and grading system for age-related maculopathy and age-related macular degeneration. The International ARM Epidemiological Study Group. Surv Ophthalmol 1995;39:367–374.
37. Bressler SB, Bressler NM, Seddon JM, Gragoudas ES, Jacobson LP. Interobserver and intraobserver reliability in the clinical classification of drusen. Retina 1988;8:102–108.
38. Sparrow JR, Parish CA, Nashimoto M, Nakanishi K. A2E, a lipofuscin fluorophore, in human retinal pigmented epithelial cells in culture. Invest Ophthalmol Vis Sci 1999;40:2988–2995.
39. Sparrow J, Fishkin N, Zhou J, et al. A2E, a byproduct of the visual cycle. Vision Res 2003;43:2983–2990.
40. Lois N, Owens SL, Coco R, Hopkins J, Fitzke FW, Bird AC. Fundus autofluorescence in patients with age-related macular degeneration and high risk of visual loss. Am J Ophthalmol 2002;133:341–349.
41. von Ruckmann A, Fitzke FW, Bird AC. Fundus autofluorescence in age-related macular disease imaged with a laser scanning ophthalmoscope. Invest Ophthalmol Vis Sci 1997;38:478–486.
42. Busuioc M, Smith RT, Chan JK, Sparow J, Koniarek J, Nagasaki T. Drusen and autofluorescence co-localization in early and late age related macular degeneration. Invest Ophthalmol Vis Sci 2004;45:E-2961.

43. Kirkpatrick JN, Spencer T, Manivannan A, Sharp PF, Forrester JV. Quantitative image analysis of macular drusen from fundus photographs and scanning laser ophthalmoscope images. Eye 1995;9:48–55.

44. Elsner AE, Burns SA, Weiter JJ, Delori FC. Infrared imaging of sub-retinal structures in the human ocular fundus. Vision Res 1996;36:191–205.

45. Beausencourt E, Remky A, Elsner AE, Hartnett ME, Trempe CL. Infrared scanning laser tomography of macular cysts. Ophthalmology 2000;107:375–385.

46. Kunze C, Elsner AE, Beausencourt E, Moraes L, Hartnett ME, Trempe CL. Spatial extent of pigment epithelial detachments in age-related macular degeneration. Ophthalmology 1999;106:1830–1840.

47. Miura M, Elsner AE, Beausencourt E, et al. Grading of infrared confocal scanning laser tomography and video displays of digitized color slides in exudative age-related macular degeneration. Retina 2002;22:300–308.

48. Zhang X, Hargitai J, Tammur J. Macular pigment and visual acuity in Stargardt macular dystrophy. Graefes Arch Clin Exp Ophthalmol 2002;240:802–809.

49. Goldbaum MH, Kouznetsova V, Cote BL, Hart WE, Nelson M. Automated registration of digital ocular fundus images for comparison of lesions. Proc SPIE 1993;187:94–99.

50. Hart WE, Goldbaum MH. Registering retinal images using automatically selected control point pairs. Image Processing, Proceedings. ICIP-94., IEEE International Conference. Nov. 1994;3:576–580.

51. Brown LG. A survey of image registration techniques. ACM Computing Surveys 1992; 24:326–376.

52. Can A, Shen H, Turner JN, Tanenbaum HL, Roysam B. Rapid automated tracing and feature extraction from retinal fundus images using direct exploratory algorithms. IEEE Trans Inf Technol Biomed 1999;3:125–138.

53. Can A, Stewart CV, Roysam B. Robust hierarchical algorithm for constructing a mosaic from images of the curved human retina. IEEE Computer Society Conference on Computer Vision and Pattern Recognition. Vol. 2, 23–25 June, 1999.

54. Can A, Stewart CV, Roysam B, Tanenbaum HL. A feature-based technique for joint, linear estimation of high-order image-to-mosaic transformations: application to mosaicing the curved human retina. IEEE Conference on Computer Vision and Pattern Recognition. June 2000; vol. 2:585–591.

55. Rivero ME, Bartsch DU, Otto T, Freeman WR. Automated scanning laser ophthalmoscope image montages of retinal diseases. Ophthalmology 1999;106:2296–2300.

56. Bogoni L. Extending dynamic range of monochrome and color images through fusion. 15th International Conference on Pattern Recognition. Sept. 2000; vol. 3:7–12.

57. Bogoni L, Hansen M, Burt P. Image enhancement using pattern-selective color image fusion. International Conference on Image Analysis and Processing. 27–29 Sept. 1999; pp. 44–49.

58. Burt PJ, Kolczynski RJ. Enhanced image capture through fusion. Fourth International Conference on Computer Vision. 11–14 May 1993; pp. 173–182.

59. Berger JW. Quantitative, spatio-temporal image analysis of fundus features in age-related macular degeneration. Proc SPIE (Ophthalmic Technologies) 1998;3246:48–53.

60. Sunness JS, Bressler NM, Tian Y, Alexander J, Applegate CA. Measuring geographic atrophy in advanced age-related macular degeneration. Invest Ophthalmol Vis Sci 1999;40:1761–1769.

61. Shin DS, Javornik NB, Berger JW. Computer-assisted, interactive fundus image processing for macular drusen quantitation [see comments]. Ophthalmology 1999;106:1119–1125.

62. Sebag M, Peli E, Lahav M. Image analysis of changes in drusen area. Acta Ophthalmologica 1991;69:603–610.

63. Morgan WH, Cooper RL, Constable IJ, Eikelboom RH. Automated extraction and quantification of macular drusen from fundal photographs. Aust New Zealand J Ophthalmol 1994;22:7–12.

64. Peli E, Lahav M. Drusen measurement from fundus photographs using computer image analysis. Ophthalmology 1986;93:1575–1580.
65. Rapantzikos K, Zervakis M, Balas K. Detection and segmentation of drusen deposits on human retina: Potential in the diagnosis of age-related macular degeneration. Med Image Anal 2003;7:95–108.
66. Sbeh B, Cohen LD, Mimoun G, Coscas G. A new approach of geodesic reconstruction for drusen segmentation in eye fundus images. IEEE Trans Med Imaging 2001;20(12):1321–1333.
67. Goldbaum MH, Katz NP, Nelson MR, Haff LR. The discrimination of similarly colored objects in computer images of the ocular fundus. Invest Ophthalmol Vis Sci 1990; 31:617–623.
68. Sivagnanavel V, Smith RT, Chong NHV. Digital drusen quantification in high-risk patients with age related maculopathy. Invest Ophthalmol Vis Sci 2003;44:E-5002.
69. Smith RT, Chan JK, Nagasaki T, et al. Automated detection of macular drusen using geometric background leveling and threshold selection. Arch Ophthalmol 2005;123:200–207.
70. Lee MS, Shin DS, Berger JW. Grading, image analysis, and stereopsis of digitally compressed fundus images. Retina 2000;20:275–281.
71. Scholl HPN, Dandekar SS, Peto T, et al. What is lost by digitizing stereoscopic fundus color slides for macular grading in age-related maculopathy and degeneration? Ophthalmology 2004;111:125–132.
72. Sajda P, Spence C, Pearson J. Learning contextual relationships in mammograms using a hierarchical pyramid neural network. IEEE Trans Med Imaging 2002;21:239–250.
73. Sajda P, Du S, Parra L, Stoyanova R, Brown T. Recovery of constituent spectra in 3D chemical shift imaging using non-negative matrix factorization. Proc. 4th International Symposium on Independent Component Analysis and Blind Signal Separation. April, 2003, Nara, Japan, pp. 71–76.
74. Sajda P, Du S, Brown TR, et al. (2004). Non-negative matrix factorization for rapid recovery of constituent spectra in magnetic resonance chemical shift imaging of the brain. IEEE Trans Med Imaging 2004;23(12):1453–1465.
75. Smith RT, Nagasaki T, Sparrow JR, Barbazetto I, Klaver CCW, Chan JK. A method of drusen measurement based on the geometry of fundus reflectance. BioMed Eng Online 2003;2:10.
76. Smith RT, Nagasaki T, Sparrow JR, Barbazetto I, Koniarek JP, Bickmann LJ. Patterns of reflectance in macular images: representation by a mathematical model. J Biomed Optics 2004;9:162–172.
77. Smith RT, Koniarek JP, Chan JK, Nagasaki T, Sparow J, Langton K. Autofluorescence characteristics of normal foveas and reconstruction of foveal autofluorescence from limited data subsets. Invest Ophthalmol Vis Sci 2005;46(8):2940–2946.
78. Smith RT, Chan JK, Nagasaki T, Sparrow JR, Barbazetto I. A method of drusen measurement based on reconstruction of fundus reflectance. B J Opthalmol 2005;89:87–91.
79. Otsu N. A threshold selection method from gray-level histograms. IEEE Trans Syst Man Cybern 1979;9:62–66.
80. Schweitzer D, Hammer M, Schweitzer F, et al. In-vivo measurement of time-resolved autofluorescence at the human fundus. J Biomed Optics 2004;9:1214–1222.
81. Carano R, Lynch JA, Redei J, et al. Multispectral analysis of bone lesions in the hands of patients with rheumatoid arthritis. Magn Reson Imaging 2004;22:505–514.
82. Ward J, Magnotta V, Andreason NC, Ooteman W, Nopoulos P, Pierson R. Color enhancement of multispectral MR images: improving the visualization of subcortical structures. J Comput Assist Tomogr 2001;25:942–949.
83. Akita K, Kuga H. A computer method of understanding ocular fundus images. Pattern Recognit 1992;15:431–433.
84. Undrill P. Towards the automatic interpretation of retinal images. Br J Opthalmol 1996; 80:937–938.

12

RPE Lipofuscin

Formation, Properties and Relevance to Retinal Degeneration

Janet R. Sparrow

CONTENTS

INTRODUCTION

The lipofuscin that accumulates in the lysosomal compartment of many nonreplicating cells, is typically derived from the autophagocytosis of modified cellular material and exhibits an autofluorescence (AF) emission in the blue (400–500 nm) region (1,2). The aging pigment of retinal pigment epithelial (RPE) cells is unique, however, because in addition to autophagy, these nonrenewing cells are burdened by nondegradable material derived from the phagocytosis of photoreceptor outer segment (OS) membrane. Some of the constituents, whether from autophagy or OS phagocytosis, may arise from proteins and lipid modified by carbonyl–amino crosslinks that form before the molecules enter the lysosome; they likely can also be generated within the lysosomal matrix. However, studies showing that RPE lipofuscin is profoundly reduced when the 11-*cis*-retinal and all-*trans*-retinal (ATR) chromophores are not circuiting the visual cycle (3–5) demonstrate that RPE lipofuscin is generated in large part from retinoid precursors. The extensive system of conjugated double bonds within these retinoid-derived fluorophores probably explains the long wavelength fluorescence emission of RPE

From: *Ophthalmology Research: Retinal Degenerations: Biology, Diagnostics, and Therapeutics*
Edited by: J. Tombran-Tink and C. J. Barnstable © Humana Press Inc., Totowa, NJ

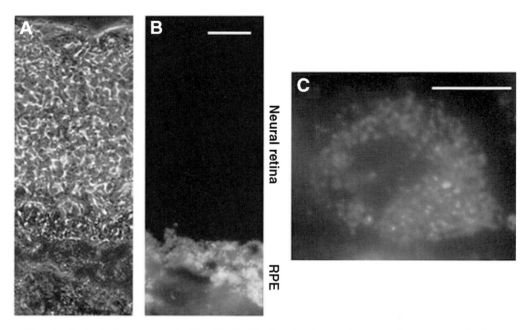

Fig. 1. Retinal pigment epithelial (RPE) lipofuscin detected as autofluorescence in human retina. A section of human retina viewed by phase contrast (**A**) and epifluorescence (**B**) imaging. The brightness at the level of RPE in **B** is caused by lipofuscin autofluorescence. In an isolated human RPE cell (**C**), RPE lipofuscin has a granule-like appearance, indicative of its location within the lysosomal compartment of the cell. Scale bar, 10 µ*M*.

lipofuscin *(2,6,7)*. Interest in the structures and modes of formation of these fluorophores stems from mounting evidence that these lipofuscin fluorophores may play a role in RPE cell dysfunction and death in some retinal disorders.

PACKAGING OF LIPOFUSCIN IN THE RPE

The lipofuscin of RPE cells is housed within membrane bound organelles—secondary lysosomes or residual bodies—that, because of their ultrastructural appearance in thin sections or after isolation by cell fractionation, are also called lipofuscin granules *(8–11)* (Fig. 1). Most of the macromolecules phagocytosed by the RPE are hydrolyzed by lysosomal enzymes to small molecules that can diffuse out of the lysosome *(9)*, but those components that are indigestible accumulate as lipofuscin. The membrane that bounds the lipofuscin-containing organelle has been visualized as a surface structure detectable by atomic force microscopy *(11)* with chloroform/methanol exposure producing irregularities in the surface owing to solubilization of phospholipids *(11)*.

Throughout life, the numbers of lipofuscin granules per RPE cell steadily increases *(12)* as does the optical density and fluorescence intensity of individual granules *(10)*. The latter observation indicates an age-related increase in the fluorophore content of these residual bodies. In young eyes, lipofuscin granules are confined to basilar portions of the RPE, whereas melanin granules are located apically *(13)*. In older eyes, however, lipofuscin granules are located throughout the entire RPE cell. Melanosomes and lipofuscin granules also undergo a poorly understood age-associated modification that involves the formation of melanin-lipofuscin complexes (melanolipofuscin granules) *(10,12)*.

CHEMICAL COMPOSITION OF RPE LIPOFUSCIN

The first comprehensive study of the composition of RPE lipofusin was performed by Eldred and Katz in 1988 *(14)*. From their chromatographical analysis of chloroform/methanol extracts of RPE, these investigators described 10 fractions. Two of these were green-emitting pigments that co-chromatographed with all-*trans*-retinol and all-*trans*-retinyl palmitate, well known retinoids of the visual cycle not suited to the definition of lipofuscin *(7,15)*. Three of the fractions were orange-emitting fluorophores with excitations in the visible range of the spectrum, features of particular interest because photoreactions in the RPE can only be initiated by visible light (>400), shorter wavelengths being absorbed by the cornea and lens. The most prominent of these three fractions was later named A2E ($C_{42}H_{58}NO$, mol wt 592) *(16)* because it could be synthesized from vitamin A aldehyde *(16,17)* and ethanolamine (2:1 ratio) and was shown to be readily detectable in extracts of human RPE by reverse phase high-performance liquid chromatography (HPLC) *(18)* (Fig. 2). The spectral characteristics of the other two orange-emitting fractions of Eldred and Katz suggest that one of them may have been iso-A2E, a less polar photoisomer of A2E *(18,19)* (Fig. 2). Whereas all of the double bonds of A2E assume the *trans* (*E*) configuration, iso-A2E has one *cis* (*Z*) olefin at the C13–14 position and its absorbance spectra is slightly blue shifted (~12 nm) relative to A2E. Under the influence of light, both in the eye and on the laboratory bench, A2E and iso-A2E are interconvertible although the *trans* configuration is favored at a ratio of 4:1 *(18)*. At least five other less abundant isomers of A2E are also detectable in hydrophobic extracts of human RPE *(19)*. These Z-isomers were originally identified because of HPLC retention times and absorbance spectra that were similar to A2E and an *m/z* (mass to charge ratio) (592) that was identical. Four of these isomers are readily accounted for by *cis* double bonds at the C9/9′–10/10′ and C11/11′–12/12′ positions whereas the fifth could contain two *cis* double bonds.

A2E is a wedge-shaped molecule with a positively charged pyridine ring and two side arms—a long and a short—each of which is derived from a molecule of ATR *(16)* (Fig. 2). Under in vivo conditions, the counterion of this pyridinium salt is presumably chloride *(18)*. Although A2E is often said to be *N*-retinylidene-*N*-retinylethanolamine, a structure initially assigned to it and then withdrawn *(17,20)*, this nomenclature is incorrect because it does not describe the pyridinium structure of the head group of A2E *(16)*.

Another constituent of RPE lipofuscin that has been structurally characterized is ATR dimer *(21,22)*, a condensation product of two ATRs that forms under conditions of release of ATR from opsin (Fig. 2). Via its aldehyde group, ATR dimer forms conjugates with various amines, one of which is phosphatidylethanolamine (PE). A protonated Schiff-base ATR dimer–PE conjugate has been detected in RPE. Besides having a novel

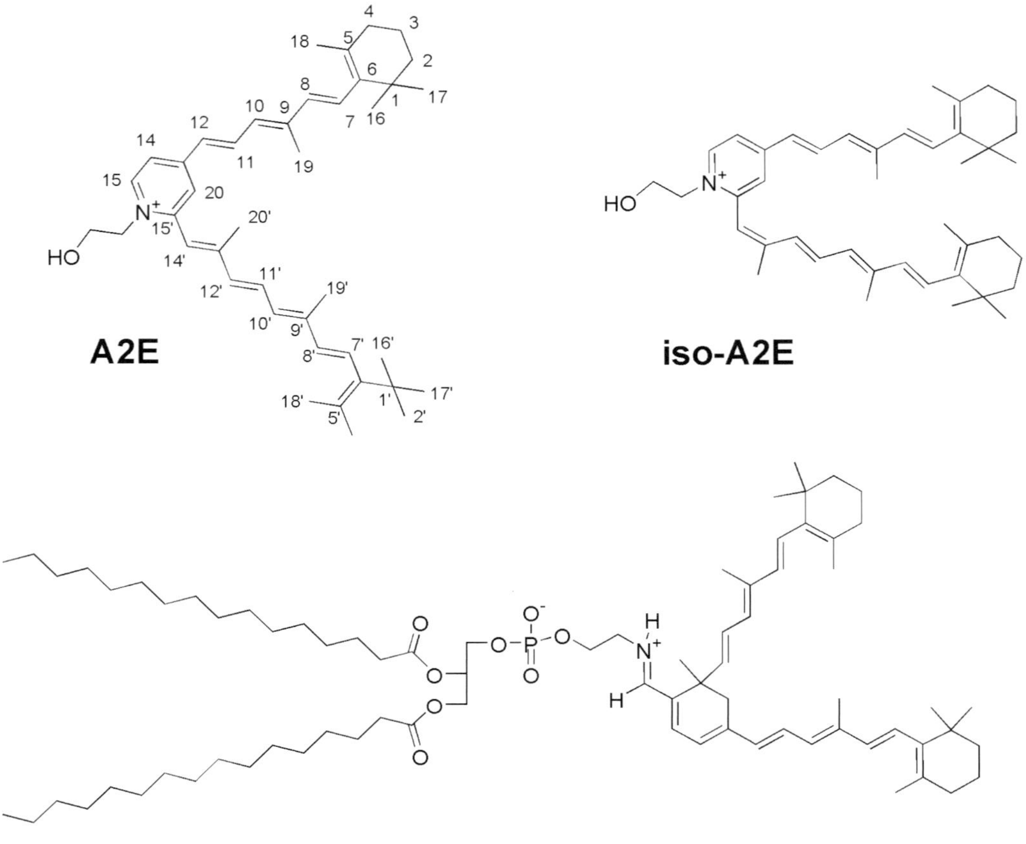

A2E **iso-A2E**

ATR dimer-PE (dipalmitoyl)

Fig. 2. Structures of some known RPE lipofuscin fluorophores, A2E; its photoisomer, iso-A2E; and all-*trans*-retinal (ATR)-dimer-phosphatylethanolamine (PE) conjugate.

structure, ATR dimer–PE has a distinctive absorbance of approx 506 nm that is red-shifted relative to A2E (~440 nm) *(22)*.

Additional components of RPE lipofuscin are generated by the photo-oxidation of A2E. These oxidative derivatives of A2E include oxiranes *(23)* and were originally identified because of the diminution in A2E fluorescence that accompanies their formation. By mass spectroscopy analysis, the addition of oxygens at carbon–carbon double bonds after blue light irradiation is detected as a series of peaks that differ in *m/z* by 16 starting from the *M*+ 592 peak attributable to A2E *(23,24)*. Intracellular A2E has been shown to undergo photo-oxidation following blue light exposure *(24)*. Moreover, it has been reported that a species with a mass of 608, corresponding to a monoepoxide of A2E, has been detected in hydrophobic extracts of human RPE cells *(25)* and A2E monoepoxides and bisepoxides have been detected in RPE isolated from Abcr$^{-/-/-}$ mice *(26)*. The addition of oxygens at olefins of A2E leads to blue shifts in A2E absorbance and to diminished A2E fluorescence; thus, an age-related increase in the proportion of A2E that is photooxidized could contribute to age-related changes in the spectral properties of RPE lipofuscin. The gradual weakening of A2E fluorescence could also account for the observation that lipofuscin fluorescence reaches a plateau or declines during the later decades of life *(27,28)*. Of course, another explanation for the latter occurrence is the loss of RPE cells. Differences in the extent to which A2E is photo-oxidized in individual lipofuscin granules may also contribute to blue shifts in some granules relative to others *(11,29)*.

The autofluorescent granules that form in RPE cultured at confluence for long periods of time likely represent inclusions derived from an autophagic source *(30)*. On the other hand, it is not known whether the fluorescent inclusions that accumulate in cultured RPE incubated with ultraviolet (UV)-irradiated or nonirradiated OS *(31–33)* are the same as native RPE lipofuscin. Photoreceptor-specific proteins, identified according to their expected molecular weight and immunoreactivity with specific antibodies, have also been detected within lipofuscin-enriched preparations *(34,35)*, but according to Feeney-Burns and colleagues, this occurs when the preparations are contaminated by phagosomes *(34)*. Other retinoid-derived fluorophores in addition to A2E, its photoisomers and photo-oxidative products and ATR dimer, may also contribute to RPE lipofuscin. For instance, ATR at high concentrations can form covalent bonds at lysine residues in rhodopsin, with the result that *bis*-retinoid adducts having emission spectra similar to A2E, are generated *(36)*. The accumulation of these adducts on rhodopsin, in vivo, could have the adverse effect of reducing light sensitivity, may necessitate the constant replacement of OS membrane *(19,37)*, and may indicate that peptides fragments bearing nondegradable fluorescent A2-moieties can be deposited in RPE cells following OS phagocytosis.

Because it was presumed for several years that RPE lipofuscin originates from peroxidized lipid, there has been considerable interest in the lipid content of RPE lipofuscin. Accordingly, a study of the lipid composition of human RPE lipofuscin granules showed that the fatty acid composition of lipofuscin granule is different than that in photoreceptor OS, the source of the precursors that form the lipofuscin38. On the other hand, it is reasonable to expect that the phospholipids that are extractable from lipofuscin granules and that increase with age in parallel with the increase in lipofuscin granules *(11,38)* originate from the phospholipids bilayer that bounds these lipofuscin-containing organelles. OS-derived products of lipid peroxidation may be present in RPE lipfuscin but are unlikely to be the major constituent because these blue-green-emitting fluorophores cannot account for the golden-yellow fluorescence of RPE lipofuscin *(6,7,39)*. Additionally, because lipofuscin granules *(40–42)* and A2E *(43)* have both been shown to mediate light-dependent lipid peroxidation, it would not be surprising if the phospholipid membrane, of lipofuscin-containing residual bodies, was to undergo lipid peroxidative damage.

A2E BIOSYNTHETIC PATHWAYS AND MODULATION OF ITS FORMATION

A2E is a pyridinium bisretinoid *(16,44)*, the formation of which begins in photoreceptor outer segments with condensation reactions between PE and first one and then a second ATR (Fig. 3). The ATR that enters the biosynthetic pathway is generated on

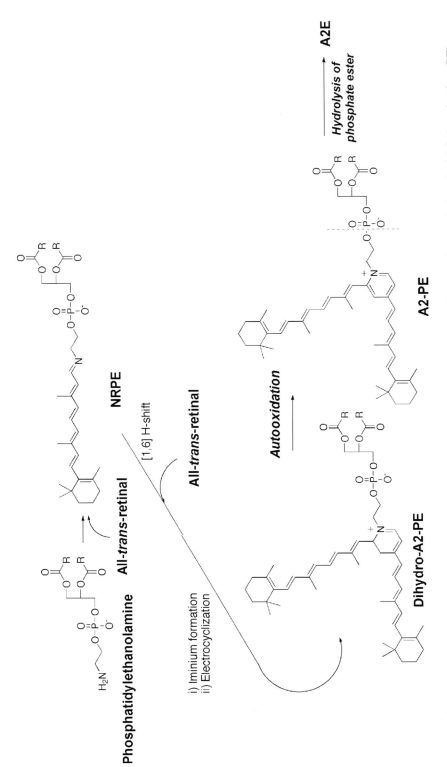

Fig. 3. Proposed biosynthetic pathway of A2E. Two molecules of all-*trans*-retinal and one molecule of phosphatidylethanolamine (PE) react in a formation cascade that generates the precursor A2-PE and after cleavage of the latter A2E is produced. Dihydro-A2-PE is an unstable intermediate that undergoes facile oxidation. NRPE, *N*-retinylidene-PE.

photoisomerization of 11-*cis*-retinal. The initial Schiff-base conjugate (*N*-retinylidene-PE [NRPE]) that forms between a single ATR and PE *(45–47)* may serve as a substrate for ABCA4 (ABCR), the photoreceptor-specific ATP-binding cassette transporter *(46,48–54)* that is the protein product of the Stargardt disease gene *(55)* (discussed below). The A2E biosynthetic pathway likely continues through a multi-step process that generates the fluorescent phosphatidyl-pyridinium bisretinoid, A2-PE *(18,19,45)*. This pigment is the precursor of A2E that accumulates in photoreceptor OS with hydrolytic cleavage of A2-PE generating A2E. The intermediate that forms prior to A2-PE is an unstable dihydropyridinium molecule that immediately undergoes autoxidation *(18,19,45)*. Because any of the intermediates formed before the oxidation is likely capable of reversal, this autoxidation step may be the last stage at which it is possible to intervene in the synthesis of A2E *(56)*.

Evidence from a number of experimental approaches has shown conclusively that A2-PE, the precursor fluorophore, forms in photoreceptor OS *(45)*, not in RPE cell lysosomes as originally thought. Nevertheless, A2-PE is not normally accrued in photoreceptor OS to detectable levels *(19,45,57)* because of continuous replacement of OS discs *(58)*. Indeed, prevention of the latter accumulation may contribute to the need for constant turnover of OS membrane. Conversely, in Royal College of Surgeons (RCS) rats, a strain characterized by an inability to phagocytosis shed OS membrane, A2-PE contributes to the orange-colored pigment and golden-yellow fluorescence that characterizes the degenerating photoreceptor outer segment debris *(59,60)* (Fig. 4). Extracts of RCS rat retina were shown to exhibit peaks with absorbances consistent with A2-PE of variable fatty acid composition, and signals detected by mass spectroscopy were attributable to A2-PE because they were of the appropriate mass and because the permanent positive charge of the quaternary nitrogen of A2-PE, by positive ionization, was definitive for A2-PE identification *(45)*. A2-PE is probably also at least one of the lipofuscin-like compounds that, because of AF, has been detected in photoreceptor cells in Stargardt disease and in some forms of retinitis pigmentosa (RP) *(61–63)*. It is unlikely that acid hydrolysis of A2-PE within RPE lysosomes can account for A2-PE cleavage *(19)*. Rather, the generation of A2E from A2-PE is probably enzyme mediated and phospholipase D has been implicated in this process *(45)*.

Only ATR that leaves the visual cycle and eludes reduction to all-*trans*-retinol by all-*trans*-retinol dehydrogenase, is available to form the ATR-derived fluorophores of RPE lipofuscin. It thus follows that conditions that increase the availability of ATR, thereby allowing random and inadvertent reactions between ATR and amine-containing molecules, enhance the opportunity for these fluorophores to form. Accordingly, because the generation of ATR in photoreceptor OS is light-activated, it is not surprising that light is a determinant of the rate of A2E formation. This assertion is supported by in vivo experiments demonstrating that the A2E precursor, A2-PE, in photoreceptor OS is augmented by exposing rats to bright light *(19)*. Additionally, dark rearing of Abcr–/– mice impedes the deposition of A2E *(57)*. Because A2E levels are not diminished if mice are raised in cyclic light and then transferred to darkness, it is also clear that once formed, A2E is not eliminated from the RPE *(57)*. Another well known factor that modulates A2E formation is the activity of ABCA4 (ABCR), the photoreceptor-specific ATP-binding cassette transporter *(49,50,51,53)* that is thought to aid in the movement of ATR to the cytosolic side of the disc membrane *(47,51,53,64,65)* where it is accessible

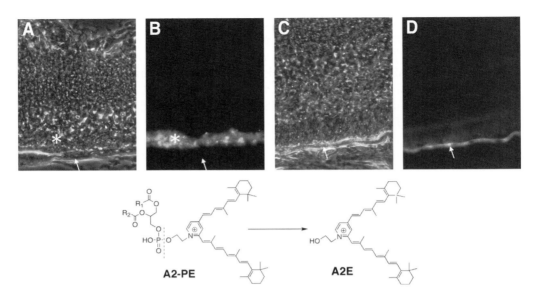

Fig. 4. Precursors of retinal pigment epithelial (RPE) lipofuscin form in photoreceptor outer segments (OS). Histological sections of retina from Royal College of Surgeons (RCS) **(A,B)** and normal **(C,D)** rats viewed as phase contrast **(A,C)** and epifluorescence **(B,D)** images. In the RCS rat, an autofluorescence (AF) is present at the level of photoreceptor OS (*), whereas AF is absent in the RPE cell layer (arrows). Conversely, in the normal rat, AF is present in the RPE cell layer (arrows). The RCS rat has a defect in the ability of RPE cells to phagocytose shed OS membrane. Consequently, the OS membrane that is discarded daily by photoreceptor cells builds up at the photoreceptor-RPE interface and an AF accumulates in this debris. Mass spectroscopy analysis *(45)* has shown that A2-PE, the precursor of A2E, accounts, at least in part, for this AF. With phagocytosis of OS debris, (normal rat), the AF material, including A2-PE, is deposited in the RPE.

to retinol dehydrogenase, the enzyme responsible for its reduction to all-*trans*-retinol *(66)*. As a consequence of the loss of Abcr activity in Abcr –/– mice, the levels of A2E in RPE cells are several-fold greater than those in normal mice *(47,67)*. In Abcr+/– mice, accumulation of A2E is approx 40% of that in the null mutant mouse *(68)*. Because the source of ATR for A2E formation is the photoisomerization of 11-*cis*-retinal, a third determinant of A2E accumulation is the kinetics of 11-*cis*-retinal regeneration. Evidence for this has come from studies of an amino acid variant in RPE65, the visual cycle protein that may have a rate-determining role in the visual cycle *(69)*. Specifically, in albino and pigmented mice in which the amino acid residue at position 450 of RPE 65 is methionine (C57BL/6J-c2J; Abcr –/– Met/Met; Abcr +/+ Met/Met) instead of leucine (BALB/cByJ; Abcr –/– Leu/Leu; Abcr +/+ Leu/Leu), recovery of the electroretinographic response following a photobleach and rhodopsin regeneration are retarded *(70–73)*, and the content of A2E in the RPE is diminished by a similar magnitude *(67)*. The acne medication isotretinoin (13-*cis*-retinoic acid; Accutane, Roche Laboratories, Nutley NJ) which was shown to reduce visual sensitivity under darkened conditions by retarding 11-*cis*-retinal regeneration74, also reduces A2E deposition in the RPE of Abcr–/– mice75. Isotretinoin was suggested to act by inhibiting

11-*cis*-retinol dehydrogenase *(75)*; however, because isotretinoin treatment causes an accumulation of both all-*trans*-retinyl esters and 11-*cis*-retinol *(74)*, perhaps the drug also acts through Rpe65.

SPECTROSCOPY AND FLUORESCENCE IMAGING

Using calibrated spectrophotometers that allowed for spectra to be corrected for the overall spectral sensitivity of the instrument, Eldred, Feeney-Burns, and colleagues showed in intact RPE cells and in extracts maintained so as to retard lipid peroxidation, that age-related RPE lipofuscin emits with a peak extending from 540 to 640 nm (yellow-orange when excited at 366 nm) and a maximum around 590 nm *(7,76)*. Similar spectra with emission maxima at 590–600 nm have been obtained from explants of adult human eyes *(77)*, although other spectra have differed *(35)*. The sieving effect associated with the use of suspensions of lipofuscin granules produces a broadening of the spectra and distinct maxima are masked; nevertheless, results obtained with these preparations have revealed excitation maxima in the blue region of the spectrum and yellow-orange emissions *(9,10)*. Blue emission peaks, thought previously to support lipid peroxidation theories of lipofuscinogenesis, were shown to be generated when fluorescent drugs have accumulated in the human tissue or when the instruments are not adequately corrected for spectral sensitivity *(7)*.

The absorbance spectrum of A2E in methanol has a major peak in the visible range at 435 nm. The excitation peak, monitored at 600 nm emission is similar in shape with a maximum at 418 nm *(78)*. The difference between the excitation and absorbance maxima of A2E has been suggested as indicating differences in the energy states of absorption and excitation *(79)*. At an excitation wavelength of 380 nm, A2E in methanol generates a yellow emission that is centered around 600 nm. This emission maximum is solvent dependent, however, with hydrophobic solvents such as *n*-butyl chloride giving a blue-shifted maximum (585 nm) as compared to emission maxima in methanol (600 nm) and buffered saline (610 nm) *(80)*. The blue shift in nonpolar solvent relative to polar media has been observed in multiple studies *(80,81)*, although a red shift in a series of solvents with decreasing polarity was reported in one *(82)*. The low fluorescence quantum yield of A2E (0.006 in chloroform; 0.002 in acetonitrile) *(82)* suggests a loss of excitation energy through nonradiative pathways such as heat or photochemical reaction. Another deactivation mechanism for A2E is the *trans-cis* isomerization *(18)*.

RPE lipofuscin is perhaps best known clinically as the source of fundus AF, the inherent fluorescence of the fundus that can be monitored noninvasively by in vivo spectrophotometry *(77,83)* and confocal laser scanning ophthalmoscopy *(84)*. Spectrophotometric measurements have shown that fundus AF is emitted across a broad band between 500 and 750 nm with a maximum of approx 590 to 630 nm *(77,85)*. The emission maximum shifts towards longer wavelengths and the spectrum narrows as the excitation wavelength is increased from 430 to 550 nm *(77)*. Although the shape of the spectra obtained at the fundus correspond well to spectra obtained from human RPE removed from the eye, optimal excitation of fundus fluorescence occurs at 510 nm, a wavelength that is red shifted relative to ex vivo RPE lipofuscin because ocular media absorption interferes with shorter wavelength excitation *(77)*.

AGE-DEPENDENCE AND SPATIAL RELATIONSHIPS

Fluorescence measurements obtained from histological material, from isolated RPE lipofuscin granules and from nonpolar extracts of human RPE have consistently demonstrated an age-related increase in fluorescent material *(9,10,13,14,27,84,86,87)*. This relationship to age has been corroborated by ultrastructural morphometric studies *(12)*. By fundus spectrophotometry, an age-associated increase in fluorescence is most pronounced at 7° to 15° from the fovea *(77)*. Estimates of the rate of lipofuscin accumulation, calculated as the ratio of levels at age 65 relative to that at age 25, have ranged from 1.1 to 4.6 *(28)*, depending on whether measurements were based on lipofuscin granule counts, in vivo fundus AF or ex vivo measurements of fluorescence *(27,28)*. In general, rates estimated from counts of lipofuscin granules have been lower than rates determined by fluorescence measurements, a difference that probably reflects the addition of lipofuscin to already existing lysosomal storage bodies. Spectrophotometric measurements of fundus AF concluded that fluorescence at age 65 was 2.8 times greater than that at age 25 *(28)*.

Several studies have concluded that the greatest accumulation occurs in the early decades of life *(12,13,27,88)* and it is likely that the accumulation reaches a plateau or decreases sometime during the 7th to 9th decade *(9,12,27,28)*. Although a leveling off of the accumulation may be the result of an age-related loss of photoreceptors *(89)*, this mechanism would not explain a decrease in fundus AF after age 70 *(28)*. The death of lipofuscin-containing RPE may be a cause of declining fluorescence in this older age group *(28,90)*. An additional contributing factor could be a change in fluorescence associated with the photooxidation of A2E *(24)*. Although the intensity of fundus AF in fellow eyes of a subject is highly correlated *(28)*, within any given age group, the intensity varies considerably amongst individuals *(28)*. This variation also appears to increase with age *(27,28)*.

Histometric studies, fundus spectrophotometry, and scanning laser ophthalmoscopy combined have shown that RPE lipofuscin is not uniformly distributed within the quadrants of the retina *(12,13,84,88,91)*. Instead, when measured 7° from the fovea, fluorescence was shown to be highest in the temporal quadrant followed by the nasal, superior and inferior quadrants, in that order *(28)*. Levels at the posterior pole are consistently found to descend to a minimum at the fovea and to reach a peak at eccentricities situated 7°– to 15° from the fovea *(12,13,84,88,91)*. From these maxima at the posterior pole, fluorescence intensity then decreases towards the periphery. Lipofuscin in the fovea is reduced approx 20–30% relative to that at an eccentricity of 7° *(28,84)*. The fluorescence profile at the fovea, when detected by fundus spectrophotometry, is influenced by the absorption of short wavelength light by macular pigment and by the distribution of RPE melanin *(77)*. Nonetheless, histological studies also reveal a lower lipofuscin content in the fovea *(13,88,91)*.

In cryostat sections of human macula, some sub-RPE deposits have a blue-green fluorescence, an emission that is blue shifted relative to RPE lipofuscin *(92)*. Nevertheless, overall AF levels at sites of drusen do not differ greatly from background *(84,93,94)*. Detailed analysis of the topography of AF over drusen has shown that for both hard and soften drusen (60–175 μm in size), there is a distinct annulus of hyperfluorescence over the outer edge of the druse and lower AF over the center. This pattern is thought to reflect drusen-associated displacement and thinning of RPE cells *(95)*. Areas that give the appearance of pigment clumping on fundus photographs, possibly because of the multilayering of the cells, can present as focal areas of highest AF and these same regions often correspond to darks areas on fluorescein angiography *(84,96)*. At sites of geographic atrophy (GA), the loss of RPE leads to a distinct absence of AF, whereas a band of increased AF at the margin of GA likely eminates from RPE cells burdened with high levels of lipofuscin *(84,93,97)*. Two observations suggest that rods make a greater contribution to lipofuscin formation than do cones. First, in the fovea where cone density is maximum and where cone density exceeds rod density, RPE lipofuscin is at a minimum. And second, the distribution of fundus AF corresponds approximately to the density of rods at eccentricities greater than $7°$ *(28)*.

PHOTOCHEMISTRY

There is abundant evidence that the photoreactivity of RPE is determined by the lipofuscin of the cell. Blue light provides the most effective irradiation and photo-dependent uptake of oxygen can be measured in both suspensions of RPE cells and in lipofuscin granules *(40,98–100)*. The photoexcited pigments in the granules form triplet states with both singlet oxygen and superoxide anion being produced *(40,41,101,102)* and the rate of oxygen uptake as well as the generation of superoxide radical anion by photoexcited lipofuscin increases with age *(40,98,100)*. The latter senescence-related increases probably reflect age-associated increases in lipofuscin. The uptake of oxygen into irradiated lipofuscin granules is also accompanied by lipid and protein oxidation *(98)*.

By detection of a 1270 nm phosphorescence and by measuring singlet-oxygen-mediated production of cholesterol hydroperoxides, it has been shown that direct photoexcitation of A2E leads to singlet oxygen production *(23,103)*. Measurements of the quantum yield of singlet-oxygen production by A2E have provided widely varying results. Not surprisingly, the lowest efficiencies (0.0008) were measured in a polar environment using cholesterol as a singlet-oxygen trap *(104)*. Using singlet oxygen phosphorescence for detection, with A2E in acetronitrile (dielectric constant, 36.6) or cyclohexane/acetone (4:1 ratio), quantum yields of 0.02 and 0.013, respectively, were reported. Because photoexcited A2E would be expected to undergo intersystem crossing from the excited singlet state to an excited triplet state with the latter transferring energy to ground state oxygen to yield high energy singlet oxygen, another approach has been to measure the efficiencies of these transformations. By photoacoustic spectroscopy, the quantum yield for intersystem crossing of A2E in methanol solution was found to be 0.0379, a value which could be expected to be higher in nonpolar solvents but which is consistent with singlet oxygen quantum yield of 0.02 (in the polar solvent acetonitrile) measured by the same investigators. The inability, in some studies, to detect triplet state A2E after photoexcitation *(82,105)* may be attributable to an absence of oxygen; as such this observation may also indicate that oxygen assists intersystem crossing in this system *(104)*.

Estimates of singlet-oxygen efficiencies in nonpolar solvents are particularly relevant because spectroscopic studies indicate that A2E associates with the hydrophobic lipid environment of intracellular membranes *(80)*, surroundings that permit a high level of oxygen solubility. As compared to quantum yields associated with compounds used in photodynamic therapy, the quantum yield of A2E is low. Nevertheless, a singlet oxygen quantum yield of 0.03 might still be considerable for a photosensitizer that accumulates in a cell over decades and that is exposed to visible light. In addition, some photosensitizers with low singlet-oxygen quantum yields (0.005) have been found to be capable of levels of destructiveness that are by orders of magnitude greater than expected on the basis of their photophysical characteristics. This destructive capability is explained on the basis of a conjugated double bond structure that serves to directly quench the singlet oxygen produced to form damaging oxidized intermediates *(106–108)*.

In oxygen-free methanol, UV-irradiated human lipofuscin extract and a synthetic lipofuscin mixture (generated by reacting PE with ATR but uncharacterized in terms of content) was shown to generate a radical resulting from either hydrogen atom or electron abstraction from the solvent *(101)*. This same synthetic lipofuscin in deuterated methanol could also generate singlet oxygen upon 355 nm excitation *(102)*. More recent studies carried out by electron paramagnet resonance (EPR) trapping suggest the formation of A2E radicals and superoxide anion following 400-nm irradiation *(103,109)*. The production of superoxide anion after photosensitization of A2E may occur because superoxide dismutase can inhibit EPR-detected free radical formation *(109)* and reduce photo-consumption of A2E (Kim and Sparrow, unpublished observations). Nevertheless, whether it is generated with low quantum efficiency *(103)* or whether it is the major species generated *(109)* remains to be determined as does mechanisms for its formation.

There is general agreement that A2E can quench singlet oxygen with considerable efficiency (rate constant of $\sim 3 \times 10^7$ $M^{-1}s^{-1}$, *[23]*; 10^8 $M^{-1}s^{-1}$ *[110]*). Indeed, it is the addition of singlet oxygen at carbon–carbon double bonds that leads to polyoxygenation of A2E. The involvement of singlet oxygen in the photoxidation of A2E is indicated by the phosphorescence detection of singlet oxygen upon 430-nm irradiation of A2E; by the potentiation of A2E photooxidation in deuterium oxide (D_2O), a solvent that that extends the life time of singlet oxygen; and by experiments demonstrating that endoperoxide-derived singlet oxygen can substitute for blue light irradiation *(23,24)*. Detection of the 608 species (592 + 16) indicates that the addition of oxygen atoms as pairs is not necessary; individual atoms of the dioxygen molecule could add to the opposite side arms of A2E and even to neighboring A2E molecules.

ADVERSE EFFECTS OF LIPOFUSCIN ACCUMULATION

Although the accumulation of lipofuscin in some cells is considered to be a benign accompaniment to aging *(1,111)*, there is considerable evidence that RPE lipofuscin has unfavorable consequences for the cell. Much of the evidence for these adverse effects has come from studies of A2E. Several unfavorable effects of A2E accumulation have been described, all of which are likely attributable to two properties of the molecule, the first of which is its amphiphilic structure. Nuclear magnetic resonance *(16)* and corroborative total chemical synthesis *(44)* have shown A2E to be a diretinoid conjugate consisting of an unprecedented pyridinium polar head group and two hydrophobic retinoid tails, an amphiphilic structure that accounts for the aggregation of A2E *(16,112)* and for its capacity to induce membrane blebbing *(37)*. The detergent-like nature of A2E is demonstrated by the tendency of A2E to permeabilize cell membranes in a concentration dependent manner *(80)* and to solubilize unilamellar vesicles *(112)*. Fluorescence anisotropy studies *(112)* indicate that the movement of A2E into the lipid bilayer may involve electrostatic interactions between the positively charged pyridinium moiety of A2E and the anionic head group of phosphatidylserine *(112)*.

Because it is within organelles of the lysosomal system that lipofuscin is housed, it would be the lysosomal membrane against which A2E might act, at least initially. Accordingly we suggested that one target of the detergent-like activity of A2E might be the membrane-bound adenosine triphosphatase that actively pumps protons into the lysosome *(80)*. Subsequently, intracellular A2E was shown to inhibit this lysosomal proton pump *(113)* resulting in alkalinization of lysosomes *(114)*. This finding is consistent with the general view that abnormal lysosomal functioning tends not to be the result of altered hydrolase activity but is more likely to be because of changes in intralysosomal pH 1.0. Questions related to the effect of A2E on enzymatic activity are significant to the notion that the failure of protein or lipid digestion, brought about by the inhibitory effects of existing lipofuscin pigments, promotes further lipofuscinogenesis. The possibility of a direct effect of A2E on lysosomal enzymes was eliminated by studies showing that the activities of lysosomal enzymes were not changed in the presence of A2E *(115)*. In cultured RPE cells that have accumulated A2E and engulfed isolated OS (ROS), the degradation rate of ROS proteins was reported to be unaltered, whereas the degradation of ROS lipids was slowed *(116)*.

Light is not only central to the formation of retinoid-derived lipofuscin constituents, it may also play a role in the mechanisms by which these fluorophores damage the RPE. For instance, in cultured RPE cells that have accumulated A2E specifically, short-wavelength light exposure can provoke a cell death program *(78,117)* that is proportional to the A2E content, that is not executed in the absence of A2E and that is implemented by cysteine-dependent proteases (caspases) and regulated by the mitochondrial protein Bcl-2 *(118)*. The potency of blue light as compared to green light reflects a wavelength dependency that is consistent with the excitation spectra of A2E *(78)* and is significant because work in animal models has shown that RPE cells demonstrate susceptibility to injury from light in the blue region of the spectrum (type 2 retinal light damage) *(119,120,120a, Busch, 1999 #70,121–123)*. DNA is a subcellular target of the photochemical reactions initiated by the illumination of A2E-laden RPE with the oxidation of guanine by the addition of oxygen at the C8 position, being one of the DNA base lesions *(124)*. Other effects include the cellular and extracellular modification of proteins by lipid peroxidation products and advanced glycation end-products and changes in gene expression *(43)*.

Evidence indicates that the generation of reactive species of oxygen upon photoexcitation of A2E is integral to blue light-mediated death of A2E-laden RPE. For instance, enhancement of cell death when assays are performed in D_2O-based media, together with the protection afforded by quenchers/scavengers (histidine, DABCO, azide) implicates singlet oxygen. Although the singlet oxygen that is generated by photosensitization of A2E is a potentially important cytotoxic agent, given its short lifetime, singlet oxygen cannot account for damage distant from the lysosome (e.g., DNA damage) *(24,124,125)*. Additionally, because much of the singlet oxygen is quenched by A2E *(23,110)* such that polyoxygenated forms of A2E are generated *(23,24)*, it may be that photooxidation products of A2E are involved.

CLINICAL IMPLICATIONS OF LIPOFUSCIN ACCUMULATION

The importance of RPE lipofuscin to the progression of recessive Stargardt disease (STGD) has long been recognized *(126–128)*. For instance, the mutations in both alleles of the ABCA4 (ABCR) gene that are the cause of this form of macular degeneration *(55,130–132)* result in levels of fundus AF that are three- to fourfold higher than in age-matched normal subjects *(133,134)*. A2E, in particular, was reported to be increased 6- to 12-fold in samples obtained from the perimacular region of human Stargardt retinas *(57)* and to be increased 10- to 20-fold in the Abcr null mutant mouse *(47,57,67,68)*. The spectral characteristics of fundus autofluoresence in patients with STGD is similar to that of age-related fundus AF, suggesting a similar composition. The one exception to this is the emission spectra on the hyperfluorescence sites known as flecks *(135)*, wherein fluorescence emission is blue shifted by 10–15 nm *(133)*. This shift may indicate a difference in the relative levels of individual fluorophores in these areas or a change in the status of these fluorophores. With regard to the latter, it may be relevant that photooxidized A2E, which is blue shifted, has been observed in Abcr–/– mice *(26)*. Fundus AF is of course absent within zones of atrophy in STGD *(133)*.

Other inherited retinal degenerations, in particular a majority of cases of autosomal recessive (AR) cone-rod dystrophy and a form of AR RP (RP19) are caused by mutations in ABCA4 and are also associated with high levels of RPE lipofuscin *(136–140)*. Indeed, mutational analysis indicates that in the Stargardt phenotype, partial functioning of the ABCR protein is retained at one or both of the mutant alleles *(55)*, whereas subjects harboring two null ABCR alleles and having the most severe clinical phenotype (RP19) are presumed to have an absence of ABCR activity *(55,141)*. A model such as this whereby the severity of the disorder is inversely correlated with residual ABCA4/ABCR activity, would also predict that RPE lipofuscin levels would vary concomitantly. However, this scenario has not been tested. Because ABCR has been linked to the visual cycle, it is reasonable to expect that the lipofuscin that accumulates in the RPE in these ABCR-associated disorders and with age is all of similar composition. Whether or not this is the case for the lipofuscin-like material that accumulates in RPE cells in vitelliform macular dystrophy (VMD; Best disease) an early onset autosomal dominant disorder *(142–144)* is not known. Bestrophin, the product of the VMD gene, is thought to participate in the generation of chloride conductances *(145,146)* but studies of the function of bestrophin have not yet clarified the link between deficient bestrophin functioning and accumulation of lipofuscin. Mechanisms involved in the high levels of lipofuscin that have also been described with point mutations (Arg172Trp) in the human RDS/peripherin gene *(134,147)* are also not understood. On the other hand, it may be that the AF material that accumulates in RPE cells in Batten disease (juvenile neuronal ceroid lipofuscinosis [JNCL]; CLN3) an AR neurodegenerative disorder, is of a different composition because these intracellular inclusions are present not only in RPE, but also in neurons of the retina and brain and one of the constituents of this material is the subunit c of mitochondrial ATP synthase *(148–150)*.

Circumstantial evidence linking RPE lipofuscin to age-related macular degeneration (AMD) has for many years caused investigators to suggest that the accumulation of this material is involved in AMD etiology *(151–157)*. The features of RPE lipofuscin accumulation that have caused it to be implicated in atrophic AMD include its age-related accumulation and its concentration within central retina. That lipofuscin levels are greater in whites than in blacks *(86,88)* may also be significant because AMD is more prevalent amongst white than amongst black persons *(158)*. Additionally, in vivo monitoring of RPE lipofuscin by ophthalmoscopic imaging of fundus AF has demonstrated that the presence of focal areas of increased AF are a risk for progression to AMD *(96)* and have established that zones of RPE exhibiting intense AF are prone to atrophy *(93,137)*. With regard to associations between AMD and RPE lipofuscin, it is also of interest that two factors known to influence the acquisition of lipofuscin by RPE cells, light exposure *(151,152,159–161)* and mutations in ABCA4 (ABCR) (discussed previously), have also been considered as risk factors for AMD. Heterozygous mutations in the ABCA4 gene are reported in some *(131,129 Allikmets, 1997 #13,162)*, but not all studies *(163)* to increase the risk of atrophic AMD in a subgroup of patients. Given evidence that ABCA4-mediated transport of PE-ATR Schiff-base adduct serves as a mechanism for removing ATR from the disc interior *(54)*, decreased ABCA4 activity owing to heterozygous mutations in the gene, could, over the course of a lifetime, lead to enhanced formation of RPE lipofuscin fluorophores. Light exposure has long been postulated as a contributor to AMD, although epidemiological studies investigating this notion have generated conflicting results *(159–161,164–167)*. The issue of an association between light exposure and AMD may have particular relevance to cataract surgery because an intraocular lens implant exposes the retina to blue light levels previously attenuated by the yellowing crystalline lens *(168,169)*. In this regard, several studies have demonstrated an association between cataract surgery and AMD *(170–173)*.

SUMMARY

RPE cells are distinctive in that they are exposed to visible light, whereas at the same time housing photodynamic molecules that aggregate as lipofuscin. These fluorophores accumulate in RPE cells not because the complement of lysosomal enzymes in these cells is defective but because the structures of the fluorophores are unusual and not amenable to degradation. Emerging evidence indicates that the lipofuscin of RPE cells is inimitable because much of this material forms as a consequence of the light capturing function of the retina. An origin from retinoids that leave the visual cycle also explains the finding that the accumulation of RPE lipofuscin is most marked in central retina, the area having the greatest accumulation of visual chromophore. Continued investigation into the composition, biogenesis, and activity of RPE lipofuscin fluorophores will reveal the extent to which the accumulation of these pigments renders the macula prone to insult and may lead to therapies that can reduce the accumulation.

REFERENCES

1. Cuervo AM, Dice JR. When lysosomes get old. Exp Gerontol 2000;35:119–131.
2. Yin D. Biochemical basis of lipofuscin, ceroid, and age pigment-like fluorophores. Free Rad Biol Med 1996;21:871–888.
3. Katz ML, Redmond TM. Effect of Rpe65 knockout on accumulation of lipofuscin fluorophores in the retinal pigment epithelium. Invest Ophthalmol Vis Sci 2001; 42:3023–3030.
4. Katz ML, Drea CM, Robison WG, Jr. Relationship between dietary retinol and lipofuscin in the retinal pigment epithelium. Mech Ageing Dev 1986;35:291–305.

 5. Katz ML, Eldred GE, Robison WG Jr. Lipofuscin autofluorescence: evidence for vitamin A involvement in the retina. Mech Ageing Dev 1987;39:81–90.

 6. Eldred GE, Katz ML. The autofluorescent products of lipid peroxidation may not be lipofuscin-like [see comments]. Free Radic Biol Med 1989;7:157–163.

 7. Eldred G, Katz ML. The lipid peroxidation theory of lipofuscinogenesis cannot yet be confirmed. Free Rad Biol Med 1991;10:445–447.

 8. Clanc CMR, Krogmeier JR, Pawlak A, Rozanowska M, Sarna T, Dunn RC, Simon JD. Atomic force microscopy and near-field scanning optical microscopy measurements of single human retinal lipofuscin granules. J Phys Chem B 2000;104:12,098–12,101.

 9. Feeney-Burns L, Eldred GE. The fate of the phagosome: conversion to 'age pigment' and impact in human retinal pigment epithelium. Trans Ophthalmol Soc UK 1983; 103:416–421.

10. Boulton M, Docchio F, Dayhaw-Barker P, Ramponi R, Cubeddu R. Age-related changes in the morphology, absorption and fluorescence of melanosomes and lipofuscin granules of the retinal pigment epithelium. Vision Res 1990;30:1291–1303.

11. Haralampus-Grynaviski NM, Lamb LE, Clancy CMR, et al. Spectroscopic and morphological studies of human retinal lipofuscin granules. Proc Natl Acad Sci USA 2003;100:3179–3184.

12. Feeney-Burns L, Hilderbrand ES, Eldridge S. Aging human RPE: morphometric analysis of macular, equatorial, and peripheral cells. Invest Ophthalmol Vis Sci 1984;25:195–200.

13. Wing GL, Blanchard GC, Weiter JJ. The topography and age relationship of lipofuscin concentration in the retinal pigment epithelium. Invest Ophthalmol Vis Sci 1978; 17:601–607.

14. Eldred GE, Katz ML. Fluorophores of the human retinal pigment epithelium: separation and spectral characterization. Exp Eye Res 1988;47:71–86.

15. Eldred GE. Vitamins A and E in RPE lipofuscin formation and implications for age-related macular degeneration. In: la Vail, MM, Anderson, RE, Hollyfield, JG, eds. Inherited and environmentally inducedd retinal degenerations, New York: Alan R. Liss, 1989:113–129.

16. Sakai N, Decatur J, Nakanishi K, Eldred GE. Ocular age pigment "A2E": an unprecedented pyridinium bisretinoid. J Am Chem Soc 1996;118:1559–1560.

17. Eldred GE, Lasky MR. Retinal age pigments generated by self-assembling lysosomotropic detergents. Nature 1993;361:724–726.

18. Parish CA, Hashimoto M, Nakanishi K, Dillon J, Sparrow JR. Isolation and one-step preparation of A2E and iso-A2E, fluorophores from human retinal pigment epithelium. Proc Natl Acad Sci USA 1998;95:14,609–14,613.

19. Ben-Shabat S, Parish CA, Vollmer HR, et al. Biosynthetic studies of A2E, a major fluorophore of RPE lipofuscin. J Biol Chem 2002;277:7183–7190.

20. Eldred GE. Age pigment structure. Nature 1993;364:396.

21. Fishkin N, Pescitelli G, Sparrow JR, Nakanishi K, Berova N. Absolute configurational determination of an all-trans-retinal dimer isolated from photoreceptor outer segments. Chirality 2004;16:637–641.

22. Fishkin NE, Pescitelli G, Itagaki Y, et al. Isolation and characterization of a novel RPE fluorophore: an all-trans-retinal dimer. Invest Ophthalmol Vis Sci 2004;45:E-abstract 1803.

23. Ben-Shabat S, Itagaki Y, Jockusch S, Sparrow JR, Turro NJ, Nakanishi K. Formation of a nona-oxirane from A2E, a lipofuscin fluorophore related to macular degeneration, and evidence of singlet oxygen involvement. Angew Chem Int Ed 2002; 41:814–817.

24. Sparrow JR, Zhou J, Ben-Shabat S, Vollmer H, Itagaki Y, Nakanishi K. Involvement of oxidative mechanisms in blue light induced damage to A2E-laden RPE. Invest Ophthalmol Vis Sci 2002;43:1222–1227.

25. Avalle LB, Wang Z, Dillon JP, Gaillard ER. Observation of A2E oxidation products in human retinal lipofuscin. Exp Eye Res 2004;78:895–898.

26. Radu RA, Mata NL, Bagla A, Travis GH. Light exposure stimulates formation of A2E oxiranes in a mouse model of Stargardt's macular degeneration. Proc Natl Acad Sci USA 2004;101:5928–5933.

27. Okubo A, Rosa RHJ, Bunce CV, et al. The relationships of age changes in retinal pigment epithelium and Bruch's membrane. Invest Ophthalmol Vis Sci 1999;40:443–449.

28. Delori FC, Goger DG, Dorey CK. Age-related accumulation and spatial distribution of lipofuscin in RPE of normal subjects. Invest Ophthalmol Vis Sci 2001;42:1855–1866.

29. Haralampus-Grynaviski NM, Lamb LE, Simon JD, et al. Probing the spatial dependence of the emission spectrum of single human retinal lipofuscin granules using near-field scanning optical microscopy. Photochem Photobiol 2001;74:364–368.

30. Burke JM, Skumatz CMB. Autofluorescent inclusions in long-term postconfluent cultures of retinal pigment epithelium. Invest Ophthalmol Vis Sci 1998;39:1478–1486.

31. Boulton M, McKechnie NM, Breda J, Bayly M, Marshall J. The formation of autofluorescent granules in cultured human RPE. Invest Ophthalmol Vis Sci 1989;30:82–89.

32. Sundelin SP, Nilsson SEG. Lipofuscin-formatoin in retinal pigment epithelial cells is reduced by antioxidants. Free Rad Biol Med 2001;31:217–225.

33. Sundelin S, Wihlmark U, Nilsson SEG, Brunk UT. Lipofuscin accumulation in cultured retinal pigment epithelial cells reduces their phagocytic capacity. Curr Eye Res 1998;17:851–857.

34. Feeney-Burns L, Gao CL, Berman ER. The fate of immunoreactive opsin following phagocytosis by pigment epithelium in human and monkey retinas. Invest Ophthalmol Vis Sci 1988;29:708–719.

35. Schutt F, Ueberle B, Schnolzer M, Holz FG, Kopitz J. Proteome analysis of lipofuscin in human retinal pigment epithelial cells. FEBS Lett 2002;528:217–221.

36. Fishkin N, Jang YP, Itagaki Y, Sparrow JR, Nakanishi K. A2-rhodopsin: a new fluorophore isolated from photoreceptor outer segments. Org Biomol Chem 2003;1:1101–1105.

37. Sparrow JR, Fishkin N, Zhou J, et al. A2E, a byproduct of the visual cycle. Vision Res 2003;43:2983–2990.

38. Bazan HE, Bazan NG, Feeney-Burns L, Berman ER. Lipids in human lipofuscin-enriched subcellular fractions of two age populations. Comparison with rod outer segments and neural retina. Invest Ophthalmol Vis Sci 1990;31:1433–1443.

39. Chowdhury PK, Halder M, Choudhury PK, et al. Generation of fluorescent adducts of malondialdehyde and amino acids: toward an understanding of lipofuscin. Photochem Photobiol 2004;79:21–25.

40. Rozanowska M, Jarvis-Evans J, Korytowski W, Boulton ME, Burke JM, Sarna T. Blue light-induced reactivity of retinal age pigment. In vitro generation of oxygen-reactive species. J Biol Chem 1995;270:18,825–18,830.

41. Rozanowska M, Wessels J, Boulton M, et al. Blue light-induced singlet oxygen generation by retinal lipofuscin in non-polar media. Free Rad Biol Med 1998;24:1107–1112.

42. Wassell J, Davies S, Bardsley W, Boulton M. The photoreactivity of the retinal age pigment lipofuscin. J Biol Chem 1999;274:23,828–23,832.

43. Zhou J, Cai B, Jang YP, Pachydaki S, Schmidt AM, Sparrow JR. Mechanisms for the induction of HNE- MDA- and AGE-adducts, RAGE and VEGF in retinal pigment epithelial cells. Exp Eye Res 2005;80:567–580.

44. Ren RF, Sakai N, Nakanishi K. Total synthesis of the ocular age pigment A2E: a convergent pathway. J Am Chem Soc 1997;119:3619–3620.

45. Liu J, Itagaki Y, Ben-Shabat S, Nakanishi K, Sparrow JR. The biosynthesis of A2E, a fluorophore of aging retina, involves the formation of the precursor, A2-PE, in the photoreceptor outer segment membrane. J Biol Chem 2000;275:29,354–29,360.

46. Sun H, Molday RS, Nathans J. Retinal stimulates ATP hydrolysis by purified and recon-stituted ABCR, the photoreceptor-specific ATP-binding cassette transporter responsible for Stargardt disease. J Biol Chem 1999;274:8269–8281.

47. Weng J, Mata NL, Azarian SM, Tzekov RT, Birch DG, Travis GH. Insights into the func-tion of Rim protein in photoreceptors and etiology of Stargardt's disease from the pheno-type in abcr knockout mice. Cell 1999;98:13–23.

48. Molday RS, Molday LL. Identification and characterization of multiple forms of rhodopsin and minor proteins in frog and bovine outer segment disc membranes. Electrophoresis, lectin labeling and proteolysis studies. J Biol Chem 1979;254:4653–4660.

49. Molday LL, Rabin AR, Molday RS. ABCR expression in foveal cone photoreceptors and its role in Stargardt macular dystrophy. Nat Genet 2000;25:257–258.

50. Papermaster DS, Schneider BG, Zorn MA, Kraehenbuhl JP. Immunocytochemical local-ization of a large intrinsic membrane protein to the incisures and margins of frog rod outer segment disks. J Cell Biol 1978;78:415–425.

51. Sun H, Nathans J. Stargardt's ABCR is localized to the disc membrane of retinal rod outer segments. Nat Genet 1997;17:15–16.

52. Sun H, Nathans J. Mechanistic studies of ABCR, the ABC transporter in photoreceptor outer segments responsible for autosomal recessive Stargardt disease. J Bioenerg Biomembrane 2001;33:523–530.

53. Sun H, Nathans J. ABCR, the ATP-binding cassette transporter responsible for Stargardt macular dystrophy, is an efficient target of all-trans retinal-mediated photo-oxidative damage in vitro: implications for retinal disease. J Biol Chem 2001;276:11,766–11,774.

54. Beharry S, Zhong M, Molday RS. N-retinylidene-phosphatidylethanolamine is the pre-ferred retinoid substrate for the photoreceptor-specific ABC transporter ABCA4 (ABCR). J Biol Chem 2004;279(52):53,972–53,979.

55. Allikmets R, Singh N, Sun H, et al. A photoreceptor cell-specific ATP-binding transporter gene (ABCR) is mutated in recessive Stargardt macular dystrophy. Nat Genet 1997; 15:236–246.

56. Sparrow JR. Therapy for macular degeneration: insights from acne. Proc Natl Acad Sci USA 2003;100:4353–4354.

57. Mata NL, Weng J, Travis GH. Biosynthesis of a major lipofuscin fluorophore in mice and humans with ABCR-mediated retinal and macular degeneration. Proc Natl Acad Sci USA 2000;97:7154–7159.

58. Young RW. The renewal of rod and cone outer segments in the rhesus monkey. J Cell Biol 1971;49:303–318.

59. Katz ML, Drea CM, Eldred GE, Hess HH, Robison WG Jr. Influence of early photorecep-tor degeneration on lipofuscin in the retinal pigment epithelium. Exp Eye Res 1986;43:561–573.

60. Eldred GE. The fluorophores of the RCS rat retina and implications for retinal degenera-tion. In: Hollyfield JG, Anderson RE, La Vail MM, eds. Retinal Degenerations. Boca Raton, Florida: CRC Press, 1991.

61. Birnbach CD, Jarvelainen M, Possin DE, Milam AH. Histopathology and immunocyto-chemistry of the neurosensory retina in fundus flavimaculatus. Ophthalmology 1994; 101:1211–1219.

62. Bunt-Milam AH, Kalina RE, Pagon RA. Clinical-ultrastructural study of a retinal dystro-phy. Invest Ophthalmol Vis Sci 1983;24:458–469.

63. Szamier RB, Berson EL. Retinal ultrastructure in advanced retinitis pigmentosa. Invest Ophthalmol Vis Sci 1977;16:947–962.

64. Ahn J, Wong JT, Molday RS. The effect of lipid environment and retinoids on the ATPase activity of ABCR, the photoreceptor ABC transporter responsible for Stargardt macular dystrophy. J Biol Chem 2000;275:20,399–20,405.

65. Illing M, Molday LL, Molday RS. The 220-kDa rim protein of retinal rod outer segments is a member of the ABC transporter superfamily. J Biol Chem 1997;272:10,303–10,310.

66. Saari JC, Garwin GG, Van Hooser JP, Palczewski K. Reduction of all-trans-retinal limits regeneration of visual pigment in mice. Vision Res 1998;38:1325–1333.

67. Kim SR, Fishkin N, Kong J, Nakanishi K, Allikmets R, Sparrow JR. The Rpe65 Leu450Met variant is associated with reduced levels of the RPE lipofuscin fluorophores A2E and iso-A2E. Proc Natl Acad Sci USA 2004;101:11,668–11,672.

68. Mata NL, Tzekov RT, Liu X, Weng J, Birch DG, Travis GH. Delayed dark adaptation and lipofuscin accumulation in abcr+/– mice: implications for involvement of ABCR in age-related macular degeneration. Invest Ophthalmol Vis Sci 2001;42:1685–1690.

69. Xue L, Gollapalli DR, Maiti P, Jahng WJ, Rando RR. A palmitoylation switch mechanism in the regulation of the visual cycle. Cell 2004;117:761–771.

70. Wenzel A, Reme CE, Williams TP, Hafezi F, Grimm C. The Rpe65 Leu450Met variation increases retinal resistance against light-induced degeneration by slowing rhodopsin regeneration. J Neurosci 2001;21:53–58.

71. Danciger M, Matthes MT, Yasamura D, et al. A QTL on distal chromosome 3 that influences the severity of light-induced damage to mouse photoreceptors. Mam Genome 2000;11:422–427.

72. Nusinowitz S, Nguyen L, Radu RA, Kashani Z, Farber DB, Danciger M. Electroretinographic evidence for altered phototransduction gain and slowed recovery from photobleaches in albino mice with a MET450 variant in RPE6. Exp Eye Res 2003;77:627–638.

73. Wenzel A, Grimm C, Samardzija M, Reme CE. The genetic modified Rpe65Leu$_{450}$: effect on light damage susceptibility in c-Fos-deficient mice. Invest Ophthalmol Vis Sci 2003;44:2798–2802.

74. Sieving PA, Chaudhry P, Kondo M, et al. Inhibition of the visual cycle in vivo by 13-cis retinoic acid protects from light damage and provides a mechanism for night blindness in isotretinoin therapy. Proc Natl Acad Sci USA 2001;98:1835–1840.

75. Radu RA, Mata NL, Nusinowitz S, Liu X, Sieving PA, Travis GH. Treatment with isotretinoin inhibits lipofuscin and A2E accumulation in a mouse model of recessive Stargardt's macular degeneration. Proc Natl Acad Sci USA 2003;100:4742–4747.

76. Eldred GE, Miller GV, Stark WS, Feeney-Burns L. Lipofuscin: resolution of discrepant fluorescence data. Science 1982;216:757–758.

77. Delori FC, Dorey CK, Staurenghi G, Arend O, Goger DG, Weiter JJ. In vivo fluorescence of the ocular fundus exhibits retinal pigment epithelium lipofuscin characteristics. Invest Ophthalmol Vis Sci 1995;36:718–729.

78. Sparrow JR, Nakanishi K, Parish CA. The lipofuscin fluorophore A2E mediates blue light-induced damage to retinal pigmented epithelial cells. Invest Ophthalmol Vis Sci 2000;41:1981–1989.

79. Lamb LE, Ye T, Haralampus-Grynaviski NM, et al. Primary photophysical properties of A2E in solution. J Phys Chem B 2001;105:11,507–11,512.

80. Sparrow JR, Parish CA, Hashimoto M, Nakanishi K. A2E, a lipofuscin fluorophore, in human retinal pigmented epithelial cells in culture. Invest Ophthalmol Vis Sci 1999;40:2988–2995.

81. De S, Sakmar TP. Interaction of A2E with model membranes. Implications to the pathogenesis of age-related macular degeneration. J Gen Physiol 2002;120:147–157.

82. Ragauskaite L, Heckathorn RC, Gaillard ER. Environmental effects on the photochemistry of A2E, a component of human retinal lipofuscin. Photochem Photobiol 2001; 74:483–488.

83. Delori FC. Spectrophotometer for noninvasive measurement of intrinsic fluorescence and reflectance of the ocular fundus. Appl Optics 1994;33:7439–7452.

84. von Rückmann A, Fitzke FW, Bird AC. Fundus autofluorescence in age-related macular disease imaged with a laser scanning ophthalmoscope. Invest Ophthalmol Vis Sci 1997;38:478–486.

85. Delori FC. Autofluorescence method to measure macular pigment optical densities fluorometry and autofluorescence imaging. Arch Biochem Biophys 2004;430:156–162.

86. Dorey CK, Wu G, Ebenstein D, Garsd A, Weiter JJ. Cell loss in the aging retina. Relationship to lipofuscin accumulation and macular degeneration. Invest Ophthalmol Vis Sci 1989;30:1691–1699.

87. Docchio F, Boulton M, Cubeddu R, Ramponi R, Barker PD. Age-related changes in the fluorescence of melanin and lipofuscin granules of the retinal pigment epithelium: a time-resolved fluorescence spectroscopy study. Photochem Photobiol 1991; 54:247–253.

88. Weiter JJ, Delori FC, Wing GL, Fitch KA. Retinal pigment epithelial lipofuscin and melanin and choroidal melanin in human eyes. Invest Ophthalmol Vis Sci 1986;27:145–151.

89. Curcio CA, Millican CL, Allen KA, Kalina RE. Aging of the human photoreceptor mosaic: evidence for selective vulnerability of rods in central retina. Invest Ophthalmol Vis Sci 1993;34:3278–3296.

90. Del Priore LV, Kuo YH, Tezel TH. Age-related changes in human RPE cell density and apoptosis proportion in situ. Invest Ophthalmol Vis Sci 2002;43:3312–3318.

91. Weiter JJ, Delori FC, Dorey CK. Central sparing in annular macular degeneration. Am J Ophthalmol 1988;106:286–290.

92. Marmorstein AD, Marmorstein LY, Sakaguchi H, Hollyfield JG. Spectral profiling of autofluorescence associated with lipofuscin, Bruch's Membrane, and sub-RPE deposits in normal and AMD eyes. Invest Ophthalmol Vis Sci 2002;43:2435–2441.

93. Holz FG, Bellman C, Staudt S, Schutt F, Volcker HE. Fundus autofluorescence and development of geographic atrophy in age-related macular degeneration. Invest Ophthalmol Vis Sci 2001;42:1051–1056.

94. Lois N, Owens SL, Coco R, Hopkins J, Fitzke FW, Bird AC. Fundus autofluorescence in patients with age-related macular degeneration and high risk of visual loss. Am J Ophthalmol 2002;133:341–349.

95. Delori FC, Fleckner MR, Goger DG, Weiter JJ, Dorey CK. Autofluorescence distribution associated with drusen in age-related macular degeneration. Invest Ophthalmol Vis Sci 2000;41:496–504.

96. Solbach U, Keilhauer C, Knabben H, Wolf S. Imaging of retinal autofluorescence in patients with age-related macular degeneration. Retina 1997;17:385–389.

97. Holz FG, Bellmann C, Margaritidis M, Schutt F, Otto TP, Volcker HE. Patterns of increased in vivo fundus autofluorescence in the junctional zone of geographic atrophy of the retinal pigment epithelium associated with age-related macular degeneration. Graefe's Arch Clin Exp Ophthalmol 1999;237:145–152.

98. Rozanowska M, Korytowski W, Rozanowska B, et al. Photoreactivitiy of aged human RPE melanosomes: a comparison with lipofuscin. Invest Ophthalmol Vis Sci 2002;43:2088–2096.

99. Pawlak A, Rozanowska M, Zareba M, Lamb LE, Simon JD, Sarna T. Action spectra for the photoconsumptioin of oxygen by human ocular lipofuscin and lipofuscin extracts. Arch Biochem Biophys 2002;403:59–62.

100. Rozanowska M, Pawlak A, Rozanowska B, et al. Age-related changes in the photoreactivity of retinal lipofuscin granules: role of chloroform-insoluble components. Invest Ophthalmol Vis Sci 2004;45:1052–1060.

101. Reszka K, Eldred GE, Wang RH, Chignell C, Dillon J. The photochemistry of human retinal lipofuscin as studied by EPR. Photochem Photobiol 1995;62:1005–1008.

102. Gaillard ER, Atherton SJ, Eldred G, Dillon J. Photophysical studies on human retinal lipofuscin. Photochem Photobiol 1995;61:448–453.

103. Pawlak A, Wrona M, Rozanowska M, et al. Comparison of the aerobic photoreactivity of A2E with its precursor retinal. Photochem Photobiol 2003;77:253–258.

104. Kanofsky JR, Sima PD, Richter C. Singlet-oxygen generation from A2E. Photochem Photobiol 2003;77:235–242.

105. Cantrell A, McGarvey DJ, Roberts J, Sarna T, Truscott TG. Photochemical studies of A2E. J Photochem Photobiol B: Biology 2001;64:162–165.

106. Bunting JR. A test of the singlet oxygen mechanism of cationic dye photosensitization of mitochondrial damage. Photochem Photobiol 1992;55:81–87.

107. Delaey E, van Laar F, De Vos D, Kamuhabwa A, Jacobs P, de Witte P. A comparative study of the photosensitizing characteristics of some cyanine dyes. J Photochem Photobiol B 2000;55:27–36.

108. Krieg M, Srichai MB, Redmond RW. Photophysical properties of 3,3'-dialkylthiacarbocyanine dyes in organized media: unilamellar liposomes and thin polymer films. Biochim Biophys Acta 1993;1151:168–174.

109. Gaillard ER, Avalle LB, Keller LMM, Wang Z, Reszka KJ, Dillon JP. A mechanistic study of the photooxidation of A2E, a component of human retinal lipofuscin. Exp Eye Res 2004;79:313–319.

110. Roberts JE, Kukielczak BM, Hu DN, et al. The role of A2E in prevention or enhancement of light damage in human retinal pigment epithelial cells. Photochem Photobiol 2002;75:184–190.

111. Schmucker DL, Sachs H. Quantifying dense bodies and lipofuscin during aging: a morphologist's perspective. Arch Gerontol Geriatr 2002;34:249–261.

112. De S, Sakmar TP. Interaction of A2E with model membranes. Implications to the pathogenesis of age-related macular degeneration. J Gen Physiol 2002;120:147–157.

113. Bergmann M, Schutt F, Holz FG, Kopitz J. Inhibition of the ATP-driven proton pump in RPE lysosomes by the major lipofuscin fluorophore A2E may contribute to the pathogenesis of age-related macular degeneration. FASEB J 2004;18:562–564.

114. Holz FG, Schutt F, Kopitz J, et al. Inhibition of lysosomal degradative functions in RPE cells by a retinoid component of lipofuscin. Invest Ophthalmol Vis Sci 1999;40: 737–743.

115. Berman M, Schutt F, Holz FG, Kopitz J. Does A2E, a retinoid component of lipofuscin and inhibitor of lysosomal degradative functions, directly affect the activity of lysosomal hydrolases. Exp Eye Res 2001;72:191–195.

116. Finneman SC, Leung LW, Rodriguez-Boulan E. The lipofuscin component A2E selectively inhibits phagolysosomal degradation of photoreceptor phospholipid by the retinal pigment epithelium. Proc Natl Acad Sci USA 2002;99:3842–3847.

117. Schutt F, Davies S, Kopitz J, Holz FG, Boulton ME. Photodamage to human RPE cells by A2-E, a retinoid component of lipofuscin. Invest Ophthalmol Vis Sci 2000;41: 2303–2308.

118. Sparrow JR, Cai B. Blue light-induced apoptosis of A2E-containing RPE: involvement of caspase-3 and protection by Bcl-2. Invest Ophthalmol Vis Sci 2001;42: 1356–1362.

119. Ham WTJ, Allen RG, Feeney-Burns L, et al. The involvement of the retinal pigment epithelium. In: Waxler M, Hitchins VM, eds. CRC Optical Radiation and Visual Health. Boca Raton, Florida: CRC Press, Inc., 1986:43–67.

120. Ham WT, Mueller HA, Ruffolo JJ, et al. Basic mechanisms underlying the production of photochemical lesions in the mammalian retina. Curr Eye Res 1984;3:165–174.

120a. Busch EM, Gorgels TGMF, Roberts JE, van Norren D. The effects of two stereoisomers of N-acetylcysteine on photochemical damage by UVA and blue light in rat retina. Photochem Photobiol 1999;70:353–358.

121. Borges J, Li Z-Y, Tso MO. Effects of repeated photic exposures on the monkey macula. Arch Ophthalmol 1990;108:727–733.

122. Putting BJ, Van Best JA, Vrensen GFJM, Oosterhuis JA. Blue-light-induced dysfunction of the blood-retinal barrier at the pigment epithelium in albino versus pigmented rabbits. Exp Eye Res 1994;58:31–40.

123. Paultler EL, Morita M, Beezley D. Reversible and irreversible blue light damage to the isolated mammalian pigment epithelium. In: La Vail MM, Anderson RE, Hollyfield JG, eds. Inherited and Environmental Induced Retinal Degeneration. New York: Alan R. Liss, 1989;555–567.

124. Sparrow JR, Zhou J, Cai B. DNA is a target of the photodynamic effects elicited in A2E-laden RPE by blue light illumination. Invest Ophthalmol Vis Sci 2003;44:2245–2251.

125. Sparrow JR, Vollmer-Snarr HR, Zhou J, et al. A2E-epoxides damage DNA in retinal pigment epithelial cells. Vitamin E and other antioxidants inhibit A2E-epoxide formation. J Biol Chem 2003;278:18,207–18,213.

126. Eagle RC, Lucier AC, Bernardino VB, Yanoff M. Retinal pigment epithelial abnormalities in fundus flavimaculatus. Ophthalmol 1980;87:1189–1200.

127. Lois N, Holder GE, Fitzke FW, Plant C, Bird AC. Intrafamilial variation of phenotype in Stargardt macular dystrophy-fundus flavimaculatus. Invest Ophthalmol Vis Sci 1999;40:2668–2675.

128. Lopez PF, Maumenee IH, de la Cruz Z, Green WR. Autosomal-dominant fundus favimaculatus. Clinicopathologic correlation. Ophthalmol 1990;97:798–809.

129. Allikmets R, Shroyer NF, Singh N, et al. Mutation of the Stargardt disease gene (ABCR) in age-related macular degeneration. Science 1997;277:1805–1807.

130. Shroyer NF, Lewis RA, Yatsenko AN, Lupski JR. Null missense ABCR (ABCA4) mutations in a family with Stargardt disease and retinitis pigmentosa. Invest Ophthalmol Vis Sci 2001;42:2757–2761.

131. Shroyer NF, Lewis RA, Yatsenko AN, Wensel TG, Lupski JR. Cosegregation and functional analysis of mutant ABCR (ABCA4) alleles in families that manifest both Stargardt disease and age-related macular degeneration. Hum Mol Genet 2001;10: 2671–2678.

132. Yatsenko AN, Shroyer NF, Lewis RA, Lupski JR. Late-onset Stargardt disease is associated with missense mutations that map outside known functional regions of ABCR (ABCA4). Hum Genet 2001;108:346–355.

133. Delori FC, Staurenghi G, Arend O, Dorey CK, Goger DG, Weiter JJ. In vivo measurement of lipofuscin in Stargardt's disease—Fundus flavimaculatus. Invest Ophthalmol Vis Sci 1995;36:2327–2331.

134. von Ruckmann A, Fitzke FW, Bird AC. In vivo fundus autofluorescence in macular dystrophies. Arch Ophthalmol 1997;115:609–615.

135. Lois N, Holder GE, Bunce CV, Fitzke FW, Bird AC. Phenotypic subtypes of Stargardt macular dystrophy-fundus flavimaculatus. Arch Ophthalmol 2001;119:359–369.

136. Rabb MF, Tso MO, Fishman GA. Cone-rod dystrophy. A clinical and histopathologic report. Ophthalmology 1986;93:1443–1451.

137. von Ruckmann A, Fitzke FW, Bird AC. Distribution of pigment epithelium autofluorescence in retinal disease state recorded in vivo and its change over time. Graefe's Arch Clin Exp Ophthalmol 1999;237:1–9.

138. Fishman GA, Stone EM, Eliason DA, Taylor CM, Liindeman M, Derlacki DJ. ABCA4 gene sequence variationsw in patients with autosomal recessive cone-rod dystrophy. Arch Ophthalmol 2003;121:851–855.

139. Klevering BJ, Maugeri A, Wagner A, et al. Three families displayinng the combination of Stargardt's disease with cone-rod dystrophy or retinitis pigmentosa. Ophthalmol 2004; 111:546–553.

140. Maugeri A, Klevering BJ, Rohrschneider K, et al. Mutations in the ABCA4 (ABCR) gene are the major cause of autosomal recessive cone-rod dystrophy. Am J Hum Genet 2000; 67:960–966.

141. Shroyer NF, Lewis RA, Allikmets R, et al. The rod photoreceptor ATP-binding cassette transporter gene, ABCR, and retinal disease: from monogenic to multifactorial. Vision Res 1999;39:2537–2544.

142. Weingeist TA, Kobrin JL, Watzke RC. Histopathology of Best's macular dystrophy. Arch Ophthalmol 1982;100:1108–1114.

143. Frangieh GT, Green WR, Fine SL. A histopathologic study of Best's macular dystrophy. Arch Ophthalmol 1982;100:1115–1121.

144. Petrukhin K, Koisti MJ, Bakall B, et al. Identification of the gene responsible for Best macular dystrophy. Nat Genet 1998;19:241–247.

145. Sun H, Tsunenari T, Yau KW, Nathans J. The vitelliform macular dystrophy protein defines a new family of chloride channels. Proc Natl Acad Sci USA 2002;99: 4008–4013.

146. Marmostein AD, Stanton JB, Yocom J, et al. A model of Best vitelliform macular dystrophy in rats. Invest Ophthalmol Vis Sci 2004;45:3733–3739.

147. Downes SM, Fitzke FW, Holder GD, et al. Clinical features of codon 172 RDS macular dystrophy. Similar phenotype in 12 families. Arch Ophthalmol 1999;117:1373–1383.

148. Hall NA, Lake BD, Dewji NN, Patrick AD. Lysosomal storage of subunit c of mitochondrial ATP synthase in Batten's disease (ceroid-lipofuscinosis). Biochem J 1991;275:269–272.

149. Haskell RE, Carr CJ, Pearce DA, Bennett MJ, Davidson BL. Batten disease: Evaluation of CLN3 mutations on protein localization and function. Hum Mol Genet 2000; 9:735–744.

150. Katz ML, Gao C, Prabhakaram M, Shibuya H, Liu P, Johnson GS. Immunochemical localization of the Batten disease (CLN3) protein in retina. Invest Ophthalmol Vis Sci 1996;38:2373–2384.

151. Young RW. Pathophysiology of age-related macular degeneration. Surv Ophthalmol 1987;31:291–306.

152. Young RW. Solar radiation and age-related macular degeneration. Surv Ophthalmol 1988;32:252–269.

153. Winkler BS, Boulton ME, Gottsch JD, Sternberg P. Oxidative damage and age-related macular degeneration. Mol Vision 1999;5:32.

154. Beatty S, Koh H-H, Henson D, Boulton M. The role of oxidative stress in the pathogenesis of age-related macular degeneration. Surv Ophthalmol 2000;45:115–134.

155. Kennedy CJ, Rakoczy PE, Constable IJ. Lipofuscin of the retinal pigment epithelium: a review. Eye 1995;9:763–771.

156. Eldred GE. Lipofuscin fluorophore inhibits lysosomal protein degradation and may cause early stages of macular degeneration. Gerontology 1995;41:15–28.

157. Mainster MA. Light and macular degeneration: A biophysical and clinical perspective. Eye 1987;1:304–310.

158. Friedman DS, O'Colmain BJ, Munoz B, et al. Prevalence of age-related macular degeneration in the United States. Arch Ophthalmol 2004;122:564–572.

159. Taylor HR, West S, Munoz B, Rosenthal FS, Bressler SB, Bressler NM. The long-term effects of visible light on the eye [see comments]. Arch Ophthalmol 1992;110:99–104.

160. Cruickshanks KJ, Klein R, Klein BEK, Nondahl DM. Sunlight and the 5-year incidence of early age-related maculopathy: the Beaver Dam Eye Study. Arch Ophthalmol 2001;119:246–250.

161. Tomany SC, Cruickshanks KJ, Klein R, Klein BEK, Knudtson MD. Sunlight and the 10-year incidence of age-related maculopathy. The Beaver Dam Eye Study. Arch Ophthalmol 2004;122:750–757.

162. Bernstein PS, Leppert M, Singh N, et al. Genotype-phenotype analysis of ABCR variants in macular degeneration probands and siblings. Invest Ophthalmol Vis Sci 2002; 43:466–473.

163. Guymer RH, Heon E, Lotery AJ, et al. Variation of codons 1961 and 2177 of the Stargardt disease gene is not associated with age-related macular degeneration. Arch Ophthalmol 2001;119:745–751.

164. Darzins P, Mitchell P, Heller RF. Sun exposure and age-related macular degeneration. An Australian case-control study. Ophthalmol 1997;104:770–776.

165. Delcourt C, Carriere I, Ponton-Sanchez A, et al. Light exposure and the risk of age-related macular degeneration. Arch Ophthalmol 2001;119:1463–1468.

166. AREDS Research Group. Risk factors for neovascular age-related macular degeneration. Arch Ophthalmol 1992;110:1701–1708.

167. AREDS Research Group. Risk factors associated with age-related macular degeneration. A case-control study in the age-related eye disease study: Age-related eye disease study report number 3. Ophthalmol 2000;107:2224–2232.

168. Mellerio J. Yellowing of the human lens: nuclear and cortical contributions. Vision Res 1987;27:1581–1587.

169. Mainster MA, Sparrow JR. How much blue light should an IOL transmit? Br J Ophthalmol 2003;87:1523–1529.

170. Liu IY, White L, LaCroix AZ. The association of age-related macular degeneration and lens opacities in the aged. Am J Public Health 1989;79:765–769.

171. Pollack A, Marcovich A, Bukelman A, Oliver M. Age-related macular degeneration after extracapsular cataract extraction with intraocular lens implantation. Ophthalmology 1996;103:1546–1554.

172. Klein R, Klein BEK, Wong TY, Tomany SC, Cruickshanks KJ. The association of cataract and cataract surgery with the long-term incidence of age-related maculopathy. Arch Ophthalmol 2002;120:1551–1558.

173. Wang JJ, Klein R, Smith W, Klein BEK, Tomany SC, Michell P. Cataract surgery and the 5-year incidence of late-stage age-related maculopathy. Pooled findings from the Beaver Dam and Blue Mountains Eye Studies. Ophthalmology 2003;110:1960–1967.

13

Genetic Modifiers That Affect Phenotypic Expression of Retinal Diseases

Malia M. Edwards, PhD, Dennis M. Maddox, PhD, Jungyeon Won, PhD, Jürgen K. Naggert, PhD, and Patsy M. Nishina, PhD

CONTENTS

INTRODUCTION

Variability in onset, progression, severity, and phenotypic expression is commonly observed in many retinal diseases (Tables 1 and 2). Although interfamily variability may be caused by environmental or allelic differences, intrafamily variability, when a common mutation is segregating, may also be due to genetic modifiers *(1–3)*. In contrast to independently acting alleles that may lead to an additive effect on disease severity or age of onset, genetic modifiers are defined as background genes that epistatically interact with a given disease genotype to affect phenotypic outcome. In general, allelic variability at modifier loci does not in itself produce a phenotype. A single gene or possibly a combination of genes in the same or parallel pathways as the mutant gene may act to create a final effect on the expression of the disease phenotype. These modifiers may enhance the effect of the mutation to cause a more severe mutant phenotype or an earlier onset, or conversely, delay or reduce the mutant phenotype even to the extent of completely restoring the wild-type (WT) condition.

Genetic modifiers have engendered excitement because the study and identification of these genes promise new insights into biological pathways that Mendelian disease genes act in and through which they cause pathologies *(4,5)*. For example, knowing the molecular basis of a genetic modifier may help in better diagnosis and treatment of a

From: *Ophthalmology Research: Retinal Degenerations: Biology, Diagnostics, and Therapeutics*
Edited by: J. Tombran-Tink and C. J. Barnstable © Humana Press Inc., Totowa, NJ

Table 1
Demonstrated or Probable Examples of Genetic Modification of Retinal Disease in Humans in which the Disease Haplotype or Genotype has been Established and Large Phenotypic Variability, Independent of Age, has been Observed

Primary mutation	Genotype	Phenotypic variability observed	Chromosomal location or identity of modifier	References
Arrestin	1147delA	Oguchi's Disease and ARRP: interfamilial variability ranging from pigmentary retinal degeneration in the midperipheral area with or without macular involvement		82
Bestrophin		Vitelliform macular dystrophy 2 (*VMD2*): phenotypes of 11 children from a homozygous parent ranged from early onset cystoid macular degeneration with accumulation of deeply and irregularly pigmented yellow macular mass to absence of clinical symptoms with pathological EOG-values		83
Cadherin23	Arg1746Gln	*USH1D*: Severity and progression of RP is highly variable		84,85
Clarin1	Asn48Lys	*USH3A*: Inter and intra-familial severity and progression of RP noted		86
Elongation of very long chain fatty acids -like4 (*ELOVL4*)	5 bp deletion	Stargardt 3 (*STGD3*): phenotypes range from Stargardt-like macular dystrophy to pattern dystrophy in related families	Co-inheritance of a mutation in (*ABC4*) increases disease severity of *STGD3*	87,88
Fascin, sea urchin, homology of, 2 (*FSCN2*)	208delG	*ADRP* or *ADMD*	ATP-binding cassette	89
Guanylate cyclase activator 1A (*GUCA1A*)	Pro50Leu	Dominant cone (*COD3*) or cone-rod dystrophy: minimal effects to macular function to cone-rod dystrophy observed in a family		90
Guanylate cyclase activator 1LB (*GUCA1B*)	Gly157Arg	Different forms or retinal diseases are observed, ADRP and ADMD		91

Gene/Protein	Mutation	Phenotype	References
Peripherin/RDS	Leu85Pro	Retinal disease: variability ranging from RP, pattern dystrophy, fundus flavimaculatus, and macular degeneration. Diallelic inheritance observed with rod outer segment membrane protein 1 (ROM1)	6,15–17
(PAP-1) protein target of Pim-1 kinase	His137Leu	Retinitis pigmentosa 9 (RP9): in the original large pedigree reported, phenotypes ranged from minimally affected with no symptoms, moderately affected with mild symptoms, abnormal ERGs, and equal loss of rod and cone function in affected areas of the retina; and severely affected with extinguished ERGs and barely detectable dark adapted static threshold sensitivities	10,92
RP1	Arg677Ter	Retinitis Pigmentosa 1 (RP1): intra and interfamilial variability in patients ranging from degeneration, initially of rods, in the far peripheral inferior nasal retina, to minimal or absence of disease	93
Retinoschisin	375-378delAGAT	X-linked retinoschisis: variability in disease progression	94
Tissue inhibitor of metalloproteinase 3 (TIMP3)	Ser181Cys	Sorsby fundus dystrophy (SFD): founder effect in which families in British Isles, Canada, United States, and South Africa with variation in phenotype including white to yellow fundus spots accompanying disciform macular degeneration, absence of fundus spots, or yellow deposits associated with atrophic macular degeneration	7–9,95

ARPP-autosomal recessive retinitis pigmentosa, EOG-electrooculography, USH-usher syndrome, ADMD-autosomal dominant mocular dystrophy.

239

Table 2
Mouse Models for Modifiers of Retinal Degeneration

Disease gene; phenotype, strain	Modifier effect	Modifier strain	Chromosome	Reference
Proposed:				
p53; vitreal opacity, retinal folds				
Fibrous retrolental tissue, C57BL/6J	No disease	129/SvJ	Not mapped	*36*
BMP4; anterior segment defects, glaucoma	Varied severity	C57BL/6J,BliA CAST/Ei,C3H AKR/J, BALB/C 129/SvEvTac	Not mapped	*96*
Myopia	Varied eye growth	50 strains	Not mapped	*97*
Chx 10; ocular retardation	Partial rescue	CASA/Rk	Not mapped	*98*
Growth factor-stimulated angiogenesis	Varied angiogenesis	129/Rej, C57BL/6J, SJL/J	Not mapped	*99*
isa, iris stromal atrophy, DBA/2J	Increased cell death	AKXD-28/Ty	Not mapped	*100*
ipd, iris pigment dispersion	Reduced phenotype	AKXD-28/Ty	Not mapped	*100*
Rho; retinal degeneration, C57BL/6J-129Sv	Varied photoreceptor apoptosis	129/Sv, C57BL/6J	Not mapped	*101*
Mapped:				
Iris atrophy, glaucoma	Interacting loci	C57BL/6J, DBA/2J	6, 4q	*38*
c-Fos; light induced photoreceptor death	Increased apoptosis		3, Rpe65	*42*
Tub; retinal degeneration, C57BL/6J	Partial rescue	AKR/J	11 (motr1) 2, 8	*55*

BMP-bone morphogenic protein; Chx10-*C. elegans* ceh-10 homeodomain containing homolog, *motr1*-modifier of retinal degeneration 1

disease, perhaps by defining a subgroup within the disease population. In addition, the identification of modifier genes may lead to new treatment modalities either by providing additional information about genetic contributions to the phenotype for which treatment may already be available or by revealing additional steps in a biological pathway that may be more amenable to treatment.

As described here previously, phenotypic variability is often observed in disease, retinal diseases notwithstanding. Although there are only a handful of cases in which genetic modifier loci have been recognized, many examples of inter- and intrafamily phenotypic variability, in which the primary mutation is the same, have been reported. These are summarized in Table 1. Presumably, more examples of modifiers and map positions of retinal phenotypes will be reported as the necessary genotype data become available.

Segregating crosses of inbred mice that allow for control of both environmental factors and allelic variability of the primary mutation have also yielded the map positions of a number of genetic modifiers and, in some cases, their identities. These are summarized in Table 2. Finally, potential strategies and examples for the identification of genetic modifiers are also discussed.

GENETIC MODIFIERS OF RETINAL DISEASES IN HUMANS

In most cases in which phenotypic variability is observed in humans, it is difficult to determine whether the effects are the result of environmental, allelic, or genetic modification or a combination of these factors. Clinicians and investigators generally suggest genetic modification when variability is observed in large pedigrees or in founder populations segregating for the same primary mutation. One of the first reports, suggesting genetic background modifier effects, was a nuclear family in which members carrying the same peripherin retinal degeneratic slow (RDS) mutation developed either retinitis pigmentosa (RP), pattern dystrophy, or fundus flavimaculatus *(6)*. Since then, genetic modification has been proposed for a number of retinal diseases including Sorsby fundus dystrophy (SFD), RP1, RP9, and primary congenital glaucoma (PCG) (*see* Table 1 for a more comprehensive summary). SFD is a macular degenerative disease characterized by submacular choroidal neovascularization. Weber et al. *(7)* identified point mutations in the gene en coding tissue inhibitor of metalloproteinase 3 (*TIMP3*), a potent angiogenesis inhibitor, as the genetic basis of SFD heritability in two pedigrees. Subsequent haplotype and mutational analysis led investigators to conclude that the Ser181Cys *TIMP3* mutation in families residing in the British Isles, Canada, United States, and South Africa was probably the result of an ancestral founder effect *(8,9)*. Therefore, the phenotypic heterogeneity in these patients, which ranges from white to yellow fundus spots accompanying disciform macular degeneration to absence of fundus spots, is not caused by genetic heterogeneity at the *TIMP3* locus, but rather by genetic or environmental modifiers, or a combination of both.

In the case of RP1, characterization of 22 patients from 10 different pedigrees (11 patients were from a single pedigree) in whom the primary mutation is a premature stop codon, Arg677Ter, in the RP1 protein demonstrated wide variability in the severity of visual field loss both within and between families. The authors suggested a role for modifier genes as well as environmental influences in the regulation of phenotypic variability.

In another example, a large pedigree segregating for RPA, an autosomal dominant RP was found to have regional retinal dysfunction with variable expressivity *(10)* . Subjects carrying the affected haplotype were reported to be asymptomatic, moderately affected with abnormal electroretinogram (ERG) responses, or severely affected with extinguished ERGs.

Mutations in *CYP1B1* are commonly found in Saudi Arabian families segregating for PCG. *CYP1B1*, which is specifically expressed in the iris, trabecular meshwork, and ciliary body, encodes cytochrome P4501B1 *(11–13)*. Absence of disease in siblings carrying the same haplotype and disease causing mutations in *CYP1B1*, has led investigators to suggest that a dominant modifier may exist in the Saudi Arabian population *(14)*.

As primary mutations are identified and more knowledge is gained about the various pathways through which genes carrying the mutations function, the number of modifiers identified has increased. For example, potential modifier variants have been identified for RP caused by mutations in peripherin/RDS and retinal pigment epithelum (RPE)65, and primary open angle glaucoma (POAG). Apfelstedt-Sylla et al. *(15)* and others showed remarkable both inter- or intrafamilial variability in the severity of the retinal disease in patients with mutations in peripherin/RDS that ranges from subtle to widespread pigmentary changes associated with choroidal neovascularization (CNV) *(16)*. Although the genes underlying these particular modifications are yet to be identified, one gene, *ROM1*, is known to interact with peripherin/RDS. Individuals heterozygous for a Leu185Pro allele of peripherin/RDS who also carry a null allele of *ROM1*, present with reduced electroretinogram amplitudes typical of RP and are said to have digenic RP *(17)*, whereas individuals carrying either allele alone have no or minimal abnormalities.

RPE65 is an abundantly expressed protein in the RPE *(18)* and in the cone photoreceptors (Prs) *(19)*. Moiseyev et al. *(21)* has recently reported that REP65 is an isomerohydrolase. Studies by Xue et al. *(20)* suggest that there are two forms of RPE65, a palmitoylated membrane associated form (mRPE65) and a nonpalmitoylated soluble form (sRPE65). The mRPE65 is proposed to be a chaperone for all-*trans*-retinyl esters and the sRPE65, a chaperone for vitamin A *(20)*. Hence, all-*trans*-retinyl esters over-accumulate in the RPE in *Rpe65* deficiency *(22)*. Mutations in *RPE65* lead to Leber congenital amaurosis type II (LCA2) *(23)* as well as autosomal recessive childhood-onset retinal dystrophy *(24)* and autosomal recessive (AR) RP in humans *(25)*. Recently, Silva et al. *(26)* reported a missense mutation in retinal guanylate cyclase 2D (*GUCY2D*) which they suggest acts as a modifier of AR LCA that results from a *RPE65* mutation. Two siblings from a consanguineous mating, who carried a homozygous nonsense mutation in *RPE65*, Glu102ter, had widely divergent disease progression. Sequencing of known LCA-causing genes in the two siblings lead to the identification of an Ile539Val allele of *GUCY2D* in the more severely affected sibling. This has lead the investigators to postulate that the mutation of the highly conserved residue in an important functional domain of GUCY2D further compromises the retinal function occurring as a result of an RPE65 deficiency. The investigators are currently confirming their results in in vivo and *in vitro* experiments.

Mutations in myocilin *(27)*, optineurin *(28)*, and, more recently, *CYP1B1 (29,30)* have been associated with POAG. Haplotype analysis of individuals with the Gln368Ter myocilin mutation suggests a founder effect. Patients carrying the same mutation show a variation both in age of onset and in the severity of the disease, from ocular hypertension to severe visual field loss to legal blindness *(31)*. A study examining subjects from a large pedigree with juvenile open-angle glaucoma for mutations in *MYOC, CYP1B1*, and *PITX2* was able to show that affected subjects carrying both Gly399Val *MYOC* and Arg368His *CYP1B1* mutations had earlier disease onset than those only carrying the *MYOC* mutation alone. These observations have led the investigators to postulate that the *CYP1B1* Arg368His allele is able to modify *MYOC* expression *(29)*. In another study of unrelated individuals with POAG, Copin et al. *(32)* provided supportive evidence two APOE promoter single-nucleotide polymorphisms (SNPs) suggesting differential modification of the POAG phenotype by. A SNP within the apolipoorotavi E (*APOE*) (-219G) promoter was associated with visual field alteration and an increase in optic nerve damage. Additionally, SNPs the *APOE* (-491T) and *MYOC* (-1000G) were shown to interact and modify intraocular pressure (IOP) and limit effectiveness of IOP-lowering treatments in patients with POAG. Although association studies are not definitive, these results allow for further hypothesis building and testing. Indeed, a follow-up study in a different population assessing the association of APOE variants to POAG phenotypes have failed to replicate the original study *(33,34)*.

GENETIC MODIFIERS OF RETINAL DISEASES IN MICE AND OTHER MODEL ORGANISMS

In mice, the presence of modifier genes was anticipated from the variation in phenotype observed when spontaneous mutations were crossed onto different inbred backgrounds to generate congenic strains. Additionally, as large numbers of targeted mutation models have been generated on mixed genetic backgrounds, mutant phenotypes change or disappear entirely as the mutations are moved from the mixed 129/B6 background to the (B6) background *(35,36)*. An example of such modification is seen when the *Tulp1* mutation on the (B6) background is crossed with the AKR/J strain (Fig. 1). Despite the fact that each F2 animal in Fig. 1 is homozygous for the *Tulp1* null allele, the degree of degeneration is highly variable. The phenomenon of phenotypic modification, which was first viewed as discouraging by many, has now become recognized as a useful tool for identifying factors that interact with genes involved in known pathways or for providing entry points to all for the elucidution of the function of novel genes.

Currently, there are three reports of mapped modifier loci involved in retinal degenerative diseases in mouse (Table 2). These investigations serve as a demonstration of the efficacy of identifying genetic interactions/modifier loci capable of modulating the progression of retinal degeneration and suggest that investigation of modifier loci for the remaining cloned and/or mapped retinal degeneration genes will provide a rich resource for future discovery.

Modification of Iris Atrophy and Glaucoma in DBA/2J

Glaucoma typically involves increased IOP and subsequent ganglion cell death, leading to blindness. In humans, glaucomas are associated disorders of the anterior segment, including pigment dispersion syndrome and iris atrophy *(37)*. In an effort to define genetic pathways through which these degenerative processes contribute to glaucoma, Chang et al. *(38)* performed crosses between C57BL/6J mice and the DBA/2J (D2) strain, which is known to develop glaucoma subsequent to the development of pigment dispersion and optic atrophy *(37)*. A genome screen of 50 mice with early onset trans-illumination defects, pigment dispersion and iris atrophy revealed that all affected mice were homozygous for the D2 alleles for one locus on chromosome 6 (termed iris pigment dispersion [*ipd*]) and for another locus on chromosome 4 (termed iris stromal atrophy [*isa*]). Histological examination demonstrated that mice homozygous for the D2 allele of *ipd* primarily demonstrated defects of the iris pigment epithelium. Mice homozygous for the D2 allele of *isa* demonstrated defects of the anterior iris stroma. Mice homozygous for the D2 allele at both loci suffered more severe defects in both the iris pigment epithelium and anterior iris stroma, with accompanying ganglion cell death. As *isa* maps near to the tyrosinase related protein 1 gene (*Tyrp1*) and

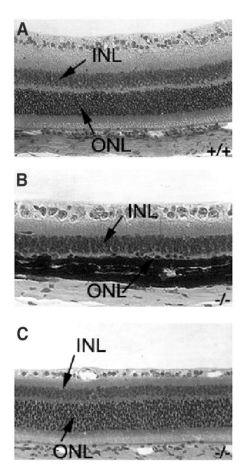

Fig. 1. Retinal degeneration is suppressed in some *tulp1–/–* mutant mice. (**A**) Wild-type retina, (**B**) *tulp1–/– ×* AKR/J mutant, (**C**) *tulp1–/– ×* AKR/J mutant with suppressed retinal degeneration. INL, inner nuclear layer; ONL, outer nuclear layer.

coat color segregates with *isa*, it was hypothesized that the D2 allele of *Tyrp1* was involved with *isa* and that biochemical processes involved in pigment production have an important role in disease etiology.

RPE65Leu$_{450}$ Variant Allele Is Able to Modify the Protection From Light-Induced Photoreceptor Damage Mediated by a c-Fos Null Allele

Danciger et al. *(39)* reported that strain C57BL/6J-c(2J) (c2J) demonstrated marked resistance to light-induced PR damage. A quantitative trait locus (QTL) genome-wide scan for genes that affected light-induced PR survival, as assessed by retinal thickness, was carried out in progeny of a (c2J X BALB/c)F1 X c2J backcross. A major QTL on chromosome 3, accounting for 50% of the protective effect, and three other weak but significant QTL on chromosomes 9, 12, and 14 were identified. The protective effect observed in the c2J background was determined to result from a single nucleotide polymorphism in the *RPE65* gene; *RPE65Leu$_{450}$* confers susceptibility to light-induced damage.

c-*Fos* binding to activator protein (AP)-1 is essential for PR apoptosis induced by bright light exposure *(40,41)*. Consequently, c-*Fos* deficient (c-*fos*–/–) mice are resistant to light-induced damage *(41)*. Presumably, the absence of c-*Fos* leads to suppression of light induced damage because c-Fos is not available for AP-1 DNA binding. However, the c-Fos deficient mice carry an *Rpe65Met$_{450}$* allele. Recently, Wenzel et al. *(42)* have reported that the *RPE65Leu$_{450}$* variant allele is capable of overcoming protection against light-induced PR apoptosis in c-*fos*–/– mice. Introduction of the *RPE65Leu$_{450}$* allele to c-*fos*–/– mice leads to increased levels of RPE65 protein and accelerated rhodopsin regeneration, a factor known to accelerate light-induced PR degeneration. Additionally, Fra-2 and FosB, Fos family members, are able to substitute for c-Fos AP-1 DNA binding and enable light-induced AP-1 activity and produce PR apoptosis in *RPE65Leu$_{450}$;c-fos*–/– mice *(42)*.

Modification of Retinal Degeneration in the Tubby (tub/tub) Mutant Mouse

Retinal degeneration 5 (*rd5*) was originally described in the tubby mouse. Tubby mice suffer retinal and cochlear degeneration with maturity onset obesity *(43–47)*. This phenotype led to the suggestion that the *rd5* mutant mouse might be an appropriate model for Usher syndrome type I *(44)*, the most common hereditary form of combined deafness and blindness in humans *(48)*. Obesity coupled with retinal degeneration and hearing loss also make tubby mice a good model for Alström syndrome and Bardet-Biedl syndrome *(49)*.

Retinal degeneration in *rd5* mutant mice is characterized by abnormal levels of cellular apoptosis of photoreceptors from postnatal day 16 (P16) until P23 *(50)*. In *rd5* mutant mice, reduction of the outer nuclear layer (ONL), with concomitant shortening and disorganization of the inner and outer segment, is observed through 8 mo of age, at which point the photoreceptor cell layer is reduced to one or two nuclei in thickness. Early stages of retinal degeneration in the *rd5* mutant mouse are characterized by the presence of membrane-bound vesicles of unknown origin in the interphotoreceptor space *(44)*. As retinal degeneration progresses, ERG readings deteriorate until they are extinguished at around 6 mo of age *(44)*.

TUB is the founding member of a small family of proteins, collectively known as the TUBBY-LIKE proteins (TULPS) *(51)*. Although the function of the TULPS is uncertain, it has been hypothesized that they may normally be involved in intracellular vesicular trafficking of opsins or other components of the visual system *(52)*, act as an intracellular adaptor molecule that is a downstream target for insulin and/or insulin-like growth factor-1 receptor *(53)*, or a member of a novel class of transcription factors *(54)*.

In order to elucidate the pathways through which TUB functions, Ikeda et al. *(55)* identified modifiers in an F2 intercross between (C57BL/6J-*tub/tub* and AKR/J). The thickness of the ONL and the number of photoreceptor nuclei were assessed, and a genome-wide scan revealed significant linkage on chromosome 11 (*motr1*) between markers D11MIT39 and D11MIT360 and suggestive linkage on chromosomes 2 and 8.

As previously mentioned, the phenotype of *tub/tub* mice is pleiotropic, including obesity and cochlear degeneration in addition to retinal degeneration. A modifier screen was performed comparing auditory brain response (ABR) threshold in F2 intercrosses between (C57BL/6J-*tub/tub* and AKR/J-+/+, CAST/Ei-+/+ or 129/Ola-+/+) *(56)*. A major modifier (*moth1*) was identified and mapped to chromosome 2, with a logarithm of the odds (LOD) score of 33.4 in the AKR/J-+/+ intercross and of 6.0 in the CAST/Ei-+/+ intercross, accounting for 57% and 43% of ABR threshold variance, respectively. Further mapping efforts (57) resulted in the identification of *moth1* as the *MAP1A* gene.

As the studies in *tub/tub* mutant mice show, different modifier loci can account for variation in different aspects of pleiotropic phenotypes. On the other hand, both *moth1* and *motr1* reside in the same QTL interval on chromosome 2, suggesting that they may represent the same gene and that modifier genes may exist that can influence more than one phenotype in a pleiotropic disease. Although we currently do not know if the *motr1* and *moth1* genes are the same, the potential exists for identifying modifiers that are capable of modifying multiple disease manifestations of a mutation. Identification of such modifiers would be of significant biological interest, as they would provide clues to common pathologies underlying disease phenotypes. Modifiers capable of ameliorating multiple defects within the same pleiotropic disease would also be incredibly valuable as therapeutic targets, as treatment could be provided for an entire syndrome rather than for a single manifestation of the disease.

STRATEGIES AND EXAMPLES OF CLONING GENETIC MODIFIERS

As illustrated in the previous sections, some genetic modifiers that affect retinal disease phenotypes have been mapped or identified. However, their number is small compared to that of diseases associated with retinal damage. Therefore, it seems probable that many additional modifiers await discovery. At this juncture, it seems appropriate that we should review methods that are being utilized in other diseases and model organisms to identify modifier genes. Application of these methods to retinal degenerative diseases may hasten the identification of retinal genetic modifiers.

Chromosomal Localization of a Genetic Modifier

In the case of mice, identification of genetic modification begins with outcrossing mice carrying a mutation to another strain. Phenotypes are scored on backcross or intercross mice carrying the disease causing mutation(s). Phenotypic data, be it a quantitative trait such as ONL thickness or absence or presence of a phenotype, is correlated to marker genotypes obtained in a genome-wide scan. To be certain that the locus identified is a genetic modifier and not the result of the interaction of the two genetic backgrounds, irrespective of the primary genetic mutation, mice that are WT or heterozygous (in the case of recessive diseases) for the primary mutation are phenotyped and genotyped as well. If the locus is truly a genetic modifier of the primary mutation, then no significant association should be observed in mice not carrying the disease mutation(s). Once a genetic modifier is mapped to a specific chromosomal region, a number of methodologies to identify the underlying genes have been applied.

Conventional Approaches to Refining the Genetic Modifier Interval: Recombinational Mapping With or Without Isolation of Modifiers as Congenics

Ultimately, it is important that the genetic modifier be identified. In order to do this, the genetic interval containing the modifier must be small enough to allow for positional candidate testing. Initially, the modifier genetic interval may be very large, encompassing regions greater than 20 cM. The method by which the interval is narrowed will depend greatly upon the contribution that the particular locus has on the observed phenotype. If a major genetic modifier is identified (e.g., >40% contribution to the phenotypic variance), a standard F2 intercross or backcross can be undertaken with progeny testing of mice recombinant within the modifier interval *(57)*. The number of meioses tested will determine the size of the genetic interval. For example, if 1000 F2 mice are tested, the region containing the modifier could be theoretically as small as 0.05 cM in size.

If, however, multiple genetic modifier loci are identified with modest effects, interval specific congenic strains may need to be constructed to isolate the various loci before approaches to narrowing the genetic interval can be applied. A congenic is a variant strain that is constructed by repeated backcrossing of a donor strain carrying the genomic region of interest to a recipient inbred strain for 10 backcross generations with subsequent intercrossing *(58)*. With each subsequent backcross generation the percentage of donor genome declines by one-half in unlinked regions, such that by 10 backcross generations, 99.9% of the unlinked genomic regions are of recipient strain origin *(58)*. The purpose of the congenic is to isolate the genetic modifier so that any phenotypic variations that are observed between the congenic and recipient parental strain must be a consequence of the gene encompassed by the modifier locus *(59)*. The congenic can then be used in traditional recombinant crosses to narrow the genetic interval encompassing the modifier and/or in combination with expression profiling, discussed on page 248.

Crosses Utilizing Multiple Inbred Strains and Application of Conserved Haplotype Block Analysis

The conventional methods described previously necessitate the generation of a large numbers of animals and significant time commitment. It is important, therefore, that these methods be made more efficient. One way the efficiency of refining modifier intervals can be improved is by performing multiple crosses with different strains in which the primary mutation is segregating *(60,61)*. Although recombinations are thought to be random in the genome, empirically, we have noted that different strain combinations may have different recombinational hot spots within a given interval. Hence, the use of multiple crosses may narrow the modifier interval by increasing the number and distribution of recombinations within a region. Additionally, detecting the same modifier region in multiple strain combinations might indicate a common ancestral allele that may be detected by an *in silico* haplotype block analysis approach *(62–64)*.

Although the use of haplotype analysis has been used in humans to identify chromosomal regions associated with increased disease risk *(65,66)*, this methodology has not been used extensively in mice. However, as in humans, the haplotype block analysis approach in mice relies on the fact that the current inbred strains are derived from a mixed but limited pool of founders *(67)*. As a consequence, chromosomal regions that are not polymorphic between two inbred strains that have been used to identify a genetic modifier are likely to be inherited from a common ancestor, and can be eliminated from consideration as regions harboring candidate modifier loci *(62)* or, conversely, haplotype blocks that are shared among a group of strains known to modify a phenotype, may be more likely to harbor the modifier in question. The current availability of genotypic data in the form of microsatellite and SNPs across multiple strains makes this approach feasible *(64)*. Additionally, the attraction of this technique is that it reduces the need to generate large cohorts of mice to refine a region. This technique has been used in mice to narrow modifier regions in studies investigating cancer *(68)*, heart failure *(69)*, and hypertension *(61)*. An example of the power of haplotype block analysis in multiple crosses has been demonstrated in a recent study by Wang et al. *(70)*, which identified *Apoa2* and the mutation within the gene that underlies a quantitative trait locus, *Hdlq5 (70)*.

Expression Analysis as a Method to Identify Potential Candidate Modifiers

There is a growing body of literature supporting the usefulness of combining the mapping of modifier loci with expression profiling, especially in the identification of complex traits *(71,72)*. Chromosomal localization identifies regions that are associated with the modification of the disease phenotype, but this region contains many genes. Expression profiling identifies genes whose expression levels differ between two populations, but whose association with the observed phenotype is unknown. By combining these two techniques, data acquired in the mapping phase can be used to filter the data acquired in the expression-profiling phase, allowing for the identification of genes that are functioning specifically to impact the biological system of interest.

In an elegant study, Dyck et al. *(73)* combined these two techniques to understand the development of gallstones in C57L/J mice that carry the *Lith1* gallstone-susceptibility locus. Microarray analysis was used to identify differences in gene expression between C57L/J mice, which are prone to gallstone formation, and resistant AKR/J mice. Numerous genes involved in fatty acid metabolism were identified. Through literature searches of common regulatory elements within antioxidant systems, the nuclear transcription factor *Nrf2*, which maps to the *Lith1* locus, was identified.

Sensitized Mutagenesis Screens to Identify Modifiers

Another method available for identifying genetic modifiers are sensitized screens in which an alteration of a disease phenotype is sought in disease carrying models that have been mutagenesized. Although this methodology has been used for years to screen for modifiers in lower organisms (such as Drosophila) *(74–76)*, it has only recently been used for this purpose in mice.

In Drosophila melanogaster

One of the strengths of utilizing *Drosophila melanogaster* as a model organism is the ability to perform large-scale forward genetic screens to identify genes that are involved in selected biological processes. With the development of modifier screens, in which modifiers are identified because of their ability to modulate the phenotype of flies carrying mutations in the biological pathways of interest, the potential for using *Drosophila* genetics to unravel questions concerning disease processes increased exponentially.

One of the first such modifier screens to take advantage of the *Drosophila* model utilized the dominant *irregular facets* (*If*) mutation *(74)*. This mutation results from a dominant gain-of-function allele of the zinc finger-type transcription factor, *Krüppel* (*Kr*). After the blastoderm stage, *Kr* is normally expressed during development of the larval visual system; however, in the *If* mutant, *Kr* is misexpressed during development of the eye imaginal disc.

A dominant modifier screen was employed to map loci interacting with *Krüppel*. The phenotypes of heterozygous *If* mutants were examined in combination with chromosomal deficiencies or lethal P-element enhancer trap insertions, and 30-modifier loci (12 enhancers and 18 suppressors) were identified. Two P-element insertions, one an enhancer and one a suppressor were identified. The enhancer was determined to be the eyelid (*eld*) gene, and the suppressor was determined to be the extra macrochaetae (*emc*) gene. Both modifier genes encode known transcription factors, and previously characterized alleles of the two loci were determined to modify the *If* mutant phenotype suggesting that these two genes function in regulation of *Krüppel* activity during normal development.

Modifier screens in *Drosophila* become even more powerful when transgenic overexpression of molecules of interest can be driven in chosen retinal cell populations. This strategy has been employed to develop sensitized modifier screens with very promising results *(75,76)*. One such screen, utilizing flies overexpressing the noncoding SCA8 triplet-repeat expansion mutation to investigate modifiers of retinal neurodegeneration was especially fruitful *(75)*. Using the retinal degenerative phenotype as a sensitized background screen, three enhancing modifiers (*staufen, muscleblind,* and *split ends*) and one suppressive modifier (*CG3249*) were identified. *CG3249* encodes a protein kinase A anchoring protein and contains a kinescin heavy chain (KH) KH-motif for RNA binding. It is hypothesized that this KH-motif allows CG3249 to bind the SCA8 repeat expansion, thus preventing harmful interactions. Although the function of *spen* is poorly understood, the other enhancers (*muscleblind* and *staufen*) are known to encode proteins that bind CUG repeat domains. These studies provide a framework to begin understanding CTG expansion in disease processes and lay groundwork for determining the pathology of in repeat expansion diseases such as SCA8 and Huntington's disease.

ENU Mutagenesis in mice

In a landmark article, Carpinelli et al. *(77)* reported a successful suppressor screen for modifiers of thrombocytopenia in mice. Thrombocytopenia, a failure of megakaryocytes to produce sufficient blood platelets, results in hemophilia. Thrombopoietin, acting through a cell surface receptor c-Mpl, is thought to be the principle cytokine-controlling production of megakaryocytes and platelets *(78,79)*. By administering *N*-ethyl-*N*-nitrosourea (ENU) to male *Mpl–/–* mice that are susceptible to thrombocytopenia, Carpinelli et al. *(77)* were able to introduce mutations into the *Mpl–/–* background. Subsequent mating and analysis of the G1 offspring revealed the presence of two dominant suppressors of the thrombocytopenic phenotype. The modifiers, termed Plt3 and Plt4, both map to chromosome 10 and are hypothesized to be differing alleles of the same gene, *c-Myb*, a gene known to be a regulator of platelet production.

Performing this type of screen in mice, which generally serve as good models for human diseases, is potentially very powerful. Additionally, genetic manipulation of the mouse is possible for both generating desired sensitized mutant models and for transgenic verification of rescue experiments once putative modifiers have been identified. With this first success in screening modifiers in mice, one is hopeful that future screens may be employed to identify therapeutic targets in mouse models of human disease.

SUMMARY AND PERSPECTIVES

Recent figures provided on a publicly available website suggest the existence of at least 158 cloned and/or mapped genes that, when mutated, lead to retinal degeneration *(80)* (Retnet: http://sph.uth.tmc.edu/Retnet/disease.htm). At the present time, modifiers have been reported for only a relatively small percentage of these genes (listed in Tables 1 and 2), although it has been conjectured that most mutations are modified, at least to some extent, on different genetic backgrounds *(81)*. As methods that improve the efficiency of identifying these genetic modifiers are applied, it is hoped that pathways and mechanisms important in function and maintenance of the visual systems will be determined. As more genes modifying the progression of retinal degeneration are discovered, it is envisioned that these modifiers will provide therapeutic targets that are more amenable to treatment than the primary mutant gene, unlocking doors to new treatment modalities.

REFERENCES

1. Nadeau JH. Modifier genes in mice and humans. Nat Rev Genet 2001;2:165–174.
2. al-Maghtheh M, Gregory C, Inglehearn C, Hardcastle A, Bhattacharya S. Rhodopsin mutations in autosomal dominant retinitis pigmentosa. Hum Mutat 1993;2:249–255.
3. Aller E, Najera C, Millan JM, et al. Genetic analysis of 2299delG and C759F mutations (USH2A) in patients with visual and/or auditory impairments. Eur J Hum Genet 2004; 12:407–410.
4. Vincent AL. Searching for modifier genes. Clin Experiment Ophthalmol 2003;31: 374–375.
5. Sontag MK, Accurso FJ. Gene modifiers in pediatrics: application to cystic fibrosis. Adv Pediatr 2004;51:5–36.
6. Weleber RG, Carr RE, Murphey WH, Sheffield VC, Stone EM. Phenotypic variation including retinitis pigmentosa, pattern dystrophy, and fundus flavimaculatus in a single family with a deletion of codon 153 or 154 of the peripherin/RDS gene. Arch Ophthalmol 1993;111:1531–1542.
7. Weber BH, Vogt G, Pruett RC, Stohr H, Felbor U. Mutations in the tissue inhibitor of metalloproteinases-3 (TIMP3) in patients with Sorsby's fundus dystrophy. Nat Genet 1994;8:352–356.
8. Wijesuriya SD, Evans K, Jay MR, et al. Sorsby's fundus dystrophy in the British Isles: demonstration of a striking founder effect by microsatellite-generated haplotypes. Genome Res 1996;6:92–101.
9. Felbor U, Benkwitz C, Klein ML, Greenberg J, Gregory CY, Weber BH. Sorsby fundus dystrophy: reevaluation of variable expressivity in patients carrying a TIMP3 founder mutation. Arch Ophthalmol 1997;115:1569–1571.
10. Kim RY, Fitzke FW, Moore AT, et al. Autosomal dominant retinitis pigmentosa mapping to chromosome 7p exhibits variable expression. Br J Ophthalmol 1995;79:23–27.
11. Sutter TR, Tang YM, Hayes CL, et al. Complete cDNA sequence of a human dioxin-inducible mRNA identifies a new gene subfamily of cytochrome P450 that maps to chromosome 2. J Biol Chem 1994;269:13,092–13,099.

12. Shimada T, Yamazaki H, Mimura M, et al. Characterization of microsomal cytochrome P450 enzymes involved in the oxidation of xenobiotic chemicals in human fetal liver and adult lungs. Drug Metab Dispos 1996;24:515–522.

13. Stoilov I, Akarsu AN, Sarfarazi M. Identification of three different truncating mutations in cytochrome P4501B1 (CYP1B1) as the principal cause of primary congenital glaucoma (Buphthalmos) in families linked to the GLC3A locus on chromosome 2p21. Hum Mol Genet 1997;6:641–647.

14. Bejjani BA, Stockton DW, Lewis RA, et al. Multiple CYP1B1 mutations and incomplete penetrance in an inbred population segregating primary congenital glaucoma suggest frequent de novo events and a dominant modifier locus. Hum Mol Genet 2000;9:367–374.

15. Apfelstedt-Sylla E, Theischen M, Ruther K, Wedemann H, Gal A, Zrenner E. Extensive intrafamilial and interfamilial phenotypic variation among patients with autosomal dominant retinal dystrophy and mutations in the human RDS/peripherin gene. Br J Ophthalmol 1995;79:28–34.

16. Kim RY, Dollfus H, Keen TJ, et al. Autosomal dominant pattern dystrophy of the retina associated with a 4-base pair insertion at codon 140 in the peripherin/RDS gene. Arch Ophthalmol 1995;113:451–455.

17. Kajiwara K, Berson EL, Dryja TP. Digenic retinitis pigmentosa due to mutations at the unlinked peripherin/RDS and ROM1 loci. Science 1994;264:1604–1608.

18. Hamel CP, Tsilou E, Pfeffer BA, Hooks JJ, Detrick B, Redmond TM. Molecular cloning and expression of RPE65, a novel retinal pigment epithelium-specific microsomal protein that is post-transcriptionally regulated in vitro. J Biol Chem 1993;268:15, 751–15,757.

19. Znoiko SL, Crouch RK, Moiseyev G, Ma JX. Identification of the RPE65 protein in mammalian cone photoreceptors. Invest Ophthalmol Vis Sci 2002;43:1604–1609.

20. Xue L, Gollapalli DR, Maiti P, Jahng WJ, Rando RR. A palmitoylation switch mechanism in the regulation of the visual cycle. Cell 2004;117:761–771.

21. Moiseyev G, Chen Y, Takahashi Y, et al. RPE65 is the isomerohydrolase in the retinoid visual cycle. Proc Natl Acad Sci USA 2005;102:12413–12418.

22. Redmond TM, Yu S, Lee E, et al. Rpe65 is necessary for production of 11-cis-vitamin A in the retinal visual cycle. Nature Genet 1998;20:344–351.

23. Marlhens F, Bareil C, Griffoin JM, et al. Mutations in RPE65 cause Leber's congenital amaurosis. Nat Genet 1997;17:139–141.

24. Gu SM, Thompson DA, Srikumari CR, et al. Mutations in RPE65 cause autosomal recessive childhood-onset severe retinal dystrophy. Nat Genet 1997;17:194–197.

25. Morimura H, Fishman GA, Grover SA, Fulton AB, Berson EL, Dryja TP. Mutations in the RPE65 gene in patients with autosomal recessive retinitis pigmentosa or leber congenital amaurosis. Proc Natl Acad Sci USA 1998;95:3088–3093.

26. Silva E, Dharmaraj S, Li YY, et al. A missense mutation in GUCY2D acts as a genetic modifier in RPE65-related Leber congenital amaurosis. Ophthalmic Genet 2004; 25: 205–217.

27. Stone EM, Fingert JH, Alward WL, et al. Identification of a gene that causes primary open angle glaucoma. Science 1997;275:668–670.

28. Rezaie T, Child A, Hitchings R, et al. Adult-onset primary open-angle glaucoma caused by mutations in optineurin. Science 2002;295:1077–1079.

29. Vincent AL, Billingsley G, Buys Y, et al. Digenic inheritance of early-onset glaucoma: CYP1B1, a potential modifier gene. Am J Hum Genet 2002;70:448–460.

30. Melki R, Colomb E, Lefort N, Brezin AP, Garchon HJ. CYP1B1 mutations in French patients with early-onset primary open-angle glaucoma. J Med Genet 2004;41:647–651.

31. Craig JE, Baird PN, Healey DL, et al. Evidence for genetic heterogeneity within eight glaucoma families, with the GLC1A Gln368STOP mutation being an important phenotypic modifier. Ophthalmology 2001;108:1607–1620.

32. Copin B, Brezin AP, Valtot F, Dascotte JC, Bechetoille A, Garchon HJ. Apolipoprotein E-promoter single-nucleotide polymorphisms affect the phenotype of primary open-angle glaucoma and demonstrate interaction with the myocilin gene. Am J Hum Genet 2002;70:1575–1581.

33. Ressiniotis T, Griffiths PG, Birch M, Keers S, Chinnery PF. The role of apolipoprotein E gene polymorphisms in primary open-angle glaucoma. Arch Ophthalmol 2004; 122:258–261.

34. Ressiniotis T, Griffiths PG, Birch M, Keers SM, Chinnery PF. Apolipoprotein E promoter polymorphisms do not have a major influence on the risk of developing primary open angle glaucoma. Mol Vis 2004;10:805–807.

35. Gong X, Agopian K, Kumar NM, Gilula NB. Genetic factors influence cataract formation in alpha 3 connexin knockout mice. Dev Genet 1999;24:27–32.

36. Ikeda S, Hawes NL, Chang B, Avery CS, Smith RS, Nishina PM. Severe ocular abnormalities in C57BL/6 but not in 129/Sv p53-deficient mice. Invest Ophthalmol Vis Sci 1999;40:1874–1878.

37. John SW, Smith RS, Savinova OV, et al. Essential iris atrophy, pigment dispersion, and glaucoma in DBA/2J mice. Invest Ophthalmol Vis Sci 1998;39:951–962.

38. Chang B, Smith RS, Hawes NL, et al. Interacting loci cause severe iris atrophy and glaucoma in DBA/2J mice. Nat Genet 1999;21:405–409.

39. Danciger M, Matthes MT, Yasamura D, et al. A QTL on distal chromosome 3 that influences the severity of light-induced damage to mouse photoreceptors. Mamm Genome 2000;11:422–427.

40. Hafezi F, Steinbach JP, Marti A, et al. The absence of c-fos prevents light-induced apoptotic cell death of photoreceptors in retinal degeneration in vivo. Nat Med 1997; 3:346–349.

41. Wenzel A, Grimm C, Marti A, et al. c-fos controls the "private pathway" of light-induced apoptosis of retinal photoreceptors. J Neurosci 2000;20:81–88.

42. Wenzel A, Grimm C, Samardzija M, Reme CE. The genetic modifier Rpe65Leu(450): effect on light damage susceptibility in c-Fos-deficient mice. Invest Ophthalmol Vis Sci 2003;44:2798–2802.

43. Coleman DL, Eicher EM. Fat (fat) and tubby (tub): two autosomal recessive mutations causing obesity syndromes in the mouse. J Hered 1990;81:424–427.

44. Heckenlively JR, Chang B, Erway LC, et al. Mouse model for Usher syndrome: linkage mapping suggests homology to Usher type I reported at human chromosome 11p15. Proc Natl Acad Sci USA 1995;92:11,100–11,104.

45. Ohlemiller KK, Hughes RM, Lett JM, et al. Progression of cochlear and retinal degeneration in the tubby (rd5) mouse. Audiol Neurootol 1997;2:175–185.

46. Noben-Trauth K, Naggert JK, North MA, Nishina PM. A candidate gene for the mouse mutation tubby. Nature 1996;380:534–538.

47. Kleyn PW, Fan W, Kovats SG, et al. Identification and characterization of the mouse obesity gene tubby: a member of a novel gene family. Cell 1996;85:281–290.

48. Petit C. Usher syndrome: from genetics to pathogenesis. Annu Rev Genomics Hum Genet 2001;2:271–297.

49. Bray GA. Human obesity and some of its experimental counterparts. Ann Nutr Aliment 1979;33:17–25.

50. Bode C, Wolfrum U. Caspase-3 inhibitor reduces apototic photoreceptor cell death during inherited retinal degeneration in tubby mice. Mol Vis 2003;9:144–150.

51. Ikeda A, Nishina PM, Naggert JK. The tubby-like proteins, a family with roles in neuronal development and function. J Cell Sci 2002;115:9–14.
52. Hagstrom SA, Adamian M, Scimeca M, Pawlyk BS, Yue G, Li T. A role for the Tubby-like protein 1 in rhodopsin transport. Invest Ophthalmol Vis Sci 2001;42:1955–1962.
53. Kapeller R, Moriarty A, Strauss A, et al. Tyrosine phosphorylation of tub and its association with Src homology 2 domain-containing proteins implicate tub in intracellular signaling by insulin. J Biol Chem 1999;274:24,980–24,986.
54. Boggon TJ, Shan WS, Santagata S, Myers SC, Shapiro L. Implication of tubby proteins as transcription factors by structure-based functional analysis. Science 1999;286:2119–2125.
55. Ikeda A, Naggert JK, Nishina PM. Genetic modification of retinal degeneration in tubby mice. Exp Eye Res 2002;74:455–461.
56. Ikeda A, Zheng QY, Rosenstiel P, et al. Genetic modification of hearing in tubby mice: evidence for the existence of a major gene (moth1) which protects tubby mice from hearing loss. Hum Mol Genet 1999;8:1761–1767.
57. Ikeda A, Zheng QY, Zuberi AR, Johnson KR, Naggert JK, Nishina PM. Microtubule-associated protein 1A is a modifier of tubby hearing (moth1). Nat Genet 2002;30:401–405.
58. Green EL. Biology of the laboratory mouse. In: Green EL, ed. Breeding systems. New York: McGraw-Hill, 1966, Chapter 3, Laboratory mice.
59. Silver L. Mouse genetics. Oxford: Oxford University Press, 1995, pp. 32–61.
60. Wang X, Paigen B. Genetics of variation in HDL cholesterol in humans and mice. Circ Res 2005;96:27–42.
61. DiPetrillo K, Tsaih SW, Sheehan S, et al. Genetic analysis of blood pressure in C3H/HeJ and SWR/J mice. Physiol Genomics 2004;17:215–220.
62. Wade CM, Kulbokas EJ, 3rd, Kirby AW, et al. The mosaic structure of variation in the laboratory mouse genome. Nature 2002;420:574–578.
63. Frazer KA, Wade CM, Hinds DA, Patil N, Cox DR, Daly MJ. Segmental phylogenetic relationships of inbred mouse strains revealed by fine-scale analysis of sequence variation across 4.6 mb of mouse genome. Genome Res 2004;14:1493–1500.
64. Pletcher MT, McClurg P, Batalov S, et al. Use of a dense single nucleotide polymorphism map for in silico mapping in the mouse. PLoS Biol 2004;2:E393.
65. Lambrechts D, Storkebaum E, Morimoto M, et al. VEGF is a modifier of amyotrophic lateral sclerosis in mice and humans and protects motoneurons against ischemic death. Nat Genet 2003;34:383–394.
66. Mirel DB, Valdes AM, Lazzeroni LC, Reynolds RL, Erlich HA, Noble JA. Association of IL4R haplotypes with type 1 diabetes. Diabetes 2002;51:3336–3341.
67. Bonhomme F, Guenet J-L, Dod B, Moriwaki K, Bulfield G. The polyphyletic origin of laboratory inbred mice and their rate of evolution. J Linn Soc 1987;30:51–58.
68. Peissel B, Zaffaroni D, Zanesi N, et al. Linkage disequilibrium and haplotype mapping of a skin cancer susceptibility locus in outbred mice. Mamm Genome 2000;11:979–981.
69. Suzuki M, Carlson KM, Marchuk DA, Rockman HA. Genetic modifier loci affecting survival and cardiac function in murine dilated cardiomyopathy. Circulation 2002;105:1824–1829.
70. Wang X, Korstanje R, Higgins D, Paigen B. Haplotype analysis in multiple crosses to identify a QTL gene. Genome Res 2004;14:1767–1772.
71. Tabakoff B, Bhave SV, Hoffman PL. Selective breeding, quantitative trait locus analysis, and gene arrays identify candidate genes for complex drug-related behaviors. J Neurosci 2003;23:4491–4498.
72. Wayne ML, McIntyre LM. Combining mapping and arraying: An approach to candidate gene identification. Proc Natl Acad Sci USA 2002;99:14,903–14,906.
73. Dyck PA, Hoda F, Osmer ES, Green RM. Microarray analysis of hepatic gene expression in gallstone-susceptible and gallstone-resistant mice. Mamm Genome 2003;14:601–610.

74. Carrera P, Abrell S, Kerber B, et al. A modifier screen in the eye reveals control genes for Kruppel activity in the Drosophila embryo. Proc Natl Acad Sci USA 1998;95:10,779–10,784.

75. Mutsuddi M, Marshall CM, Benzow KA, Koob MD, Rebay I. The spinocerebellar ataxia 8 noncoding RNA causes neurodegeneration and associates with staufen in Drosophila. Curr Biol 2004;14:302–308.

76. Therrien M, Morrison DK, Wong AM, Rubin GM. A genetic screen for modifiers of a kinase suppressor of Ras-dependent rough eye phenotype in Drosophila. Genetics 2000;156:1231–1242.

77. Carpinelli MR, Hilton DJ, Metcalf D, et al. Suppressor screen in Mpl–/– mice: c-Myb mutation causes supraphysiological production of platelets in the absence of thrombopoietin signaling. Proc Natl Acad Sci USA 2004;101:6553–6558.

78. Kaushansky K. Thrombopoietin: the primary regulator of megakaryocyte and platelet production. Thromb Haemost 1995;74:521–525.

79. Kaushansky K, Drachman JG. The molecular and cellular biology of thrombopoietin: the primary regulator of platelet production. Oncogene 2002;21:3359–3367.

80. RetNet: Cloned and/or Mapped Genes Causing Retinal Diseases. Listed by chromosome, Vol. 2005. http://www.sph.uth.tmc.edu/Retnet/disease.htm.

81. Nadeau JH. Modifier genes and protective alleles in humans and mice. Curr Opin Genet Dev 2003;13:290–295.

82. Nakazawa M, Wada Y, Tamai M. Arrestin gene mutations in autosomal recessive retinitis pigmentosa. Arch Ophthalmol 1998;116:498–501.

83. Nordstrom S, Thorburn W. Dominantly inherited macular degeneration (Best's disease) in a homozygous father with 11 children. Clin Genet 1980;18:211–216.

84. Astuto LM, Bork JM, Weston MD, et al. CDH23 mutation and phenotype heterogeneity: a profile of 107 diverse families with Usher syndrome and nonsyndromic deafness. Am J Hum Genet 2002;71:262–275.

85. Bolz H, von Brederlow B, Ramirez A, et al. Mutation of CDH23, encoding a new member of the cadherin gene family, causes Usher syndrome type 1D. Nat Genet 2001;27:108–112.

86. Ness SL, Ben-Yosef T, Bar-Lev A, et al. Genetic homogeneity and phenotypic variability among Ashkenazi Jews with Usher syndrome type III. J Med Genet 2003;40:767–772.

87. Bernstein PS, Tammur J, Singh N, et al. Diverse macular dystrophy phenotype caused by a novel complex mutation in the ELOVL4 gene. Invest Ophthalmol Vis Sci 2001;42:3331–3336.

88. Zhang K, Kniazeva M, Hutchinson A, Han M, Dean M, Allikmets R. The ABCR gene in recessive and dominant Stargardt diseases: a genetic pathway in macular degeneration. Genomics 1999;60:234–237.

89. Wada Y, Abe T, Itabashi T, Sato H, Kawamura M, Tamai M. Autosomal dominant macular degeneration associated with 208delG mutation in the FSCN2 gene. Arch Ophthalmol 2003;121:1613–1620.

90. Downes SM, Payne AM, Kelsell RE, et al. Autosomal dominant cone-rod dystrophy with mutations in the guanylate cyclase 2D gene encoding retinal guanylate cyclase-1. Arch Ophthalmol 2001;119:1667–1673.

91. Sato M, Nakazawa M, Usui T, Tanimoto N, Abe H, Ohguro H. Mutations in the gene coding for guanylate cyclase-activating protein 2 (GUCA1B gene) in patients with autosomal dominant retinal dystrophies. Graefes Arch Clin Exp Ophthalmol 2005;243:235–242.

92. Keen TJ, Hims MM, McKie AB, et al. Mutations in a protein target of the Pim-1 kinase associated with the RP9 form of autosomal dominant retinitis pigmentosa. Eur J Hum Genet 2002;10:245–249.

93. Jacobson SG, Cideciyan AV, Iannaccone A, et al. Disease expression of RP1 mutations causing autosomal dominant retinitis pigmentosa. Invest Ophthalmol Vis Sci 2000; 41:1898–1908.

94. Tantri A, Vrabec TR, Cu-Unjieng A, Frost A, Annesley WH, Jr, Donoso LA. X-linked retinoschisis: report of a family with a rare deletion in the XLRS1 gene. Am J Ophthalmol 2003;136:547–549.

95. Hamilton WK, Ewing CC, Ives EJ, Carruthers JD. Sorsby's fundus dystrophy. Ophthalmology 1989;96:1755–1762.

96. Hong HK, Lass JH, Chakravarti A. Pleiotropic skeletal and ocular phenotypes of the mouse mutation congenital hydrocephalus (ch/Mf1) arise from a winged helix/forkhead transcriptionfactor gene. Hum Mol Genet 1999;8:625–637.

97. Zhou G, Williams RW. Eye1 and Eye2: gene loci that modulate eye size, lens weight, and retinal area in the mouse. Invest Ophthalmol Vis Sci 1999;40:817–825.

98. Bone-Larson C, Basu S, Radel JD, et al. Partial rescue of the ocular retardation phenotype by genetic modifiers. J Neurobiol 2000;42:232–247.

99. Rohan RM, Fernandez A, Udagawa T, Yuan J, D'Amato RJ. Genetic heterogeneity of angiogenesis in mice. FASEB J 2000;14:871–876.

100. Anderson MG, Smith RS, Savinova OV, et al. Genetic modification of glaucoma associated phenotypes between AKXD-28/Ty and DBA/2J mice. BMC Genet 2001;2:1.

101. Humphries MM, Kiang S, McNally N, et al. Comparative structural and functional analysis of photoreceptor neurons of Rho–/– mice reveal increased survival on C57BL/6J in comparison to 129Sv genetic background. Vis Neurosci 2001;18:437–443.

14

X-Linked Retinal Dystrophies and Microtubular Functions Within the Retina

Alan F. Wright, PhD, FRCP, FRSE and Xinhua Shu, PhD

INTRODUCTION

RP is one of the most heterogeneous genetic disorders known in man *(1)*. There are currently about 40 genes known or identified in this group of disorders *(2)*. Most cases result from one of a series of monogenic disorders inherited in an autosomal, X-linked or mitochondrial manner. The extent to which it includes a subset of oligogenic or even polygenic conditions is unclear. Oligogenic inheritance has been established in a small proportion of RP families, for example, a combination of mutations in the *ROM1* and RDS/peripherin *(PRPH2)* genes *(3)*. A significant excess of RP simplex cases (single-affected individual within a family) has been reported in segregation analyses, suggesting that 12 to 40% of all RP results from nongenetic causes, new mutations, or complex inheritance *(4,5)*. The early literature also found a significant excess of affected males relative to females *(6,7)* (e.g., Nettleship *[6]* found a ratio of 1.6:1) and that males were less likely than females to transmit the disease to their offspring *(7)*—both suggesting the possibility of X-linkage. This question needs to be revisited in the light of recent molecular findings, as discussed in the section on the *RPGR* gene.

RP was first described clinically by van Trigt in 1853 *(8)* and although several X-linked RP (XLRP) pedigrees were reported in the early 1900s, they were not recognised as such until a paper by Usher in 1935 *(9)*, which described an X-linked recessive RP pedigree.

The prevalence of XLRP is in the region of 1 in 10,000–15,000 in most populations of European origin. This figure is based first on RP population prevalence studies, which generally report figures of 1 in 3000–7000 *(10,11)*, although it may be less

From: *Ophthalmology Research: Retinal Degenerations: Biology, Diagnostics, and Therapeutics*
Edited by: J. Tombran-Tink and C. J. Barnstable © Humana Press Inc., Totowa, NJ

common in some Mediterranean and African countries. However, two extremely thorough population-based studies, covering multiple ascertainment sources in populations of more than 1 million, found prevalences for RP in the region of 1 in 1500 people, which is probably the most accurate figure *(12,13)*. In the United States, Fishman found that 16% of his sample of patients with RP showed X-linkage *(14)* compared with 14–16% in two UK studies *(4,15)*. Assuming therefore that 15% of all RP cases are caused by X-linkage (which compares well with 15–20% estimated from molecular data) then XLRP has a population prevalence of 1 in 9400. Assuming a lower percentage of X-linked cases, such as 10%, then the prevalence of XLRP would be 1 in 15,000. A prevalence study that fails to examine female family members for the carrier state or to account for excess simplex males or male multiplex sibships are likely to underestimate significantly the proportion of all RP caused by X-linked disease.

CLINICAL MANIFESTATIONS OF XLRP

The earliest clinical manifestation of XLRP in males is generally night blindness, with onset in the first decade, progressing to reduction in visual fields in the second decade, a reduction in visual acuity by age 20 and severe visual loss (<20/200) by age 40 *(16)*. More than half of all XLRP hemizygotes (affected males) are symptomatic by age 10 and only 16% retain useful vision by age 40 *(15,16)*. The average age at onset in XLRP has been reported to be 7.2 ± 1.7 yr *(17)*, so that this is one of the most consistently severe forms of RP. Other clinical features have been noted, including some that are seen in all types of RP and others that show at least some predilection for the X-linked subtype. The former include the characteristic bone spicule fundus deposits, attenuation of retinal arterioles, optic disk pallor, posterior subcapsular lens opacities, and absent or subnormal electroretinogram (ERG) amplitudes. The signs that are more indicative of X-linked RP (but by no means diagnostic) include macular or foveal lesions, impaired color vision (blue-yellow defect), and a spherical refractive error of −2.00 diopters or greater *(18)*. However, XLRP cannot be distinguished clinically from other severe forms of retinal dystrophy.

Carrier females in XLRP can show some relatively characteristic features *(16,18)*. The most common manifestations in carriers are however rather non-specific, such as late-onset night blindness, both pigment epithelial changes and a few pigmentary deposits in the peripheral retina, associated with full-field ERG abnormalities, such as reduced amplitude to white light or delayed cone-wave implicit times *(16,19,20)*. The flicker ERG may also be abnormal in XLRP, implicating a reduced signal-to-noise ratio in the rod system compared with normal *(21)*. The signs that show greater specificity for the carrier state include, first, the tapetal reflex, noted by Frost first in 1902 *(22)* and later described in detail by Falls and Cotterman *(23)*. It is a golden metallic sheen in the macular region, best seen on direct ophthalmoscopy, named after the similar appearance seen in many mammals (but not humans) that have a tapetum lucidum (a reflective layer of the choroid), when a light is shone into the eye at night. However, the sign is only present in a minority of carrier families *(16)*. More recently, Lorenz and co-workers described patchy loss of rod and cone sensitivity in XLRP carriers by two-color threshold perimetry, with rods more severely affected than cones *(24)*. However, a new and more specific finding was an abnormal radial pattern of fundus autofluorescence in 80% of carriers. The authors suggested that the radial pattern could be explained by random X-inactivation in early embryogenesis and a radial and centrifugal pattern of cell growth in the developing retina *(24)*. If confirmed, this would provide a useful and relatively simple and specific test for the X-linked carrier state. Because XLRP carriers under the age of 40 yr are often asymptomatic and yet are at 1 in 2 risk of having affected sons, this type of test would be useful for genetic counselling. Recognizing XLRP pedigrees is often not straightforward because family sizes are commonly small. Recognizing the carrier state in mothers of single-affected males or male siblings with such a test would help to recognize XLRP and initiate genetic testing.

The key pedigree features that alert the clinician to the possibility of XLRP are a classical X-linked inheritance pattern with affected males and unaffected or more mildly affected but transmitting females and absence of male-to-male transmission. What is often confusing and results in mislabeling of XLRP pedigrees as autosomal dominant ones is the presence of severely affected females in a pedigree in which there is no male-to-male transmission. There has long been debate as to whether X-linked dominant or intermediate compared with recessive inheritance occurs in XLRP and the matter is still not fully resolved. There do appear to be some pedigrees in which female carriers are more consistently or severely affected than others. This probably mirrors the severity of the responsible mutation but chance skewing of X-inactivation and ascertainment bias probably also contribute.

Early genetic linkage studies established the presence of two major XLRP loci (RP3, RP2), situated 16–25 cM apart on the short arm of the X chromosome *(25,26)*. The results indicated that the RP3 locus in Xp21.1 accounted for 60 to 75% of the XLRP families analysed and the remainder mapped to the RP2 locus in the Xp11.2-p11.3 region. This led to prolonged and difficult positional cloning efforts, which finally led, firstly, to the identification of the gene responsible for RP3, the RP GTPase Regulator or *RPGR* gene *(27,28)*.

RP3 TYPE X-LINKED RP AND THE *RPGR* GENE

The *RPGR* gene is located in chromosomal region Xp21.1 and spans 172 kb *(29)*. There are multiple alternatively spliced transcripts, all of which encode an amino (N)-terminal RCC1-like domain (RLD) that is structurally similar to the RCC1 protein, a guanine nucleotide exchange factor for the small GTP-binding protein, Ran *(27,29,30)*. The X-ray crystallographic structure of RCC1 consists of a seven-bladed propeller formed from internal repeats of 51–68 residues per blade *(30)*. The RLD of RPGR interacts with at least two proteins, RPGRIP1 *(31–33)* and a 17 kD prenyl binding protein called PDED *(34)*. RPGRIP1 has multiple isoforms that are of unknown function but contains a long N-terminal coiled-coil domain, a Ca_2 phospholipid binding domain, and a carboxyl C-terminal RPGR interaction domain *(31–33)*. Mutations in *RPGRIP1* were subsequently shown to cause a form of congenital retinal blindness, Leber's congenital amaurosis, in 5–10% of patients *(35,36)* and a subtype of cone-rod dystrophy (CRD) *(37)*. The function of the 17 kD prenyl binding protein is also unclear but it binds prenylated photoreceptor proteins such as opsin kinase (GRK1, GRK7), rod cGMP phosphodiesterase (PDE6) subunits, and the small GTPase Rab8 *(38)*. It is implicated in the transport and membrane targeting of prenylated proteins destined for photoreceptor OS and shows increased labeling near the connecting cilia of bovine photoreceptors, at the junction of IS and OS (38).

The RPGR transcript that was initially identified is widely expressed and contains 19 exons (RPGR^{ex1-19}), encoding a predicted 90 kDa protein (27,28). Exons 1–11 encode the RLD, whereas exons 12–19 encode a C-terminal domain rich in acidic residues and ending in an isoprenylation anchorage signal (Fig. 1A). Subsequently, alternative transcripts were found, the most important of which is RPGR^{ORF15} (29), which shows highest expression in photoreceptors, and is the only transcript known to be involved in retinal disease. Human RPGR^{ORF15} contains exons 1–14 of RPGR^{ex1-19} plus a large alternatively spliced C-terminal exon, ORF15, encoding 567 amino acids (Fig. 1A). The full-length human RPGR^{ORF15} isoform encodes a 1152 amino acid protein. Exon ORF15 encodes a repetitive glycine and glutamic acid-rich domain of unknown function and a basic C-terminal domain (ORF15^{C2}),which is evolutionarily conserved and binds the multifunctional chaperone nucleophosmin (NPM) at centrosomes and mitotic spindle poles (39).

RPGR Function

RPGR is a component of centrioles, ciliary axonemes, and microtubular transport complexes, although its precise function is unknown. It co-localizes with RPGRIP1 at the axonemes of connecting cilia in rod and cone photoreceptors (40) by binding to RPGRIP1$_2$ because this localization is lost in Rpgrip1 knockout (KO) mice. Rpgr KO mice develop a slow retinal degeneration, with features resembling a cone-rod degeneration—cone photoreceptors degenerate faster than rods and there is partial mislocalization of cone opsins (41). Some residual Rpgr^{ORF15} expression has been reported in this model (42).

RPGR has been shown to co-immunoprecipitate in retinal extracts with a number of different axonemal, basal body and microtubular transport proteins (42). These include nephrocystin-5 and calmodulin, which localize to photoreceptor connecting cilia; the microtubule-based transport proteins, kinesin II (KIF3A, KAP3 subunits), dynein (DIC subunit), SMC1, and SMC3; and two regulators of cytoplasmic dynein, p150^{Glued} and p50-dynamitin, which tether cargoes to the dynein motor. Inhibition of dynein by overexpressing p50-dynamitin abrogates the localization of RPGR^{ORF15} to basal bodies. RPGR^{ORF15} can be co-immunoprecipitated from retinal extracts with other basal body proteins, including NPM, IFT88, 14-3-3ε, and γ-tubulin (39,42). RPGR^{ORF15} and RPGRIP1 co-localize at centrosomes in a wide variety of nonciliated cells and at basal bodies in ciliated cells (39). Both proteins are core components of centrioles and basal bodies (39). In summary, RPGR^{ORF15} appears to have a role in microtubule-based transport to and from the basal bodies and within photoreceptor axonemes, perhaps concerned with movement of cargoes between IS and OS.

RPGR^{ORF15} is predominantly expressed in photoreceptor connecting cilia and basal bodies but expression has also been reported in OS in some species (43), although this has been disputed (40). RPGR is also expressed in the transitional zone of motile cilia in the epithelial lining of human bronchi and sinuses (RPGR^{ex1-19} only) and within the human and monkey cochlea (40,44).

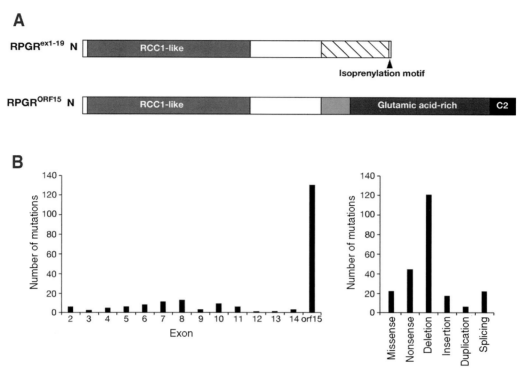

Fig. 1. (A) Domain structure of two major RPGR isoforms (RPGR[ex1–19] and RPGR[ORF15]), as discussed in the text. (B) The distribution of reported mutations in the *RPGR*[ORF15] isoform within each exon, showing the large excess of mutations within the C-terminal exon ORF15. The types of mutation are also shown, again indicating the excess of deletions due to the ORF15 mutational "hot spot."

Involvement of RPGR in Disease

240 different mutations in *RPGR* have been reported which account for a number of different retinopathies. Numerically, the most important group is XLRP because up to 75% of cases in most series result from mutations in RPGR[ORF15] *(29,45–47)*. Lower figures are reported in which X-linkage is uncertain. Mutations can occur in any part of the protein but the glycine/glutamic-acid rich domain of exon ORF15 is a mutation "hotspot," accounting for up to 80% of RPGR mutations *(29,45–47)* (Fig. 1B). Because of its high-mutation rate, RPGR accounts for disease in up to 20% of all patients with RP *(45)*. The highest frequency of mutations is associated with a central repetitive, purine-rich region of ORF15, which contains only 2–3% pyrimidines. The purine-rich sequences may promote polymerase arrest and slipped strand mispairing because the majority of mutations are out-of-frame deletions of 1–5 bp within short repetitive regions. These are predicted to produce truncated proteins of varying length, and can include novel amino acid sequences that change the charge of the domain from acidic to basic. Severe disease is correlated with longer regions of such abnormal sequence *(47)*. Because ORF15 is a C-terminal exon, it is not subject to nonsense-mediated decay, so that truncated RPGR[ORF15] may accumulate and cause a gain-of-function or

toxic phenotype *(48,49)*. On the other hand, the majority of missense mutations within the RLD are nonconservative and affect highly conserved residues, suggesting a loss-of-function mechanism. Some of the phenotypic variability found with RPGR mutations (in hemizygotes and heterozygotes) may relate to such differences in the mutant protein.

In addition to unconditional and conditional KO mouse models of XLRP, there are naturally occurring animal models in mouse (rd9) and dogs. Canine X-linked progressive retinal atrophy (XLPRA) is subdivided by mutational type into XLPRA1 (Siberian husky, Samoyed), which have the same 5-bbp deletion in exon ORF15, and XLPRA2, which has a 2-bp deletion in ORF15 (mixed breed) *(48)*. The XLPRA1 mutation results in a frameshift and immediate premature stop codon, and shows normal photoreceptor development and function until about 6 mo of age, after which a slow degeneration of rod and cone photoreceptors occurs, with cones being less severely affected. In contrast, the frameshift in XLPRA2 shows a premature stop codon 71 amino acids downstream, including 34 additional basic residues. The XLPRA2 phenotype is very severe and manifests during retinal development, with disorganized and disoriented rod and cone OS, which is worst in rods, and faster degeneration. The XLPRA2 protein aggregates in the endoplasmic reticulum of transfected cells, in contrast to the normal and XLPRA1 proteins *(48)*.

RPGR^{ORF15} mutations can give rise to central or macular dystrophies, including X-linked forms of CRD *(50,51)*, cone dystrophy *(29,52)*, and atrophic macular degeneration *(53)*. In the X-linked CRD families, a parafoveal ring of fundus autofluorescence, similar to that described in XLRP carriers *(24)*, was evident in younger affected males (but not in carriers). Mutations that are closer to the 3′ end of RPGR^{ORF15} and that lack abnormal (basic) sequence tend to be associated with CRD and milder forms of RP *(47)*. Similarly, loss of the final third (180 amino acids) of RPGR^{ORF15} results in X-linked atrophic macular degeneration *(53)*. A histopathological study of a CRD carrier showed a bull's eye maculopathy, focally absent macular retinal pigment epithelium and absent perifoveal cones and rods *(54)*. Elsewhere in the macula and in the peripheral retina, cone but not rod photoreceptors were reduced in numbers and both rod and cone OS shortened.

Mutations in *RPGR* can give rise to a syndrome combining features of a primary ciliary dyskinesia, sensorineural hearing loss, and RP *(55–58)*. This syndrome is most commonly associated with deletion, missense or splice site mutations within the RLD of RPGR and can be confused with Usher syndrome.

RP2-TYPE XLRP AND THE *RP2* GENE

The second XLRP locus, which mapped to Xp11.2-p11.3 by linkage studies *(25,26,59)* was positionally cloned in 1998 and named the *RP2* gene *(60)*. *RP2* consists of five exons and encompasses 3.8 kb of DNA. Exon 2 encodes a cofactor-C homologous domain (CFCHD) and a microtubule-associated protein homologous domain (MAPHD). Exons 3 and 4 encode a NM23 homologous domain (NM23HD) *(61)* (Fig. 2A). NM23 belongs to a large family of structurally and functionally conserved proteins, consisting of four to six identically folded subunits of approx 16–20 kDa in size. These oligomeric proteins show nucleoside diphosphate kinase (NDPK) activity

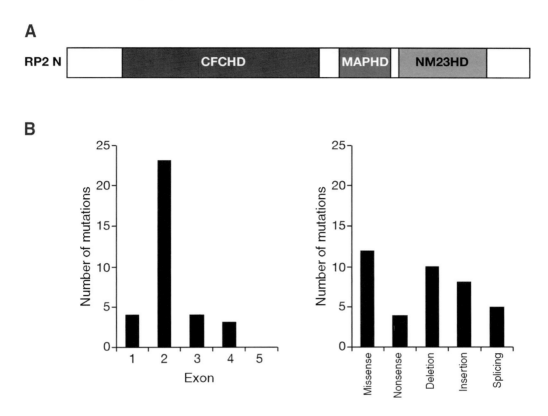

Fig. 2. (**A**) Domain structure of the RP2 protein, as discussed in the text. CFCHD, cofactor-C homologous domain; MAPHD, microtubule-associated protein homologous domain; NM23HD, NM23 homologous domain. (**B**) The distribution of reported mutations in *RP2* in each exon, showing the excess of mutations within exon 2, which encodes the CFCHD and MAPHD domains. The types of mutation are also shown, indicating a more even distribution than with *RPGR*.

that catalyzes nonsubstrate specific conversions of nucleoside diphosphates to nucleoside triphosphates. Many NM23 proteins bind DNA. In vivo, NM23-NDPKs regulate a diverse range of cellular events, including growth and development *(62)*.

To date, there are 39 *RP2* mutations identified in XLRP families. The distribution and type of mutations are shown in Fig. 2B. Most mutations have been found in exon 2, which encodes the CFCHD and MAPHD domains. In three studies in which both *RPGR* and *RP2* were screened for mutation, *RP2* mutations accounted for 7–8% of patients with XLRP whereas *RPGR* accounted for 55–73% of mutations *(45–47)*.

RP2 protein is widely expressed at low levels in human tissues, and has been reported to be targeted to the plasma membrane by myristoylation and palmitoylation of its N-terminal amino acids *(63)*. There is a potential *N*-myristoyl transferase recognition site at the N-terminus of RP2. The N-terminal 15 amino acids appear to be sufficient for targeting RP2 to the membrane. A single amino acid deletion of Serine residue 6 in a patient with RP2 resulted in failure of the protein to associate with the plasma membrane *(63)*. It was therefore proposed that XLRP is prevented from reaching the correct cellular location in sufficient amounts for normal activity *(63)*. C-terminal truncation mutations, which account for two-thirds of pathogenic RP2 variants, led to misfolding and subsequent degradation of the resultant nonfunctional proteins. Loss of the protein and/or its aberrant intracellular distribution might therefore be a common basis for the photoreceptor cell degeneration occurring in patients with RP2 *(64)*.

RP2 and the tubulin-specific chaperone cofactor C share 53% similarity and 29% identity over a domain of approx 200 amino acids. Both proteins stimulate the GTPase activity of native tubulin. Their functions are overlapping but not identical, because only cofactor C participates in the heterodimerization of newly folded tubulin subunits. The adenosine diphosphate (ADP)-ribosylation factor-like 2 (Arl2) protein regulates the tubulin-GTPase activating protein (GAP) activity of cofactor C and D. Arl3 does not affect the tubulin-GAP activity of RP2 or cofactor D. RP2 binds to Arl3 in a nucleotide and myristoylation-dependent manner. An R118H pathogenic mutation, which does not affect the normal plasma membrane localization of RP2 *(63)*, implies that this residue acts as an "arginine finger" to trigger the tubulin-GAP activity *(65)*.

RP2 was localized to the plasma membrane of cells throughout the human retina, whereas Arl3 and cofactor C localized predominantly to the connecting cilium of rod and cone photoreceptor cells. The localization of RP2 and its interacting proteins to this region suggest that RP2 may be involved in vesicular transport *(66)*, providing an intriguing parallel with RPGR. Much remains to be learned about the precise functions of RP2 in the retina but at this stage it is certainly possible that, like RPGR, it has a role in some aspect of microtubular function relating IS and OS of photoreceptors.

X-linked retinitis pigmentosa (RP) is a genetically heterogeneous disorder in which two major genes have been identified (*RPGR, RP2*) that account for the disease in the majority of patients. RPGR joins a growing number of proteins implicated in ciliary function and may be concerned with microtubule-based transport between inner and outer segments (IS and OS, respectively) of rod and cone photoreceptors. The function of RP2 is less clear but it appears to have a role in microtubular assembly or function in photoreceptor IS.

ACKNOWLEDGMENTS

We would like to acknowledge the financial support of the Medical Research Council, The British Retinitis Pigmentosa Society, the Foundation Fighting Blindness, and EVI-GENORET (European Union) for financial support.

REFERENCES

1. Rattner A, Sun H, Nathans J. Molecular genetics of human retinal disease. Annu Rev Genet 1999;33:89–131.
2. Retina Information Network. http://www.sph.uth.tmc.edu/Retnet/ Date accessed: September 2005.
3. Kajiwara K, Berson EL, Dryja TP. Digenic retinitis pigmentosa due to mutations at the unlinked peripherin/RDS and ROM1 loci. Science 1994;264:1604–1608.
4. Jay M. On the heredity of retinitis pigmentosa. Br J Ophthalmol 1982;66:405–416.
5. Najera C, Millan JM, Beneyto M, Prieto F. Epidemiology of retinitis pigmentosa in the Valencian community (Spain). Genet Epidemiol 1995;12:37–46.

6. Nettleship E. On retinitis pigmentosa and allied diseases. R Lond Ophthal Hosp Rep 1908;17:333–426.
7. Bell J. Retinitis pigmentosa and allied diseases. In: Pearson K, ed. Treasury of Human Inheritance, vol. 2. Cambridge University Press, Cambridge, UK 1922;1–28.
8. Van Trigt AC. De oogspiegel. Nederlandisch Lancet, third series, Utrecht, 1852–1853;2d: 417–509.
9. Usher CH. On a few hereditary eye affections. Trans Ophthal Soc UK 1935;55:164–245.
10. Pagon RA. Retinitis pigmentosa. Surv Ophthalmol 1988;33:137–177.
11. Humphries P, Kenna P, Farrar GJ. On the molecular genetics of retinitis pigmentosa. Science 1992;25:804–808.
12. Bundey S, Crews SJ. A study of retinitis pigmentosa in the city of Birmingham. I Prevalence. J Med Genet 1984;21:417–420.
13. Puech B, Kostrubiec B, Hache JC, Francois P. Epidemiology and prevalence of hereditary retinal dystrophies in the Northern France. J Fr Ophtalmol 1991;14:153–164.
14. Fishman GA. Retinitis pigmentosa. Genetic percentages. Arch Ophthalmol 1978;96: 822–826.
15. Bundey S, Crews SJ. A study of retinitis pigmentosa in the city of Birmingham. II Clinical and genetic heterogeneity. J Med Genet 1984;21:421–428.
16. Bird AC. X-linked retinitis pigmentosa. Br J Ophthalmol 1975;59:177–199.
17. Maumenee IH, Pierce ER. Bias WB, Schleutermann DA. Linkage studies of typical retinitis pigmentosa and common markers. Am J Hum Genet 1975;27:505–508.
18. Fishman GA, Farber MD, Derlacki DJ. X-linked retinitis pigmentosa. Profile of clinical findings. Arch Ophthalmol 1988;106:369–375.
19. Fishman GA, Weinberg AB, McMahon TT. X-linked recessive retinitis pigmentosa. Clinical characteristics of carriers. Arch Ophthalmol 1986;104:1329–1335.
20. Berson EL, Rosen JB, Simonoff EA. Electroretinographic testing as an aid in detection of carriers of X-chromosome-linked retinitis pigmentosa. Am J Ophthalmol 1979;87:460–468.
21. Ernst W, Clover G, Faulkner DJ. X-linked retinitis pigmentosa: reduced rod flicker sensitivity in heterozygous females. Invest Ophthalmol Vis Sci 1981;20:812–816.
22. Frost WA. Usually well marked retinla reflex. Trans Ophthalmol Soc UK 1902;22:208–209.
23. Falls HF, Cotterman C. Choroidal degeneration. A sex-linked form in which heterozygous women exhibit a tapetal-like retinal reflex. Arch Ophthalmol 1948;40:683–703.
24. Wegscheider E, Preising MN, Lorenz B. Fundus autofluorescence in carriers of X-linked recessive retinitis pigmentosa associated with mutations in RPGR, and correlation with electrophysiological and psychophysical data. Graefes Arch Clin Exp Ophthalmol 2004;242:501–511.
25. Ott J, Bhattacharya S, Chen JD, et al. Localizing multiple X chromosome-linked retinitis pigmentosa loci using multilocus homogeneity tests. Proc Natl Acad Sci USA 1990; 87:701–704.
26. Teague PW, Aldred MA, Jay M, et al. Heterogeneity analysis in 40 X-linked retinitis pigmentosa families. Am J Hum Genet 1994;55:105–111.
27. Meindl A, Dry K, Herrmann K, et al. A gene (RPGR) with homology to the RCC1 guanine nucleotide exchange factor is mutated in X-linked retinitis pigmentosa (RP3). Nat Genet 1996;13:35–42.
28. Roepman R, van Duijnhoven G, Rosenberg T, et al. Positional cloning of the gene for X-linked retinitis pigmentosa 3: homology with the guanine-nucleotide-exchange factor RCC1. Hum Mol Genet 1996;5:1035–1041.
29. Vervoort R, Lennon A, Bird AC, et al. Mutational hot spot within a new RPGR exon in X-linked retinitis pigmentosa. Nat Genet 2000;25:462–466.
30. Renault L, Nassar N, Vetter I, et al. The 1.7 A crystal structure of the regulator of chromosome condensation (RCC1) reveals a seven-bladed propeller. Nature 1998;392:97–101.

31. Boylan JP, Wright AF. Identification of a novel protein interacting with RPGR. Hum Mol Genet 2000;9:2085–2093.

32. Roepman R, Bernoud-Hubac N, Schick E, et al. The retinitis pigmentosa GTPase regulator (RPGR) interacts with novel transport-like proteins in the outer segments of rod photoreceptors. Hum Mol Genet 2000;9:2095–2105.

33. Hong DH, Yue G, Adamian M, Li T. Retinitis pigmentosa GTPase regulator (RPGRr)-interacting protein is stably associated with the photoreceptor ciliary axoneme and anchors RPGR to the connecting cilium. J Biol Chem 2002;276:12,091–12,099.

34. Linari M, Ueffing M, Manson F, Wright A, Meitinger T, Becker J. The retinitis pigmentosa GTPase regulator, RPGR, interacts with the delta subunit of rod cyclic GMP phosphodiesterase. Proc Natl Acad Sci USA 1999;96:1315–1320.

35. Dryja TP, Adams SM, Grimsby JL, et al. Null RPGRIP1 alleles in patients with Leber congenital amaurosis. Am J Hum Genet 2001;68:1295–1298.

36. Gerber S, Perrault I, Hanein S, et al. Complete exon-intron structure of the RPGR-interacting protein (RPGRIP1) gene allows the identification of mutations underlying Leber congenital amaurosis. Eur J Hum Genet 2001;9:561–571.

37. Hameed A, Abid A, Aziz A, Ismail M, Mehdi SQ, Khaliq S. Evidence of RPGRIP1 gene mutations associated with recessive cone-rod dystrophy. J Med Genet 2003;40:616–619.

38. Norton AW, Hosier S, Terew JM, et al. Evaluation of the 17-kDa prenyl-binding protein as a regulatory protein for phototransduction in retinal photoreceptors. J Biol Chem 2005;280:1248–1256.

39. Shu X, Fry AM, Tulloch B, et al. RPGR ORF15 isoform co-localizes with RPGRIP1 at centrioles and basal bodies and interacts with nucleophosmin. Hum Mol Genet 2005;14:1183–1197.

40. Hong DH, Pawlyk B, Sokolov M, et al. RPGR isoforms in photoreceptor connecting cilia and the transitional zone of motile cilia. Invest Ophthalmol Vis Sci 2003;44:2413–2421.

41. Hong DH, Pawlyk BS, Shang J, Sandberg MA, Berson EL, Li T. A retinitis pigmentosa GTPase regulator (RPGR)-deficient mouse model for X-linked retinitis pigmentosa (RP3). Proc Natl Acad Sci USA 2000;97:3649–3654.

42. Khanna H, Hurd TW, Lillo C, et al. RPGR-ORF15, Which is mutated in retinitis pigmentosa, associates with SMC1, SMC3, and microtubule transport proteins. J Biol Chem 2005;280:33,580–33,587.

43. Mavlyutov TA, Zhao H, Ferreira PA. Species-specific subcellular localization of RPGR and RPGRIP isoforms: implications for the phenotypic variability of congenital retinopathies among species. Hum Mol Genet 2002;11:1899–1907.

44. Iannaccone A, Wang X, Jablonski MM, et al. Increasing evidence for syndromic phenotypes associated with RPGR mutations. Am J Ophthalmol 2004;137:785–786.

45. Breuer DK, Yashar BM, Filippova E, et al. A comprehensive mutation analysis of RP2 and RPGR in a North American cohort of families with X-linked retinitis pigmentosa. Am J Hum Genet 2002;70:1545–1554.

46. Bader I, Brandau O, Achatz H, et al. X-linked retinitis pigmentosa: RPGR mutations in most families with definite X linkage and clustering of mutations in a short sequence stretch of exon ORF15. Invest Ophthalmol Vis Sci 2003;44:1458–1463.

47. Sharon D, Sandberg MA, Rabe VW, Stillberger M, Dryja TP, Berson EL. RP2 and RPGR mutations and clinical correlations in patients with X-linked retinitis pigmentosa. Am J Hum Genet 2003;73:1131–1146.

48. Zhang Q, Acland GM, Wu WX, et al. Different RPGR exon ORF15 mutations in Canids provide insights into photoreceptor cell degeneration. Hum Mol Genet 2002;11:993–1003.

49. Hong DH, Pawlyk BS, Adamian M, Li T. Dominant, gain-of-function mutant produced by truncation of RPGR. Invest Ophthalmol Vis Sci 2004;45:36–41.

50. Demirci FY, Rigatti BW, Wen G, et al. X-linked cone-rod dystrophy (locus COD1): identification of mutations in RPGR exon ORF15. Am J Hum Genet 2002;70:1049–1053.

51. Ebenezer ND, Michaelides M, Jenkins SA, et al. Identification of novel RPGR ORF15 mutations in X-linked progressive cone-rod dystrophy (XLCORD) families. Invest Ophthalmol Vis Sci 2005;46:1891–1898.

52. Yang Z, Peachey NS, Moshfeghi DM, et al. Mutations in the RPGR gene cause X-linked cone dystrophy. Hum Mol Genet 2002;11:605–611.

53. Ayyagari R, Demirci FY, Liu J, et al. X-linked recessive atrophic macular degeneration from RPGR mutation. Genomics 2002;80:166–171.

54. Aguirre GD, Yashar BM, John SK, et al. Retinal histopathology of an XLRP carrier with a mutation in the RPGR exon ORF15. Exp Eye Res 2002;75:431–443.

55. Dry KL, Manson FD, Lennon A, Bergen AA, Van Dorp DB, Wright AF. Identification of a 5′ splice site mutation in the RPGR gene in a family with X-linked retinitis pigmentosa (RP3). Hum Mutat 1999;13:141–145.

56. Iannaccone A, Breuer DK, Wang XF, et al. Clinical and immunohistochemical evidence for an X linked retinitis pigmentosa syndrome with recurrent infections and hearing loss in association with an RPGR mutation. J Med Genet 2003;40:e118.

57. Zito I, Downes SM, Patel RJ, et al. RPGR mutation associated with retinitis pigmentosa, impaired hearing, and sinorespiratory infections. J Med Genet 2003;40:609–615.

58. Koenekoop RK, Loyer M, Hand CK, et al. Novel RPGR mutations with distinct retinitis pigmentosa phenotypes in French-Canadian families. Am J Ophthalmol 2003;136:678–687.

59. Bhattacharya SS, Wright AF, Clayton JF, et al. Close genetic linkage between X-linked retinitis pigmentosa and a restriction fragment length polymorphism identified by recombinant DNA probe L1.28. Nature 1984;309:253–255.

60. Schwahn U, Lenzner S, Dong J, et al. Positional cloning of the gene for X-linked retinitis pigmentosa 2. Nat Genet 1998;19:327–332.

61. Miano MG, Testa F, Filippini F, et al. Identification of novel RP2 mutations in a subset of X-linked retinitis pigmentosa families and prediction of new domains. Hum Mutat 2001;18:109–119.

62. Postel EH. NM23-NDP kinase. Int J Biochem Cell Biol 1998;30:1291–1295.

63. Chapple JP, Hardcastle AJ, Grayson C, Spackman LA, Willison KR, Cheetham ME. Mutations in the N-terminus of the X-linked retinitis pigmentosa protein RP2 interfere with the normal targeting of the protein to the plasma membrane. Hum Mol Genet 2000;9:1919–1926.

64. Schwahn U, Paland N, Techritz S, Lenzner S, Berger W. Mutations in the X-linked RP2 gene cause intracellular misrouting and loss of the protein. Hum Mol Genet 2001; 10:1177–1183.

65. Bartolini F, Bhamidipati A, Thomas S, Schwahn U, Lewis SA, Cowan NJ. Functional overlap between retinitis pigmentosa 2 protein and the tubulin-specific chaperone cofactor C. J Biol Chem 2002;277:14,629–14,634.

66. Grayson C, Bartolini F, Chapple JP, et al. Localization in the human retina of the X-linked retinitis pigmentosa protein RP2, its homologue cofactor C and the RP2 interacting protein Arl3. Hum Mol Genet 2002;11:3065–3074.

15

Synaptic Remodeling in Retinal Degeneration

You-Wei Peng and Fulton Wong

CONTENTS

INTRODUCTION: COMMON CONSEQUENTIAL EVENTS OF MUTATION-INDUCED ROD-CONE PHOTORECEPTOR DEGENERATION

Retinitis pigmentosa (RP) is a group of hereditary retinal degenerative diseases with a complex molecular etiology. Hundreds of RP-inducing mutations, involving dozens of genes, have been identified in patients (*see* references in other chapters in this book; a list of identified mutations that cause retinal degeneration is updated at www.sph.uth.tmc.edu/RetNet). Despite this genetic heterogeneity, patients with RP tend to have a common pattern of vision loss. Typically, patients experience loss of night vision early in life as a result of degeneration of rod photoreceptors. Some loss of cone photoreceptor function may be detected early as well. Nevertheless, the majority of cones survives and remains functional, and hence daytime vision persists. Over years and decades,

From: *Ophthalmology Research: Retinal Degenerations: Biology, Diagnostics, and Therapeutics*
Edited by: J. Tombran-Tink and C. J. Barnstable © Humana Press Inc., Totowa, NJ

however, these cones progressively degenerate, leading ultimately to blindness *(1)*. The link between the myriad of mutations to the disease mechanisms underlying the clinical course of rod-cone degeneration in RP remains unknown; meanwhile, there is no cure or effective treatment broadly available for RP.

The combination of vast genetic heterogeneity and a relatively well-defined, "stereotypic phenotype" argues strongly for a decisive role in photoreceptor death of some "final common pathway," irrespective of the specific RP-inducing mutation. This interpretation is supported by the fact that many RP-inducing mutations are rod photoreceptor specific. In other words, the RP-inducing mutation may involve a gene that is not expressed in cone photoreceptors and yet, the cones eventually die. Death of the "genetically normal" cones must be caused by mechanisms secondary to the immediate effects initiated by the mutation that leads to rod degeneration because neither the normal nor the mutated gene in question is expressed in cones. In RP, it is the death of the cones that forces patients to dependency and thus the most practical and critical issue in finding an effective treatment for RP is to identify methods that would prolong the survival and function of the remaining cones in a diseased retina. In this regard, one of the prevailing views is that an RP-inducing mutation triggers a sequence of "downstream" cellular events that lead eventually to rod and cone photoreceptor death *(2–7)*. Although still unidentified, the pursuit of these elusive—sometimes even seemingly abstract—cellular events *(5)* has yielded some tangible results that may lead to the future identification of the precise cellular mechanisms causing pan-retinal degeneration, including death of the cones, in RP.

Common approaches in RP research either aim at finding ways to rescue photoreceptors, focusing on therapeutic outcomes, or seek to identify the immediate (cell-autonomous) mechanisms linking a specific mutation to the death of the cells that express the mutant gene (*see* other chapters in this book). Our approach of seeking to understand the disease mechanisms underlying RP, in contrast, has led us to shift the research focus from simply the survival of photoreceptors to their contacts with the postsynaptic neurons, especially the bipolar cells which receive synaptic input from the photoreceptors. Serendipitously, we found changes in these synaptic connections in RP animal models that might be common consequential events of mutation-induced photoreceptor degeneration. These events occur at the very early stages of the disease process and hence may contribute to pathogenesis. In this chapter, we shall review the discovery of ectopic synaptogenesis in degenerating retinas of RP animal models and briefly discuss the implications of these results for RP disease mechanisms.

SYNAPTIC ORGANIZATION IN THE OPL AND SIGNALING PATHWAYS OF MAMMALIAN RETINAS

Synaptic organization of the mammalian retina has been well-characterized and described *(8–11)*. Accordingly, there is a rich background of information for comparison and to detect even subtle changes in the outer plexiform layer (OPL) of diseased retinas in animal models of rod-cone degeneration. Herein, we briefly summarize the salient features of the normal synaptic structure between photoreceptors and their postsynaptic partners, the bipolar cells and horizontal cells.

When viewed by electron microscopy (EM), the morphology of rod and cone photoreceptor synaptic terminals shows distinguishing landmarks. The rod synaptic terminal, sometimes called spherule, has a spherical form. The cone synaptic terminal, at times called a pedicle, has a flatter and broader shape and dimension than the rod terminal, in addition to being less electron dense. In the mammalian retina, cone pedicles are located, usually in one or two layers, at the innermost part of the OPL. Because there are many more rods than cones, their nuclei and soma are arranged in several layers, making up almost the full thickness of the outer nuclear layer (ONL). Similarly, the rod spherules are located and arranged in multiple layers in the OPL. Therefore, processes of the postsynaptic cells, such as the rod bipolar cell dendrites, often penetrate deep into the distal part of the OPL where they synapse with the rod spherules located in the outer margin of the OPL.

Each rod spherule has only one central invaginating ribbon synaptic complex. For historical reasons, the complex is sometimes described as the "triad" structure *(12)*. In the spherule, a synaptic ribbon points toward the central part of the triad. Occasionally, more than one ribbon can be observed. In the triad, there are two lateral components coming from the processes of horizontal cells. The horizontal cell processes can be differentiated by their slightly larger and more transparent synaptic vesicles. Facing the synaptic ribbon and in the central part of the triad, there are two (at times one may be seen) *(12)* invaginating component from the dendrite of a rod bipolar cell, which usually is void of synaptic vesicles *(11,12)*. Because each postsynaptic component has its own cell membrane, two juxtaposing plasma membranes can be observed between the cytoplasm of the photoreceptor terminal and the postsynaptic components. Most of the rod spherules do not have any conventional flat synaptic contacts with the postsynaptic cells. However, in the normal adult mammalian retina, approxi 5–20% of rod terminals make conventional flat synaptic contacts with dendrites of OFF-cone bipolar cells *(13,14)*.

Unlike rod spherules, each cone pedicle contains multiple invaginating ribbon synaptic complexes. However, their locations, and hence the extent of the invagination, are more superficial than those observed in the rod terminals. Although all rod bipolar cells are the ON-type, there are ON- and OFF-cone bipolar cells. Nevertheless, all invaginating cone bipolar cells are ON-cone bipolar cells. There are also numerous conventional flat, or basal, contact synapses with cone bipolar cell dendrites on the broad basal surface of the cone pedicle. These flat contact synapses have clearly observable pre- and postsynaptic thickenings. Most flat contact synapses are with OFF-cone bipolar cell dendrites, although cone photoreceptors also make flat contact synapses with some ON-cone bipolar cells *(15–17)*.

In the mammalian retina, signals from photoreceptors are transmitted through two separate and parallel processing pathways: the ON and OFF pathways. The rod photoreceptors, responsible for scotopic vision, activate the rod bipolar cells (all ON-type), which do not directly make synapses with the ganglion cells. Instead, the rod bipolar cells transmit visual information through AII amacrine cells, which provide input to cone bipolar cells that in turn provide input to the ganglion cells. The AII amacrine cells make gap junctions (excitatory input) with ON-cone bipolar cells and conventional inhibitory synapses (inhibitory input) with OFF-cone bipolar cells.

The cone photoreceptors, responsible for photopic vision, transmit signals to both the ON- and OFF-cone bipolar cells, which make synaptic connections with and provide signals to amacrine cells and ganglion cells *(8,18–22)*.

In addition to these main pathways, there are other "rod-to-cone" neural circuits in the mammalian retina. These include the direct rod-cone photoreceptor coupling through gap junctions and the "third rod pathway," namely, the direct link from rods to OFF-cone bipolar cells *(14,22–24)*. In normal mice, the third rod pathway may involve up to 20% of all rod terminals *(14)*. It is important to emphasize that in normal mammalian retina, cones almost never make direct synapses with rod bipolar cells *(18)*.

RETINAL DEGENERATION IN THE RHODOPSIN P347L TRANSGENIC PIG

Animal models, which manifest rod-cone degeneration similar to human RP, have been important for gaining the current understanding of these diseases *(25)*. There are many animal models of RP; nevertheless, the transgenic pig has emerged as a critical model for researching the disease mechanisms underlying rod-cone degeneration because, compared to other rod-dominated retinas, the porcine retina has a higher percentage of cones. The average ratio of rods to cones in the porcine retina is 8:1, and this ratio is fairly homogeneous across the retina. However, in the central posterior area, just above the optic disc, the rod-to-cone ratio can reach 5:1 *(26)*. These numbers compare favorably with the rod-to-cone ratio in the human macula and thus the pig retina would be a good model for studying cone degeneration if its rods can be triggered to degenerate, for example, by an RP-inducing mutation.

A pig model expressing the P347L mutant rhodopsin transgene has been created and established *(6,27–31)*. These pigs manifest a phenotype reminiscent of human RP caused by the same mutation in the human rhodopsin gene. In the P347L transgenic pig retina, rod numbers are normal at birth, but the outer segments (OS) are shorter than normal and disorganized. Instead of restricted to the OS, rhodopsin is delocalized, identified by immunocytochemistry throughout the entire photoreceptor. In 2-wk-old retinas, the number of rod nuclei in the ONL decreased slightly; by 3 wk postnatal, rod degeneration becomes clearly noticeable. At the age of 4 mo, most of the rods, and some cones, have died; yet, a large number (about 85%) of cones survive. These remaining cones have normal-length OS and, judging from the electroretinogram (ERG), are functioning quite well. However, these surviving cones gradually degenerate. By age 20 mo, there is only a single layer of cones remaining in the retina. The majority of these remaining cones have grossly abnormal morphology, although the photopic ERG was still recordable, albeit much reduced in amplitude.

ECTOPIC SYNAPTOGENESIS IN RETINA OF RHODOPSIN P347L TRANSGENIC PIG

Although the remaining cones undergo a long period of protracted degeneration after the complete loss of rods, their abundance and persistent function in the P347L transgenic pig retina create a unique opportunity to study the events that take place in the slowly degenerating retina. In so doing, we noted *(32)* that components of the rod-mediated signaling pathway, the rod bipolar cells, and the AII amacrine cells, mostly remained long after the loss of input signals from the rods. This observation led us to hypothesize that the rod bipolar cells might be receiving synaptic input from the surviving cones *(32)*.

Most Rod Bipolar Cells Survived Rod Degeneration

Although its function in the rod bipolar cell remains unclear, protein kinase Cα (PKCα) is known to express abundantly in these cells. Hence, the many commercially available antibodies against PKCα provide a convenient tool to label and identify the rod bipolar cell *(33–37)*. In the P347L transgenic porcine retina, most of the rods have degenerated by age 4 mo; after 9 mo, there are virtually no rhodopsin-positive cells identifiable by immunocytochemistry *(28)*. Nevertheless, in a 10-mo-old P347L transgenic porcine retina, we have identified many rod bipolar cells using immunocytochemistry with anti-PKCα antibodies *(6)*. In addition to the well-established specificity of PKCα as a marker for rod bipolar cells, the location of their cell bodies in the distal part of the inner nuclear layer (INL) and the termination of their axon terminals as large varicosities in the innermost layer of the inner plexiform layer (IPL) are characteristic features of rod bipolar cells that are seen in the labeled cells *(34,37,38)*. Morphometric analysis of both peripheral and central retinas revealed that the densities of rod bipolar cells are comparable between normal and P347L transgenic porcine retinas from 10-mo-old animals; whereas, there are approx 10% fewer rod bipolar cells than normal in a 22-mo-old P347L transgenic pig *(6)*. Clearly, the majority of rod bipolar cells continue to exist in the absence of rod photoreceptors. What function might they perform in a retina where the only photoreceptors are cones? Despite the nearly normal cell number and the appearance of the surviving rod bipolar cells, we noted some subtle changes in their dendritic structure.

In a normal retina, dendrites of the rod bipolar cells often penetrate deep into the distal part of the OPL where they synapse with the multiple layers of rod spherules. When labeled with anti-PKCα antibodies, these dendrites appear erect, showing a fine candelabrum-like appearance. Instead of this erect appearance, the dendrites of rod bipolar cells in the P347L transgenic porcine retina appear flattened, extending only to the inner part of the OPL, where the cone pedicles are located. This abnormal dendritic appearance could be detected even in a 6-d-old P347L transgenic porcine retina; by age 4 mo, almost all the rod bipolar cell dendrites are restricted to the innermost part of the OPL.

Synaptophysin as a Marker for Cone Synaptic Terminals

Just like PKCα serving as a specific marker for rod bipolar cells, the synaptic vesicle protein, synaptophysin, is a useful marker for labeling cone synaptic terminals. In the retina, synaptophysin is present in both the OPL and the IPL. In the OPL, it is specifically localized in the terminals of both rods and cones, located in separate but adjacent layers in the OPL *(39)*. Accordingly, in a 10-mo-old porcine retina, immunostaining using antibodies against synaptophysin labels a thick multilayered band; the innermost layer corresponds to the cone pedicles and the distal layers correspond to the rod spherules. Double immunostaining using antibodies against synaptophysin and a cone-specific marker, the cone transducin γ-subunit *(40,41)*, confirmed this interpretation. In the P347L transgenic porcine retina, as retinal degeneration progresses, the band labeled by anti-synaptophysin antibodies becomes thinner; by age 10 mo, labeling in the OPL no longer appears as a continuous band, but rather as discrete and sporadic "spots," corresponding to the pedicles of surviving cones.

Dendrites of Rod Bipolar Cells Make Ectopic Synaptic Contacts With Cone Terminals

In the 10-mo-old P347L transgenic porcine retina, double immunostaining using antibodies against PKCα and synaptophysin reveal that the rod bipolar cell dendrites extend laterally, appearing to seek out isolated cone pedicles and arborize below them. Serially overlapping optical sections of double labeled 10- and 22-mo-old P347L transgenic porcine retinas clearly indicate co-localization of dendritic tips of rod bipolar cells and the cone synaptic terminals. Using anti-PKCα immunocytochemistry conducted at the EM level, ectopic synaptic contacts between rod bipolar cell dendrites and cone pedicles are revealed. The synaptic contacts have pre- and postsynaptic thickenings, filamentous cross-bridges, and presynaptic vesicles that are typical of functioning synapses *(6)*. Therefore, based on morphological evidence, these ectopic synapses between the cones and the rod bipolar cells are most likely functional. These ectopic synapses are found throughout the entire retina, from periphery to center, of 4-, 10- and 22-mo-old P347L transgenic porcine retinas. In a parallel study of the normal porcine retina, we identify 230 cone pedicles in retinal sections; of these, none show any synaptic contacts with rod bipolar cell dendrites. Morphometric analysis based on EM micrographs leads to the conclusion that in the 10-mo-old transgenic porcine retina greater than 70% of cone terminals make synaptic contacts with rod bipolar cell dendrites.

In a normal retina, the dendrites of rod bipolar cells make exclusively invaginating synapses with rod terminals *(8,11,38,42)*. In contrast, most of the ectopic synapses found are of the flat contact type. In a study of EM sections of a 10-mo-old P347L retina, 240 of 280 cone pedicles counted make ectopic synaptic contacts with rod bipolar cells; out of the 240 synapses identified, there are 228 flat contact and 12 invaginating synapses. Therefore, at this stage of retinal degeneration, the rod bipolar cell dendrites make almost exclusively flat contact with cone synaptic terminals.

The Surviving Rod Bipolar Cells Likely Remain ON-Bipolar Cells

As described in an earlier section, all rod bipolar cells are normally the ON-type and form invaginating synapses with rods; whereas, there are ON- and OFF-cone bipolar cells. Most of the ON-cone bipolar cells have invaginating synapses with cones. In contrast, all the OFF-cone bipolar cells make flat contact synapses with cones. In addition, ON-bipolar cells (both rod and cone bipolar cells) express metabotropic glutamate receptors (mGluR6) and G protein, G_o; whereas, OFF-bipolar cells express ionotropic glutamate receptors *(8,11,16,22,42–46)*. Because the ectopic cone-to-rod bipolar cell synapses are the flat contact type, consistent with one of the properties of OFF-bipolar cells, it seems possible that the rod bipolar cells may have become OFF-type bipolar cells.

Results of our studies, however, indicate that the surviving rod bipolar cells likely remain the ON-type. First, the dendrites of surviving rod bipolar cells in P347L transgenic retinas still express the G_o G protein, which is associated with the signal transduction pathway mediated by the mGluR6 receptor and thus found only in ON-bipolar cells *(45,47)*. Second, all the axons of the surviving rod bipolar cells could be traced to the innermost part of the IPL. Termination of axons in this location, sub-lamina b, is one of the most important criteria for identifying ON-bipolar cells *(11)*.

Cones May Provide Input to Both Rod- and Cone-Mediated Signaling Pathways

For the sake of simplicity, the following discussion will center on the cone-mediated ERG of the pig elicited by a bright white flash when the rod response is suppressed by a bright background light (photopic condition). As reported previously, the response recorded from the normal pig is a transient negative (hyperpolarizing) a-wave, followed by a transient positive (depolarizing) b-wave that returns to baseline after a series of oscillations *(27)*. According to a study of the primate retina *(48)*, the phasic waveform of the photopic cone ERG is thought to be largely shaped by the summation of two opposing field potentials: a depolarizing potential contributed by the ON-cone bipolar cells and a hyperpolarizing potential contributed by the OFF-cone bipolar cells and horizontal cells *(49,50)*. It is likely that, in broad terms, the conclusions of this study can be applied to the photopic ERG recorded from the pig as well.

According to the model described above and under photopic conditions, the rod function in a normal retina is saturated and transmission of signal through the rod bipo-lar cell and AII amacrine cell to the cone bipolar cell is rendered ineffective. However, if the cones are providing input to the rod bipolar cells in the P347L transgenic porcine retina, one would predict that the combination of excitatory input to the ON-cone path-way and inhibitory input to the OFF-cone pathway, both via the AII amacrine cells, would add an extra depolarizing component to the photopic ERG. This prediction is fulfilled in that, compared to the ERG recorded from the normal pig under the same conditions, the photopic ERG recorded from the P347L transgenic pig shows an anom-alous waveform; it appears as though an additional depolarizing wave, coinciding with the series of oscillations, is superimposed on the normal waveform *(27)*.

In another study of the P347L transgenic pig using a different paradigm and based solely on analyses of ERG, the cone photoreceptor physiology appears normal for the first several postnatal months before the delayed onset of cone cell death. However, even during this period of apparent normal physiology, abnormal cone post-receptoral function is detected. The defect is localized to hyperpolarizing cells postsynaptic to the middle wavelength-sensitive cones (the dominant type in the porcine retina) and the abnormality is attributed to a failure of cone circuitry maturation *(29)*. Because of the inherent difficulties in the extraction of cellular information from ERG recordings, however, this study does not pinpoint the cellular mechanisms underlying the defect. In contrast, with the benefits of the current morphological evidence, alternative explana-tions of the abnormal ERG that would be consistent with the cellular mechanism being proposed here can be made. For example, the abnormal connections formed between cones and rod bipolar cells may have caused the specific defects detected in the pho-topic ERG. This may occur if rod bipolar cells inhibit OFF-cone bipolar cells through

AII amacrine cells when the cones provide input to the (former) rod-mediated signaling pathway in addition to the cone-mediated pathway.

In sum, although a rigorous test of the proposed hypothesis will require detailed physiological analyses at the cellular level, nevertheless, based on currently accepted models of signaling pathways in the retina and the sources of electrical signals which may underlie the ERG, observed features of the anomalous cone-mediated ERG in the P347L transgenic pig do support the hypothesis that the ectopic cone-to-rod bipolar cell synapses are functional and participating in the modified cone-mediated signaling pathway.

ECTOPIC SYNAPTOGENESIS IS A COMMON CONSEQUENTIAL EVENT IN ANIMAL MODELS OF MUTATION-INDUCED RP

Ectopic Synaptogenesis in rd Mice

Because the mechanisms leading to rod degeneration in the rhodopsin P347L transgenic porcine retina are not known and, compared to other animal models, the cones survive for a long period after early severe rod degeneration, it may be argued that ectopic synaptogenesis is a consequence unique to the phenotype of this porcine model and hence without the merit of generality. To rule out this argument, we have performed similar investigations in one of the best established murine models of RP—the *retinal degeneration* (*rd*) mouse, which has a mutation in another rod-specific protein: the phosphodiesterase β-subunit (PDE6B) *(51,52)*.

This naturally occurring murine model of RP manifests rapid and severe photoreceptor degeneration. By postnatal 18 d, most of the rods in the *rd* retina have degenerated and disappeared; only a few cell bodies belonging to cones remain. However, almost all the rod bipolar cells survive the initial photoreceptor degeneration. Morphometric analysis indicates that the density of rod bipolar cells is comparable between 18-d-old normal and *rd* mouse retinas. Double immunostaining using antibodies against synaptophysin and the cone transducin γ-subunit have confirmed that at this age all the remaining photoreceptor terminals in the OPL belong to cones. Analysis of EM micrographs have revealed that most of these remaining cone terminals, like those in the rhodopsin P347L transgenic porcine retina, have synaptic contacts with the dendrites of the rod bipolar cells; almost all of these ectopic synapses are of the flat contact type (Fig. 1). Nevertheless, the surviving rod bipolar cells of the *rd* mouse still express the G_o G protein, and their axons terminate at the innermost border of the IPL, suggesting that they likely remain ON-bipolar cells *(6)*.

In both the porcine and murine models studied, the morphology of the ectopic cone-to-rod bipolar cell synapse is consistent with the characteristics expected of a functional synapse. In addition, documented alterations in the photopic ERG recorded from the rhodopsin P347L transgenic pigs are circumstantial evidence, which supports the idea that cones may provide input to the former rod-mediated signaling pathway. Further supporting evidence is obtained from studying the *rd* retina using the anti-PKCα monoclonal antibody, Ab-2, which specifically recognizes the hinge region of the enzyme *(53)*. Like other anti-PKCα antibodies, Ab-2 labels rod bipolar cells in the mouse retina. However, it does so only in dark-adapted retinas; PKCα-immunoreactivity (ir) is lost

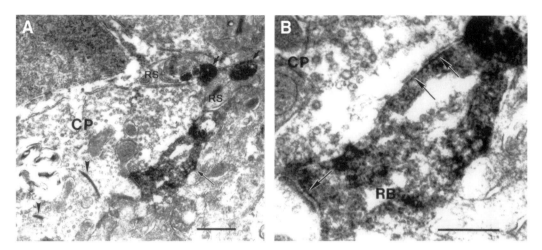

Fig. 1. Electron micrographs of 18-d-old *rd* retina immunostained with anti-PKCα antibody show that rod bipolar cell dendrites make ectopic flat contact synapses with cone terminals. In (**A**) the long arrow indicates a PKCα-ir rod bipolar cell dendritic tip, which forms ectopic synaptic contacts with a cone pedicle (CP). The CP is recognized by its electron density, size, shape, and multiple synaptic ribbons (arrowheads). There are two rod spherules (RS), on the right side of the cone pedicle, which have invaginating synapses with PKCα-ir rod bipolar cell dendritic tips (short arrows). In (**B**) higher magnification of the ectopic synapses in (**A**). These ectopic synapses have pre- and postsynaptic thickening and presynaptic vesicles. Arrows indicate the postsynaptic thickening. Scale bar = 0.5 μm. Reproduced from ref. *6*, with permission.

in animals that have been exposed to light for 1 h. Although the underlying mechanism of this phenomenon is still under investigation, PKCα-ir detected by Ab-2 is a useful tool for monitoring any potential "light-induced activity" in rod bipolar cells.

In 4-mo-old *rd* retinas, in which all the rods have long since degenerated and only sporadic cones are identified, Ab-2 could detect light adaptation dependent changes of PKC-ir reversibly in rod bipolar cells. In our studies, all the rod bipolar cells are labeled by Ab-2 if the retina is dark adapted; whereas, light adaptation abolishes the Ab-2 detected PKCα-ir. The simplest interpretation of this observation is that the rod bipolar cells are getting light-induced signals from the cones *(54)*.

The morphological evidence of ectopic cone-to-rod bipolar cells and the functional evidence of cone input to the residual rod-mediated signaling pathway in 4-mo-old *rd* retinas argue strongly against the notion that ectopic synaptogenesis in degenerating retinas is a transient phenomenon. On the contrary, data would support the interpretation that the ectopic cone-to-rod bipolar cell synapse is the cellular basis of an anomalous cone-rod bipolar cell-signaling pathway, the functional significance of which, especially in a diseased retina devoid of rod photoreceptors, remains to be determined.

Comments on Other Studies of the **rd** Retina Relevant to Ectopic Synaptogenesis

The concept that RP research may yield additional benefits by broadening its focus from simply the survival of photoreceptors to include the potential effects that death of the photoreceptors may have on the whole retina has been pursued independently by various groups of investigators. A study on the *rd* mouse published by Strettoi and Pignatelli *(55)* is especially relevant to the subject and their conclusions are germane to a discussion of the discovery of ectopic synaptogenesis in degenerating retinas *(6)*.

Based on their light microscopic observations of (1) the absence of normal rod bipolar cell dendritic arborization and lack of axonal processes originating from the horizontal cells and (2) changes in the pattern of immunoreactivity of the glutamate receptor mGluR6 in rod bipolar cells, Strettoi and Pignatelli *(55)* conclude that "healthy photoreceptors are required to ensure the normal development of second-order neurons." Although Strettoi and Pignatelli *(55)* have not provided any data that specifically address the issues of formation and function of ectopic cone to rod bipolar cell synapses in degenerating *rd* retinas, a casual reading of the authors' other statements could lead to the erroneous impression that their conclusions contradict the existence of ectopic cone to rod bipolar cell synapses in the *rd* retina at early stages of degeneration *(56)*.

In PKCα-stained vertical sections of a 20-d-old *rd* retina (Fig. 4E from ref. *55*), the authors describe "rudimentary [rod bipolar cell] dendrites of abnormal morphology, resembling club-shaped processes directed toward photoreceptors" and "limited to a rudimental border." In adult (2- to 3-mo-old) *rd* mice, these authors report rod bipolar cell are totally devoid of processes in the OPL. Thus, the authors conclude that they "never observed normally developed dendritic arborizations in the OPL"—a conclusion that is consistent with a previous report *(57)*. All their studies are conducted at the light microscopic level, without any EM investigation of the abnormal, club-shaped rod bipolar cell dendrites directing toward photoreceptors; none of their statements address potential opportunities for ectopic cone-to-rod bipolar cell synapses. However, the image of the 20-d-old *rd* retina presented in Fig. 4E of ref. *55* is remarkably similar to the image of an 18-d-old *rd* retina shown in Fig. 7 of ref. *6* and from which immunocytochemistry performed at the EM level has revealed that the altered (or rudimentary) rod bipolar cell dendrites can make ectopic synapses with cone pedicles (Fig. 8 of ref. *6* reproduced as Fig. 1 in this chapter and Fig. 2).

Similar to the statements concluded from analysis of PKCα-ir, a casual reading of the authors' statements regarding the loss of normal distribution of mGluR6-ir in the *rd* retina (Fig. 5 of ref. *55*) may lead to the incorrect generalization that a major redistribution of glutamate receptors in rod bipolar cells may have rendered these cells incompetent to signal in response to changes in glutamate concentration. It should be pointed out that the mGluR6-ir data presented by Strettoi and Pignatelli *(55)* are from studies of 2- to 3-mo-old *rd* retinas in which photoreceptors, including the cones, have mostly disappeared. At these late stages of degeneration in *rd* retinas, the rod bipolar cells undergo more overt morphological changes. The authors do not show any data on mGluR6-ir in younger *rd* retinas, such as at age 18 d old, where a substantial fraction of cones still remain. On the other hand, we have used the same anti-mGluR6 antibody obtained from Dr. S. Nakanishi and observed mGluR6-ir in the OPL, including the dendrites of rod bipolar cells, at stages of retinal degeneration that ectopic synapses are observed in the *rd* retina *(58)*.

Although the interpretation of their respective data (obtained by different techniques) by the two groups differ significantly in details, the conclusion of Strettoi and

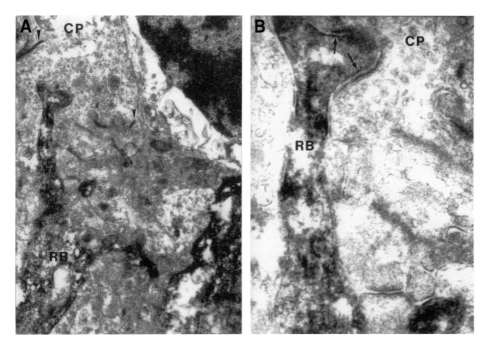

Fig. 2. Electron micrographs of 18-d-old *rd* retina immunostained with anti-PKCα antibody. **(A)** Rod bipolar cell dendrite make ectopic synapses with cone terminal (CP). Arrowheads indicate the synaptic ribbons. **(B)** Higher magnification of the ectopic synapses in **(A)**. Arrows indicate ectopic synaptic contacts.

Pignatelli *(55)* that "photoreceptor degeneration appears to trigger dramatic changes in the morphology of second-order neurons that predict even more serious functional consequences" is broadly consistent with and supportive of the notion that ectopic synapses form and function in the degenerating *rd* retina *(6)* .

Ectopic Synaptogenesis in RCS Rat

It can be argued that both the transgenic pig model and the *rd* mouse involve rod photoreceptor-specific genes and, hence, it is possible that ectopic synaptogenesis of the kind observed is a result of specific rod cell-autonomous mechanisms owing to the mutations rather than a general consequence of photoreceptor degeneration. In other words, in retinas in which the mutated gene is not expressed in rod photoreceptors, ectopic synapses may not form. To address this possibility, we have investigated one of the most extensively studied rodent models of RP—the Royal College Surgeon (RCS) rat *(7)*. In the RCS rat, a mutation in the receptor tyrosine kinase gene, *Mertk*, causes failure of the retinal pigment epithelial (RPE) cells to phagocytose shed photoreceptor OS *(59)*. The nonphagocytic phenotype of the RCS rat is RPE cell autonomous. As a consequence of the failed RPE function, the normal rod and cone photoreceptors in the RCS rat die *(60,61)*.

The ONL thickness of a 20-d-old RCS rat retina is normal. Signs of apoptotic photoreceptor death are detectable by approximately postnatal day 25 *(60,62)*. By postnatal day 35, the thickness of the ONL has decreased significantly. At this stage, many rod photoreceptors still have long OS. In contrast to animal models expressing mutations in rhodopsin *(6,28,63)*, the RCS retina does not show abnormal localization of rhodopsin in rod cell bodies and terminals because the immunostaining of rhodopsin is concentrated in the rod outer segments. Therefore, delocalization of rhodopsin may not be the trigger of ectopic synaptogenesis.

Comparing the percentages of remaining cell bodies at different postnatal days, the number of rod bipolar cells decreases much slower than the number of rods. The results also indicate, despite the much slower rate of cone loss during the ages studied, the RCS retina remains rod dominated; for example, in the 35-d-old RCS retina, the rod to cone ratio is approx 8.6:1. At this stage of degeneration in RCS rat retinas, consistent with the loss of rods and cones, the band of terminals in the OPL stained by antisynaptophysin has thinned. The estimated number of labeled terminals is 62.4% of normal, indicating synaptic abnormalities in about one-third of the rod terminals. Similar to findings in P347L pigs and *rd* mice, most rod bipolar cell dendrites in a 35-d-old RCS retina are no longer erect but appear to extend laterally. When the RCS retina is double labeled with antisynaptophysin and anti-PKCα antibodies, the results indicate that some rod bipolar cell dendrites still penetrate the band of cone pedicles to terminate near the outer border of the OPL, where the rod spherules are located. Furthermore, when double labeled with antisynaptophysin and anticone transducin γ antibodies, only some of the terminals are double labeled, indicating the presence of synaptophysin in some of the surviving rods. These results suggest that in the 35-d-old RCS retina, many surviving rods and cones may still have functioning synapses.

Consistent with this view, the b-wave amplitude of a single-flash, mixed rod-cone ERG of the 35-d-old RCS rats is approx 25.5% that of the non-dystrophic rat. By 45 d, the averaged b-wave amplitude reduces to 15.3%, and by 55 d, only 7.2%, of the nondystrophic rat. Examination of 35-d-old RCS retinas by EM shows that a small number of surviving rods still have intact invaginating synapses from rod bipolar cell dendrites. Even in a 55-d-old RCS retina, 4 out of 88 rod terminals examined (4.6%) still have invaginating synapses. Therefore, functional and abnormal rod synapses co-exist in the degenerating RCS retina.

Of the 154 cone terminals examined in a 35-d-old RCS retina, 54 (35.1%) have ectopic synaptic contacts with rod bipolar cell dendrites; all of them are the flat contact type. Some rod bipolar cell dendrites could make flat contact synapses with two cone terminals simultaneously. In addition to the cone to rod bipolar cell ectopic synapses, there are abnormal flat contact synapses on some rod terminals; such flat-contact synapses at rod terminals are not observed in nondystrophic retinas and, therefore, are likely ectopic synaptic contacts with rod bipolar cell dendrites. As demonstrated by EM immunocytochemistry, PKCα-stained rod bipolar cell dendrites indeed make flat contact ectopic synapses with rod terminals. Of the 252 rod terminals counted in the 35-d-old RCS retina, 50 (19.8%) have flat-contact synapses with rod bipolar cell dendrites. Most of these rod terminals show signs of degeneration *(64)*, lacking intact synaptic ribbons and triad structure.

Identification of ectopic synapses in the RCS rat supports the conclusion that ectopic synaptogenesis is a general consequence of photoreceptor degeneration; and this event may take place independent of a particular disease-inducing cell-specific gene mutation. Furthermore, as shown in the RCS rat, rather than being a default developmental outcome, ectopic synapses involving rod and cone photoreceptors and rod bipolar cell dendrites are the results of "neural re-wiring," as a consequence of mutation-induced photoreceptor loss.

SYNAPTIC REMODELING IN OTHER RETINAL DEGENERATION ANIMAL MODELS AND THE UNDERLYING PRINCIPLES OF SYNAPTIC PARTNERING IN THE RETINA

The aggregate of published results from studying ectopic synaptogenesis in a variety of RP animal models, from different species (and with mutations in different genes), including the P347L transgenic pig (rhodopsin), the *rd* mouse (PDE6B), and the RCS rat (*Mertk*), strongly suggests that the phenomenon is a common feature (downstream event) of mutation-induced retinal degeneration. Additional results from our laboratory further support this view. For example, in rhodopsin P347L transgenic mice, which express the same porcine rhodopsin P347L transgene as the P347L transgenic pig, EM analysis of anti-PKCα labeling has revealed similar cone to rod bipolar cell synapses (Fig. 3). Parallel experiments performed on analogous models, such as mice and pigs each expressing the porcine rhodopsin P347S transgene, have led to the same conclusion. Our preliminary studies of a *rod-cone dysplasia 1 (rcd1)* dog, which like the *rd* mouse, has a mutation in the PDE6B gene *(65)*, have yielded results that support the notion that ectopic synaptogenesis occurs during degeneration in this animal model as well (Peng and Wong, unpublished results).

From the standpoint of potential postdeafferentation responses of the rod bipolar cells, which may include protein redistribution, synaptic remodeling, and degeneration, ectopic synaptogenesis involving cones and rod bipolar cells in the degenerating retina are certainly consistent with the expectations of general neuronal behavior *(66–69)*.

As it was pointed out previously *(6)*, formation of abnormal cone-to-rod bipolar cell synapses in animal models of mutation-induced photoreceptor degeneration demonstrates that the rod bipolar cell dendrites have the capability to make alternative connections when the preferred contacts are apparently not available. Hence, the rules that govern synaptic partnering between rods and rod bipolar cells and between cones and cone bipolar cells are not absolute; furthermore, the molecular mechanisms used by rods and cones to choose their synaptic partners share some common features. These concepts of the principles underlying synaptic partnering in the retina have been illustrated in a different context by a study of the neural retina leucine zipper (*Nrl*) gene knock out (KO) mice. In these mice, the rods fail to form developmentally and hence all the photoreceptors are cones *(70)*. In these cone-only retinas, the rod bipolar cells form synaptic connections with cones *(71)*. Accordingly, it seems that the principles guiding the formation of photoreceptor and bipolar cell synapses in the retina, as illustrated in the *Nrl* KO mouse, would predict ectopic synaptogenesis in degenerating retinas just as we have previously reported.

From these premises, it is a small leap of faith to hypothesize that even though there may not be any remaining rods, the residual rod-mediated signaling pathway persists and functions in the diseased retina. Although direct evidence is still lacking, the

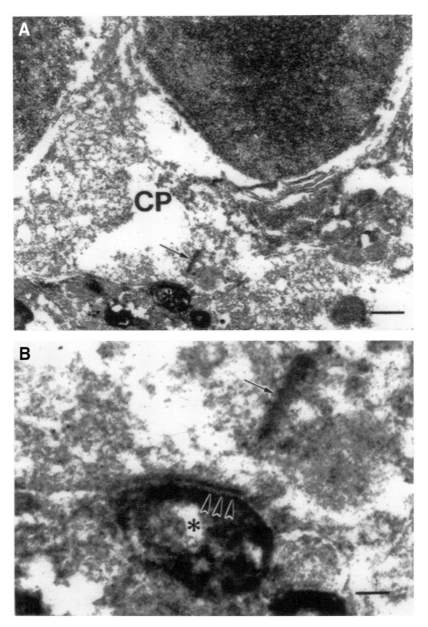

Fig. 3. Electron micrograph of a 10-wk-old P347L transgenic mouse retina immunostained with PKCα antibody. Asterisk marks a labeled rod bipolar cell dendritic tip making ectopic synapse with a cone pedicle (CP). **(B)** Shows the higher magnification of the ectopic synaptic contact (arrowheads) in **(A)**. Arrows indicate synaptic ribbon. Bar = 0.5 μm **(A)** and 0.1 μm **(B)**.

collective morphological evidence and the indirect evidence summarized in this chapter are consistent with this hypothesis, e.g., ERG in the P347L transgenic pigs and light-induced changes in PKCα-ir detected by monoclonal antibody Ab-2 in rod bipolar cells of the degenerated *rd* retina. Furthermore, even though the electron microscopic details of the purported ectopic synapses have not been presented, morphological evidence collected at the light microscopic level and the parallel ERG analysis conducted in the rhodopsin P23H transgenic rat by Cuenca et al. *(72)* provide indirect evidence in support of the notion that after severe rod degeneration, the cone-mediated ERG has acquired a component that may result from cone input to the former rod-mediated signaling pathway. In short, in the rhodopsin P23H transgenic rat, just like in other animal models described in this chapter, ectopic cone-to-rod bipolar cell synapses may exist and via these synapses, the remaining cones may provide input to both the residual rod-mediated signaling pathway as well as the cone-mediated pathway.

IMPLICATIONS FOR PATHOGENESIS OF LATERAL EXTENSIONS OF ROD BIPOLAR CELL DENDRITES

It has been demonstrated a decade ago in the peripheral retina of specimens collected from patients with RP that rods extend long processes into the inner retina *(73)*. These rod neurites usually bypass the OPL, where dendrites and processes of bipolar cells and horizontal cells—the usual synaptic targets of rod synaptic terminals—are located *(56)*. Instead, the rod neurites extend to the INL and inner IPL, reaching even the inner limiting membrane. There is no evidence, however, suggesting that these rod neurites form any synapses with neurons in the inner retina *(74)*; however, later studies reported presence of synaptic protein SV-2 ir in the terminals of these neurites, hinting at the potential presence of synaptic vesicles *(75)*. In humans, such neurite sprouting is not observed in the macular region *(73,74,76)*. Nonetheless, recently, Fei *(77)* has observed sprouting neurites from cones in the retinas of *rd* mice, starting from early stages of rod degeneration. Some of these sprouting neurites extend laterally and appear to make contacts with rod bipolar cell, although their synaptic fate is not known.

Rod bipolar cell dendrites undergo analogous "sprouting" under various conditions *(78–80)*. Especially relevant to the topic under discussion is the lateral extensions of these dendrites in animal models of rod-cone degeneration. As illustrated in Fig. 4 (4-mo-old *rd* mouse) and Fig. 5 (9-mo-old P347S transgenic mouse), the dendrites of the rod bipolar cells extend laterally and make synapses with the terminals of neighboring cones. Furthermore, the rod bipolar cell soma have dislocated to the ONL. At the light microscopic level, the synaptic contacts appear to be on the cone cell bodies. Similar extensions of the rod bipolar cell dendrites have been documented independently by other investigators, e.g., in *rd* mouse81 and P23H transgenic rat *(72)*.

The images presented in Figs. 4 and 5 suggest a scenario that the rod bipolar cell may actively initiate synaptic contacts with the surviving cones via its extensive dendritic tree. If the autonomous behavior of the rod bipolar cell dendrites is manifested very early, before extensive rod death, such action may in fact contribute to pathogenesis. For example, during the initial phase of rod death, when a few rod bipolar cell dendrites may have lost their synaptic input from the degenerated rods, the rod bipolar cell may withdraw some of its other dendrites from functioning rod-to-rod bipolar cell synapses. In so doing, the rod bipolar cell would accelerate the death of the rods. Morphological evidence consistent with such withdrawal by rod bipolar cell dendrites from functioning rod spherules indeed exists *(82)* and the speculative model of rod bipolar cell initiated rod death is currently under investigation.

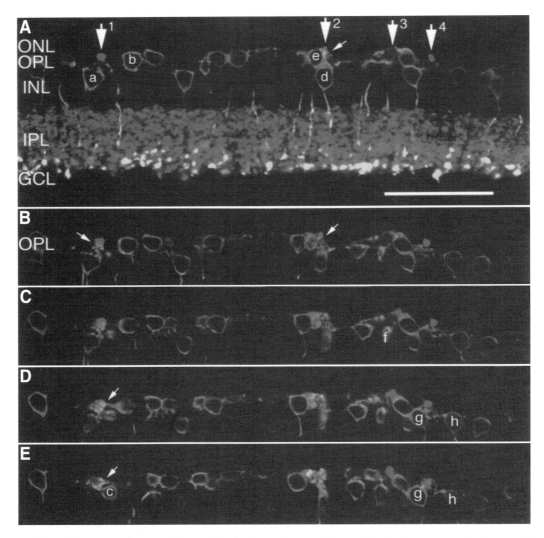

Fig. 4. Cross sections at different focal planes obtained by confocal microscopy of a 4-mo-old *rd* mouse retina double-immunostained with PKCα and antisynaptophysin antibodies. Large arrows (1–4) indicate different remaining cone terminals (labeled with antisynaptophysin antibody in red) form ectopic synapses (yellow spots) with different rod bipolar cell dendrites (a–h, labeled with anti-PKCα antibody in green). Small arrows indicate extended rod bipolar cell dendrites. In this series of confocal images, it can be seen that a synaptophysin labeled cone terminal (large arrow 1 in panel **A**, red) makes synaptic contacts (yellow spots, small arrows in panels **B**, **D**, and **E**) with three different rod bipolar cells (a, b, and c, green), respectively. In panel **D**, rod bipolar cell b is seen extending its dendrite laterally to make synaptic contact (yellow spot, small arrow) with the cone terminal (red). In panel **A**, large arrow 2 indicates another synaptophysin labeled cone terminal (red) forming synaptic contact (yellow spot, small arrow) with the dendrite of rod bipolar cell d (green). The same cone terminal (large arrow 2) is contacted by rod bipolar cell e (small arrow in panel **B**). Analogous synaptic contacts between other cone terminals (arrows 3 and 4) and rod bipolar cells f, g, and h can be traced through the same series of potical sections. ONL, outer nuclear layer; OPL, outer plexiform layer; INL, inner nuclear layer; IPL, inner plexiform layer; GCL, ganglion cell layer. Scale bar = 50 μm.

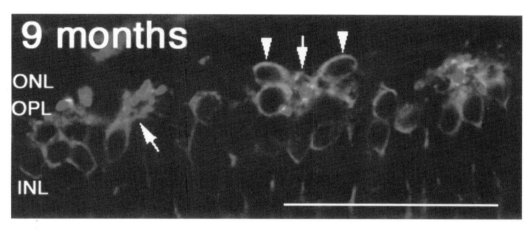

Fig. 5. Cross section of a 9-mo-old P347S mouse retina double immunostained with anti-PKCα (green) and antisynaptophysin (red) antibodies. Arrows indicate laterally extended rod bipolar cell dendrites. Arrowheads indicate dislocated rod bipolar cell somas in the ONL. These rod bipolar cells extended dendrites laterally toward the surviving synaptophysin labeled cone terminals (arrow, red). Labels as in Fig. 4.

CONCLUDING REMARKS

In this chapter, we have reviewed the discovery of ectopic synaptogenesis in several animal models of rod-cone degeneration. The major conclusions are: (1) ectopic synaptogenesis reflects a set of common downstream consequential events, triggered by a multitude of mutations that cause retinal degeneration; (2) ectopic synaptogenesis might occur early and could persist until the very late stages of degeneration; and (3) long after the death of rods, the residual rod-mediated signaling pathway persists and receives input from the surviving cones. The significance of the persistent anomalous wiring is not known.

We have restricted our discussions of synaptic remodeling to the OPL because what occurs in the inner retina, especially in the late stages of degeneration, has been reviewed by others *(56)*. The "neuronal remodeling" or "negative cell remodeling" observed in the inner retina certainly would impact significantly some of the current theories regarding therapeutic approaches to treat patients in late stages of retinal degeneration. In parallel, retina-wide changes in the synaptic connections between photoreceptors and the second order neurons that occur early in the disease process, such as those summarized in this chapter, are expected to lead to more obvious and dramatic changes, including consequential events that would occur in the inner retina *(83,84)*. In other words, early synaptic changes in the OPL and the changes in the inner retina observed at late stages of retinal degeneration may be causally related. Although this possibility remains to be investigated, the discovery of ectopic synaptogenesis has led us to a new hypothesis regarding the role of the rod bipolar cell in this phenomenon. Careful examination of the speculative model outlined in this chapter may lead us to a better understanding of the potential disease mechanisms underlying mutation-induced rod-cone degeneration.

REFERENCES

1. Berson EL. Retinitis pigmentosa. The Friedenwald lecture. Invest Ophthalmol Vis Sci 1993;34:1659–1676.
2. Wong F. Visual pigments, blue cone monochromasy, and retinitis pigmentosa. Arch Ophthalmol 1990;108:935–936.
3. Chang GQ, Hao Y, Wong F. Apoptosis: final common pathway of photoreceptor death in rd, rds, and rhodopsin mutant mice. Neuron 1993;11:595–605.
4. Huang PC, Gaitan AE, Hao Y, Petters RM, Wong F. Cellular interactions implicated in the mechanism of photoreceptor degeneration in transgenic mice expressing a mutant rhodopsin gene. Proc Natl Acad Sci USA 1993;90:8484–8488.
5. Wong F. Investigating retinitis pigmentosa: a laboratory scientist's perspective. Prog Ret Eye Res 1997;16:353–373.
6. Peng YW, Hao Y, Petters RM, Wong F. Ectopic synaptogenesis in the mammalian retina caused by rod photoreceptor-specific mutations. Nat Neurosci 2000;3:1121–1127.
7. Peng YW, Senda T, Hao Y, Matsuno K, Wong F. Ectopic synaptogenesis during retinal degeneration in the royal college of surgeons rat. Neuroscience 2003;119:813–820.
8. Wassle H, Boycott BB. Functional architecture of the mammalian retina. Physiol Rev 1991;71:447–480.
9. Wassle H. Parallel processing in the mammalian retina. Nat Rev Neurosci 2004;5:747–757.
10. Rodieck RW. The first steps in seeing. Sunderland, MA: Sihauer Associates, 1998.
11. Sterling P. Retina. In: the synaptic organization of the brain. New York: Oxford University Press, 1998;205–253.
12. Rao-Mirotznik R, Harkins AB, Buchsbaum G, Sterling P. Mammalian rod terminal: architecture of a binary synapse. Neuron 1995;14:561–569.
13. Hack I, Peichl L, Brandstatter JH. An alternative pathway for rod signals in the rodent retina: rod photoreceptors, cone bipolar cells, and the localization of glutamate receptors. Proc Natl Acad Sci USA 1999;96:14,130–14,135.
14. Tsukamoto Y, Morigiwa K, Ueda M, Sterling P. Microcircuits for night vision in mouse retina. J Neurosci 2001;21:8616–8623.
15. Calkins DJ, Tsukamoto Y, Sterling P. Foveal cones form basal as well as invaginating junctions with diffuse ON bipolar cells. Vision Res 1996;36:3373–3381.
16. Chun MH, Grunert U, Martin PR, Wassle H. The synaptic complex of cones in the fovea and in the periphery of the macaque monkey retina. Vision Res 1996;36:3383–3395.
17. Hopkins JM, Boycott BB. The cone synapses of cone bipolar cells of primate retina. J Neurocytol 1997;26:313–325.
18. Dacheux RF, Raviola E. The rod pathway in the rabbit retina: a depolarizing bipolar and amacrine cell. J Neurosci 1986;6:331–345.
19. Wassle H, Grunert U, Chun MH, Boycott BB. The rod pathway of the macaque monkey retina: identification of AII-amacrine cells with antibodies against calretinin. J Comp Neurol 1995;361:537–551.
20. Boycott B, Wassle H. Parallel processing in the mammalian retina: the Proctor lecture. Invest Ophthalmol Vis Sci 1999;40:1313–1327.
21. Sharpe LT, Stockman A. Rod pathways: the importance of seeing nothing. Trends Neurosci 1999;22:497–504.
22. Bloomfield SA, Dacheux RF. Rod vision: pathways and processing in the mammalian retina. Prog Retin Eye Res 2001;20:351–384.
23. Vaney DI. Neuronal coupling in rod-signal pathways of the retina. Invest Ophthalmol Vis Sci 1997;38:267–273.
24. Soucy E, Wang Y, Nirenberg S, Nathans J, Meister M. A novel signaling pathway from rod photoreceptors to ganglion cells in mammalian retina. Neuron 1998;21:481–493.

25. Chader GJ. Animal models in research on retinal degenerations: past progress and future hope. Vision Res 2002;42:393–399.

26. Gerke CG Jr, Hao Y, Wong F. Topography of rods and cones in the retina of the domestic pig. Hong Kong Med J 1995;1:302–308.

27. Petters RM, Alexander CA, Wells KD, et al. Genetically engineered large animal model for studying cone photoreceptor survival and degeneration in retinitis pigmentosa. Nat Biotechnol 1997;15:965–970.

28. Li ZY, Wong F, Chang JH, et al. Rhodopsin transgenic pigs as a model for human retinitis pigmentosa. Invest Ophthalmol Vis Sci 1998;39:808–819.

29. Banin E, Cideciyan AV, Aleman TS, et al. Retinal rod photoreceptor-specific gene mutation perturbs cone pathway development. Neuron 1999;23:549–557.

30. Blackmon SM, Peng YW, Hao Y, et al. Early loss of synaptic protein PSD-95 from rod termnals of rhodopsin P347L transgenic porcine retina. Brain Res 2000;885:53–61.

31. Tso MO, Li WW, Zhang C, et al. A pathologic study of degeneration of the rod and cone populations of the rhodopsin Pro347Leu transgenic pigs. Trans Am Ophthalmol Soc 1997;95:467–479.

32. Peng YW, Mahmoud TH, Oliveira LB, et al. Rhodopsin mutation induces ectopic cone-rod bipolar cell synaptic connections in transgenic swine. Mol Biol Cell (suppl) 1999;10:76a.

33. Negishi K, Kato S, Teranishi T. Dopamine cells and rod bipolar cells contain protein kinase C-like immunoreactivity in some vertebrate retinas. Neurosci Lett 1988;94:247–252.

34. Greferath U, Grunert U, Wassle H. Rod bipolar cells in the mammalian retina show protein kinase C-like immunoreactivity. J Comp Neurol 1990;301:433–442.

35. Grunert U, Martin PR, Wassle H. Immunocytochemical analysis of bipolar cells in the macaque monkey retina. J Comp Neurol 1994;348:607–627.

36. Muller B, Peichl L. Rod bipolar cells in the cone-dominated retina of the tree shrew Tupaia belangeri. Vis Neurosci 1991;6:629–639.

37. Wassle H, Yamashita M, Greferath U, Grunert U, Muller F. The rod bipolar cell of the mammalian retina. Vis Neurosci 1991;7:99–112.

38. Chun MH, Han SH, Chung JW, Wassle H. Electron microscopic analysis of the rod pathway of the rat retina. J Comp Neurol 1993;332:421–432.

39. Brandstatter JH, Lohrke S, Morgans CW, Wassle H. Distributions of two homologous synaptic vesicle proteins, synaptoporin and synaptophysin, in the mammalian retina. J Comp Neurol 1996;370:1–10.

40. Peng YW, Robishaw JD, Levine MA, Yau KW. Retinal rods and cones have distinct G protein beta and gamma subunits. Proc Natl Acad Sci USA 1992;89:10,882–10,886.

41. Ong OC, Yamane HK, Phan KB, et al. Molecular cloning and characterization of the G protein gamma subunit of cone photoreceptors. J Biol Chem 1995;270:8495–8500.

42. Hopkins JM, Boycott BB. Synapses between cones and diffuse bipolar cells of a primate retina. J Neurocytol 1995;24:680–694.

43. Masu M, Iwakabe H, Tagawa Y, et al. Specific deficit of the ON response in visual transmission by targeted disruption of the mGluR6 gene. Cell 1995;80:757–765.

44. Vardi N, Morigiwa K, Wang TL, Shi YJ, Sterling P. Neurochemistry of the mammalian cone 'synaptic complex.' Vision Res 1998;38:1359–1369.

45. Vardi N. Alpha subunit of Go localizes in the dendritic tips of ON bipolar cells. J Comp Neurol 1998;395:43–52.

46. Morigiwa K, Vardi N. Differential expression of ionotropic glutamate receptor subunits in the outer retina. J Comp Neurol 1999;405:173–184.

47. Dhingra A, Jiang M, Wang TL, et al. The light response of ON bipolar neurons requires G[alpha]o. J Neurosci 2000;20:9053–9058.

48. Sieving PA, Murayama K, Naarendorp F. Push-pull model of the primate photopic electroretinogram: a role for hyperpolarizing neurons in shaping the b-wave. Vis Neurosci 1994;11:519–532.

49. Knapp AG, Schiller PH. The contribution of on-bipolar cells to the electroretinogram of rabbits and monkeys. A study using 2-amino-4-phosphonobutyrate (APB). Vision Res 1984;24:1841–1846.

50. Stockton RA, Slaughter MM. B-wave of the electroretinogram. A reflection of ON bipolar cell activity. J Gen Physiol 1989;93:101–122.

51. Bowes C, Danciger M, Baxter LC, Applebury ML, Farber DB. Retinal degeneration in the rd mouse is caused by a defect in the beta subunit of rod cGMP-phosphodiesterase. Nature 1990;347:677–680.

52. Farber DB, Flannery JG, Bowes-Rickman C. The rd mouse story: seventy years of research on an animal model of inherited retinal degeneration. Prog Ret Eye Res 1994;13:31–64.

53. Young S, Rothbard J, Parker PJ. A monoclonal antibody recognising the site of limited proteolysis of protein kinase C. Inhibition of down-regulation in vivo. Eur J Biochem 1988;173:247–252.

54. Peng YW, Hao Y, Oka K, Wong F. Light-induced changes of PKC alpha immunoreactivity in rod bipolar cells of normal and rd Mice. Invest Ophthalmol Vis Sci 2002;Suppl 43: ARVO e-abstract 741.

55. Strettoi E, Pignatelli V. Modifications of retinal neurons in a mouse model of retinitis pigmentosa. Proc Natl Acad Sci USA 2000;97:11,020–11,025.

56. Marc RE, Jones BW, Watt CB, Strettoi E. Neural remodeling in retinal degeneration. Prog Retin Eye Res 2003;22:607–655.

57. Blanks JC, Adinolfi AM, Lolley RN. Photoreceptor degeneration and synaptogenesis in retinal-degenerative (rd) mice. J Comp Neurol 1974;156:95–106.

58. Peng YW, Hao Y, Wong F. Metabotropic glutamate receptors mGluR5 and mGluR6 are localized to the dendrites of rd mouse rod bipolar cells. Soc Neurosci Abs 2002;28.

59. D'Cruz PM, Yasumura D, Weir J, et al. Mutation of the receptor tyrosine kinase gene Mertk in the retinal dystrophic RCS rat. Hum Mol Genet 2000;9:645–651.

60. LaVail MM. Legacy of the RCS rat: impact of a seminal study on retinal cell biology and retinal degenerative diseases. Prog Brain Res 2001;131:617–627.

61. Vollrath D, Feng W, Duncan JL, et al. Correction of the retinal dystrophy phenotype of the RCS rat by viral gene transfer of Mertk. Proc Natl Acad Sci USA 2001;98:12,584–12,589.

62. Tso MO, Zhang C, Abler AS, et al. Apoptosis leads to photoreceptor degeneration in inherited retinal dystrophy of RCS rats. Invest Ophthalmol Vis Sci 1994;35:2693–2699.

63. Olsson JE, Gordon JW, Pawlyk BS, et al. Transgenic mice with a rhodopsin mutation (Pro23His): a mouse model of autosomal dominant retinitis pigmentosa. Neuron 1992;9: 815–830.

64. Eisenfeld AJ, LaVail MM, LaVail JH. Assessment of possible transneuronal changes in the retina of rats with inherited retinal dystrophy: cell size, number, synapses, and axonal transport by retinal ganglion cells. J Comp Neurol 1984;223:22–34.

65. Suber ML, Pittler SJ, Qin N, et al. Irish setter dogs affected with rod/cone dysplasia contain a nonsense mutation in the rod cGMP phosphodiesterase beta-subunit gene. Proc Natl Acad Sci USA 1993;90:3968–3972.

66. Baekelandt V, Arckens L, Annaert W, Eysel UT, Orban GA, Vandesande F. Alterations in GAP-43 and synapsin immunoreactivity provide evidence for synaptic reorganization in adult cat dorsal lateral geniculate nucleus following retinal lesions. Eur J Neurosci 1994;6:754–765.

67. Gilbert CD. Adult cortical dynamics. Physiol Rev 1998;78:467–485.

68. Fiala JC, Spacek J, Harris KM. Dendritic spine pathology: cause or consequence of neurological disorders? Brain Res Rev 2002;39:29–54.

69. Rubel EW, Fritzsch B. Auditory system development: primary auditory neurons and their targets. Annu Rev Neurosci 2002;25:51–101.

70. Mears AJ, Kondo M, Swain PK, et al. Nrl is required for rod photoreceptor development. Nat Genet 2001;29:447–452.

71. Strettoi E, Mears AJ, Swaroop A. Recruitment of the rod pathway by cones in the absence of rods. J Neurosci 2004;24:7576–7582.

72. Cuenca N, Pinilla I, Sauve Y, Lu B, Wang S, Lund RD. Regressive and reactive changes in the connectivity patterns of rod and cone pathways of P23H transgenic rat retina. Neuroscience 2004;127:301–317.

73. Milam AH, Li ZY, Fariss RN. Histopathology of the human retina in retinitis pigmentosa. Prog Ret Eye Res 1998;17:175–205.

74. Li ZY, Kljavin IJ, Milam AH. Rod photoreceptor neurite sprouting in retinitis pigmentosa. J Neurosci 1995;15:5429–5438.

75. Fariss RN, Li ZY, Milam AH. Abnormalities in rod photoreceptors, amacrine cells, and horizontal cells in human retinas with retinitis pigmentosa. Am J Ophthalmol 2000; 129:215–223.

76. Milam AH, Li ZY, Cideciyan AV, Jacobson SG. Clinicopathologic effects of the Q64ter rhodopsin mutation in retinitis pigmentosa. Invest Ophthalmol Vis Sci 1996;37:753–765.

77. Fei Y. Cone neurite sprouting: an early onset abnormality of the cone photoreceptors in the retinal degeneration mouse. Mol Vis 2002;8:306–314.

78. Lewis GP, Linberg KA, Fisher SK. Neurite outgrowth from bipolar and horizontal cells after experimental retinal detachment. Invest Ophthalmol Vis Sci 1998;39:424–434.

79. Dick O, tom Dieck S, Altrock WD, et al. The presynaptic active zone protein bassoon is essential for photoreceptor ribbon synapse formation in retina. Neuron 2003;37:775–786.

80. Haeseleer F, Imanishi Y, Maeda T, et al. Essential role of Ca2+-binding protein 4, a Cav1.4 channel regulator, in photoreceptor synaptic function. Nat Neurosci 2004;7:1079–1087.

81. Strettoi E, Pignatelli V, Rossi C, Porciatti V, Falsini B. Remodeling of second-order neurons in the retina of rd/rd mutant mice. Vision Res 2003;43:867–877.

82. Peng Y-W, Hao Y, Gaitan A, Zhang W, Wong F. Loss of invaginating rod bipolar cell dendrites in degenerating rod terminals. Invest Ophthalmol Vis Sci 2004;(Suppl) 45:ARVO e-abstract 5360.

83. Marc RE, Jones BW. Retinal remodeling in inherited photoreceptor degenerations. Mol Neurobiol 2003;28:139–147.

84. Park SJ, Lim EJ, Oh SJ, et al. Ectopic localization of putative AII amacrine cells in the outer plexiform layer of the developing FVB/N mouse retina. Cell Tissue Res 2004; 315:407–412.

IV
DEVELOPING THERAPEUTIC STRATEGIES
FOR RETINAL DEGENERATIVE DISEASES

16

On The Suppression of Photoreceptor Cell Death in Retinitis Pigmentosa

Fiona Kernan, PhD, Alex G. McKee, G. Jane Farrar, and Peter Humphries

CONTENTS

INTRODUCTION

A hereditary degenerative disease of the retina, retinitis pigmentosa (RP), is the leading cause of visual handicap among working populations in developed countries, with an estimated 1.5 million patients worldwide *(1,2)*. Clinically, RP is characterized by night blindness (nyctalopia) as a result of initial death of rod photoreceptors, followed by progressive visual loss owing to secondary degeneration of cone cells *(3)*. Mutations underlying RP reveal a genetically heterogenous condition, which can be inherited in an autosomal dominant (AD), recessive, X-linked recessive, digenic, or mitochondrial mode, with around 40 known or predicted genes implicated in disease pathology (http://www.sph.uth.tmc.edu/RetNet/). Of those genes with known functions, some encode proteins involved in the visual transduction cascade, e.g., rhodopsin, others in maintenance of photoreceptor structure, e.g., peripherin, and others involved in regeneration of the rhodopsin chromophore (11-*cis*-retinal) in the visual cycle,

From: *Ophthalmology Research: Retinal Degenerations: Biology, Diagnostics, and Therapeutics*
Edited by: J. Tombran-Tink and C. J. Barnstable © Humana Press Inc., Totowa, NJ

e.g., retinal pigment epithelial protein (RPE) 65 *(4)*. Interestingly, a number of RP genes are widely expressed but only cause disease pathology within the retina, highlighting the unique and complex biochemistry of photoreceptor cells. Included in the latter category are the genes *HPRP3* and *PRPC8* encoding pre-messenger RNA splicing factors and the gene encoding inosine monophosphate dehydrogenase type 1 (*IMPDH1*), the rate-limiting enzyme of the *de novo* pathway of guanine nucleotide biosynthesis *(5–7)*. Despite such genetic heterogeneity, photoreceptors degenerate in RP, and indeed in other inherited retinal degenerations, by a common form of cell death, apoptosis *(8,9)*. Apoptosis is a regulated mode of cell death that is essential for normal development and homeostasis *(10,11)*. However, abnormal regulation of apoptosis contributes to many disease pathologies, including cancer, autoimmune disorders, and neurodegenerative diseases, for example, Alzheimer's disease and amyotrophic lateral sclerosis (ALS) *(12–16)*. Numerous studies in cell culture, and in various animal models of retinal degeneration, including inherited and light-induced models of retinal damage support the initial observation by Chang et al. *(8)* that photoreceptors die by apoptosis in retinal degenerations *(17–20)*. One of the key aims of RP research is the development of effective therapeutics, and modulation of apoptosis clearly represents a potential therapeutic approach. It is unlikely that each RP mutation initiates an equivalent number of separate apoptotic pathways, so what is more probable is that such events converge and progress via one, or a limited number of apoptotic cascades, providing an alternative therapeutic approach to targeting the underlying primary mutations. Therapeutic invervention for primary mutations may involve either gene replacement for autosomal recessive (AR) RP or alternatively some form of gene suppression for ADRP. With respect to ADRP, targeting of primary mutations presents a particularly formidable challenge, since multiple mutations are routinely encountered in any given disease-causing gene, e.g., more than 100 different rhodopsin mutations have been identified *(4)*. On the other hand, the goal of inhibiting apoptosis is to modulate the course of the disease in an entirely mutation independent fashion, providing therapeutic benefit by targeting a common pathway. In addition to modulating apoptotic programs, other therapeutic strategies may include promoting photoreceptor survival using neurotrophic factors *(21–23)* or replacing lost photoreceptor cells by retinal transplantation or stem cell therapy *(24,25)*. For recent reviews of RP therapy, *see* Delyfer, et al. and Doonan et al. *(26,27)*. None of these therapeutic approaches is mutually exclusive and indeed it is likely that a combination of therapies may ultimately be used to treat this group of conditions. A summary of therapeutic strategies for RP is provided in Table 1.

The focus of this chapter is on how apoptosis can be modulated for potential therapeutic benefit in RP, including the inhibition of key proteases involved in mediating apoptosis and the reduction of reactive oxygen species (ROS) that may play a role in photoreceptor degeneration. Recent exciting developments in the area of cell survival factors will also be discussed. In addition, the role of light in apoptosis will be reviewed: how light-induced animal models of retinal degeneration have provided insights into mechanisms of degeneration in models of RP and how such discoveries may impact on the development of therapeutic strategies. Finally, it is clear from studies of the segregation pattern of genetic disorders in humans and from studies in animal models that so

Table 1
Strategies for Rescue in Models of Retinal Degeneration

Target	Therapy	Species/model	Effect of treatment	Reference
Caspase 3	Ac-DEVD-CHO	*rd* mouse	Transient protection	56
Caspase 3	p35 (transgene)	Drosophila	Protection of structure and function	57
Caspase 3	p35 (transgene)	661W cone cell line	Protects against FADD-induced death	17
Calpains/calcium channel	D-*cis*-diltiazem	Light induced	Prevents occurrence of TUNEL positive cells in ONL	70
Calpains/calcium channel	D-*cis*-diltiazem	*rd* mouse	Photoreceptor rescue	134
Calpains/calcium channel	D-*cis*-diltiazem	*rd* mouse	No protection	102,104
Calpains/calcium channel	D-*cis*-diltiazem	P23H	No protection	103
Calpains/calcium channel	D-*cis*-diltiazem	*rcd1* dog	No protection	105
Calpains/calcium channel	D-*cis*-diltiazem	RCS rat	No protection	106
Calpains	CR6 (ROS scavenger)	661W cone cell line	Protects against chemical induced death	109
NOS	L-NAME	Light-induced	Prevents occurrence of TUNEL positive cell in ONL	54
NOS	L-NAME	Light-induced, P23H, S334ter rat	Partial protection of morphology but not function in LI model, No protection in transgenics	148
Calpains (and caspase-3)	calpain inhibitor SJA6017	661W cone cell line	Protects against chemical induced death	51
Mitochondria (MOMP)	Bcl-2 (transgene)	*rd* mouse S334ter mouse	Increased PR survival for 2-4 weeks	77
Mitochondria (MOMP)	Bcl-2 and Bcl-XL (transgenes)	*rd* mouse	No protection	79
Mitochondria (MOMP)	Bcl-2 (Ad)	*rd* mouse	Rescue lasting 6 weeks	78
Mitochondria (MOMP)	Bcl-2 and BAG-1 (transgenes)	S334ter mouse	Rescue lasting 7–9 weeks	81
Mitochondria (MOMP)	MITO-4565	S334ter rat	Inhibits apoptosis	96

(Continued)

Table 1 *(Continued)*

Target	Therapy	Species/model	Effect of treatment	Reference
Reactive oxygen	DMTU	Light-induced in P23H and S334ter rats	Protection	*111,112*
Reactive oxygen	PBN	Light-induced, P23H, S334ter rats	Protection in LI but not transgenics	*112*
Reactive oxygen	Thioredoxin	Light-induced	Protection	*113*
nd, not determined	FGF2	RCS rat	Slower degeneration	*120,122, 123*
nd	FGF2	Light-induced	Slower degeneration	*121,124*
nd	FGF2	S334ter rat	Partial protection of morphology, not function	*124*
nd	EPO (transgene)	Light-induced, *rd*1, VPP mouse	Protection in LI but not transgenics	*149*
nd	CNTF (Ad)	*rd* mouse	Transient protection	*125*
nd	CNTF (intravitreal injection)	Q344ter, VPP, S334ter, P23H, mouse	No protection	*126*
nd	CNTF (Ad)	*rds* mouse	Protection and increase in ERG response	*127*
nd	CNTF and BDNF	*rd* mouse explants	PRs are rescued, but rod differentiation is depressed	*128*
nd	CNTF (AAV)	*Rho-/-* mouse	Protection of morphology, but not function	*129*
nd	CNTF (Ad)	P216L *rds* mouse	Protection of morphology, but not function	*130*
nd	CNTF (AAV)	*rd2* mouse	Protection of morphology, ERG lower than untreated eyes	*131*
nd	CNTF (ECT device)	*rcd1* dog	Protection or morphology, but function not evaluated	*21*
nd	Cardiotrophin-1 (repeated intravitreal injection)	S334ter rat	Protection of morphology	*133*
nd	GDNF (subretinal injection)	*rd* mouse	Protection and detectable ERG	*134*
nd	GDNF (AAV)	RCS rat, *rd2* mouse	Protection of morphology and function	*173*
nd	GDNF (AAV)	S334ter rats	Protection of morphology and function	*135*

(Continued)

Table 1 *(Continued)*

Target	Therapy	Species/model	Effect of treatment	Reference
nd	BDNF (transgenic cell transplant)	RCS rat	Protection of morphology	22
nd	RdCVF	*rd1* mouse	40% increase in cone survival	23

AAV, adeno-associated virus; Ac-DEVD-CHO, N-Ac-Asp-Glu-Val-Asp-CHO; Ad, adenovirus; BAG-1, Bcl-2 associated anthogene-1; Bcl-2, B-cell leukemia/lymphoma 2; Bcl-XL, homologue of Bcl-2; BNDF, brain-derived neurotrophic factor; CNTF, ciliary neurotrophic growth factor; CR-6, 3,4-dihydro-6-hydroxy-7-methoxy-2,2-dimethy1-1(2*H*)-benzopyran; DMTU, dimethylnitrourea; ECT, encapsulated cell technology; EPO, erythropoietin; ERG, electroretinogram; FADD, FAS-associating death domain-containing protein; FGF2, fibroblast growth factor-2; GDNF, glial-derived neurotrophic factor; LI, light-induced; L-NAME, *N*(G)-nitro-L-arginine methyl ester; NOS, nitric oxide synthase; ONL, outer nuclear layer; p35, baculoviral anti-apoptotic protein; PBN, phenyl-*N*-tert-butyInitrone; PR, photoreceptor; *rcdl*, rod-cone-dysplasia type 1; RCS, Royal College of Surgeons; *rd,* retinal degeneration; *rdl,* retinal degeneration 1 (same as *rd*); *rd2,* retinal degeneration 2 (previously known as *rds*); RdCVF, rod-derived cone viability factor; *rds*, rential degeneration slow; ROS, reactive oxygen species; TUNEL, terminal dUTP nick-end labeling; VPP, mutant transgene for opsin (V20G, P23H, P27L).

Adapted from ref. *170.*

called genetic modifiers influence progression of the disease. Identification of such modifiers, some of which are likely to regulate pathways of apoptosis and cell survival, may possibly illuminate novel therapeutic targets.

APOPTOSIS

On the Mechanism of Apoptosis

Apoptosis can be mediated by caspases, a group of cysteine-aspartyl-specific proteases *(28–31)*. To date, 14 mammalian caspases have been identified, a subset of which are involved in apoptosis, whereas the remainder are involved in processing pro-inflammatory cytokines *(32)*. Apoptotic caspases fall broadly into two categories, initiators and effectors. Initiator caspases, such as caspase-8, -10, and -12 are the first to be activated in response to a death stimulus, which in turn activate the effector caspases, namely caspase-3, -6, and-7 *(33)*. Once activated, these caspases mediate cell destruction by degrading a broad range of structural and regulatory proteins *(34)*. Apoptosis can be initiated from both outside and within the cell, depending on the pro-apoptotic stimulus. The extrinsic pathway is triggered via the activation of cell surface death receptors, e.g., Fas (or CD95) receptor and tumor necrosis factor receptor 1 (TNFR1), which in turn, activate caspase-8 within the cell *(35,36)*. The intrinsic pathway can be activated by a variety of stimuli, including ultraviolet light, chemotherapeutic agents, or growth factor deprivation, which trigger mitochondrial outer membrane permeabilization (MOMP), releasing cytochrome-*c* and other pro-apoptotic factors *(37–39)*. MOMP is a central event in cell death, and is tightly regulated by the Bcl-2 family of proteins, comprising both pro- and anti-apoptotic members *(40,41)*. An intrinsic pathway that centres on the endoplasmic reticulum (ER) has also been identified, in which insults that induce ER stress including misfolded proteins and oxidative stress, lead to caspase-12 activation *(42,43)*. Thus, it is clear that apoptosis is a complex process with numerous potential points for modulation. In the context of therapeutic development, initial targets to be explored were the caspases, in which inhibition of these proteins was used as a therapeutic approach for conditions including neurodegenerative disorders, myocardial infarction, and acute brain injury *(44–47)*. In this context, there have been notable successes, and several drugs are now at the clinical trial stage of development. For example, novel caspase inhibitors (Idun Pharmaceuticals, Inc., IDN-1965 and IDN-6556) have been shown to be protective in instances of heart and liver injury *(47,48)*.

Caspase-Dependent Mechanisms of Photoreceptor Cell Death

The possible involvement of caspases in RP has been explored to assess whether caspase inhibition is a potential therapeutic strategy for the disease. For example, there is substantial evidence to support the activation of caspase-3 in various models of retinal degeneration, the *rd* mouse *(49–51)*, *tubby* mouse *(19)*, ser334ter rhodopsin mutant rat *(18)*, and in chemically induced models of retinal degeneration *(52)*. In contrast, however, results from other studies suggest that caspase-independent apoptosis may be occurring *(53,54)*. Although there is significant evidence to support caspase-3 activation, the impact of caspase-3 ablation in knockout (KO) mice has been shown to provide only minimal protection against photoreceptor degeneration in the *rd* model of retinal degeneration *(55)*. This supports a transient protective effect previously observed using the caspase-3 inhibitor Ac-DEVD-CHO in the *rd* mouse *(56)*. Clearly, caspase-3 is activated in such systems but it may not play a critical role in the mediation of apoptosis, its function perhaps being compensated for by other caspases. Considering the complex nature of the pathways that lead from the numerous primary mutations encountered in RP to the death of photoreceptor cells, it is premature at this stage to discount caspase inhibition as a therapeutic strategy.

In contrast to these aforementioned studies involving caspase-3 inhibitors, successes have been achieved with pan-caspase inhibitors, most notably the p35 protein. p35 was originally identified in baculoviruses and is a pan-caspase inhibitor targeting both initiator (caspase-2, -8, and -10,) and effector (caspase-3, -6, and -7) caspases and it has been shown to rescue photoreceptor degeneration in Drosophila models of retinal degeneration *(57,58)*. Furthermore, p35 has also been shown to protect against chemically induced apoptosis in the cone photoreceptor cell line, 661W *(17)*. p35 inhibits several caspases, in contrast to the specific caspase-3 inhibitors, possibly explaining its greater protective effect. Clearly further evaluation in animal models will be required before any conclusions can be made regarding the use of caspase inhibitors as therapeutic agents in RP.

Caspase-Independent Mechanisms of Photoreceptor Cell Death

Caspase-mediated apoptosis may not be the only pathway of photoreceptor degeneration in RP. Caspases were long considered the key executioners of apoptosis, but research has shown that caspase-independent mechanisms of cell death exist, where dying cells retain many morphological characteristics of apoptosis *(59)*. Caspase-independent pathways have been demonstrated in neuronal systems in response to ischaemia, traumatic brain injury and in neurodegenerative diseases such as Huntington's and Alzheimer's diseases *(60–63)*. Proteases involved in caspase-independent pathways of cell death include cathepsins, calpains, and serine proteases such as granzyme B *(64–66)*. Calpains are a family of calcium-dependent proteases, comprising at least 15 members, the best characterized of which are μ- and m-calpain *(65)*. Although much remains to be learned about the regulation and function of calpains, these proteases have been implicated in the pathogenesis of cell death in cerebral ischaemia *(67)*, cataract formation *(68)*, and neurodegenerative disorders including Huntington's disease *(69)*. In reference to photoreceptor cell death, calpain activation has been shown in light-induced and inherited models of retinal degeneration *(51,70,71)*. In one study, a calpain inhibitor prevented calcium-induced death in cone photoreceptor-derived 661W cells, further supporting a possible role for calpains in photoreceptor cell death *(51)*. The successful inhibition of cell death in 661W cells cal-pains using a calpain inhibitor, warrants further exploration of calpains as novel thera-peutic targets for modulation of apoptosis in degenerative retinopathies. It is notable that in the previously mentioned study caspase-3 activation was also detected, indicating cross talk between the two proteolytic systems of caspases and calpains. Interaction between these different proteases has been demonstrated in previous studies, including the activation of caspase-3 and -12 by calpains *(72,73)*. Activation of both systems in photoreceptor degeneration suggests a possible explanation for the limited success of caspase inhibitors in preventing apoptosis in models of RP. However, results from other studies suggested no caspase activation in the *rd* inherited and light-induced model of retinal degeneration, indicating that the complex pathways of cell death in RP remain to be fully elucidated *(53,54)*.

A recent study by the same group showed that although treatment of *rd* retinal explants with a calpain inhibitor successfully inhibited calpain-induced alpha-fodrin cleavage, it did not protect against photoreceptor degeneration, suggesting the involve-ment of multiple cell death pathways *(171)*. Lohr and colleagues reached a similar con-clusion by comparing three photoreceptor degenerations caused by different events: calcium overload (*rd* mouse), structural defects (*rds* mouse), and light induced retinal degeneration *(172)*. By comparing caspase, lysozyme and cathepsin activity, as well as the expression of several other apoptotic marker genes, they concluded that multiple parallel cell death mechanisms are involved in retinal cell death *(172)*. Until a common upstream initiator of cell death can be determined, each of these components must be addressed for successful inhibition of photoreceptor degeneration.

Mitochondria

Caspases and calpains represent some possible therapeutic targets, in respect of pho-toreceptor protection, but there are several others within the apoptotic pathway, most notably those centring on the mitochondria. Apoptosis proceeding through the mito-chondria represents an important pathway of cell death, which is characterized by a central event, that of MOMP *(39)*. Following MOMP, factors mediating apoptosis are released including cytochrome-*c*, apoptosis inducing factor, and Smac/Diablo *(74–76)*. As a result, the mitochondrial potential is dissipated and the essential functions of the mitochondria are lost. Initiation of this process is tightly regulated by the Bcl-2 family of proteins, comprising both pro- and anti-apoptotic members and they modulate the formation of permeability transition (PT) pores on the surface of the outer membrane. Anti-apoptotic members block MOMP by preventing the formation of the PT pores, whereas pro-apoptotic members facilitate opening of the pores. Modulation of this process has been evaluated as a therapeutic approach for RP, with varying results. For example, overexpression of Bcl-2 was shown to provide transient protection in mouse models of retinal degeneration *(77,78)*. In contrast, however, other studies have reported no protection by either Bcl-2 or Bcl-XL expression *(79)*. Furthermore, different combinations of Bcl-2 anti-apoptotic proteins have been evaluated and again, although some protection has been observed in photoreceptors, effects were transient *(80,81)*. In support of this approach, overexpression of Bcl-XL has been shown to protect against lead-induced photoreceptor apoptosis up to postnatal day 90 (P90) in mice *(82)*. Overall, however, results from studies of exploring Bcl-2 family members as potential therapeutics for RP have not been encouraging. This view is supported by a recent observation in which Bax, the target of Bcl-2, was found to be downregulated during normal retinal development, possibly explaining, at least in part, the lack of protection provided by overexpression of Bcl-2 *(81)*.

It is notable, however, that there are alternative ways to inhibit the formation of the mitochondrial PT pore. For example, cyclosporin A blocks the loss of membrane potential by targeting proteins involved in PT function including cyclophilin D *(83,84)*. It has been shown to be protective in models of Alzheimer's disease, Parkinson's disease and ALS *(85–87)*. In addition, this agent was shown to decrease the death of cortical neurons in a model of focal ischemic stroke *(88)* and has been shown to protect against calcium-induced apoptosis in isolated rat retina *(89)*. Another example of an agent, which has been found to be protective at the level of mitochondria is tauroursodeoxycholic acid (TUDCA), an endogenously produced hydrophilic bile. TUDCA acts, in part, by inhibiting the translocation of pro-apoptotic Bax to the mitochondria and in addition has antioxidant properties *(90)*. It has been shown to be neuroprotective in animal models of Huntington's disease *(91)* and also reduced apoptosis in RPE cells *(92)*. Other agents that protect by a similar mode of action are minocyclin and rasagiline, which have been shown to be of therapeutic benefit in models of Parkinson's disease and ALS *(93,94)*.

Another agent targeting mitochondria, MITO 4565, has been evaluated in a rat model of RP expressing a dominant mutation (Ser344ter) within the rhodopsin gene *(95)*. MITO 4565 is a novel oestrogen analogue that does not inhibit the formation of the PT pore, but rather stabilizes the mitochondrial membrane. MITO 4565 was injected into the left retinas of the Ser344ter model at PD9, and by PD20 the loss of outer nuclear layer (ONL) thickness was shown to be significantly reduced compared to the control right retinas *(96)*. MITO 4565 is believed to intercalate into the mitochondrial membrane, terminating lipid peroxidation and thus maintaining mitochondrial membrane potential.

CALCIUM INVOLVEMENT IN APOPTOSIS

Elevated calcium levels probably play a key role in photoreceptor apoptosis. Calcium overload has been observed in various models of inherited and chemical-induced retinal degenerations *(97–99)*. The role of calcium is well characterized in the *rd* mouse, which has a mutation in the gene encoding the β-subunit of cGMP phosphodiesterase *(100)*. When cGMP levels rise, channels regulated by cGMP remain open, resulting in the build up of toxic levels of calcium within the photoreceptors. A recent study demonstrated in 661W cone photoreceptor cells, that calcium-induced apoptosis is mediated by calpain activation, resulting in caspase-3 dependent cell death and that this apoptotic pathway could be inhibited by a calpain inhibitor, SJA6017. In the same study, activation of calpain and caspase-3 were observed in the retinas of the *rd* mouse, indicating that the pathway of apoptosis observed in 661W cells is the possible mechanism of photoreceptor degeneration in the *rd* model *(51)*. As elevated calcium has been shown to play a key role in mediating photoreceptor apoptosis, modulation of calcium levels is another potential strategy for slowing cell death in the retina. Calcium enters the cell via voltage-dependent calcium channels, and channel blockers have been evaluated as a potential therapy, with varying results. D-*cis*-diltiazem, one such channel blocker, has been reported to be protective in the *rd* mouse model and in a light-induced model of retinal degeneration *(70,101)*. In contrast, D-*cis*-diltiazem was not found to be protective in other studies using the *rd* mouse, in a Pro23His rat model, or in a canine model of retinal degeneration *(102–105)*. Similar results have been obtained with nilvadipine, which has been shown to be protective in the *rd* mouse retina, and in another model of retinal degeneration, the Royal College of Surgeons (RCS) rat *(104,106)*. Micoarray analysis of gene expression in the *rd* mouse following nilvadipine administration suggests that protection is likely mediated through suppression of caspases, and upregulation of fibroblast growth factor (FGF), a neuroprotective cytokine *(104)*. Considering the observation of therapeutic benefit in a number of studies, calcium-channel blockers warrant further investigation as potential therapeutic agents for degenerative retinopathies.

OXIDATIVE STRESS INVOLVEMENT IN APOPTOSIS

Another factor suggested to be involved in photoreceptor degeneration is oxidative stress, resulting from the generation of damaging ROS within retinal tissue. As the retina is one of the highest oxygen-consuming tissues in the body, it is particularly sensitive to oxidative stress *(107)*. Oxidative stress in photoreceptor apoptosis has been studied predominantly using light-induced models of retinal degeneration, in which short exposure to bright light induces retinal damage *(20)*. The role of oxidative stress in mediating apoptosis in light-induced models is supported by various *in vitro* and *in vivo* studies demonstrating that increased levels of ROS represents an early event in photoreceptor apoptosis, which can be inhibited by antioxidants *(108,109)*.

In addition to indicating an involvement of ROS in retinal degeneration, these results suggested a possible therapeutic strategy for RP because light has been shown to be a cofactor accelerating disease progression *(110)*. Antioxidants such as dimethylthiourea (DMTU) and phenyl-*N*-tert-butylnitrone (PBN) have been evaluated using inherited rodent models of retinal degeneration exposed to damaging levels of light *(111,112)*. Protection from the deleterious effects of light has been observed with DMTU in Pro23His and Ser344ter transgenic rats and with PBN in the Pro23His rat model. However, PBN was found to have no effect on the rate of degeneration of the photoreceptors in either model in the absence of additional light insult, indicating that such therapies may be of benefit in limited cases of RP, in which light-accelerated damage is more significant. Another potential protective agent in the context of photodegeneration is thioredoxin, an endogenous protein with various activities including elimination of ROS and regulation of the apoptotic pathway. Thioredoxin has been shown to protect against light-induced retinal degeneration in several studies, but has yet to be evaluated in inherited models of retinal degeneration *(113,114)*.

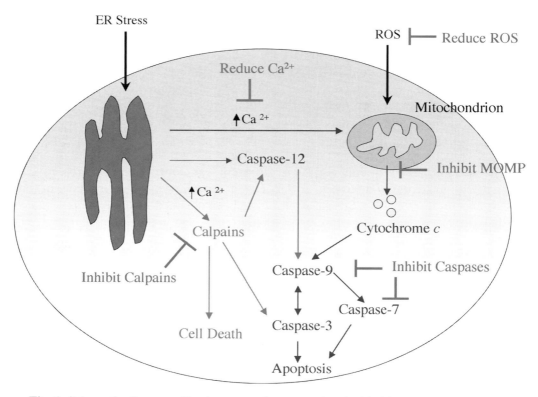

Fig. 1. Schematic diagram of basic proposed caspase- (marked in blue) and calpain- (marked in green) pathways of cell death in photoreceptor cells with potential therapeutic strategies (marked in red). Apoptotic pathways in RP are ill defined to date but ER stress is known to play a significant role, particularly in cases of RP as a result of mutated rhodopsin *(168,169)*. Other factors that may be involved in degeneration are calcium and ROS *(100,108,109)*.

Oxidative stress may also have a more fundamental role in retinal degenerations, in that it may be responsible for the secondary loss of cone photoreceptors, which almost invariably follows initial rod cell death in conditions such as RP. Understanding processes involved in cone degeneration is of vital importance because cone loss is responsible for the main visual handicap in RP. Several theories have been advanced to explain cone photoreceptor loss, including oxidative damage *(115)*. Markers for oxidative stress, indicating damage to proteins, lipid, and DNA were detected in cone photoreceptors of the transgenic pig model of RP, with a Pro347Leu rhodopsin mutation *(116)*. The data suggest that as the rod photoreceptors degenerate, there is a reduction in oxygen consumption, resulting in increased oxygen levels within the retina (hyperoxia), oxidative stress, and finally cone degeneration. The hypothesis is supported by studies demonstrating increased oxygen levels in models of retinal degeneration *(117,118)*. It remains to be seen whether antioxidants may slow down cone photoreceptor loss in RP, but the potential therapeutic benefit observed in animals supports further evaluation of agents that reduce oxidative stress (*see* Fig. 1).

TROPHIC FACTORS AND THE POTENTIAL FOR GENE THERAPY

A parallel therapeutic approach to inhibiting cell death is promoting cell survival. One way of achieving this is through use of neurotrophic factors that modulate neuronal growth during development to maintain existing cells and aid recovery of injured neurons. In developing retinal neurons, correct synaptic connections are reinforced by trophic factors, whereas cells with inappropriate connections receive no trophic support and die by apoptosis *(119)*. This observation has lead to the premise that because the removal of neurotrophic factors stimulates cell death, the addition of exogenous trophic factors may have neuroprotective effects in the retina.

In the first study demonstrating that growth factors might protect against photoreceptor degeneration the use of basic FGF (bFGF or FGF-2) was explored in a rat model of inherited retinal dystrophy *(120)*. However, despite the observation that FGF-2 treatment rescued degenerating photoreceptors *(121–124)*, FGF-2 has also been shown to trigger pathological retinal neovascularisation, making it unacceptable for human therapy.

Other trophic factors have also been shown to protect against photoreceptor degeneration. The most well characterized of these is ciliary neurotrophic factor (CNTF), a member of the interleukin (IL)-6 family of cytokines, which has been shown to delay photoreceptor apoptosis in several models of retinal degeneration *(125–131)*. However, CNTF therapy also has its drawbacks. Despite inducing a morphological rescue, functional analysis by electroretinogram (ERG) showed a decreased response in CNTF-treated retinas *(129,130)*. In contrast a study by Cayouette et al. *(127)* showed a significant preservation of the ERG response in the *rds* mouse after adenovirus-mediated gene delivery of a gene-encoding CNTF and it has been suggested that this might be in part because of a lower level of expression, whereas previous studies were suspected to have delivered toxic dose levels of CNTF. To provide controlled, continuous, long-term delivery of CNTF, Tao et al. *(21)* developed an encapsulated cell therapy (ECT) device, specifically designed for intraocular implantation. This involved loading a polymer membrane capsule with mammalian cells genetically engineered to secrete CNTF that was surgically implanted into the vitreous of 7-wk-old *rcd1* dogs. After 7 wk, the ECT treated eyes had significantly higher levels of nuclei in the ONL, but retinal function was not evaluated in the study *(14)*. To study the effects of dose of CNTF on normal retinal function, ECT devices secreting a high or low dose of CNTF were implanted into white albino rabbits *(132)*. Low (5 ng/d) doses had no adverse effects, whereas the higher (22 ng/day) dose showed morphological changes in the ONL. but caused no reduction in the ERG, leading the authors to suggest that low-therapeutic doses are not toxic.

Another neurotrophic factor, cardiotrophin (CT)-1, also a member of the IL-6 family, has been shown to protect photoreceptors of the S334ter transgenic rat *(133)*. Repeated intravitreal injection of CT-1 every 4 or 5 d resulted in a significant rescue of ONL cells of the retina. The biological effects of CT-1 are mediated thorough a signal transducer and activator of transcription 3 (STAT3) pathway and the marked increase in phosphorylated STAT3 observed in Müller cells suggests that these cells probably mediate the protective effect *(61)*.

Glial derived neurotrophic factor (GDNF) has also been shown to slow photoreceptor degeneration and preserve visual function in an *rd* mouse model. Intraocular injections of GDNF were efficient at delaying photoreceptor cell death and detectable ERG responses were recorded in 4 out of 10 GNDF-treated animals *(134)*. Use of a recombinant adeno-associated virus vector to deliver GDNF has been shown to significantly increase rod photoreceptor survival and substantially increase the amplitude of the ERG response in Ser334ter transgenic rats *(135)*. GDNF-replacement therapy has also been shown to slow cell death and enhance retinal function in the *rd2* mouse and the RCS rat in combination with gene replacement therapy *(173)*.

Despite the fact that the mechanisms underlying cone cell death remain relatively unknown, it is important to note that degeneration of cones occurs in patients with rhodopsin mutations after rod cell death exceeds 75% *(110)*. This has also been observed in several animal models of retinal degeneration *(136)*. As the cones are not directly affected by the mutation, they therefore degenerate in a "non-cell autonomous" mechanism, which results in progressive loss of cone function. One proposed hypothesis to account for the degeneration of cones suggests that rod photoreceptors release a trophic factor or factors that are essential for cone cell preservation. Evidence for the loss of trophic support theory has been described in an elegant set of experiments by Thierry Leveillard and colleagues *(23)*.

Mohand-Said et al. *(137)* demonstrated that retinas from *rd* mice cultured in the presence of normal retinas showed significantly (15–20%) greater numbers of surviving cones compared with controls. These data suggested the existence of a diffusable trophic factor released by rods that protects cones *(137)*. This factor was eventually identified using an expression cloning approach, and named rod-derived cone viability factor (RdCVF). RdCVF is a truncated thioredoxin-like protein, expressed and secreted specifically by rod photoreceptors *(23)*. These data suggest a novel mode of trophic interactions and should in principle allow for the development of unique therapies aimed at preventing secondary cone cell death and subsequent loss of central vision in degenerative retinopathies. Because only 5% of the normal complement of cone cells is still compatible with visual discrimination and orientation, and 50% cone survival is compatible with normal, 20/20 visual acuity, RdCVF could, in principle, provide substantial therapeutic benefit even if delivered at a relatively advanced stage of disease.

ON THE SIGNIFICANCE OF LIGHT-INDUCED RETINAL APOPTOSIS AS A MODEL OF RP

Although many transgenic mouse lines that mimic human retinal disease are now available *(138–140)*, much of the research carried out to date on photoreceptor apoptosis has used light-induced models. Although lifelong exposure to bright light is known to be a contributing element to retinal disease progression, the light-induced model of retinal degeneration does not necessarily mimic the human condition and potentially activates alternative apoptotic pathways. However, this model does have the advantages of being significantly quicker and more convenient than using transgenic models. It takes only hours of light exposure to significantly alter the appearance of the photoreceptors, whereas with many transgenic animal models degeneration may often not occur for several months.

It has been firmly established that light-induced apoptosis in mice is dependent on a functional visual cycle. Photoreceptors lacking rhodopsin are completely protected against light-induced apoptosis *(141)* and mice with a Leu450Met mutation in the *Rpe65* gene with slow rhodopsin regeneration kinetics are more resistant to light damage than wild-type (WT) mice *(142)*. Rhodopsin-deficient mice can also be generated by knocking out genes needed for the synthesis of 11-*cis*-retinal, such as *Rpe65*, and as expected, mice lacking *Rpe65* are also completely protected against light-induced apoptosis. This is also consistent with the finding that vitamin A deficiency, which also prevents rhodopsin synthesis, also protects against light damage *(143)*.

Another gene known to play an important role in retinal light damage is the *c-fos* gene, which codes for a proto-oncoprotein. Although it has no direct role in the visual cycle, aberrant expression of the *c-fos* transcript was observed in the *rd* mouse *(144)*. This led to the investigation of its role in light-induced damage, and the subsequent discovery that *c-fos* KO mice were protected against light damage *(145)*. This implicated the activator protein (AP)-1 transcription factor, of which c-fos is a component, in the light-induced cell death cascade. However, c-fos does not protect against photoreceptor cell death in the rhodopsin KO mouse, which suggests that the AP-1 pathway has no role in some, or possibly all mutation-induced forms of photoreceptor apoptosis *(146)*. A study by Hao et al. *(147)* has demonstrated that at least two different biochemical pathways mediated light damage, one pathway for bright light, and a second for low-level light. *Gnat1* KO mice lacking the gene coding for the α-subunit of rod transducin were used to distinguish between the roles of activated rhodopsin and phototransduction. In this way, it was possible to investigate whether light-induced activation alone leads to apoptosis, or if a downstream event in the phototransduction cascade might be involved. Both transducin-deficient and WT mice were equally susceptible to light damage induced by bright light, whereas absence of phototransduction was protective against low-light intensities. These results define a second phototransduction-dependent light-induced mechanism of photoreceptor cell death.

The direct relevance of the light-induced model of retinal degeneration to genetic models of retinal disease has still to be fully established. In this regard, it is notable that in a number of studies *(102–104,106,148,149)* neuroprotective agents that were found to be protective in light-induced models of retinal degeneration did not translate through to neuroprotection in genetic models of retinal disease.

ON THE INFLUENCE OF GENETIC AND ENVIRONMENTAL FACTORS ON PHOTORECEPTOR CELL DEATH

Light is also known to modify the severity of disease progression in human retinal degenerations. A human case report of two families with Pro23His rhodopsin mutations provided the first indication that light phototoxicity may be an accelerator of RP *(150)*. Studies in transgenic rodent models *(111,151)* with rhodopsin mutations lent support to this theory, showing that light activation of rhodopsin contributes to the severity of the disease, leading to the suggestion that minimizing exposure to light might delay retinal degenerations arising from rhodopsin mutations.

A recent report using a naturally occurring canine model of ADRP caused by a rhodopsin mutation again tested this hypothesis, exploiting the similarity in eye-size and preretinal light transmission characteristics between dog and human *(152)*. Investigation into the illuminating effects of clinical retinal photography led to the surprising observation of circular-degenerating areas of retina that matched the pattern of the light flashes. Changes in retinal tissue were visible minutes after light exposure. By titrating transient increases in neuronal stress by light exposure, a dose–response relationship between light exposure and long-term outcomes of these early alterations was established. High doses of light caused a rapid degeneration of neurons, whereas low doses revealed mechanisms acting over several weeks or months that repair the damage. This study represents the first report of repair of retinal injury in an inherited retinal degeneration and establishes a useful in vivo assay to study the balance between pro- and anti-apoptotic signaling and repair-compensation mechanisms. Taken together, these findings suggest that limiting light exposure in patients with rhodopsin mutations may slow disease progression.

Another factor contributing to the variability in RP phenotypes is the influence of genetic modifier loci. In principle, modifiers cause variation in phenotype by interacting in the same or a parallel biological pathway as that of the disease gene. This modifier effect may suppress the mutant phenotype even to the extent of completely restoring the WT phenotype. Alternatively, expression of the modifier gene may lead to a more severe mutant phenotype or affect the pleiotropy of a given disease resulting in a different combination of traits. In addition, combinations of modifier genes may act together to create cumulative effects on the expression of a phenotype *(153)*. Clearly, the further characterization of genetic modifier loci should provide insights into the biological pathways in which these genes act to cause disease as well as providing novel therapeutic targets. For such reasons, interest in modifier genes has grown rapidly.

MODIFIERS OF RP IN HUMANS

A classic example of the effect of a strong modifier in RP was the occurrence of three separate phenotypes within a single nuclear family with a novel three-base deletion of codon 153 or 154 in the peripherin/*rds* gene. These patients developed either RP, pattern dystrophy, or *fundus flavimaculatus (154)*. Although the genetic modifier locus or loci underlying these particular modifications have yet to be identified, one gene product known to interact with peripherin/*rds* is ROM1. Rare individuals heterozygous for a Leu85Pro allele of peripherin/*rds* who also carry a null *ROM1* allele develop digenic RP, whereas individuals with either mutation alone are unaffected *(155)*. In another study of 1941 probands with a clinical diagnosis of RP (who had previously been screened for mutations in rhodopsin and peripherin/*rds* and found to be negative), 17 families were shown to harbor mutations in the *RP1* gene. Patients with a premature stop codon, Arg677ter, demonstrated wide variability in the severity of visual field loss both within and between families, which led the authors to suggest an important role for modifier genes or environmental factors in RP1 releated disease *(156)*.

In a study of RP9, the mutation causing ADRP was shown to lead to regional retinal dysfunction with greatly variable expressivity. Family members with this mutation were reported to be either minimally effected, with normal electrophysiological responses, moderately affected with abnormal ERG responses, or severely effected with no ERG response *(157)*. RP9 has since been shown to be caused by a mutation in the splicing factor gene, *PAP-1 (158,159)*.

Despite a wealth of studies reporting phenotypic differences between individuals with the same genotype, very few modifier loci have been chromosomally localized. In humans, this difficulty may be largely the result of the genetic variation in the population. In contrast, this problem can be reduced significantly in inbred mice as discussed earlier.

MODIFIERS OF RP IN MOUSE MODELS

The existence of modifiers in mice was originally recognized when the spontaneous obesity mutations *Lep^ob* and *Lep^db* were shown to cause diabetes on a C57BL/KsJ, but not on a C57BL/6J genetic background *(160)*. Recent advances in generating targeted mutant mouse models have also revealed many important examples of modifiers. Such mutants are usually propagated in stem cells derived from 129 mice that were made chimeric with the C57BL/6J strain and then crossed onto a specific genetic background. Alterations in the initial targeted phenotypes have been reported when mice with retinopathies were back crossed onto specific genetic backgrounds *(161,162)*. Modifier effects of RP can result from a single gene at a locus independent of the disease gene *(163–165)* or can be caused by the combined effects of several genes at different loci, as is typical of quantitative trait loci (QTL) *(165)*. The first report of variable phenotypic expression in a mouse model of retinal degeneration was observed in the *rd*3 mouse. A mutation at the *rd3* locus led to a unique retinal degeneration whereby photoreceptor cell death starts at 3 wk postnatally and is complete at 5 wk *(166)*. However, significant variation in the onset and progression of the disease was observed when mice were crossed onto different background strains. The ocular phenotypes of mice carrying a targeted disruption of the p53 tumor suppressor gene (p53–/– mice) have also been shown to be radically different on C57BL/6J and 129 genetics backgrounds. p53 KO mice bred onto a C57 background, but not on the 129 background, exhibited vitreal opacities, retinal folds, and vitreal neovascularisation, possibly as a result of abnormal developmental retinal apoptosis; although, as pointed out by the authors, angiogenic factors could also be involved in modulation of the p53 phenotype *(161)*. Mice with a targeted disruption in rhodopsin (*Rho*–/– mice) have also been reported to be protected by modifiers on the C57BL/6J background when compared to the 129 background *(162)*. In the latter investigation, C57BL/6J mice were found to have a significantly greater number of ONL nuclei by 3 months of age and TUNEL staining, over various time points, showed more positive labeling in the ONL of 129 retinas. Both amplitude and waveform features of electroretinographic analysis were remarkably different in the two strains.

Taken together, these results suggest the presence of genetic modifiers on the C57BL/6J background that significantly protect photoreceptors against the retinopathy; however, whether such modifiers directly influence apoptotic mechanisms remains to be established. Two recent studies by Danciger et al. *(163,167)* have resulted in localization of QTL that contribute to the protection of photoreceptor cells against damage induced by constant light. In the first of these studies, a genome-wide scan on the progeny from backcrossed mice using the thickness of the ONL as the quantitative trait reflecting retinal damage revealed a strong QTL on mouse chromosome 3 that contributes almost 50% of the protective effect *(163)*. A high LOD score linked the *Rpe65* gene to the apex of this QTL and sequencing revealed a single base change in codon 450, coding for a methionine in c2J mice and a leucine in the BALB/c strain *(163)*. In a second study, rhodopsin was measured spectrophotometrically subsequent to light-induced apoptosis, and this was used as the quantitative trait to reveal new QTL on mouse chromosomes

1 and 4, with suggestive QTL on chromosomes 6 and 2 *(167)*. More recently, the Rpe65 gene was determined to be a modifier for an inherited retinal degeneration in the VPP mouse model with the authors suggesting that the variation in the gene may modulate rhodopsin regeneration kinetics, therefore affecting light-damage susceptibility *(174)*. Identification of such QTL and the associated modifiers may provide important information needed to further understand human retinal degenerations. However this will be challenging. In a second study rhodopsin was measured spectrophotometrically subsequent to light-induced apoptosis, and this was used as the quantitative trait to reveal significant QTL on mouse chromosomes 1 and 4, with suggestive QTL on chromosomes 6 and 2 *(167)*. Identification of such QTL and the associated modifiers may provide important information needed to further understand human retinal degenerations. However, this will be challenging, especially if more than one gene contributes to the modification of a given phenotype. Nonetheless, the real promise of substantial therapeutic potential remains once the functions of modifier genes associated with a suppression of photoreceptor cell death are elucidated.

CONCLUSION

In the light of significant recent progress, it is tempting to speculate that several therapies for RP may be available within the next few years. Such therapies will be based either on direct intervention at the genetic level, using the technique of gene replacement or suppression of transcripts derived from dominant-acting genes, either through using techniques aimed at suppressing secondary molecular pathological effects, such as apoptosis, or by enhancing neuroprotection. Although significant progress in being made in both gene replacement and in suppression of mutant transcripts *(4)*, targeting of apoptotic or survival mechanisms holds much appeal in the sense that such strategies will be largely independent of the vast number of mutations now known to cause RP-related conditions. The most readily attainable goal of RP research is the elucidation and functional evaluation of all RP genes. Up to 40 RP genes are known to date, but it is possible that many more remain to be identified (http://www.sph.uth.tmc.edu/RetNet/). Why mutations in such genes lead to photoreceptor cell death, sometimes many years after birth, is as yet an unresolved question. It is highly unlikely that there are many different gene-specific pre-apoptotic pathways, all individually activating apoptosis. A more probable scenario is that a smaller number of such pathways, shared by many RP loci, converge toward a few pre-apoptotic initiators. A major endeavor for future RP research will be to identify molecules and interactions in such pathways, and to understand the "switch" that occurs from normal aging to that of disease. Never before have so many avenues been available through which therapeutic interventions for this group of conditions might be achieved.

REFERENCES

1. Berson EL. Retinitis pigmentosa: unfolding its mystery. Proc Natl Acad Sci USA 1996;93:4526–4528.
2. Weleber RG, Gregory-Evans K. Retinitis pigmentosa and allied disorders. In: Ryan SJ. ed. Retina. St. Louis, MO: Mosby, 2001:362–470.

3. Kalloniatis M, Fletcher EL. Retinitis pigmentosa: understanding the clinical presentation, mechanisms and treatment options. Clin Exp Optom 2004;87:65–80.

4. Farrar GJ, Kenna PF, Humphries P. On the genetics of retinitis pigmentosa and on mutation-independent approaches to therapeutic intervention. Embo J 2002;21:857–864.

5. Chakarova CF, Hims MM, Bolz H, et al. Mutations in HPRP3, a third member of pre-mRNA splicing factor genes, implicated in autosomal dominant retinitis pigmentosa. Hum Mol Genet 2002;11:87–92.

6. McKie AB, McHale JC, Keen TJ, et al. Mutations in the pre-mRNA splicing factor gene PRPC8 in autosomal dominant retinitis pigmentosa (RP13). Hum Mol Genet 2001;10: 1555–1562.

7. Aherne A, Kennan A, Kenna PF, et al. On the molecular pathology of neurodegeneration in IMPDH1-based retinitis pigmentosa. Hum Mol Genet 2004;13:641–650.

8. Chang GQ, Hao Y, Wong F. Apoptosis: final common pathway of photoreceptor death in rd, rds, and rhodopsin mutant mice. Neuron 1993;11:595–605.

9. Portera-Cailliau C, Sung CH, Nathans J, Adler R. Apoptotic photoreceptor cell death in mouse models of retinitis pigmentosa. Proc Natl Acad Sci USA 1994;91:974–978.

10. Jacobson MD, Weil M, Raff MC. Programmed cell death in animal development. Cell 1997;88:347–354.

11. Danial NN, Korsmeyer SJ. Cell death: critical control points. Cell 2004;116:205–219.

12. Green DR, Evan GI. A matter of life and death. Cancer Cell 2002;1:19–30.

13. Todaro M, Zeuner A, Stassi G. Role of apoptosis in autoimmunity. J Clin Immunol 2004;24:1–11.

14. Eldadah BA, Faden AI. Caspase pathways, neuronal apoptosis, and CNS injury. J Neurotrauma 2000;1:811–829.

15. Marques CA, Keil U, Bonert A, et al. Neurotoxic mechanisms caused by the Alzheimer's disease-linked Swedish amyloid precursor protein mutation: oxidative stress, caspases, and the JNK pathway. J Biol Chem 2003;278:28,294–28,302.

16. Raoul C, Estevez AG, Nishimune H, et al. Motoneuron death triggered by a specific pathway downstream of Fas. potentiation by ALS-linked SOD1 mutations. Neuron 2002; 35:1067–1083.

17. Tuohy G, Millington-Ward S, Kenna PF, Humphries P, Farrar GJ. Sensitivity of photoreceptor-derived cell line (661W) to baculoviral p35, Z-VAD.FMK, and Fas-associated death domain. Invest Ophthalmol Vis Sci 2002;43:3583–3589.

18. Liu C, Li Y, Peng M, Laties AM, Wen R. Activation of caspase-3 in the retina of transgenic rats with the rhodopsin mutation s334ter during photoreceptor degeneration. J Neurosci 1999;19:4778–4785.

19. Bode C, Wolfrum U. Caspase-3 inhibitor reduces apototic photoreceptor cell death during inherited retinal degeneration in tubby mice. Mol Vis 2003;9:144–150.

20. Reme CE, Grimm C, Hafezi F, Marti A, Wenzel A. Apoptotic cell death in retinal degenerations. Prog Retin Eye Res 1998;17:443–464.

21. Tao W, Wen R, Goddard MB, et al. Encapsulated cell-based delivery of CNTF reduces photoreceptor degeneration in animal models of retinitis pigmentosa. Invest Ophthalmol Vis Sci 2002;43:3292–3298.

22. Lawrence JM, Keegan DJ, Muir EM, et al. Transplantation of Schwann cell line clones secreting GDNF or BDNF into the retinas of dystrophic Royal College of Surgeons rats. Invest Ophthalmol Vis Sci 2004;45:267–274.

23. Leveillard T, Mohand-Said S, Lorentz O, et al. Identification and characterization of rod-derived cone viability factor. Nat Genet 2004;36:755–759.

24. Lund RD, Ono J, Keegan DJ, Lawrence JM. Retinal transplantation: progress and problems in clinical application. J Leukoc Biol 2003;74:151–160.

25. Ahmad I. Stem cells: new opportunities to treat eye diseases. Invest Ophthalmol Vis Sci 2001;42:2743–2748.
26. Delyfer MN, Leveillard T, Mohand-Said S, Hicks D, Picaud S, Sahel JA. Inherited retinal degenerations: therapeutic prospects. Biol Cell 2004;96:261–269.
27. Doonan F, Cotter TG. Apoptosis: A Potential Therapeutic Target for Retinal Degenerations. Curr Neurovasc Res 2004;1:41–53.
28. Thornberry NA, Lazebnik Y. Caspases: enemies within. Science 1998;281:1312–1316.
29. Nicholson DW. Caspase structure, proteolytic substrates, and function during apoptotic cell death. Cell Death Differ 1999;6:1028–1042.
30. Earnshaw WC, Martins LM, Kaufmann SH. Mammalian caspases: structure, activation, substrates, and functions during apoptosis. Annu Rev Biochem 1999;68:383–424.
31. Shi Y. Mechanisms of caspase activation and inhibition during apoptosis. Mol Cell 2002;9:459–470.
32. Creagh EM, Conroy H, Martin SJ. Caspase-activation pathways in apoptosis and immunity. Immunol Rev 2003;193:10–21.
33. Slee EA, Adrain C, Martin SJ. Serial killers: ordering caspase activation events in apoptosis. Cell Death Differ 1999;6:1067–1074.
34. Slee EA, Adrain C, Martin SJ. Executioner caspase-3, -6, and -7 perform distinct, non-redundant roles during the demolition phase of apoptosis. J Biol Chem 2001;276:7320–7326.
35. Muzio M, Chinnaiyan AM, Kischkel FC, et al. FLICE, a novel FADD-homologous ICE/CED-3-like protease, is recruited to the CD95 (Fas/APO-1) death—inducing signaling complex. Cell 1996;85:817–827.
36. Ashkenazi A, Dixit VM. Death receptors: signaling and modulation. Science 1998;28:1305–1308.
37. Green DR. Apoptotic pathways: the roads to ruin. Cell 1998;94:695–698.
38. Kroemer G, Reed JC. Mitochondrial control of cell death. Nat Med 2000;6:513–519.
39. Green DR, Kroemer G. The pathophysiology of mitochondrial cell death. Science 2004;305:626–629.
40. Gross A, McDonnell JM, Korsmeyer SJ. BCL-2 family members and the mitochondria in apoptosis. Genes Dev 1999;13:1899–1911.
41. van Loo G, Saelens X, van Gurp M, MacFarlane M, Martin SJ, Vandenabeele P. The role of mitochondrial factors in apoptosis: a Russian roulette with more than one bullet. Cell Death Differ 2002;9:1031–1042.
42. Rao RV, Hermel E, Castro-Obregon S, et al. Coupling endoplasmic reticulum stress to the cell death program. Mechanism of caspase activation. J Biol Chem 2001;276:33,869–33,874.
43. Morishima N, Nakanishi K, Takenouchi H, Shibata T, Yasuhiko Y. An endoplasmic reticulum stress-specific caspase cascade in apoptosis. Cytochrome c-independent activation of caspase-9 by caspase-12. J Biol Chem 2002;277:34,287–34,294.
44. Bilsland J, Harper S. Caspases and neuroprotection. Curr Opin Investig Drugs 2002;3:1745–1752.
45. Brunner T, Mueller C. Apoptosis in disease: about shortage and excess. Essays Biochem 2003;39:119–130.
46. Philchenkov A. Caspases: potential targets for regulating cell death. J Cell Mol Med 2004;8:432–444.
47. Kreuter M, Langer C, Kerkhoff C, et al. Stroke, myocardial infarction, acute and chronic inflammatory diseases: caspases and other apoptotic molecules as targets for drug development. Arch Immunol Ther Exp (Warsz) 2004;52:141–155.

48. Hoglen NC, Chen LS, Fisher CD, Hirakawa BP, Groessl T, Contreras PC. Characterization of IDN-6556 (3-[2-(2-tert-butyl-phenylaminooxalyl)-amino]-propionylamino]-4-oxo-5-(2,3,5,6-tetrafluoro-phenoxy)-pentanoic acid): a liver-targeted caspase inhibitor. J Pharmacol Exp Ther 2004;309:634–640.

49. Jomary C, Neal MJ, Jones SE. Characterization of cell death pathways in murine retinal neurodegeneration implicates cytochrome c release, caspase activation, and bid cleavage. Mol Cell NeuroSci 2001;18:335–3346.

50. Kim DH, Kim JA, Choi JS, Joo CK. Activation of caspase-3 during degeneration of the outer nuclear layer in the rd mouse retina. Ophthalmic Res 2002;34:150–157.

51. Sharma AK, Rohrer B. Calcium-induced calpain mediates apoptosis via caspase-3 in a mouse photoreceptor cell line. J Biol Chem 2004;279:35,564–35,572.

52. Yoshizawa K, Nambu H, Yang J, et al. Mechanisms of photoreceptor cell apoptosis induced by N-methyl-N-nitrosourea in Sprague-Dawley rats. Lab Invest 1999;79:1359–1367.

53. Doonan F, Donovan M, Cotter TG. Caspase-independent photoreceptor apoptosis in mouse models of retinal degeneration. J NeuroSci 2003;23:5723–5731.

54. Donovan M, Carmody RJ, Cotter TG. Light-induced photoreceptor apoptosis in vivo requires neuronal nitric-oxide synthase and guanylate cyclase activity and is caspase-3-independent. J Biol Chem 2001;276:23,000–23,008.

55. Zeiss CJ, Neal J, Johnson EA. Caspase-3 in postnatal retinal development and degeneration. Invest Ophthalmol Vis Sci 2004;45:964–970.

56. Yoshizawa K, Kiuchi K, Nambu H, et al. Caspase-3 inhibitor transiently delays inherited retinal degeneration in C3H mice carrying the rd gene. Graefes Arch Clin Exp Ophthalmol 2002;240:214–219.

57. Davidson FF, Steller H. Blocking apoptosis prevents blindness in Drosophila retinal degeneration mutants. Nature 1998;391:587–591.

58. Alloway PG, Howard L, Dolph PJ. The formation of stable rhodopsin-arrestin complexes induces apoptosis and photoreceptor cell degeneration. Neuron 2000;28:129–138.

59. Leist M, Jaattela M. Four deaths and a funeral: from caspases to alternative mechanisms. Nat Rev Mol Cell Biol 2001;2:589–598.

60. Zhu C, Qiu L, Wang X, et al. Involvement of apoptosis-inducing factor in neuronal death after hypoxia-ischemia in the neonatal rat brain. J Neurochem 2003;86:306–317.

61. Zhang X, Chen J, Graham SH, et al. Intranuclear localization of apoptosis-inducing factor (AIF) and large scale DNA fragmentation after traumatic brain injury in rats and in neuronal cultures exposed to peroxynitrite. J Neurochem 2002;82:181–191.

62. Wang X, Zhu S, Drozda M, et al. Minocycline inhibits caspase-independent and -dependent mitochondrial cell death pathways in models of Huntington's disease. Proc Natl Acad Sci USA 2003;100:10,483–10,487.

63. Selznick LA, Zheng TS, Flavell RA, Rakic P, Roth KA. Amyloid beta-induced neuronal death is bax-dependent but caspase-independent. J Neuropathol Exp Neurol 2000;59:271–279.

64. Foghsgaard L, Wissing D, Mauch D, et al. Cathepsin B acts as a dominant execution protease in tumor cell apoptosis induced by tumor necrosis factor. J Cell Biol 2001;153:999–1010.

65. Goll DE, Thompson VF, Li H, Wei W, Cong J. The calpain system. Physiol Rev 2003;83:731–801.

66. Stenson-Cox C, FitzGerald U, Samali A. In the cut and thrust of apoptosis, serine proteases come of age. Biochem Pharmacol 2003;66:1469–1474.

67. Rami A. Ischemic neuronal death in the rat hippocampus: the calpain-calpastatin-caspase hypothesis. Neurobiol Dis 2003;13:75–88.

68. Takeuchi N, Ito H, Namiki K, Kamei A. Effect of calpain on hereditary cataractous rat, ICR/f. Biol Pharm Bull 2001;24:1246–1251.

69. Gafni J, Ellerby LM. Calpain activation in Huntington's disease. J NeuroSci 2002;22: 4842–4849.

70. Donovan M, Cotter TG. Caspase-independent photoreceptor apoptosis in vivo and differential expression of apoptotic protease activating factor-1 and caspase-3 during retinal development. Cell Death Differ 2002;9:1220–1231.

71. Azarian SM, Williams DS. Calpain activity in the retinas of normal and RCS rats. Curr Eye Res 1995;14:731–735.

72. Nakagawa T, Yuan J. Cross-talk between two cysteine protease families. Activation of caspase-12 by calpain in apoptosis. J Cell Biol 2000;150:887–894.

73. Blomgren K, Zhu C, Wang X, et al. Synergistic activation of caspase-3 by m-calpain after neonatal hypoxia-ischemia: a mechanism of "pathological apoptosis"? J Biol Chem 2001;276:10,191–10,198.

74. Goldstein JC, Waterhouse NJ, Juin P, Evan GI, Green DR. The coordinate release of cytochrome c during apoptosis is rapid, complete and kinetically invariant. Nat Cell Biol 2000;2:156–162.

75. Susin SA, Lorenzo HK, Zamzami N, et al. Molecular characterization of mitochondrial apoptosis-inducing factor. Nature 1999;397:441–446.

76. Du C, Fang M, Li Y, Li L, Wang X. Smac, a mitochondrial protein that promotes cytochrome c-dependent caspase activation by eliminating IAP inhibition. Cell 2000;102:33–42.

77. Chen J, Flannery JG, LaVail MM, Steinberg RH, Xu J, Simon MI. bcl-2 overexpression reduces apoptotic photoreceptor cell death in three different retinal degenerations. Proc Natl Acad Sci USA 1996;93:7042–7047.

78. Bennett J, Zeng Y, Bajwa R, Klatt L, Li Y, Maguire AM. Adenovirus-mediated delivery of rhodopsin-promoted bcl-2 results in a delay in photoreceptor cell death in the rd/rd mouse. Gene Ther 1998;5:1156–1164.

79. Joseph RM, Li T. Overexpression of Bcl-2 or Bcl-XL transgenes and photoreceptor degeneration. Invest Ophthalmol Vis Sci 1996;37:2434–2446.

80. Eversole-Cire P, Concepcion FA, Simon MI, Takayama S, Reed JC, Chen J. Synergistic effect of Bcl-2 and BAG-1 on the prevention of photoreceptor cell death. Invest Ophthalmol Vis Sci 2000;41:1953–1961.

81. Eversole-Cire P, Chen J, Simon MI. Bax is not the heterodimerization partner necessary for sustained anti-photoreceptor-cell-death activity of Bcl-2. Invest Ophthalmol Vis Sci 2002;43:1636–1644.

82. He L, Perkins GA, Poblenz AT, et al. Bcl-xL overexpression blocks bax-mediated mitochondrial contact site formation and apoptosis in rod photoreceptors of lead-exposed mice. Proc Natl Acad Sci USA 2003;100:1022–1027.

83. Galat A, Metcalfe SM. Peptidylproline cis/trans isomerases. Prog Biophys Mol Biol 1995;63:67–118.

84. Liu J, Farmer JD Jr, Lane WS, Friedman J, Weissman I, Schreiber SL. Calcineurin is a common target of cyclophilin-cyclosporin A and FKBP-FK506 complexes. Cell 1991; 66:807–815.

85. Cassarino DS, Swerdlow RH, Parks JK, Parker WD Jr, Bennett JP Jr. Cyclosporin A increases resting mitochondrial membrane potential in SY5Y cells and reverses the depressed mitochondrial membrane potential of Alzheimer's disease cybrids. Biochem Biophys Res Commun 1998;248:168–173.

86. Matsuura K, Makino H, Ogawa N. Cyclosporin A attenuates the decrease in tyrosine hydroxylase immunoreactivity in nigrostriatal dopaminergic neurons and in striatal dopamine content in rats with intrastriatal injection of 6-hydroxydopamine. Exp Neurol 1997;146:526–535.

87. Kirkinezos IG, Hernandez D, Bradley WG, Moraes CT. An ALS mouse model with a permeable blood–brain barrier benefits from systemic cyclosporine A treatment. J Neurochem 2004;88:821–826.

88. Yoshimoto T, Siesjo BK. Posttreatment with the immunosuppressant cyclosporin A in transient focal ischemia. Brain Res 1999;839:283–291.

89. Fox DA, Poblenz AT, He L, Harris JB, Medrano CJ. Pharmacological strategies to block rod photoreceptor apoptosis caused by calcium overload: a mechanistic target-site approach to neuroprotection. Eur J Ophthalmol 2003;13 Suppl 3:S44–S56.

90. Rodrigues CM, Fan G, Wong PY, Kren BT, Steer CJ. Ursodeoxycholic acid may inhibit deoxycholic acid-induced apoptosis by modulating mitochondrial transmembrane potential and reactive oxygen species production. Mol Med 1998;4:165–178.

91. Keene CD, Rodrigues CM, Eich T, Chhabra MS, Steer CJ, Low WC. Tauroursodeoxycholic acid, a bile acid, is neuroprotective in a transgenic animal model of Huntington's disease. Proc Natl Acad Sci USA 2002;99:10,671–10,676.

92. Do VT, Nickerson JM, Boatright JH. Prevention of Apoptosis in an RPE Carcinoma Cell Line by Bile Acids. Fort Lauderdale, FL: Association for Research in Vision and Opthalmology, 2003.

93. Akao Y, Maruyama W, Shimizu S, et al. Mitochondrial permeability transition mediates apoptosis induced by N-methyl(R)salsolinol, an endogenous neurotoxin, and is inhibited by Bcl-2 and rasagiline, N-propargyl-1(R)-aminoindan. J Neurochem 2002;82:913–923.

94. Zhu S, Stavrovskaya IG, Drozda M, et al. Minocycline inhibits cytochrome c release and delays progression of amyotrophic lateral sclerosis in mice. Nature 2002;417:74–78.

95. Steinberg RH, Flannery JG, Naash M, Oh P, Matthes MT, Yasumura D. Transgenic rat models of inherited retinal degeneration caused by mutant opsin genes. Association for Research in Vision and Opthalmology. Investigative Opthalmology and Visual Science Fort Lauderdale 1996;37:S698, abstract no. 3190.

96. Dykens JA, Carroll AK, Wiley S, et al. Photoreceptor preservation in the S334ter model of retinitis pigmentosa by a novel estradiol analog. Biochem Pharmacol 2004; 68:1971–1984.

97. Edward DP, Lam TT, Shahinfar S, Li J, Tso MO. Amelioration of light-induced retinal degeneration by a calcium overload blocker. Flunarizine. Arch Ophthalmol 1991; 109:554–562.

98. He L, Poblenz AT, Medrano CJ, Fox DA. Lead and calcium produce rod photoreceptor cell apoptosis by opening the mitochondrial permeability transition pore. J Biol Chem 2000;275:12,175–12,184.

99. Fox DA, Poblenz AT, He L. Calcium overload triggers rod photoreceptor apoptotic cell death in chemical-induced and inherited retinal degenerations. Ann NY Acad Sci 1999;893:282–285.

100. Bowes C, Li T, Danciger M, Baxter LC, Applebury ML, Farber DB. Retinal degeneration in the rd mouse is caused by a defect in the beta subunit of rod cGMP-phosphodiesterase. Nature 1990;347:677–680.

101. Frasson M, Sahel JA, Fabre M, Simonutti M, Dreyfus H, Picaud S. Retinitis pigmentosa: rod photoreceptor rescue by a calcium-channel blocker in the rd mouse. Nat Med 1999;5:1183–1187.

102. Pawlyk BS, Li T, Scimeca MS, Sandberg MA, Berson EL. Absence of photoreceptor rescue with D-cis-diltiazem in the rd mouse. Invest Ophthalmol Vis Sci 2002;43:1912–1915.

103. Bush RA, Kononen L, Machida S, Sieving PA. The effect of calcium channel blocker diltiazem on photoreceptor degeneration in the rhodopsin Pro23His rat. Invest Ophthalmol Vis Sci 2000;41:2697–2701.

104. Takano Y, Ohguro H, Dezawa M, et al. Study of drug effects of calcium channel blockers on retinal degeneration of rd mouse. Biochem Biophys Res Commun 2004;313: 1015–1022.

105. Pearce-Kelling SE, Aleman TS, Nickle A, et al. Calcium channel blocker D-cis-diltiazem does not slow retinal degeneration in the PDE6B mutant rcd1 canine model of retinitis pigmentosa. Mol Vis 2001;7:42–47.

106. Yamazaki H, Ohguro H, Maeda T, et al. Preservation of retinal morphology and functions in royal college surgeons rat by nilvadipine, a Ca(2+) antagonist. Invest Ophthalmol Vis Sci 2002;43:919–926.

107. Yu DY, Cringle SJ. Oxygen distribution and consumption within the retina in vascularised and avascular retinas and in animal models of retinal disease. Prog Retin Eye Res 2001;20:175–208.

108. Carmody RJ, McGowan AJ, Cotter TG. Reactive oxygen species as mediators of photoreceptor apoptosis in vitro. Exp Cell Res 1999;248:520–530.

109. Sanvicens N, Gomez-Vicente V, Masip I, Messeguer A, Cotter TG. Oxidative stress-induced apoptosis in retinal photoreceptor cells is mediated by calpains and caspases and blocked by the oxygen radical scavenger CR-6. J Biol Chem 2004;279:39,268–39,278.

110. Cideciyan AV, Hood DC, Huang Y, et al. Disease sequence from mutant rhodopsin allele to rod and cone photoreceptor degeneration in man. Proc Natl Acad Sci USA 1998;95:7103–7108.

111. Organisciak DT, Darrow RM, Barsalou L, Kutty RK, Wiggert B. Susceptibility to retinal light damage in transgenic rats with rhodopsin mutations. Invest Ophthalmol Vis Sci 2003;44:486–492.

112. Ranchon I, LaVail MM, Kotake Y, Anderson RE. Free radical trap phenyl-N-tert-butylnitrone protects against light damage but does not rescue P23H and S334ter rhodopsin transgenic rats from inherited retinal degeneration. J NeuroSci 2003;23:6050–6057.

113. Tanito M, Masutani H, Nakamura H, Ohira A, Yodoi J. Cytoprotective effect of thioredoxin against retinal photic injury in mice. Invest Ophthalmol Vis Sci 2002;43:1162–1167.

114. Tanito M, Kwon YW, Kondo N, et al. Cytoprotective effects of geranylgeranylacetone against retinal photooxidative damage. J NeuroSci 2005;25:2396–2404.

115. Shen J, Yan, X, Dong A, et al. Oxidative damage is a potential cause of cone cell death in retinitis pigmentosa. J Cell Physiol 2005;203:457–464.

116. Petters RM, Alexander CA, Wells KD, et al. Genetically engineered large animal model for studying cone photoreceptor survival and degeneration in retinitis pigmentosa. Nat Biotechnol 1997;15:965–970.

117. Yu DY, Cringle SJ, Su EN, Yu PK. Intraretinal oxygen levels before and after photoreceptor loss in the RCS rat. Invest Ophthalmol Vis Sci 2000;41:3999–4006.

118. Yu DY, Cringle S, Valter K, Walsh N, Lee D, Stone J. Photoreceptor death, trophic factor expression, retinal oxygen status, and photoreceptor function in the P23H rat. Invest Ophthalmol Vis Sci 2004;45:2013–2019.

119. Crespo D, O'Leary DD, Cowan WM. Changes in the numbers of optic nerve fibers during late prenatal and postnatal development in the albino rat. Brain Res 1985;351:129–134.

120. Faktorovich EG, Steinberg RH, Yasumura D, Matthes MT, LaVail MM. Photoreceptor degeneration in inherited retinal dystrophy delayed by basic fibroblast growth factor. Nature 1990;347:83–86.

121. LaVail MM, Unoki K, Yasumura D, Matthes MT, Yancopoulos GD, Steinberg RH. Multiple growth factors, cytokines, and neurotrophins rescue photoreceptors from the damaging effects of constant light. Proc Natl Acad Sci USA 1992;89:11,249–11,253.

122. Akimoto M, Miyatake S, Kogishi J, et al. Adenovirally expressed basic fibroblast growth factor rescues photoreceptor cells in RCS rats. Invest Ophthalmol Vis Sci 1999;40: 273–279.

123. Uteza Y, Rouillot JS, Kobetz A, et al. Intravitreous transplantation of encapsulated fibroblasts secreting the human fibroblast growth factor 2 delays photoreceptor cell degeneration in Royal College of Surgeons rats. Proc Natl Acad Sci USA 1999;96: 3126–3131.

124. Lau D, McGee LH, Zhou S, et al. Retinal degeneration is slowed in transgenic rats by AAV-mediated delivery of FGF-2. Invest Ophthalmol Vis Sci 2000;41:3622–3633.

125. Cayouette M, Gravel C. Adenovirus-mediated gene transfer of ciliary neurotrophic factor can prevent photoreceptor degeneration in the retinal degeneration (rd) mouse. Hum Gene Ther 1997;8:423–430.

126. LaVail MM, Yasumura D, Matthes MT, et al. Protection of mouse photoreceptors by survival factors in retinal degenerations. Invest Ophthalmol Vis Sci 1998;39:592–602.

127. Cayouette M, Behn D, Sendtner M, Lachapelle P, Gravel C. Intraocular gene transfer of ciliary neurotrophic factor prevents death and increases responsiveness of rod photoreceptors in the retinal degeneration slow mouse. J NeuroSci 1998;18:9282–9293.

128. Caffe AR, Soderpalm AK, Holmqvist I, van Veen T. A combination of CNTF and BDNF rescues rd photoreceptors but changes rod differentiation in the presence of RPE in retinal explants. Invest Ophthalmol Vis Sci 2001;42:275–282.

129. Liang FQ, Aleman TS, Dejneka NS, et al. Long-term protection of retinal structure but not function using RAAV.CNTF in animal models of retinitis pigmentosa. Mol Ther 2001;4:461–472.

130. Bok D, Yasumura D, Matthes MT, et al. Effects of adeno-associated virus-vectored ciliary neurotrophic factor on retinal structure and function in mice with a P216L rds/peripherin mutation. Exp Eye Res 2002;74:719–735.

131. Schlichtenbrede FC, da Cruz L, Stephens C, et al. Long-term evaluation of retinal function in Prph2Rd2/Rd2 mice following AAV-mediated gene replacement therapy. J Gene Med 2003;5:757–764.

132. Bush RA, Lei B, Tao W, et al. Encapsulated cell-based intraocular delivery of ciliary neurotrophic factor in normal rabbit: dose-dependent effects on ERG and retinal histology. Invest Ophthalmol Vis Sci 2004;45:2420–2430.

133. Song Y, Zhao L, Tao W, Laties AM, Luo Z, Wen R. Photoreceptor protection by cardiotrophin-1 in transgenic rats with the rhodopsin mutation s334ter. Invest Ophthalmol Vis Sci 2003;44:4069–4075.

134. Frasson M, Picaud S, Leveillard T, et al. Glial cell line-derived neurotrophic factor induces histologic and functional protection of rod photoreceptors in the rd/rd mouse. Invest Ophthalmol Vis Sci 1999;40:2724–2734.

135. McGee Sanftner LH, Abel H, Hauswirth WW, Flannery JG. Glial cell line derived neurotrophic factor delays photoreceptor degeneration in a transgenic rat model of retinitis pigmentosa. Mol Ther 2001;4:622–629.

136. Mohand-Said S, Hicks D, Leveillard T, Picaud S, Porto F, Sahel JA. Rod-cone interactions: developmental and clinical significance. Prog Retin Eye Res 2001;20:451–467.

137. Mohand-Said S, Deudon-Combe A, Hicks D, et al. Normal retina releases a diffusible factor stimulating cone survival in the retinal degeneration mouse. Proc Natl Acad Sci USA 1998;95:8357–8362.

138. Chader GJ. Animal models in research on retinal degenerations: past progress and future hope. Vision Res 2002;42:393–399.

139. Dejneka NS, Rex TS, Bennett J. Gene therapy and animal models for retinal disease. Dev Ophthalmol 2003;37:188–198.

140. Pacione LR, Szego MJ, Ikeda S, Nishina PM, McInnes RR. Progress toward understanding the genetic and biochemical mechanisms of inherited photoreceptor degenerations. Annu Rev NeuroSci 2003;26:657–700.

141. Grimm C, Wenzel A, Hafezi F, Yu S, Redmond TM, Reme CE. Protection of Rpe65-deficient mice identifies rhodopsin as a mediator of light-induced retinal degeneration. Nat Genet 2000;25:63–66.

142. Wenzel A, Reme CE, Williams TP, Hafezi F, Grimm C. The Rpe65 Leu450Met variation increases retinal resistance against light-induced degeneration by slowing rhodopsin regeneration. J NeuroSci 2001;21:53–58.

143. Noell WK, Albrecht R. Irreversible effects on visible light on the retina: role of vitamin A. Science 1971;172:76–79.

144. Rich KA, Zhan Y, Blanks JC. Aberrant expression of c-Fos accompanies photoreceptor cell death in the rd mouse. J Neurobiol 1997;32:593–612.

145. Hafezi F, Steinbach JP, Marti A, et al. The absence of c-fos prevents light-induced apoptotic cell death of photoreceptors in retinal degeneration in vivo. Nat Med 1997;3:346–349.

146. Hobson AH, Donovan M, Humphries MM, et al. Apoptotic photoreceptor death in the rhodopsin knockout mouse in the presence and absence of c-fos. Exp Eye Res 2000;71:247–254.

147. Hao W, Wenzel A, Obin MS, et al. Evidence for two apoptotic pathways in light-induced retinal degeneration. Nat Genet 2002;32:254–260.

148. Kaldi I, Dittmar M, Pierce P, Anderson RE. L-NAME protects against acute light damage in albino rats, but not against retinal degeneration in P23H and S334ter transgenic rats. Exp Eye Res 2003;76:453–461.

149. Grimm C, Wenzel A, Stanescu D, et al. Constitutive overexpression of human erythropoietin protects the mouse retina against induced but not inherited retinal degeneration. J NeuroSci 2004;24:5651–5658.

150. Heckenlively JR, Rodriguez JA, Daiger SP. Autosomal dominant sectoral retinitis pigmentosa. Two families with transversion mutation in codon 23 of rhodopsin. Arch Ophthalmol 1991;109:84–91.

151. Naash ML, Peachey NS, Li ZY, et al. Light-induced acceleration of photoreceptor degeneration in transgenic mice expressing mutant rhodopsin. Invest Ophthalmol Vis Sci 1996;37:775–782.

152. Cideciyan AV, Jacobson SG, Aleman TS, et al. In vivo dynamics of retinal injury and repair in the rhodopsin mutant dog model of human retinitis pigmentosa. Proc Natl Acad Sci USA 2005;102:5233–5238.

153. Haider NB, Ikeda A, Naggert JK, Nishina PM. Genetic modifiers of vision and hearing. Hum Mol Genet 2002;11:1195–1206.

154. Weleber RG, Carr RE, Murphey WH, Sheffield VC, Stone EM. Phenotypic variation including retinitis pigmentosa, pattern dystrophy, and fundus flavimaculatus in a single family with a deletion of codon 153 or 154 of the peripherin/RDS gene. Arch Ophthalmol 1993;111:1531–1542.

155. Kajiwara K, Berson EL, Dryja TP. Digenic retinitis pigmentosa due to mutations at the unlinked peripherin/RDS and ROM1 loci. Science 1994;264:1604–1608.

156. Jacobson SG, Cideciyan AV, Iannaccone A, et al. Disease expression of RP1 mutations causing autosomal dominant retinitis pigmentosa. Invest Ophthalmol Vis Sci 2000;41:1898–1908.

157. Kim RY, Fitzke FW, Moore AT, et al. Autosomal dominant retinitis pigmentosa mapping to chromosome 7p exhibits variable expression. Br J Ophthalmol, 1995;79:23–27.

158. Keen TJ, Hims MM, McKie AB, et al. Mutations in a protein target of the Pim-1 kinase associated with the RP9 form of autosomal dominant retinitis pigmentosa. Eur J Hum Genet 2002;10:245–249.

159. Maita H, Kitaura H, Keen TJ, Inglehearn CF, Ariga H, Iguchi-Ariga SM. PAP-1, the mutated gene underlying the RP9 form of dominant retinitis pigmentosa, is a splicing factor. Exp Cell Res 2004;300:283–296.

160. Hummel KP, Coleman DL, Lane PW. The influence of genetic background on expression of mutations at the diabetes locus in the mouse. I. C57BL-KsJ and C57BL-6J strains. Biochem Genet 1972;7:1–13.

161. Ikeda S, Hawes NL, Chang B, Avery CS, Smith RS, Nishina PM. Severe ocular abnormalities in C57BL/6 but not in 129/Sv p53-deficient mice. Invest Ophthalmol Vis Sci 1999;40:1874–1878.

162. Humphries MM, Kiang S, McNally N, et al. Comparative structural and functional analysis of photoreceptor neurons of Rho-/- mice reveal increased survival on C57BL/6J in comparison to 129Sv genetic background. Vis Neurosci 2001;18:437–443.

163. Danciger M, Matthes MT, Yasamura D, et al. A QTL on distal chromosome 3 that influences the severity of light-induced damage to mouse photoreceptors. Mamm Genome 2000;11:422–427.

164. Ikeda A, Zheng QY, Rosenstiel P, et al. Genetic modification of hearing in tubby mice: evidence for the existence of a major gene (moth1) which protects tubby mice from hearing loss. Hum Mol Genet 1999;8:1761–1767.

165. Ikeda A, Naggert JK, Nishina PM. Genetic modification of retinal degeneration in tubby mice. Exp Eye Res 2002;74:455–461.

166. Heckenlively JR, Chang B, Peng C, Hawes NL, Roderick TH. Variable expressivity of rd-3 retinal degeneration dependent on back-ground strain. In: Hollyfield JG, Anderson RE, LaVail MM, ed. Retinal Degeneration. New York: Plenum Press, 1993:273–280.

167. Danciger M, Lyon J, Worrill D, et al. New retinal light damage QTL in mice with the light-sensitive RPE65 LEU variant. Mamm Genome 2004;15:277–283.

168. Sung CH, Schneider BG, Agarwal N, Papermaster DS, Nathans J. Functional heterogeneity of mutant rhodopsins responsible for autosomal dominant retinitis pigmentosa. Proc Natl Acad Sci USA 1991;88:8840–8844.

169. Sung CH, Davenport CM, Nathans J. Rhodopsin mutations responsible for autosomal dominant retinitis pigmentosa. Clustering of functional classes along the polypeptide chain. Proc Natl Acad Sci USA 1993;268:26,645–26,649.

170. Wenzel A, Grimm C, Samardzija M, Reme C.E. Molecular mechanisms of light-induced photoreceptor apoptosis and neuroprotection for retinal degeneration. Prog Retin Eye Res 2005;24:275–306.

171. Doonan F, Donovan M, Cotter TG. Activation of multiple pathways during photoreceptor apoptosis in the rd mouse. Invest Ophthalmol Vis Sci 2005;46(10):3530–3538.

172. Lohr HR, Kuntchithapautham K, Sharma AK, Rohrer B. Multiple, parallel cellular suicide mechanisms participate in photoreceptor cell death. Exp Eye Res 2006;83(2):380–389.

173. Buch PK, Maclaren RE, Duran Y, Balaggan KS, Macneil A, Schlichtenbrede FC, Smith AJ, Ali RR. In Contrast to AAV-Mediated Cntf Expression, AAV-Mediated Gdnf Expression Enhances Gene Replacement Therapy in Rodent Models of Retinal Degeneration. Mol Ther; 2006 (in press).

174. Samardzija M, Wenzel A, Naash M, Reme CE, Grimm C. Rpe65 as a modifier gene for inherited retinal degeneration. Eur J Neurosci 2006;23(4):1028–1034.

17

Cell-Based Therapies to Restrict the Progress of Photoreceptor Degeneration

Raymond D. Lund, PhD and Shaomei Wang, PhD

Contents

INTRODUCTION

The first studies showing the potential of retinal cell transplantation to alleviate the progress of blindness in an animal model of retinal disease, the Royal College of Surgeons (RCS) rat, were focused on the fact that in this animal there was a known defect in the retinal pigment epithelium (RPE) that resulted in secondary loss of photoreceptors. It seemed logical to effect cell replacement, by introducing into the subretinal space, normal RPE cells to replace the affected ones. A series of studies showed that the procedure did indeed rescue photoreceptors, presumably by replacing the deficiently functioning cells with ones that function normally. However, RPE cells have many functions, including phagocytosis of outer segments (OS), visual pigment recycling, maintenance of Bruch's membrane, and transport of materials in and out of the retina *(1–5)*. How many of these several roles are replicated by the grafted cells was not explored in the early studies—indeed very little other than photoreceptor rescue *per se* was ever measured and, in the absence of suitable labels, it was not even clear how long the donor cells actually survived. Because it was thought that in age-related macular degeneration (AMD), dysfunction, or depletion of RPE cells might be a cause of the photoreceptor loss, the successes in the laboratory were quickly applied to patients with advanced AMD. Irrespective of the causes of AMD, one clear role for RPE cell transplantation was after removing choroidal neovascularization

From: *Ophthalmology Research: Retinal Degenerations: Biology, Diagnostics, and Therapeutics*
Edited by: J. Tombran-Tink and C. J. Barnstable © Humana Press Inc., Totowa, NJ

(CNV) membranes that invade the space between Bruch's membrane and the RPE, RPE cells would inevitably be removed too. Repopulation with new cells might correct that deficiency. Results were disappointing: there was suggestion of graft rejection and although little hope of functional improvement might have been expected in the advanced stage patients that formed the subjects of these investigations, attempts to demonstrate visual improvement were equivocal at best. What these studies did show, however, was not that the approach was flawed, but rather that the steps that needed to be taken to achieve a viable clinical treatment are many. These include, at the laboratory level, choice of the optimal donor cell type and age, defining exactly what grafts are doing at the cellular and molecular level, effects on inner retinal and vascular integrity, questions relating to the condition of the underlying Bruch's membrane, safety, longevity of effect, controlling immune and inflammatory events that might compromise graft viability, and functional efficacy. In translation to clinic, further issues relate to the need for homology between rodent models and human disease states, the intermediate steps between rodent and human (is it necessary for example to have a larger animal model?), the suitable patient population (is AMD the right starting point?), the stage in the progress of the disease for any hope of vision rescue, good objective functional tests of efficacy, and proper controls for sham effects associated with surgery. Needless to say, any future clinical studies should be done under conditions of a formally phased clinical trial structure.

In this review, we will summarize (1) progress in the field of cell transplantation to rescue the photoreceptors from progressive degeneration and (2) the evolution in thinking as this field has developed from the initial experiments. Given the range of possibilities presented by various donor cells, and the implication that "transplantation" is restricted to homologous cell replacement, it is perhaps better to use the term "cell-based therapy" to characterize this field of endeavor. In reviewing work that has or should have been done, we are not so much exposing the shortcomings of those working in the field (including our own studies), but rather trying to emphasize the complexity of the subject: how so few people have developed the approach to a stage in which several hundred patients have received various cell-based therapies and how many questions that seemed unimportant in the early experiments have emerged and need answering.

ANIMAL MODELS

The majority of studies involved in controlling the progress of degeneration has relied on rodents, and in particular the RCS rat. In this animal, there is rapid loss of photoreceptors between 21 and 90 d of age (Fig. 1B) with a slower loss over many months, and some cells still being present at 1 yr of age *(6–8)*. The defect is associated with the inability of the RPE cells to phagocytose, shed OS material at a normal rate, because of a mutation in the *Mertk* gene *(9)*. The defect can be partially corrected by gene therapy by introducing a normal copy *(10)* and by delivery of growth factors *(11,12)*, as well as by introduction of suitable cells in the subretinal space *(13–17)*. These procedures have generally been done at an early stage in the progress of degeneration. Secondary changes occur at a relatively early stage, affecting the cells of the inner retina *(18,19)*: these are first seen even before there is significant loss of photoreceptors and fall into three main phases. The first phase

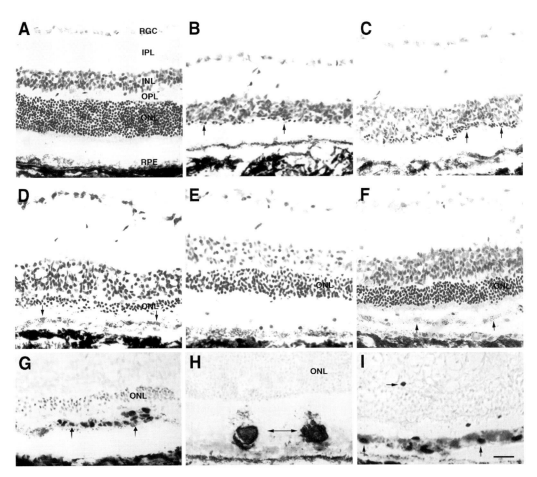

Fig. 1. (A–F) Retinal sections stained with cresyl violet. **(A)** Normal retina at P90 showing retinal layers; **(B)** RCS retina at P120, with only a single layer of cells remaining in the ONL (arrows); **(C)** RCS retina with sham injection at P120, showing local ONL rescue (arrows); **(D)** RCS retina with ARPE-19 graft at P120, showing three to four layers of rescued cells in the ONL, arrows point to an extra layer of cells with pigment granules: this extra layer can be stained by human specific antibody (*see* **G**), indicating it is composed of donor cells; **(E)** RCS retina with human Schwann cell graft at P120, showing four to six layers of rescued cells in the ONL; **(F)** RCS retina with forebrain progenitor cell graft at P150, showing six to eight layers of rescued cells in the ONL, arrows point to an extra layer of cells with pigment granules, which can be revealed by human specific antibody (arrows in **I**); **(G–I)** Donor cells labeled with human nuclear marker **(G,I)** or human Schwann cell marker-P75 **(H)**, showing donor cells after ARPE-19 **(G)**, human Schwann cell **(H)** and forebrain-derived progenitor **(I)** cell injections. Note in **I** that a few donor cells are found within the retina: these cells express nestin. Scale bar = 20 μm.

is of retraction of rod bipolar dendrites, reduction in the density of synaptic markers in the outer plexiform layer, which is evident by postnatal day 21 (P21) and progresses over 3 mo. Subsequent atrophic events involve neurons distributing in the inner plexiform layer. The second phase involves sprouting of processes of bipolar cells, horizontal cells, and Müller cells into the debris zone formed by shed, but undigested OS material. The sprouting subsides as the debris zone disappears. At late stages, vascular disorders develop associated with the formation of vascular complexes adjacent to Bruch's membrane and with migration of RPE cells into the retina: these result in a further set of events including major laminar disruption of the inner retina, ligation of ganglion cell axons with loss of ganglion cells over segments of the retina *(20,21)*, and migration of neurons from their normal location in the inner retina. The events described here relate to the RCS rat, but there are many studies in a range of animals showing that, although details may differ, inner retinal changes are the common consequence of photoreceptor degeneration *(22–28)*.

Other possible rat models include the transgenic P23H and S334ter rats. With respect to cell-based therapies, the latter has been studied more for replacement of lost photoreceptors rather than rescue studies *(29)*. A wide range of mouse models of photoreceptor degenerative diseases exists ranging from homologous models of retinitis pigmentosa (RP) to transgenics homologous to specific forms of RP *(25,30–34)*. Although these are very valuable for genetic analysis, cell biology of disease, and gene therapy studies, they can be problematic for transplantation experiments in which cells are introduced into the subretinal space because of the small size of the eye, in which minor distortion can significantly affect the optics of the eye and hence visual function. Furthermore, their poor vision, even compared with rats, makes it harder to titrate visual outcomes in treated vs control animals. Nevertheless, transplants have been achieved in mice *(35,38)* and for AMD in particular, two transgenic mice presently offer the only animal models of this disease *(39,40)*, other than a recent report of AMD primates *(41)*.

Other large animals, such as cats and dogs with defined retinal degeneration conditions and transgenic pigs, are available *(42–46)*, but have been used very little for examination of cell-based therapies with the intent of slowing the progress of photoreceptor loss.

MEASURES OF EFFICACY OF TREATMENT REGIMENS

The predominant measure of success in cell-based therapies, as well as other approaches designed to rescue photoreceptors has depended on the thickness of the outer nuclear layer (ONL). Although this is an important baseline measure, it does not take into account the organization and functioning of the OS (in the RCS rat defects here precede loss of photoreceptor cell bodies by more than 1 wk *[47–50]*, including the intimate relation between outer segments and RPE). In rodents, the animal used in most of studies, rods comprise more than 95% of photoreceptors *(51,52)* and, as a result, any count of the ONL will inevitably focus attention toward rod viability. In human studies, the continued efficacy of cones is perhaps the most important goal. An additional concern is the role of changes in the inner retina and in the retinal vasculature—how are these affected by photoreceptor loss and how might cell-based therapies affect these independent of or in concert with photoreceptor rescue, especially if the donor cells may deliver growth factors? This is especially pertinent because it has recently been shown *(53)* that receptors for ciliary neurotrophic factor (CNTF), a potent factor for promoting photoreceptor rescue, are widely expressed among cells of the retina (although enigmatically in rodents where neurotrophic effects on photoreceptors have been demonstrated, they are not found in these cells). For cell-based therapies, it is clearly important to know where the donor cells are located, how they relate to cells of the host retina, what is the exact role they play, and indeed how many survive and for how long. Although morphology is clearly important, perhaps of more significance is how vision is affected either during degeneration or rescue. There are several functional measures that have been used in recent work.

First is electroretinogram (ERG), which focuses on the ability of the retina to develop electrical responses to visual signals and relay them through the retina. Full-field ERG to single flashes can tell much about the integrity of the phototransduction process within the photoreceptors themselves and the relays through the retina. Further analysis can elucidate the relative efficacy of rod and cone pathways within the retina using double flash or cone-specific stimuli. Somewhat surprisingly, penetrating analysis of ERG responses has until recently *(47,54)* received little attention in transplant studies. However, an issue that may compromise the value of ERG recordings is that when cells are introduced into the subretinal space, they do not usually protect the whole retinal area: full-field ERG will average the response from protected and unprotected retina, so will not give a good measure of the magnitude of focal rescue. This may be achieved better by multifocal ERG, but the technique can be quite difficult to apply in small rodents *(55,56)*. An alternative is to measure physiological responsiveness across the visual field. This is best done recording from regions of the central nervous system, such as the superior colliculus, where there is a coherent map of the visual world. Such an approach has the additional advantage that substrates of centrally mediated vision can also be assessed. Two measures are very useful. First light and dark adaptation responses give indication of rod and cone function under physiological conditions *(57)*. Second, recording threshold responses under mesopic conditions produces data that compare closely with those collected from a Humphrey perimeter and show the level of protection across the retina *(58,59)*. The superior colliculus is particularly valuable as a site for collecting such data because the retinal input is relatively unfiltered and there is a well-ordered spatial map of the contralateral retina. In addition to these recordings, by characterizing single-unit responses in the visual cortex, it is possible to show at the cellular level the tuning properties of individual neurons that collectively provide the substrates of conscious vision *(60,61)*. One further physiological assessment that is important, but has not been explored rigorously *(62,63)*, is to examine how response properties of single retinal ganglion cells are affected by photoreceptor degeneration and rescue. Such information is important to see what part the inner retinal changes might play in modulating retinal output functions and whether centrally mediated visual functions might compensate for deficiencies in the retina.

Visual performance in rodents has been measured two different ways. One involves a two-choice discrimination, presenting on one display panel stripes of varying spatial frequencies against a gray panel to achieve an acuity threshold. Although graft efficacy has been demonstrated using a terrestrial two-choice test box *(64)*, significantly higher resolution can be achieved with a visual water task *(65)*, a combination of the standard two-choice method, and the Morris water maze. With that method, the progress of visual loss and its rescue with transplants can be accurately assessed over time *(66,67)*.

Although data can be collected that gives acuity measures that correlate with optimal performance using other more complex methods of data acquisition *(68)*, the approach does require careful management and is time consuming, as it requires animals to be shaped up to perform the test before experimental data can be collected. An alternative method, which depends on an automatic visual response, is that of head tracking to a moving stripe pattern. In its basic form, an animal is placed in the center of a rotating drum and, in the case of rodents, the movement of the head measured in synchrony with the stripe movement *(16,69)*. Although the approach has been used to good effect in a number of transplant studies, the threshold acuity recorded is low compared to the visual water task; the data is noisy and the responses deteriorate to undetectable within a limited time frame. A development of the methodology, the Optomotry testing method, uses a computer display of a virtual drum instead of the rotating drum, which therefore allows for a quicker assessment of acuity, and the ability with the same approach to collect contrast sensitivity data *(70)*. Even unoperated RCS rats can give positive data for prolonged periods and variance within and among animals is considerably reduced *(71)*. The one concern with the use of this method is that threshold acuity levels are still somewhat below those recorded in the visual water test (0.6 compared with 1.0 cycle/degree in nondystrophic rats). Recent work *(72)* has suggested that the optomotor response may require intact rods for execution of the response and, as such, may be measuring a different aspect of vision from the visual water task. Nevertheless, its value is in providing a quick test for assessing relative changes in spatial resolution, but because it certainly does not record absolute acuity, the figures obtained should be referred to as "relative acuity."

EXPERIMENTAL PROTOCOL

In a typical study, cells are isolated and either transplanted into the subretinal space relatively soon thereafter or stored for later treatment. For rodents, most such studies have involved introduction of cells as a suspension *(16,17,64)*. In other studies, cells have been introduced into the vitreous either in a loose suspension *(73–75)* or in an encapsulated device *(76,77)*. The former approach has the inherent problems in that cells may spread to cover the surface of the lens, so limiting vision. And, when cells do invade the retina, they may come to rest preferentially in the inner retina rather than where they are needed in the region of photoreceptors and RPE. The encapsulated delivery method avoids the consequences of possible immunological mismatch and provides a level of protection against abnormal donor cell behavior: but the cells are still some distance from where they are needed. However this approach has proven effective in the rcd-1 canine model of RP in which the ARPE-19 RPE cell line transfected to produce additional CNTF rescues photoreceptors *(76)*. Recently, a stage 1 clinical trial has begun to explore this strategy in humans *(78)*. Introduction of cells to the subretinal space is the preferable way of placement close to degenerating photoreceptors or dysfunctional RPE, as long as the safety concerns can be addressed. The procedure does create a retinal detachment, which in itself can be problematic, but this normally resolves within a few days. Early animal studies took no precaution to protect against immune disparity between donor and host, relying on the known "immunological privilege" afforded the eye. However, because no immune consequences were reported, most of these studies used relatively short survival times. In later work, it became clear that graft cells can undergo rejection *(79–82)*, but that immune-related loss was much less florid than that seen after introducing foreign cells outside the eye *(83)*. It became clear that even with allografts *(84)*, it might be necessary to provide protection using an immunosuppressant, sometime in combination with an anti-inflammatory agent, but even this may not always be sufficient in some circumstances *(85,86)*. Although syngeneic or autologous grafts are possible in a number of situations, it is only recently that longevity of survival and efficacy of such grafts has been examined in the absence of immunosuppressive regimens *(87)*.

Experimental studies introducing cells to preserve photoreceptors are usually designed for a "best case scenario," involving introduction of cells early in the course of degeneration: although a suitable model for clinical application would ideally require introduction of cells at a later stage in the progress of the disease, this has yet to be explored systematically.

DONOR CELLS

Freshly Harvested RPE Cells

Animal Studies

The transplantation of RPE into the subretinal space to slow the progress of photoreceptor loss was first examined by Li and his colleagues in the RCS rat *(13,88)* and by Gouras' group in a series of studies in a range of animals including RCS rats *(15,63)*, and Abyssinian cats (89); normal rabbits *(90,91)* and monkeys *(92,93)* were studied to examine graft survival. In the work on RCS rats, it was found that the normal photoreceptor loss that was symptomatic of this animal was limited considerably at least up to the time points of 3 mo. Localization of donor cells was not possible in that work because of the absence of adequate donor cell markers. Subsequent work *(83)* indicated that if survival times were extended, allogeneic grafts were no longer effective in rescuing photoreceptors by 4.25-mo survival without immunosuppression.

Seaton and Turner *(94)* also hinted that secondary vascular changes might be halted by the transplantation. Unfortunately, they grafted cells into the dorsal retina and used the ventral retina as a control. There is an asymmetry in the development of vascular complexes such that at the time they sacrificed their animals, these complexes would not yet have developed in the dorsal retina *(21)*, although they would have been very evident ventrally. More recent work has indicated that the grafts may indeed limit the progress of the secondary vascular changes not only in the immediate region of the grafted cells, but over a large area of retina *(18,95)*.

Human Studies

The success of subretinal grafts in animals prompted exploration of clinical application. This has taken three forms

1. RPE cells harvested from a donor eye were introduced to correct the deficit encountered in advanced AMD. The rationale for this is twofold. The first is the presumption that in AMD the primary affected cell is the RPE cell and, therefore, simple cell replacement might

correct the defect. This expectation might have to be revised in the light of recent work, which shows that in more than 40% of AMD patients, complement factor H mutations involving choroidal vasculature can be identified *(96–98)* and, therefore, the RPE cells may be secondarily involved at best. Second, in removing CNV membranes to alleviate the damage associated with "wet" AMD, RPE cells will inevitably also be removed, so repopulation with new cells might correct that problem *(99–102)* and serve as an adjunct to therapies managing the neovascular events. Results have been somewhat disappointing. There are many explanations. The first is that there is a clear indication of immune rejection *(103)*. In addition, mature cells, unlike immature ones, do not attach easily to Bruch's membrane without secondary manipulations *(104,105)*. Furthermore, a damaged Bruch's membrane is much less attractive for optimal attachment than an intact one *(106,107)*. Beyond this is the possibility that symptoms of AMD resulting from different etiologies might respond differently to cell supplementation or that the level of damage to the retina is too severe to expect effective tissue restoration and recovery of visual function.

2. Removal of cells from a peripheral region of the retina and introduction of these cells to a more central region might avoid immune complications. An exhaustive recent study using cells taken from a peripheral region of the same eye did show either a trend or significant improvement over controls when monitored using several functional indices *(108)*. The failure to get more dramatic improvement may with the exception of potential immune considerations be similar to those described here previously, but these are not necessarily unresolvable issues.

3. Retinal translocation, apposing the retina to a new region of RPE away from the macular might also serve to improve vision *(109)*. Although not technically involving RPE transplantation, the translocation should effectively prove the principal that defective RPE can be substituted and improve vision. In best cases, substantial improvement in visual performance has been recorded, but sustaining efficacy does present a problem *(110,111)*.

Alternative Sources of RPE Cells

Even if it could be shown that fresh RPE is effective in rescuing photoreceptors from degeneration, the previous discussion indicates that the best results are likely to be obtained with immature cells: logistic and ethical concerns, however, are likely to limit availability of such cells. It may be possible to manipulate the surface molecular properties of more mature cells and this could provide a suitable solution *(112)*. Nevertheless, the potential of immune rejection is still a problem, although this may be resolved by careful tissue typing and storing RPE cells obtained from eye bank donor eyes *(113)*.

Another approach to cell sourcing involves the use of immortalized cells. These can be generated using a transformation technology or they can arise spontaneously. Transformed RPE cells have been generated using SV-40 T transformation—two rodent lines *(114)* and two human lines have been generated in this way *(115)*. The ARPE19 cell line is one of several spontaneously generated cell lines. It has been studied in considerable detail and is well characterized *(116,117)*. It is clear from detailed analysis of expression patterns that although it does express a range of properties expected of healthy RPE cells, it is not normal, but changes can be achieved by manipulating the substrate *(118)*, and clearly the environment of the subretinal space may also modulate expression patterns and behavior. For one thing, cells become pigmented *(16)*, but it is not known whether biological activity is substantially modified.

Recent work has shown that RPE cells can be generated from embryonic stem (ES) cells. A full characterization indicates that they assume a much more normal RPE phenotype than ARPE19 cells *(116)*. Finally it has been reported that RPE-like cells can be generated from forebrain progenitor cells *(119)*.

Presently, efficacy has only been explored in detail using immortalized cells—either generated using SV-40 T transformation or the spontaneously generated ARPE-19 cell: most of the work has been conducted with the ARPE-19 cell line.

ARPE-19 Grafts

Morphologically, ARPE-19 cells can be identified posttransplantation in whole mounts of retina, where they attach to the retina when it is dissected from the underlying host RPE. In histological sections, they can be differentiated from host RPE using a human nuclear marker *(120)*. At early survival times after transplantation, they are usually seen as a focal area of cells close to the injection site, but with time they form a monolayer or bilayer of cells (Fig. 1G), sometimes adjacent to host RPE and sometimes interpolated among host RPE cells. At the electron microscopic level, the cells are distinguishable from host RPE by the different disposition of pigment granules. In such material, it can be seen that even when lying over host RPE, they send processes down between cells to the Bruch's membrane. They can survive as a monolayer for as long as 7 mo, providing suitable immunosuppression is maintained.

Such grafts ensure photoreceptor rescue, which is greatest in the region of the donor RPE cells (Fig. 1D), but continues beyond their area of distribution *(120)*. Sham-operated animals (medium alone) by contrast have a small area of rescue immediately around the injection site (Fig. 1C). At 15-wk survival, a photoreceptor layer of five to six cells thick is characteristically seen in areas of best rescue, reducing to two to three cells thick further away. With time, there may be further reduction of thickness. Because the majority of cells in a rat retina are rods, the rescue seen in simple cell stains largely reflects rod rescue, but staining with specific rod and cone antibodies shows that both cell types are among the rescued cells. If more rods are rescued, the cone morphology appears less pathological.

It is evident that, along with protecting photoreceptors, grafts are also effective in containing much of the secondary inner retinal changes and also slow the development of vascular changes.

In unoperated dystrophic rats, ERG recordings show that the a-wave response disappears between 50 and 60 d of age *(121,122)*. A rod contribution to the b-wave can be recorded as late as P74, whereas a cone contribution persists for around P180 *(47,54)*. Both a- and b-wave responses are sustained for a period by ARPE-19 cell grafts. a-waves can often be identified as late as P120 and rod-derived b-waves can be recorded beyond that time point. Another measure of rod activity, the oscillatory potential, can be sustained for prolonged periods.

Adaptation studies can be achieved in rats by setting the animal in a physiological recording apparatus and placing a hemisphere in front of it *(123)*. The whole hemisphere is diffusely illuminated to provide a background luminance level 0.02 candela (cd)/m^2 and a small spot of light is projected on one part of the visual field. Threshold responses can be recorded in the superior colliculus to the spot presented over a background luminance, which is progressively increased. Data collected by such studies provide a light-adaptation curve. Although a normal rat shows absolute dark-adapted threshold at −3.5 log cd/m^2, which starts to rise progressively after reaching a background luminance of −6.0 log cd/m^2, RCS rats even at 21 d of age show a threshold about 1 log unit higher than normal rats and this threshold does not change with

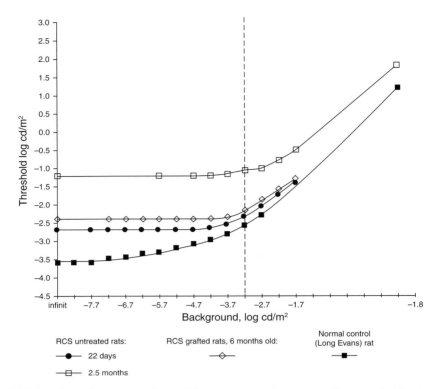

Fig. 2. Light adaptation curves obtained from a group of rats, recording threshold unit responses in the superior colliculus to a spot of light presented with varying background levels. The dotted line indicates to level chosen for mesopic recordings in other studies a) ■ non-dystrophic rat, b) ● unoperated P21 dystrophic RCS rat, c) ⊞ unoperated P60 dystrophic RCS rat, d) ◇ P120 dystrophic RCS rat with subretinal ARPE-19 graft. Note that grafting does not recover the normal curve but it does prevent the elevation of the threshold baseline.

increased background luminance until a level of about –4.0 log cd/m² is reached, when in both normal and dystrophics the curve rises more acutely—this corresponds to activation of cones. In progressively older RCS rats, the threshold levels become progressively higher. The converse record can be obtained once full luminance is achieved —namely, of measuring how long it takes to reach fully dark-adapted state and at what luminance that level is achieved. Such dark-adaptation curves show that dystrophic RCS rats do not respond at lower luminance levels, suggesting that from an early age, rods are dysfunctional and the animals are effectively "night blind"—the increased thresholds levels with time may reflect gradual deterioration of cone efficacy. After introducing ARPE-19 cells to the subretinal space, there is no improvement of function at low luminance levels, but the later elevation of thresholds level with age is prevented, suggesting it may reflect preservation of cone function (Fig. 2).

If a stable background luminance level of 0.02 cd/m² is maintained, it is possible to map threshold levels across the whole visual field by recording different points across the superior colliculus to projection of a spot of light *(124)*. This approach avoids the need for lengthy dark adaptation between tests and makes it feasible to record as many as 70 points or more in an individual colliculus. The approach has been used to record deterioration in threshold levels with time in RCS rats and is a very effective way of examining how retinal sensitivity is rescued by grafting. With this approach, it was found that although in normal rats, the thresholds were around 0.2 log units above background, dystrophic RCS rats began to show deterioration at around P30 and there was gradual deterioration to a level of around 4 log units above background at P120. Beyond that, cells in the colliculus become progressively unresponsive to focal stimulation. In dystrophic rats receiving ARPE-19 grafts, thresholds of less than 1.0 log units above background can be sustained for prolonged periods in best cases. There is deterioration with time, but recordings as late as 8 mo of age still give figures of around 1.0 log units above background over a selected area of the colliculus.

Physiological recording of single unit responses in the visual cortex also show well-tuned responses as long as 7 mo after grafting at P23 *(61)*.

In accordance with the physiological studies, the Visual Water Task also shows considerable rescue of acuity by grafting. Nondystrophic rats give a figure of around 1 cycle/degree (c/d), whereas at 1 mo of age, a figure of around 0.82 c/d is recorded in dystrophic rats. This deteriorates to 0.32 c/d at 4 mo of age and more slowly over the next 11 mo to 0.02c/d, a point by which the rats are effectively blind *(66)*. After ARPE-19 grafts, performance is sustained at around 0.69 c/d at 4 mo, with the best-performing animals giving a figure of 0.72 c/d *(87)* (Fig. 3A).

In the Optomotry test, normal rats have a threshold response to moving stripes of 0.6 c/d. Dystrophic rats aged P30 have a threshold at 0.37 c/d deteriorating to 0.30 c/d at 3 mo. After ARPE-19 grafts, an average is sustained at 0.4 c/d, with the best-performing animals having figures in the order of 0.49 c/d (TM Holmes, unpublished work).

The work described here previously has explored efficacy after grafting early in the course of degeneration. Very little has been done with grafts introduced at later ages, although from a clinical perspective, this might be the more important need. One early study showed that fresh RPE cells introduced at P38 in RCS rats are ineffective in preserving photoreceptors, when examined at P90 *(125)*. More recently, it has been shown that RCS rats receiving ARPE-19 grafts at P60 at a time when the ONL is around three cells deep preserve the remaining cells for as much as P150: luminance and optomotor threshold responses are also preserved at the levels typical of the age at which the transplants were introduced *(126)*.

Several points emerge from this series of studies:

1. Introduction of a suspension of ARPE-19 cells into the subretinal space can preserve rod photoreceptors from degeneration and sustain relatively healthy cones.
2. The cells also preserve a range of visual responses for prolonged periods.
3. Although best cases in some measures can be sustained at levels surprisingly similar to normal, there is usually deterioration with time. This is most likely caused by inadequate immune protection, although other factors cannot be overlooked.
4. There is disparity among the various measures, most notably adaptation responses and ERG. Adaptation responses suggest that rescued rods are not functioning normally, whereas ERG recordings show continued rescue of rod function in the early stages. However, ERG responses show significant deterioration in response amplitudes between P60 and P100 while other functional tests indicate sustained visual responsiveness.

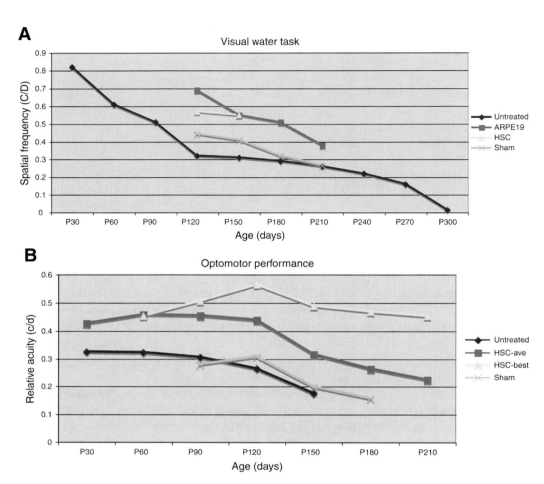

Fig. 3. (A) Effect of subretinal ARPE-19 and human Schwann cell (hSC) grafts on the grating acuity of dystrophic RCS rats compared with sham injection; measured in the Visual Water Task. Both ARPE19 and hSC groups performed significantly better than the sham and untreated groups. It is noted that the performance of hSC group maintained almost same from P120 to P150, **(B)** Effect on subretinal human Schwann cell graft on the relative acuity of dystrophic RCS rats compared with sham injection, measured in an optomotor test. hSC grafted animals performed significantly better than sham and untreated groups.

One obvious question is how the grafts exert their effect. Do they completely substitute for the existing defective RPE cells or do they have more restricted functions? This issue has not been addressed critically, but available evidence suggests the latter. The area of rescued photoreceptors appears to extend beyond the area of distribution of donor RPE cells. The indication that rods may not function at low luminance may argue for failure of the rod phototransduction process.

However the grafts do appear to have a supportive role in rod survival and this in itself may secondarily ensure rescue of cone function by a mechanism such as that proposed by Sahel and his colleagues *(127)*. In this context it should be noted that rods need to establish a close relation with RPE cells for photopigment recycling, whereas for cones, this process can be mediated via Müller cells making them less vulnerable to RPE disarray *(128)*.

Given that grafts delay the progress of vascular deterioration, the possibility exists that efficacy of grafted RPE cells may lie in their ability to produce PEDF as a diffusible factor that not only prevents photoreceptor degeneration, but also has antineovascular properties *(129–133)*. A study has shown that pigment epithelium derived factor (PEDF) injections into the vitreous similarly limit the development of vascular anomalies in RCS rats *(134)*. Other factors may also be effective in promoting photoreceptor survival. Because, as summarized below, similar patterns of rescue can be achieved with non-RPE cells, it may not be sufficient to assume that ARPE-19 grafts are simply replacing all aspects of those that are functioning abnormally, but rather that they may be functioning more simply perhaps by delivering diffusible factors.

Iris Pigment Epithelial Cells

Iris pigment epithelial (IPE) cells harvested from the iris have also been studied in some detail as an alternative cell source to RPE. The cells have similar embryological origin to RPE cells and, although they do show some characteristics in common with RPE, including the ability to phagocytose OS membranes *(135)*, they also differ sufficiently to raise question as to whether they could be effective substitutes. However, a number of studies have examined their efficacy in RCS rats. One found no significant improvement over sham-injected rats *(136)*, but subsequent work *(137)* showed some improvement even if grafts were located in the choroid, raising the possibility of an action associated with delivery of a diffusible factor. Further work *(138,139)* was able to show some rescue in the light damage model. These studies have prompted the exploration of IPE as a potential autologous grafting procedure for cell replacement in AMD *(140–143)*. These studies found that the grafts are tolerated and that they do not have obvious deleterious effects, but, although there are indications that they are effective in preventing recurrence of CNVs, the effect on rescuing vision over that seen in controls was not significant. A development of these studies in RCS rats *(144)* has transfected IPE cells with PEDF and the work suggests that efficacy can be enhanced and that the anti-angiogenic properties of PEDF may be additionally beneficial.

Schwann Cells

The rationale of exploring the use of Schwann cells (SC) derives from the observations of LaVail and colleagues that injection of growth factors into the vitreous can slow photoreceptor loss *(145)*. Subsequent work has shown that introduction of growth factors using gene therapy via viral vectors will also ensure photoreceptor survival in a range of animal models of retinal degeneration *(146–153)*. Schwann cells make a number of the growth factors that are known to keep photoreceptors alive, including CNTF, glial cell-derived neurotrophic factor (GDNF), basic fibroblast growth factor, and brain-derived neurotrophic factor (BDNF) *(154–156)*. Consequently, the possibility that they might serve to provide continuous delivery of a galaxy of factors at physiological levels deserved attention. A further strength of the approach was that it has been indicated that single factors may not be sufficient for optimal rescue but rather simultaneous delivery of more than one factor may be more beneficial *(157,158)*, However, unlike RPE or IPE cells, Schwann cells do not normally reside in the eye; furthermore, in situations in which they have been introduced into the eye, they were seen to myelinate optic nerve axons *(159)*. Both of these points might mitigate against application of these cells for a photoreceptor neurotrophic role.

In a series of studies, we have explored how these cells behave when introduced into the subretinal space, and whether they are effective in promoting photoreceptor rescue.

In the first studies, in which Schwann cells were harvested from immature rat sciatic nerves and introduced as a suspension into the subretinal space of RCS rats, we found that substantial rescue of photoreceptors can be achieved *(17)*. Most important in these studies, the cells do not appear to show any untoward behavior, such as myelinating neural profiles in the outer retina. However, if they are introduced into the vitreous cavity and come to lie on the inner limiting membrane, myelinated ganglion cell axon bundles are seen.

Subsequent work explored the ability of Schwann cells derived from human nerve roots and in these studies efficacy was measured in range of ways. First, it was found that donor cells could survive for as much as 7 mo, appearing unlike RPE cell grafts, as small clumps of cells (Fig. 1H) *(120)*. Second, a substantial level of photoreceptor rescue is achieved and this can be sustained for many months (Fig. 1E). Despite this, dark-adaptation studies still show no rescue of low-luminance vision *(123)*. Threshold response studies recorded from the SC under mesopic conditions indicate generally high levels of rescue over a large area of the visual field. Acuity measures also show good responses sustained over many months (Fig. 3A) *(67)*. In the best cases, figures as high as 0.72 c/d were obtained at 5 mo, more than 4 mo after transplantation. An average acuity of 0.36 c/d was recorded in the fibroblast-injected rats. In the optomotor test, although acuity thresholds are always lower than those recorded in the Visual Water Task, average figures of 0.45 c/d with best results of 0.56 c/d at 4 mo (compared with 0.6 c/d in normals and 0.30 c/d in shams *[160]*) were recorded (Fig. 3B).

Recent work monitoring visual performance over time after introducing syngeneic Schwann cell grafts to RCS rats at 3 wk of age has shown that even at 27 wk, there is no deterioration in optomotor thresholds *(87)*, which is in contrast to allogeneic grafts in which substantial reduction in performance was registered. This adds support to the idea that the deterioration seen in human to rat grafting is caused by immunological factors, not protected by cyclosporine.

Finally, one study has explored the use of transplantation to rescue photoreceptors in another model of retinal degeneration, the rhodopsin knockout mouse, and shown rescue occurring for a limited period *(38)*. Clearly the use of other animal models is indicated.

The observations of anatomical rescue occurring over a larger area than the actual donor cell distribution again suggest that the graft rescue effect may be through diffusible agents. The hypothesis that growth factor delivery is a key element was given support by a study using Schwann cell lines transfected to produce additional levels of growth factors and using head tracking performance to monitor efficacy *(161)*. In that study, it was found that the untransfected Schwann cell line is ineffective in rescuing photoreceptors or optomotor responses in RCS rats grafted at 3 wk of age. On the other hand, BDNF-transfected cells show a trend towards improved rescue, but GDNF-transfected cells show significant rescue of function and correlated anatomical preservation. Recent work *(144)* showing improved efficacy after gene transfection of IPE cells adds to the value of this approach.

Olfactory Ensheathing Cells

The olfactory ensheathing cell (OEC) resides in the olfactory epithelium, olfactory nerve, and bulb and is important in the continued ability of olfactory nerves to regenerate throughout life. It can be harvested from either olfactory epithelium or bulb and has been shown either alone or in partnership with other cell types to promote regeneration of severed central neural axons *(162,163)*. Like Schwann cells, OECs have been shown to produce a range of growth factors, including BDNF, CNTF, GDNF, and nerve growth factor *(164,165)*. On transplantation to the subretinal space, such cells have been shown to effect retinal rescue and sustain some visual functions *(166)*. Photoreceptor rescue is much more local than that seen after either ARPE-19 or Schwann cell grafts. There is no evidence of sustained ERG response, perhaps because of the local nature of rescue. However, optomotor responses are quite robust up to at least 4 mo and threshold luminance levels recorded from the superior colliculus under mesopic background conditions are significantly better than sham-injected animals.

These observations show that another factor-producing cell, one not normally resident in the retina, can be effective in preserving photoreceptors. Whether the more local effect is the result of reduced ability to migrate across the retina, to the different factor-producing profile, or to the amount of factor produced is not clear.

Neural Progenitor Cells

Several studies using neural progenitor cells derived from forebrain have shown that they integrate into the retina without much effect on photoreceptor rescue *(36,167–169)*, whereas ones derived from retina have been shown to be capable of providing a range of differentiated cells including new photoreceptors as well as leading to some improvement in visual performance over unoperated animals *(37)*. Whether the functional improvement, maintaining circadian rhythms, is owing to new connections, to a rescue effect, or to modulation of other phototransduction pathways is not altogether clear. Recent work has shown that forebrain-derived cells can transform into RPE-like cells and under these circumstances substantial morphological and functional rescue can be achieved in RCS rats *(119)* (Fig. 1F,I).

Stem Cells

The potential of stem cells as a cell source to provide rescue of photoreceptors has been little explored at present. Studies using manipulated ES cells have achieved populations of RPE cells, which, when characterized in detail, more closely resemble RPE cells in culture than does the immortalized cell line, ARPE-19 *(116)*. Efficacy of this and other cell types derived from various stem cell populations adds an important new development in the exploration of alternative cells for prevention of retinal degeneration. One set of studies has shown that RPE cells derived from an ES cell line was found to promote rescue in RCS rats, but disturbingly in a follow-up study, 50% of the retinas developed teratomas *(170)*. Whether this was an unfortunate isolated circumstance or whether this might be a problem associated with management of ES cell-derived cells remains to be seen. Certainly there are other studies using ES cells where teratomas have not occurred *(171)*.

SUMMARY

The studies reviewed here show that a range of very different cell types can rescue photoreceptors and associated visual function in experimental animal studies. It is likely that the list is far from complete. Several of the cells used do not appear to function by simply replacing cells afflicted with the primary gene defect, and some like Schwann cells and OEC are not normal constituents of the retina. Even the ARPE-19 cells which have much in common with RPE do not appear to be capable of rescuing normal rod function even though anatomically, rod survival is clearly improved over untreated animals. A common feature of the donor cells explored here is that they do produce a number of different growth factors, raising the possibility that they are indeed functioning as cell-based growth factor delivery units. A development of this approach, which is beginning to be explored, is to transfect cells ex vivo with a specific growth factor to test the role of particular growth factors in a continuous delivery system or to use them to enhance the efficacy of the particular cell type in rescuing photoreceptors. This has considerable potential in this area of research.

Although attention has been heavily directed toward rod photoreceptor rescue, because this is the predominant cell of the ONL in rodents, it is clear from the studies described here, that rescued rods may still not function normally, if at all. This is important to be aware of especially when ONL thickness is the sole index of success. More important maybe is to explore how the particular treatment affects a range of visual functions and to develop this as the important endpoint assay for efficacy. Clearly, morphology should always be examined, to identify any events that might compromise optimal recovery, including regressive, reactive, and pathological events and to ensure that structural and functional features have a logical correlate. In this context, it is important to know whether treatments, especially those involving potential growth factor release, might affect the inner retina directly, particularly the ganglion cells, since they have been shown to be sensitive to many of the same factors that play a role is sustaining photoreceptors. In addition, the factors identified may also affect the retinal vasculature directly. This has been investigated for ARPE-19 cells and for PEDF-transfected cells, where it was shown that vascular abnormalities seen in RCS rats do not develop until much later than normal, but little is know about potential problems using other cell types or added growth factors.

Use of alternative cells may well have a role clinically, especially for continuous release of growth factors. Although this is already being explored using encapsulated cells, therefore avoiding safety concerns, cells introduced to the subretinal space without such protection do present problems: they can carry a range of infective agents, including viruses and prions, and they may be immunogenic if harvested from a nonsyngeneic donor. Careful screening prior to injecting into the eye is absolutely essential. For cell lines, evidence of senescence is important to screen for and ideally it should be possible to produce large cohorts of similar cells so that they can be rigorously screened and are available for commercial application. For stem cells, circumstances that lead to untoward growth patterns such as teratomas must be carefully scrutinized. There are a number of syngeneic cells that avoid problems, presented by allografts and by manufactured cell lines. Schwann cells are particularly attractive because they could be harvested from a peripheral sensory nerve of the patient requiring photoreceptor rescue: such autologous grafts would largely circumvent issues associated with immune incompatibility and transmitted infective agents, serial surgeries in the same eye are avoided and secondary manipulations are not needed. Another syngeneically derived cell type that may be valuable is the IPE cell, although for optimally efficacy, ex vivo gene transfection may be desirable. Some stem cells do seem to be less immunogenic and, if they can be maintained as long-term replicating lines, they can be carefully screened for any infective agents. With the ability to maintain large repositories of such cells, they may well represent the future direction of work to find therapies to contain photoreceptor degeneration.

Clearly, the work of the past few years has enlarged the scope of transplantation into a much broader cell-based therapy approach, which, with the introduction of stem cells and of ex vivo gene transfection, introduces possibilities not conceived in the early studies of RPE transplantation.

ACKNOWLEDGMENTS

Personal work reported in this review was supported by National Institutes of Health (EY14038), Foundation Fighting Blindness, Wynn Foundation, and Walsh Foundation. Research to Prevent Blindness (RPB) and National Eye Institute (P30 EY014800) provided core support for the work. We thank the many colleagues who have participated in these studies.

REFERENCES

1. Bok D. Processing and transport of retinoids by the retinal pigment epithelium. Eye 1990;4 (Pt 2):326–332.
2. Bok D. Retinal photoreceptor-pigment epithelium interactions. Friedenwald lecture. Invest Ophthalmol Vis Sci 1985;26:1659–1694.
3. Bok D. The retinal pigment epithelium: a versatile partner in vision. J Cell Sci Suppl 1993;17:189–195.
4. Bosch E, Horwitz J, Bok D. Phagocytosis of outer segments by retinal pigment epithelium: phagosome-lysosome interaction. J Histochem Cytochem 1993;41:253–263.
5. Bok D. Photoreceptor "retinoid pumps" in health and disease. Neuron 1999;23:412–414.
6. Bourne MC, Campbell D A, Tansley K. Hereditary degeneraiton of the rat retina. Br J Ophthalmol 1938;22:613–623.
7. Dowling JE, Sidman RL. Inherited retinal dystrophy in the rat. J Cell Biol 1962;14:73–107.
8. LaVail MM. Legacy of the RCS rat: impact of a seminal study on retinal cell biology and retinal degenerative diseases. Prog Brain Res 2001;131:617–627.
9. D'Cruz PM, Yasumura D, Weir J, et al. Mutation of the receptor tyrosine kinase gene Mertk in the retinal dystrophic RCS rat. Hum Mol Genet 2000;9:645–651.
10. Vollrath D, Feng W, Duncan JL, et al. Correction of the retinal dystrophy phenotype of the RCS rat by viral gene transfer of Mertk. Proc Natl Acad Sci USA 2001;98:12,584–12,589.
11. Faktorovich EG, Steinberg RH, Yasumura D, Matthes MT, LaVail MM. Photoreceptor degeneration in inherited retinal dystrophy delayed by basic fibroblast growth factor. Nature 1990;347:83–86.
12. Perry J, Du J, Kjeldbye H, Gouras P. The effects of bFGF on RCS rat eyes. Curr Eye Res 1995;14:585–592.
13. Li LX, Turner JE. Inherited retinal dystrophy in the RCS rat: prevention of photoreceptor degeneration by pigment epithelial cell transplantation. Exp Eye Res 1988;47:911–917.

14. Sheedlo HJ, Li L, Turner JE. Photoreceptor cell rescue in the RCS rat by RPE transplantation: a therapeutic approach in a model of inherited retinal dystrophy. Prog Clin Biol Res 1989;314:645–658.

15. Lopez R, Gouras P, Kjeldbye H, et al. Transplanted retinal pigment epithelium modifies the retinal degeneration in the RCS rat. Invest Ophthalmo Vis Sci 1989;30:586–588.

16. Lund RD, Adamson P, Sauve Y, et al. Subretinal transplantation of genetically modified human cell lines attenuates loss of visual function in dystrophic rats. Proc Natl Acad Sci USA 2001;98:9942–9947.

17. Lawrence JM, Sauve Y, Keegan DJ, et al. Schwann cell grafting into the retina of the dystrophic RCS rat limits functional deterioration. Royal College of Surgeons. Invest Ophthalmol Vis Sci 2000;41:518–528.

18. Wang S, Lu B, Lund RD. Morphological changes in the Royal College of Surgeons rat retina during photoreceptor degeneration and after cell-based therapy. J Comp Neurol 2005;491:400–417.

19. Cuenca N, Pinilla I, Sauve Y, Lund R. Early changes in synaptic connectivity following progressive photoreceptor degeneration in RCS rats. Eur J Neurosci 2005; 22:1057–1072.

20. Villegas-Perez MP, Lawrence JM, Vidal-Sanz M, LaVail MM, Lund RD. Ganglion cell loss in RCS rat retina: a result of compression of axons by contracting intraretinal vessels linked to the pigment epithelium. J Comp Neurol 1998;392:58–77.

21. Wang S, Villegas-Perez MP, Holmes T, et al. Evolving neurovascular relationships in the RCS rat with age. Curr Eye Res 2003;27:183–196.

22. Strettoi E, Pignatelli V. Modifications of retinal neurons in a mouse model of retinitis pigmentosa. Proc Natl Acad Sci USA 2000;97:11,020–11,025.

23. Strettoi E, Porciatti V, Falsini B, Pignatelli V, Rossi C. Morphological and functional abnormalities in the inner retina of the rd/rd mouse. J Neurosci 2002;22:5492–5504.

24. Pignatelli V, Cepko CL, Strettoi E. Inner retinal abnormalities in a mouse model of Leber's congenital amaurosis. J Comp Neurol 2004;469:351–359.

25. Claes E, Seeliger M, Michalakis S, Biel M, Humphries P, Haverkamp S. Morphological characterization of the retina of the CNGA3(–/–)Rho(–/–) mutant mouse lacking functional cones and rods. Invest Ophthalmol Vis Sci 2004;45:2039–2048.

26. Cuenca N, Pinilla I, Sauve Y, Lu B, Wang S, Lund RD. Regressive and reactive changes in the connectivity patterns of rod and cone pathways of P23H transgenic rat retina. Neuroscience 2004;127:301–317.

27. Jones BW, Watt CB, Frederick JM, et al. Retinal remodeling triggered by photoreceptor degenerations. J Comp Neurol 2003;464:1–16.

28. Jones BW, Marc RE. Retinal remodeling during retinal degeneration. Exp Eye Res 2005;81:123–137.

29. Sagdullaev BT, Aramant RB, Seiler MJ, Woch G, McCall MA. Retinal transplantation-induced recovery of retinotectal visual function in a rodent model of retinitis pigmentosa. Invest Ophthalmol Vis Sci 2003;44:1686–1695.

30. Bowes C, Li T, Danciger M, Baxter LC, Applebury L, Farber DB. Retinal degeneration in the rd mouse is caused by a defect in the B subunit of rod cGMP-phosphodiesterase. Nature 1990;347:677–680.

31. Naash MI, Hollyfield JG, al-Ubaidi MR, Baehr W. Simulation of human autosomal dominant retinitis pigmentosa in transgenic mice expressing a mutated murine opsin gene. Proc Natl Acad Sci USA 1993;90:5499–5503.

32. Gao J, Cheon K, Nusinowitz S, et al. Progressive photoreceptor degeneration, outer segment dysplasia, and rhodopsin mislocalization in mice with targeted disruption of the retinitis pigmentosa-1 (Rp1) gene. Proc Natl Acad Sci USA 2002;99:5698–5703.

33. Lem J, Flannery JG, Li T, Applebury ML, Farber DB, Simon MI. Retinal degeneration is rescued in transgenic rd mice by expression of the cGMP phosphodiesterase beta subunit. Proc Natl Acad Sci USA 1992;89:4422–4426.

34. Chang B, Hawes NL, Hurd RE, Davisson MT, Nusinowitz S, Heckenlively JR. Retinal degeneration mutants in the mouse. Vision Res 2002;42:517–525.

35. Kwan AS, Wang S, Lund RD. Photoreceptor layer reconstruction in a rodent model of retinal degeneration. Exp Neurol 1999;159:21–33.

36. Lu B, Kwan T, Kurimoto Y, Shatos M, Lund RD, Young MJ. Transplantation of EGF-responsive neurospheres from GFP transgenic mice into the eyes of rd mice. Brain Res 2002;943:292–300.

37. Klassen HJ, Ng TF, Kurimoto Y, et al. Multipotent retinal progenitors express developmental markers, differentiate into retinal neurons, and preserve light-mediated behavior. Invest Ophthalmol Vis Sci 2004;45:4167–4173.

38. Keegan DJ, Kenna P, Humphries MM, et al. Transplantation of syngeneic Schwann cells to the retina of the rhodopsin knockout (rho(–/–)) mouse. Invest Ophthalmol Vis Sci 2003;44:3526–3532.

39. Zhang K, Kniazeva M, Han M, et al. A 5-bp deletion in ELOVL4 is associated with two related forms of autosomal dominant macular dystrophy. Nat Genet 2001;27:89–93.

40. Yang Z, Alvarez BV, Chakarova C, et al. Mutant carbonic anhydrase 4 impairs pH regulation and causes retinal photoreceptor degeneration. Hum Mol Genet 2005;14:255–265.

41. Umeda S, Suzuki MT, Lkamoto H, et al. Molecular composition of drusen and possible imvolvement of anti-retinal autoimmunity in two different forms of macular degeneration in cynomolgus monkey (Macaca fascicularis). FASEB J 2005;19:1683–1685.

42. Narfstrom K, Nilsson SE. Progressive retinal atrophy in the Abyssinian cat. J Heredity 1987;74:273–276.

43. Veske A, Nilsson SE, Narfstrom K, Gal A. Retinal dystrophy of Swedish briard/briard-beagle dogs is due to a 4-bp deletion in RPE65. Genomics 1999;57:57–61.

44. Li ZY, Wong F, Chang JH, et al. Rhodopsin transgenic pigs as a model for human retinitis pigmentosa. Invest Ophthalmo Vis Sci 1998;39:808–819.

45. Petters RM, Alexander CA, Wells KD, et al. Genetically engineered large animal model for studying cone photoreceptor survival and degeneration in retinitis pigmentosa. Nat Biotechnol 1997;15:965–970.

46. Acland GM, Aguirre GD, Ray J, et al. Gene therapy restores vision in a canine model of childhood blindness. Nat Genet 2001;28:92–95.

47. Pinilla I, Lund RD, Sauve Y. Contribution of rod and cone pathways to the dark-adapted electroretinogram (ERG) b-wave following retinal degeneration in RCS rats. Vision Res 2004;44:2467–2474.

48. DiLoreto DA Jr, del Cerro C, Cox C, del Cerro M. Changes in visually guided behavior of Royal College of Surgeons rats as a function of age: a histologic, morphometric, and functional study. Invest Ophthalmo Vis Sci 1998;39:1058–1063.

49. Kovalevsky G, Diloreto D Jr, Wyatt J, del Cerro C, Cox C, del Cerro M. The intensity of the pupillary light reflex does not correlate with the number of retinal photoreceptor cells. Exp Neurol 1995;133:43–49.

50. Perlman I. Dark-adaptation in abnormal (RCS) rats studied electroretinographically. J Physiol 1978;278:161–175.

51. Peichl L. Diversity of mammalian photoreceptor properties: Adaptations to habitat and lifestyle? Anat Rec A Discov Mol Cell Evol Biol 2005;287:1001–1012.

52. Ahnelt PK, Kolb H. The mammalian photoreceptor mosaic-adaptive design. Prog Retin Eye Res 2000;19:711–777.

53. Beltran WA, Rohrer H, Aguirre GD. Immunolocalization of ciliary neurotrophic factor receptor alpha (CNTFRalpha) in mammalian photoreceptor cells. Mol Vis 2005; 11:232–244.

54. Pinilla I, Lund RD, Lu B, Sauve Y. Measuring the cone contribution to the ERG b-wave to assess function and predict anatomical rescue in RCS rats. Vision Res 2005;45:635–641.

55. Seeliger MW, Weber BH, Besch D, Zrenner E, Schrewe H, Mayser H. MfERG waveform characteristics in the RS1h mouse model featuring a 'negative' ERG. Doc Ophthalmol 2003;107:37–44.

56. Paskowitz DM, Nune G, Yasumura D, et al. BDNF reduces the retinal toxicity of verteporfin photodynamic therapy. Invest Ophthalmol Vis Sci 2004;45:4190–4196.

57. Girman SW, Wang S, Lund RD. Time course of deterioration of rod and cone function in RCS rat and the effects of subretinal cell grafting: a lignt- and dark-adaptation study. Vis Res 2005;45:343–354.

58. Sauvé Y, Girman SV, Wang S, Keegan DJ, Lund RD. Preservation of visual responsiveness in the superior colliculus of RCS rats after retinal pigment epithelium cell transplantation. Neuroscience 2002;114:389–401.

59. Lund RD, Kwan AS, Keegan DJ, Sauve Y, Coffey PJ, Lawrence JM. Cell transplantation as a treatment for retinal disease. Prog Retin Eye Res 2001;20:415–449.

60. Girman SV, Sauvé Y, Lund RD. Receptive field properties of single neurons in rat primary visual cortex. J Neurophysiol 1999;82:301–311.

61. Girman SV, Wang S, Lund RD. Cortical visual functions can be preserved by subretinal RPE cell grafting in RCS rats. Vision Res 2003;43:1817–1827.

62. Radner W, Sadda SR, Humayun MS, et al. Light-driven retinal ganglion cell responses in blind rd mice after neural retinal transplantation. Invest Ophthalmol Vis Sci 2001;42:1057–1065.

63. Yamamoto S, Du J, Gouras P, Kjeldbye H. Retinal pigment epithelial transplants and retinal function in RCS rats. Invest Ophthalmol Vis Sci 1993;34:3068–3075.

64. Coffey PJ, Girman S, Wang SM, et al. Long-term preservation of cortically dependent visual function in RCS rats by transplantation. Nat Neurosci 2002;5:53–56.

65. Prusky GT, West PW, Douglas RM. Behavioral assessment of visual acuity in mice and rats. Vision Res 2000;40:2201–2209.

66. McGill TJ, Douglas RM, Lund RD, Prusky GT. Quantification of spatial vision in the Royal College of Surgeons rat. Invest Ophthalmol Vis Sci 2004;45:932–936.

67. McGill TJ, Lund RD, Douglas RM, Wang S, Lu B, Prusky GT. Preservation of vision following cell-based therapies in a model of retinal degenerative disease. Vision Res 2004;44:2559–2566.

68. Dean P. Visual pathways and acuity hooded rats. Behav Brain Res 1981;3:239–271.

69. Coffey PJ, Whiteley SJ, Lund RD. Preservation and restoration of vision following transplantation. Prog Brain Res 2000;127:489–499.

70. Prusky GT, Alam NM, Beekman S, Douglas RM. Rapid quantification of adult and developing mouse spatial vision using a virtual optomotor system. Invest Ophthalmol Vis Sci 2004;45:4611–4616.

71. Holmes TM, Silver B, Douglas RM, Lund RD, Prusky GT. Rapid assessment of vision in the Royal College of Surgeons & long evens rats using optomotor: a virtual reality visual screening system. Invest Ophthalmol Vis Sci 2004; Abstract no. 5156.

72. Schmucker C, Seeliger M, Humphries P, Biel M, Schaeffel F. Grating acuity at different luminances in wild-type mice and in mice lacking rod or cone function. Invest Ophthalmol Vis Sci 2005;46:398–407.

73. Jordan JF, Semkova I, Kociok N, Welsandt GR, Krieglstein GK, Schraermeyer U. Iris pigment epithelial cells transplanted into the vitreous accumulate at the optic nerve head. Graefes Arch Clin Exp Ophthalmol 2002;40:403–407.

74. Wongpichedchai S, Weiter JJ, Weber P, Dorey CK. Comparison of external and internal approaches for transplantation of autologous retinal pigment epithelium. Invest Ophthalmol Vis Sci 1992;33:3341–3352.

75. Otani A, Dorrell MI, Kinder K, et al. Rescue of retinal degeneration by intravitreally injected adult bone marrow-derived lineage-negative hematopoietic stem cells. J Clin Invest 2004;114:765–774.

76. Tao W, Wen R, Goddard MB, et al. Encapsulated cell-based delivery of CNTF reduces photoreceptor degeneration in animal models of retinitis pigmentosa. Invest Ophthalmol Vis Sci 2002;43:3292–3298.

77. Uteza Y, Rouillot JS, Kobetz A, et al. Intravitreous transplantation of encapsulated fibroblasts secreting the human fibroblast growth factor 2 delays photoreceptor cell degeneration in Royal College of Surgeons rats. Proc Natl Acad Sci USA 1999;96:3126–3131.

78. Sieving PA, Caruso RG, Coleman HR, Thompson DJ, Fullmer KR, Bush RA. Ciliary neurotrophic factor (CNTF) for human retinal degeneration: phase I trial of CNTF delivered by encapsulated cell intraocular implants. Proc Natl Acad Sci USA 2006;103:3896–3901

79. Grisanti S, Ishioka M, Kosiewicz M, Jiang LQ. Immunity and immune privilege elicited by cultured retinal pigment epithelial cell transplants. Invest Ophthalmo Vis Sci 1997;38:1619–1626.

80. Jiang LQ, Jorquera M, Streilein JW, Ishioka M. Unconventional rejection of neural retinal allografts implanted into the immunologically privileged site of the eye. Transplantation 1995;59:1201–1207.

81. Jiang LQ, Jorquera M, Streilein JW. Subretinal space and vitreous cavity as immunologically privileged sites for retinal allografts. Invest Ophthalmol Vis Sci 1993;34:3347–3354.

82. Kohen L, Enzmann V, Faude F, Wiedemann P. Mechanisms of graft rejection in the transplantation of retinal pigment epithelial cells. Ophthalmic Res 1997;29:298–304.

83. Zhang X, Bok D. Transplantation of retinal pigment epithelial cells and immune response in the subretinal space. Invest Ophthalmol Vis Sci 1998;39:1021–1027.

84. Zamiri P, Zhang Q, Streilein JW. Vulnerability of allogeneic retinal pigment epithelium to immune T-cell-mediated damage in vivo and in vitro. Invest Ophthalmol Vis Sci 2004;45:177–184.

85. Jiang LQ, Streilein JW. Enhancement of survival of intraocular neural retinal grafts by prior antigen-specific immune deviation. Transplant Proc 1992;24:2883–2884.

86. Del Priore LV, Ishida O, Johnson EW, et al. Triple immune suppression increases short-term survival of porcine fetal retinal pigment epithelium xenografts. Invest Ophthalmol Vis Sci 2003;44:4044–4053.

87. McGill TJ, Lund RD, Douglas RM, et al. Subretinal Syngeneic and Allogeneic Rat Schwann Cell Transplants Preserve Vision in Dystrophic RCS Rat. Soc Neurosci 2005; Abstract no. 9775.

88. Sheedlo HJ, Li LX, Turner JE. Functional and structural characteristics of photoreceptor cells rescued in rpe-cell grafted retinas of rcs dystrophic rats. Exp Eye Res 1989;48:841–854.

89. Ivert L, Gouras P, Naeser P, Narfstrom K. Photoreceptor allografts in a feline model of retinal degeneration. Graefes Arch Clin Exp Ophthalmol 1998;236:844–852.

90. Gouras P, Lopez R, Brittis M, Kjeldbye H. The ultrastructure of transplanted rabbit retinal epithelium. Graefes Arch Clin Exp Ophthalmol 1992;230:468–475.

91. Lopez R, Gouras P, Brittis M, Kjeldbye H. Transplantation of cultured rabbit retinal epithelium modifies to rabbit retina using a closed-eye method. Invest Ophthalmo Vis Sci 1987;28:1131–1137.

92. Gouras P, Flood MT, Kjeldbye H, Bilek MK, Eggers H. Transplantation of cultured human retinal epithelium to bruch's membrane of the owl monkey's eye. Curr Eye Res 1985; 4:253–265.

93. Berglin L, Gouras P, Sheng YH, et al. Tolerance of human fetal retinal pigment epithelium xenografts in monkey retina. Graefes Arch Clin Exp Ophthalmol 1997;235:103–110.

94. Seaton AD, Sheedlo HJ, Turner JE. A primary role for RPE transplants in the inhibition and regression of neovascularization in the RCS rat. Invest Ophthalmol Vis Sci 1994;35:162–169.

95. Holmes TM, Lu B, Lund RD. Human RPE cell transplants retard the secondary vascular events of RCS rat retinal degeneration. Invest Ophthalmol Vis Sci 2003; Abstract no. 4972.

96. Klein RJ, Zeiss C, Chew EY, et al. Complement factor H polymorphism in age-related macular degeneration. Science 2005;308:385–389.

97. Haines JL, Hauser MA, Schmidt S, et al. Complement factor H variant increases the risk of age-related macular degeneration. Science 2005;308:419–421.

98. Edwards AO, Ritter R 3rd, Abel KJ, Manning A, Panhuysen C, Farrer LA. Complement factor H polymorphism and age-related macular degeneration. Science 2005;308:421–424.

99. Algvere P, Berglin L, Gouras P, et al. Transplantation of RPE in age-related macular degeneration: Observations in disciform lesions and dry RPE atrophy. Graefes Arch Clin Exp Ophthalmol 1997;235:149–158.

100. Bindewald A, Roth F, Van Meurs J, Holz FG. Transplantation of retinal pigment pithelium (RPE) following CNV removal in patients with AMD. Techniques, results, outlook. Ophthalmologe 2004;101:886–894.

101. Stanga PE, Kychenthal A, Fitzke FW, et al. Retinal pigment epithelium translocation and central visual function in age related macular degeneration: preliminary results. Int Ophthalmol 2001;23:297–307.

102. Tsukahara I, Ninomiya S, Castellarin A, Yagi F, Sugino IK, Zarbin MA. Early attachment of uncultured retinal pigment epithelium from aged donors onto Bruch's membrane explants. Exp Eye Res 2002;74:255–266.

103. Algvere PV, Gouras P, Dafgard Kopp E. Long-term outcome of RPE allografts in non-immunosuppressed patients with AMD. Eur J Ophthalmol 1999;9:217–230.

104. Gullapalli VK, Sugino IK, Van Patten Y, Shah S, Zarbin MA. Retinal pigment epithelium resurfacing of aged submacular human Bruch's membrane. Trans Am Ophthalmol Soc 2004;102:123–138.

105. Zarbin MA. Analysis of retinal pigment epithelium integrin expression and adhesion to aged submacular human Bruch's membrane. Trans Am Ophthalmol Soc 2003;101:499–520.

106. Tezel TH, Del Priore LV. Repopulation of different layers of host human Bruch's membrane by retinal pigment epithelial cell grafts. Invest Ophthalmol Vis Sci 1999;40:767–774.

107. Tezel TH, Kaplan HJ, Del Priore LV. Fate of human retinal pigment epithelial cells seeded onto layers of human Bruch's membrane. Invest Ophthalmol Vis Sci 1999;40:467–476.

108. Binder S, Krebs I, Hilgers RD, et al. Outcome of transplantation of autologous retinal pigment epithelium in age-related macular degeneration: a prospective trial. Invest Ophthalmol Vis Sci 2004;45:4151–4160.

109. Machemer R, Steinhorst UH. Retinal separation, retinotomy, and macular relocation: II. A surgical approach for age-related macular degeneration? Graefes Arch Clin Exp Ophthalmol 1993;231:635–641.

110. Cahill MT, Mruthyunjaya P, Bowes Rickman C, Toth CA. Recurrence of retinal pigment epithelial changes after macular translocation with 360 degrees peripheral retinectomy for geographic atrophy. Arch Ophthalmol 2005;123:935–938.

111. Khurana RN, Fujii GY, Walsh AC, Humayun MS, de Juan E Jr, Sadda SR. Rapid recurrence of geographic atrophy after full macular translocation for nonexudative age-related macular degeneration. Ophthalmology 2005;112:1586–1591.

112. Ishida M, Lui GM, Yamani A, Sugino IK, Zarbin MA. Culture of human retinal pigment epithelial cells from peripheral scleral flap biopsies. Curr Eye Res 1998;17:392–402.

113. Engelmann K, Valtink M. RPE cell cultivation. Graefes Arch Clin Exp Ophthalmol 2004;242:65–67.

114. Greenwood J, Pryce G, Devine L, et al. SV40 large T immortalised cell lines of the rat blood-brain and blood-retinal barriers retain their phenotypic and immunological characteristics. J Neuroimmunol 1996;71:51–63.

115. Kanuga N, Winton HL, Beauchene L, et al. Characterization of genetically modified human retinal pigment epithelial cells developed for in vitro and transplantation studies. Invest Ophthalmol Vis Sci 2002;43:546–555.

116. Klimanskaya I, Hipp J, Rezai KA, West M, Atala A, Lanza R. Derivation and comparative assessment of retinal pigment epithelium from human embryonic stem cells using transcriptomics. Cloning Stem Cells 2004;6:217–245.

117. Dunn KC, Aotaki-Keen AE, Putkey FR, Hjelmeland LM. ARPE-19, a human retinal pigment epithelial cell line with differentiated properties. Exp Eye Res 1996;62:155–169.

118. Turowski P, Adamson P, Sathia J, et al. Basement membrane-dependent modification of phenotype and gene expression in human retinal pigment epithelial ARPE-19 cells. Invest Ophthalmol Vis Sci 2004;45:2786–2794.

119. Gamm D, Capowski B, Lu B, et al. Localization and rescue effect of human forebrain progenitor cells injected into the subretinal space of a rat photoreceptor degeneration model. Soc Neurosci 2005; Abstract no. 9777.

120. Wang S, Lu B, Wood P, Lund RD. Grafting of ARPE-19 and Schwann cells to the subretinal space in RCS rats. Invest Ophthalmol Vis Sci 2005;46:2552–2560.

121. Bush RA, Hawks KW, Sieving PA. Preservation of inner retinal responses in the aged Royal College of Surgeons rat. Evidence against glutamate excitotoxicity in photoreceptor degeneration. Invest Ophthalmol Vis Sci 1995;36:2054–2062.

122. Sauvé Y, Lu B, Lund RD. The relationship between full field electroretinogram and perimetry-like visual thresholds in RCS rats during photoreceptor degeneration and rescue by cell transplants. Vision Res 2003;20:100–112.

123. Girman SV, Wang S, Lund RD. Time course of deterioration of rod and cone function in RCS rat and the effects of subretinal cell grafting: a light- and dark-adaptation study. Vision Res 2005;45:343–354.

124. Sauvé Y, Girman SV, Wang S, Lawrence JM, Lund RD. Progressive visual sensitivity loss in the Royal College of Surgeons rat: perimetric study in the superior colliculus. Neuroscience 2001;103:51–63.

125. Li L, Turner JE. Optimal conditions for long-term photoreceptor cell rescue in RCS rats: the necessity for healthy RPE transplants. Exp Eye Res 1991;52:669–679.

126. Wang S, Girman S, Holmes T, Sauvé Y, Lu B, Lund RD. Morphological and functional rescue in RCS rats after RPE cell line transplantation at later stage of degeneration. Invest Ophthalmol Vis Sci 2005; Abstract no. 4145.

127. Leveillard T, Mohand-Said S, Lorentz O, et al. Identification and characterization of rod-derived cone viability factor. Nat Genet 2004;36:755–759.

128. Mata NL, Radu RA, Clemmons RC, Travis GH. Isomerization and oxidation of vitamin a in cone-dominant retinas: a novel pathway for visual-pigment regeneration in daylight. Neuron 2002;36:69–80.

129. Liu H, Ren JG, Cooper WL, Hawkins CE, Cowan MR, Tong PY. Identification of the anti-vasopermeability effect of pigment epithelium-derived factor and its active site. Proc Natl Acad Sci USA 2004;101:6605–6610.

130. Yamagishi S, Amano S, Inagaki Y, Okamoto T, Takeuchi M, Inoue H. Pigment epithelium-derived factor inhibits leptin-induced angiogenesis by suppressing vascular endothelial

growth factor gene expression through anti-oxidative properties. Microvasc Res 2003; 65:186–190.

131. Duh EJ, Yang HS, Suzuma I, et al. Pigment epithelium-derived factor suppresses ischemia-induced retinal neovascularization and VEGF-induced migration and growth. Invest Ophthalmol Vis Sci 2002;43:821–829.

132. Chader GJ. PEDF: Raising both hopes and questions in controlling angiogenesis. Proc Natl Acad Sci USA 2001;98:2122–2124.

133. Dawson DW, Volpert OV, Gillis P, et al. Pigment epithelium-derived factor: a potent inhibitor of angiogenesis. Science 1999;285:245–248.

134. Holmes TM, Lawrence JM, Butt RP, Lund RD. Development of retinal vascular complexes in the Royal College of Surgeons (RCS) rat and the effects of pharmaceutical intervention with pigment epothelial derived factor (PEDF) and echistatin. Invest Ophthalmol Vis Sci 2002; Abstract no. 34–86.

135. Dintelmann TS, Heimann K, Kayatz P, Schraermeyer U. Comparative study of ROS degradation by IPE and RPE cells in vitro. Graefes Arch Clin Exp Ophthalmol 1999;237:830–839.

136. Schraermeyer U, Kociok N, Heimann K. Rescue effects of IPE transplants in RCS rats: short-term results. Invest Ophthalmo Vis Sci 1999;40:1545–1556.

137. Schraermeyer U, Kayatz P, Thumann G, et al. Transplantation of iris pigment epithelium into the choroid slows down the degeneration of photoreceptors in the RCS rat. Graefes Arch Clin Exp Ophthalmol 2000;238:979–984.

138. Kano T, Abe T, Tomita H, Sakata T, Ishiguro S, Tamai M. Protective effect against ischemia and light damage of iris pigment epithelial cells transfected with the BDNF gene. Invest Ophthalmol Vis Sci 2002;43:3744–3753.

139. Hojo M, Abe T, Sugano E, et al. Photoreceptor protection by iris pigment epithelial transplantation transduced with AAV-mediated brain-derived neurotrophic factor gene. Invest Ophthalmol Vis Sci 2004;45:3721–3726.

140. Abe T, Yoshida M, Tomita H, et al. Auto iris pigment epithelial cell transplantation in patients with age-related macular degeneration: short-term results. Tohoku J Exp Med 2000;191:7–20.

141. Lappas A, Weinberger AW, Foerster AM, Kube T, Rezai KA, Kirchhof B. Iris pigment epithelial cell translocation in exudative age-related macular degeneration. A pilot study in patients. Graefes Arch Clin Exp Ophthalmol 2000;238:631–641.

142. Thumann G, Aisenbrey S, Schraermeyer U, et al. Transplantation of autologous iris pigment epithelium after removal of choroidal neovascular membranes. Arch Ophthalmol 2000;118:1350–1355.

143. Lappas A, Foerster AM, Weinberger AW, Coburger S, Schrage NF, Kirchhof B. Translocation of iris pigment epithelium in patients with exudative age-related macular degeneration: long-term results. Graefes Arch Clin Exp Ophthalmol 2004;242:638–647.

144. Semkova I, Kreppel F, Welsandt G, et al. Autologous transplantation of genetically modified iris pigment epithelial cells: a promising concept for the treatment of age-related macular degeneration and other disorders of the eye. Proc Natl Acad Sci USA 2002;99:13,090–13,095.

145. LaVail MM, Faktorovich EG, Hepler JM, et al. Basic fibroblast growth factor protects photoreceptors from light-induced degeneration in albino rats. Ann NY Acad Sci 1991;638:341–347.

146. Akimoto M, Miyatake S, Kogishi J, et al. Adenovirally expressed basic fibroblast growth factor rescues photoreceptor cells in RCS rats. Invest Ophthalmol Vis Sci 1999;40:273–279.

147. Wu WC, Lai CC, Chen SL, et al. Gene therapy for detached retina by adeno-associated virus vector expressing glial cell line-derived neurotrophic factor. Invest Ophthalmol Vis Sci 2002;43:3480–3488.

148. Hauswirth WW, Li Q, Raisler B, et al. Range of retinal diseases potentially treatable by AAV-vectored gene therapy. Novartis Found Symp 2004;255:179–194.

149. McGee Sanftner LH, Abel H, Hauswirth WW, Flannery JG. Glial cell line derived neurotrophic factor delays photoreceptor degeneration in a transgenic rat model of retinitis pigmentosa. Mol Ther 2001;4:622–629.

150. Machida S, Tanaka M, Ishii T, Ohtaka K, Takahashi T, Tazawa Y. Neuroprotective effect of hepatocyte growth factor against photoreceptor degeneration in rats. Invest Ophthalmol Vis Sci 2004;45:4174–4182.

151. Lau D, McGee LH, Zhou S, et al. Retinal degeneration is slowed in transgenic rats by AAV-mediated delivery of FGF-2. Invest Ophthalmol Vis Sci 2000;41:3622–3633.

152. Cayouette M, Smith SB, Becerra SP, Gravel C. Pigment epithelium-derived factor delays the death of photoreceptors in mouse models of inherited retinal degenerations. Neurobiol Dis 1999;6:523–532.

153. Lambiase A, Aloe L. Nerve growth factor delays retinal degeneration in C3H mice. Graefes Arch Clin Exp Ophthalmol 1996;234 Suppl 1:S96–S100.

154. Neuberger TJ, De Vries GH. Distribution of fibroblast growth factor in cultured dorsal root ganglion neurons and Schwann cells. II. Redistribution after neural injury. J Neurocytol 1993;22:449–460.

155. Sendtner M, Stockli KA, Thoenen H. Synthesis and localization of ciliary neurotrophic factor in the sciatic nerve of the adult rat after lesion and during regeneration. J Cell Biol 1992;118:139–148.

156. Meyer M, Matsuoka I, Wetmore C, Olson L, Thoenen H. Enhanced synthesis of brain-derived neurotrophic factor in the lesioned peripheral nerve: different mechanisms are responsible for the regulation of BDNF and NGF mRNA. J Cell Biol 1992;119:45–54.

157. Ogilvie JM, Speck JD, Lett JM. Growth factors in combination, but not individually, rescue rd mouse photoreceptors in organ culture. Exp Neurol 2000;161:676–685.

158. Caffe AR, Soderpalm AK, Holmqvist I, Van Veen T. A combination of CNTF and BDNF rescues rd photoreceptors but changes rod differentiation in the presence of RPE in retinal explants. Invest Ophthalmo Vis Sci 2001;42:275–282.

159. Perry VH, Hayes L. Lesion-induced myelin formation in the retina. J Neurocytol 1985; 14:297–307.

160. Holmes T, Lu B, Sauvé Y, et al. Schwann cell therapy sustains long-term visual function in the Royal College of Surgeons rat. Invest Ophthalmol Vis Sci 2005; Abstract no. 1658.

161. Lawrence JM, Keegan DJ, Muir EM, et al. Transplantation of Schwann cell line clones secreting GDNF or BDNF into the retinas of dystrophic Royal College of Surgeons rats. Invest Ophthalmol Vis Sci 2004;45:267–274.

162. Li Y, Decherchi P, Raisman G. Transplantation of olfactory ensheathing cells into spinal cord lesions restores breathing and climbing. J Neurosci 2003;23:727–731.

163. Cao L, Liu L, Chen ZY, et al. Olfactory ensheathing cells genetically modified to secrete GDNF to promote spinal cord repair. Brain 2004;127:535–549.

164. Boruch AV, Conners JJ, Pipitone M, et al. Neurotrophic and migratory properties of an olfactory ensheathing cell line. Glia 2001;33:225–229.

165. Lipson AC, Widenfalk J, Lindqvist E, Ebendal T, Olson L. Neurotrophic properties of olfactory ensheathing glia. Exp Neurol 2003;180:167–171.

166. Lund R, Budko E, Lu B, et al. Assessment of olfactory ensheathing cells as a cell-based therapy to prevent photoreceptor degeneraion in the Royal College of Surgeons rat. Invest Ophthalmol Vis Sci 2005; Abstract no. 1653.

167. Mizumoto H, Mizumoto K, Whiteley SJ, Shatos M, Klassen H, Young MJ. Transplantation of human neural progenitor cells to the vitreous cavity of the Royal College of Surgeons rat. Cell Transplant 2001;10:223–233.

168. Young MJ, Ray J, Whiteley SJ, Klassen H, Gage FH. Neuronal differentiation and morphological integration of hippocampal progenitor cells transplanted to the retina of immature and mature dystrophic rats. Mol Cell Neurosci 2000;16:197–205.
169. Takahashi M, Palmer TD, Takahashi J, Gage FH. Widespread integration and survival of adult-derived neural progenitor cells in the developing optic retina. Mol Cell Neurosci 1998;12:340–348.
170. Arnhold S, Klein H, Semkova I, Addicks K, Schraermeyer U. Neurally selected embryonic stem cells induce tumor formation after long-term survival following engraftment into the subretinal space. Invest Ophthalmol Vis Sci 2004;45:4251–4255.
171. Banin E, Obolensky A, Idelson M, et al. Retinal incorporation and differentiation of neural precursors derived from human embryonic stem cells. Stem Cells 2006;24:246–257.

18

Current Status of IPE Transplantation and Its Potential as a Cell-Based Therapy for Age-Related Macular Degeneration and Retinal Dystrophies

Makoto Tamai, MD, PhD

CONTENTS

INTRODUCTION

The vital properties of the neural retina that ensures good vision are maintained by the highly specialized functions of the retinal pigment epithelium (RPE) cells. Among these functions are a physical barrier with tight junctions, absorption of stray light, metabolic/biochemical phagocytosis and vitamin A metabolism, developmental/trophic support (cytokines), and transport of ion, amino acids, and vitamin A. In addition, the microenvironment of the subretinal space between the photoreceptor cells of the sensory retina and the choriocapillaris (1) is maintained by RPE cells. For example, each RPE cell is estimated to phagocytize about 3×10^8 discs during a 70-yr life span, and RPE cells are the scavenger operating under oxidative stress during continuous light stimulation (2). Abnormalities in any of these functions of the RPE cells will lead to disturbances of the microenvironment and result in the programmed cell death of the photoreceptor cells (3).

Genomic mutations of genes controlling structural or functional proteins that are present in either the photoreceptors or RPE cells are responsible for hereditary retinal degenerations (4,5). In addition, oxidative stress can lead to the invasion of choroidal

From: *Ophthalmology Research: Retinal Degenerations: Biology, Diagnostics, and Therapeutics*
Edited by: J. Tombran-Tink and C. J. Barnstable © Humana Press Inc., Totowa, NJ

neovascularization (CNV) membranes into the subretinal space in eyes with age-related macular degeneration (AMD). Both of these conditions, retinal degeneration and AMD, are the most prevalent hereditary and age-related causes of severe loss of vision in the developed countries.

At present, there are no established and effective treatments for these diseases but partial success in rescuing photoreceptor cells have been attained by the transplantation of RPE cells *(6,7)* and iris pigment epithelial (IPE) cells in experimental animals *(8)*. These experiments have been conducted on such animals as *rd* and *rds* mice, Royal College of Surgeon (RCS) and Fisher-344 rats, and the results suggested that transplantation of RPE cells may be an alternative treatment for human retinal diseases. Interestingly, the transplantation of pigmented cells or neural retina has already been performed on patients with retinitis pigmentosa (RP) or AMD.

In this chapter, I will review the basic and clinical studies that are related to the transplantation of IPE cells, and the results we have obtained with IPE transplantation as a developing therapy for RP and AMD.

EVIDENCE FOR RPE PROLIFERATION/MIGRATION OF RPE CELLS IN SUBRETINAL SPACE AND FUNCTIONAL RECOVERY

Clinical Observations of a Case

A 43-yr-old man was referred to Tohoku University Hospital because of a sudden development of a visual field defect in his right eye on November 16, 1991. His best corrected visual acuity was 0.5 ocular dexter (OD) and 1.5 ocular sinister (OS). The anterior segment of the right eye showed no signs of inflammation but a severe serous detachment was observed in the inferior half of the retina including macular area. His vision soon recovered to 1.5 OD, but a large RPE tear was found in the temporal posterior retina (Fig. 1A) under the detached retina that showed diffuse fluorescein leakage. The fovea centralis was spared, so his vision was still 1.5 OD. The detached RPE curled into a tight circular bundle as can be seen in Fig. 1A–G. Follow-up examination by ophthalmoscopy and fluorescein angiography showed a gradual increase of pigmentation and a decrease of fluorescein leakage, especially in the marginal zone with bared RPE. His vision was good, so we were able to follow his visual field changes precisely by Humphrey perimetry. Initially, the sensitivity corresponding to the bared area decreased, but there was a progressive recovery during the following 6 mo (Fig. 1B,E,H). We suggest that the functional recovery in this patient was the result of the proliferation and migration of RPE cells into the bared area just as in experimental animals *(9)*.

RPE tears are not rare and can be found in patients with multiple posterior pigment epitheliopathy, and quite often in eyes with age-related RPE detachment. Repair of areas bared by laser photocoagulation have been observed in clinical cases *(10)*. Because of the capricious recovery of the functional processes, an accurate visual prognosis is difficult in such cases.

A disturbance of Bruch's membrane caused by the accumulation of debris, such as neutral lipids or phospholipids *(11)*, has been suggested to be one of the factors affecting the repair of RPE cells. If an RPE tear happens in an elderly person, the visual

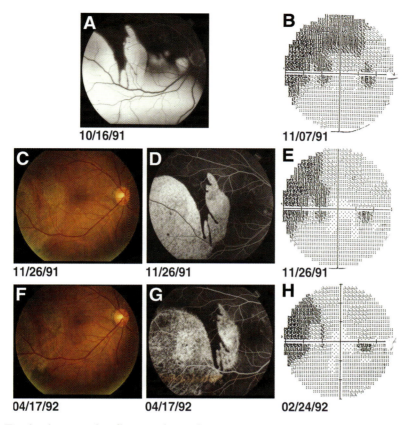

Fig. 1. Funds photographs, fluorescein angiograms, and Humphrey visual fields recorded at the follow-up periods indicated under each picture. Just after a large RPE tear developed, active fluorescein leakage with serous retinal detachment was observed in the bared area **(A)**. About 3 wk later, perimetry showed a large visual defect corresponding to the RPE tear with better sensitivity at the RPE-remaining area **(B)**. Follow-up examinations of fluorescein angiography **(D,G)** and perimetry **(E,H)** show the recovery of the barrier function of RPE-bared area and reduction of scotoma indicated by arrows and enlargement and better sensitivity in the area indicated by arrowheads.

prognosis is usually poor, but in younger persons, as shown here, reconstruction of the blood–retinal barrier and recovery of the retinal function of the torn area can occur.

Rearrangement and Function of Transplanted RPE Cells Under the Neural Retina: Discrepancy Between Basic and Clinical Observations

RPE cells transplanted into the subretinal space can survive (monkey *[12]*), regenerate and proliferate (rabbit *[9,13]*; monkey *[14]*), rescue photoreceptor cells (RCS rats *[15]*), and delay the progression of age-related cell death (Fisher-344 rat *[7]*). Within 7 d after RPE ablation, the blood–retinal barrier function and tight junction complexes are restored by the infiltration of no-pigmented fibroblast-like cells in rabbits *(8)*, and the density of regenerated RPE cells is increased by more than four times the preoperative level. These cells also secrete neurotrophic (NT) factors such as basic fibroblast growth factor (bFGF) *(13)*. The regenerated cells lack melanin and are undifferentiated, but after 9 mo, the repopulated RPE cells became pigmented. These observations indicate that RPE cells in the subretinal space can differentiate with a longer time period *(14)*.

The surgical removal of subfoveal hemorrages and/or CNV membranes is performed as an alternative AMD treatment among vitreous surgeons in the developed countries. However, as the RPE cells constitute part or the margin of the CNV *(16,17)*, surgical removal of a CNV must result in the loss of RPE cells. The mean number of cells lost has been calculated to be 1.52×10^4 cells *(18,19*; Fig. 2A). A functional recovery of the retina has been observed in some of these cases, but in many, the recovery of vision was to 0.1 or less (Fig. 2B). Moreover, the pigmented areas were not necessarily used as the fixation area as determined by microperimetry, and scotomas were present in some patients. These observations suggest that the proliferation and migration of residual RPE cells do not always lead to functional recovery under these clinical conditions, and the necessity of supplemental RPE cells or other types of cells may be necessary to obtained better vision.

COMPARISONS OF CHARACTERISTICS OF RPE AND IPE CELLS AND RESULTS OF SIMULTANEOUS SUBMACULAR SURGERY AND CULTURED AUTOLOGOUS IPE TRANSPLANTATION IN EYES WITH AMD

Based on these basic and clinical observations, RPE cells were isolated from eye bank eyes or fetal human eyes and cultured for allograft transplantation in patients with AMD *(20,21)*. Following transplantation, ophthalmoscopy showed a thickening in the transplanted area, and fluorescein angiography (FA) showed fluorescein leakage. With time, there was a gradual decrease of vision suggesting a host-graft rejection, and these patients required careful monitoring for tissue rejection *(22,23)*.

On the other hand, the transplantation of autologous RPE cells into eyes with exudative AMD was reported to improve the visual acuity *(24,25)*. These results indicate that the subretinal space was not a completely immunologically privileged site *(26)*, and systemic immunity can exert a significant influence on nonautologous transplanted cells into the subretinal space when there is a breakdown of the blood–retinal barrier *(27)*. Thus, allograft transplantation can elicit host-graft rejection in humans even if the intraocular space is considered to be immunologically privileged.

To avoid or minimize the host-graft rejection in patients, autologous cells should be used, and IPE cells may be the best alternative to RPE cells, because IPE cells have the same embryonic origin and sufficient numbers can be easily obtained by peripheral iridectomy.

Comparisons of RPE and IPE Cells

Methods of Culture and Proliferation Rates

The isolation and culturing of RPE and IPE cells have been reported in detail *(28,29)*. In brief, after removing the anterior segment and vitreous from eyes, the eyecups are incubated in a calcium- and magnesium-free Hank's balanced salt solution (HBSS) supplemented with trypsin (0.05%)/ethylenediaminetetraaminicacid (EDTA; 0.53 m*M*) for 40 min at 37°C in a 5% CO_2 incubator. Human IPE cells were separated from the

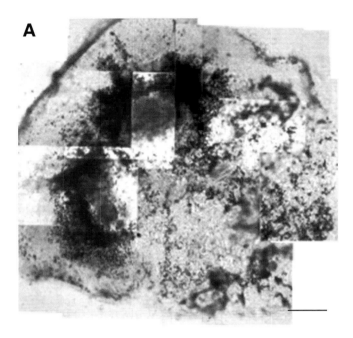

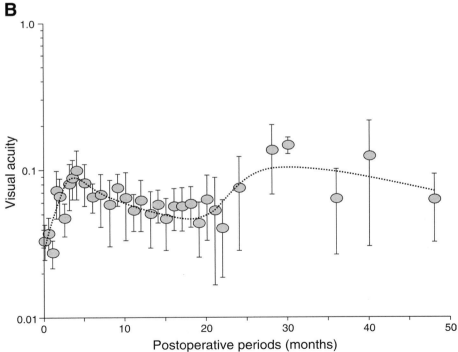

Fig. 2. (A) Removal of a choroidal neovascular membrane by submacular surgery. A large amount of RPE and melanin-laden cells are removed. (Bar: 200 μm). **(B)** Visual acuity (*y*-axis) and follow-up periods after choroidal neovascular membrane removal (*x*-axis) of 45 patients. During the 3 mo after surgery, the visual acuity improved rapidly but then was stationary.

iris stroma by incubating the iris tissue in 2.5% trypsin/0.05% EDTA for 20 min. After incubation, the RPE or IPE cells were mechanically detached from the Bruch's membrane or the iris stroma, and individual cells were separated by pipetting through a glass pipet. The RPE or IPE cells were collected in 20% fetal bovine serum and F-12 medium and seeded into modified polystyrene dishes at 37°C in 5% CO_2. The maximum enhancement of cell proliferation was obtained with an autologous serum concentration of 10 to 15%.

The proliferation rates of monkey IPE and human RPE cells that were passaged under the same conditions were both exponential, but the proliferation rate was higher for RPE cells ($R = 0.970$ to 0.986) than for IPE cells ($R = 0.883$ to 0.933). The proliferation rate of human IPE was also exponential ($R = 0.822$), and we were able to collect about 60,000 IPE cells after about 4 to 5 wk (Fig. 3A,B). *(28)*

mRNA Expression of Melanogenesis, Cytokines,
and Their Receptors, and Melanin Concentration

Tyrosinase and tyrosinase-related protein-1 and -2 are critical for regulating the synthesis of melanin pigments. The melanogenesis-related genes in fresh adult RPE were amplified by reverse transcriptase-polymerase chain reaction (RT-PCR) except for tyrosinase, which is expressed only in fetal RPE cells *(30,31)*. The expression level of these genes was higher in bFGF-supplemented growth medium.

Human adult IPE cells expressed these melanogenesis-related genes including tyrosinase in vitro for more passages than human RPE cells *(30)*. Their expression level was enhanced by bFGF supplementation. The level of expression of these genes in human IPE cells decreased after passage 8, and the amount of melanin decreased rapidly with successive passages.

RPE cells also synthesize a number of cytokines and their receptors which are important for the proper functioning and the creation of the microenvironment of RPE cells. The messenger RNA (mRNA) expressions of cytokines and their receptors in cultured IPE cells were compared with those of RPE cells *(32)*. No qualitative or quantitative differences were found in 94% of 36 cytokines or their receptors in cultured adult IPE and RPE cells. But the mRNA expression levels of vascular endothelial growth factor (VEGF) and its receptor 2 were lower in IPE than in RPE cells.

The mRNA of cellular retinaldehyde binding protein (CRALBP) was expressed in RPE, but not in IPE cells *(31)*. Low levels of VEGF and its receptor may be beneficial, but the lack of expression of CRALBP is not a good property of IPE cells for transplantation for the treatment of AMD.

Phagocytosis of Photoreceptor Outer Segments and Activities of Lysosomal Enzymes,
Cathepsin D, and Cathepsin S

The removal and elimination of shed outer segments are essential for the renewal process of the photoreceptor membranes. Thus, the ability and capacity of the transplanted cells to phagocytose the photoreceptor outer segments is very important especially if IPE cells are transplanted. To determine if cultured IPE cells are able to phagocytose outer segments, isolated photoreceptor outer segments were labeled with Alexa 488 dye and fed to cultured IPE cells. After 24 h of exposure, the excess outer segments were washed out, and the amount of outer segments phagocytized was determined by examining the culture under a fluorescence microscope with fluorescein

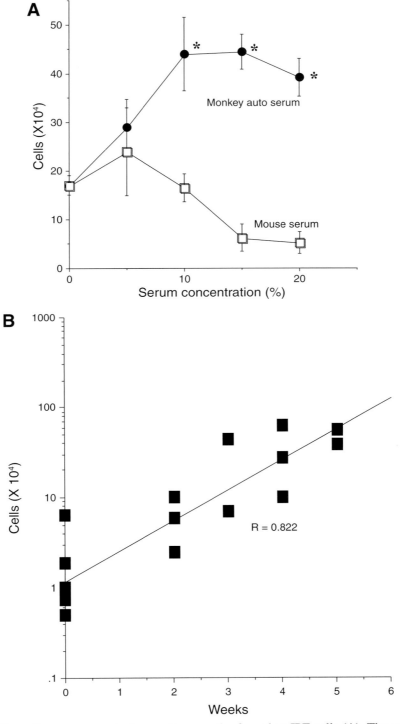

Fig. 3. Effect of serum concentration on the growth of monkey IPE cells (**A**). The y-axis represents the number of cells. The optimal serum concentration was around 10–15% (*$p < 0.05$, compare to 0% serum). The growth curve of human IPE cells obtained from 10 patients (**B**).

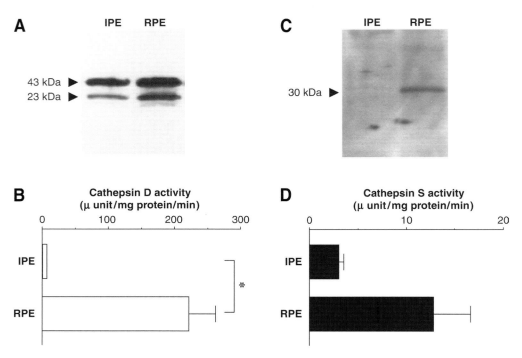

Fig. 4. Western blot analysis of cathepsin D **(A)** and S **(B)** in cultured rat IPE and RPE cells. Cathepsin D **(C)** and S **(D)** activities were measured using fluorogenic substrates. (Student's *t*-test, *$p < 0.05$, **$p < 0.01$) Error bars represent SEM ($n = 5$–10).

isothiocyanate. Indeed, the IPE cells were able to phagocytose outer segments, and their capacity was about 70% of that of RPE cells *(33,34)*. The level of phagocytosis was enhanced by the stable expression of neurotrophines such as bFGF as shown in Fig. 4A.

The lysosomal functions of the cultured IPE cells were compared to that of RPE cells by measuring the mRNA expression of cathepsin D and S enzyme activities and protein levels *(35)* (Fig. 4). Lysosomal activities were present in IPE cells, but both cathepsin D and S activities were significantly lower than those in RPE cells. These results supported our findings that the ability to digest ROS by the IPE cells is lower than that by RPE cells.

Survival of Transplanted Cells in Subretinal Space

Cultured autologous IPE cells were transplanted into the subretinal space of monkey eyes. After the transplantation, fluorescein leakage was not detected, and a thickening of the retina was not observed. We were able to identify the autologous IPE cells in the subretinal space 6 mo after transplantation by histological and electron microscopic examinations. These cells appeared to be less pigmented and to have less mitochondria than did the host RPE cells, and some of the grafted cells were seen to have phagocytized photoreceptor outer segments *(17)*. These observations suggest that transplanted autologous cells can survive for relatively long periods in the subretinal space and can affect the surrounding environment.

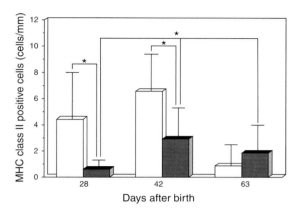

Fig. 5. The number of MHC class-II positive cells in RCS rat retinas during the developmental stage. Rat RPE cells were transplanted into the subretinal space in RCS rats (closed column). Sham operation was performed in RCS rats (white column). $*p < 0.01$; Welch t-test.

Host-Graft Rejection

The ability of transplanted cells to rescue photoreceptor cells was evaluated in dystrophic RCS rats. RPE cells were used for iso- or allografts *(36,37)*, IPE cells for allograft *(9)*, and human RPE cells for xenograft *(38)*. In most of these experiments, the transplanted cells were found to rescue the photoreceptor cells, and no evidence of rejection was detected *(36,37)*. However, after the xenograft transplantation of RPE cells grown on collagen sheets into the subretinal space or anterior chamber of rabbits, the retinas appeared to be well-preserved histochemically, but there was evidence of rejection and decreased electroretinographic responses. These results suggest a graft rejection of nonself cells *(31,39)* even in the intraocular environment. Zhang and Bok *(27)* also observed the absence of an acute immune rejection, but presented evidence for chronic rejection. They warned of an imperfect immunologically privileged status of the subretinal space, and this needs to be considered in future trials of allograft transplantation of RPE cells in humans.

The major histocompatibility complex (MHC) is the genetic region where immunological events, such as immune responses and rejections following transplantation, are regulated in all mammals. Quantitative analyses were performed with immunohistochemical methods for MHC class II antigen (OX6), a cytoplasmic antigen in bone marrow-derived macrophages (ED1), and a microglia/macrophage marker (OX42) in the RPE cells following allograft transplantation. Immunohistologically positive cells were counted on postnatal days (P)10 to 140 in animals without transplantation and at P80 and 2 mo after transplantation (Fig. 5). The results showed that MHC class II-positive cells appeared in the outer nuclear layer and debris of outer segments, and the numbers increased with increasing days after transplantation. But the number of positive cells were significantly ($p < 0.05$) less with RPE cell transplantation *(31,40)*. These results suggest a short-term rescue effect even with allograft transplantation.

Cytokine Gene Expressions in Transplanted Cells

The results of these transplantation studies showed that photoreceptor cells from dystrophic animals can be rescued by the transplantation of RPE cells, but there was

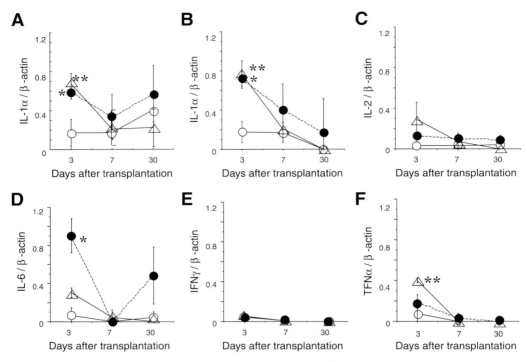

Fig. 6. Semiquantitative RT-PCR of rat for IL-1α (**A**), -1β (**B**), -2 (**C**), -6 (**D**), IFN-γ (**E**), and TNF (**F**) in cultured human RPE cells. The expression of each cytokine gene is expressed as ratios of the internal control, β-actin. Closed circles represent cultured human RPE cells; slashed triangles; cultured rat IPE; and open circles, vehicle only.

some evidence for the presence of inflammatory cytokines at the transplanted site with allografts *(41)*. These observations indicated that cytokines may play an important role in the success of the preservation of the photoreceptor cells.

To determine whether cytokines were influencing the results, we measured the expression of the cytokine genes with RT-PCR and semi-quantitative PCR at the transplanted sites in RCS rats. For these experiments, approx 8×10^4 cultured human RPE cells in a 2-μL suspension was injected through a scleral incision in the superior part of the eye of RCS rats. Cultured RPE cells were injected into the subretinal space; human RPE cells as xenografts, Long Evans rat IPE cells as allografts, and culture medium as control. The level of expression of the genes for rat interleukin (IL)-1α, -1β, -2, -6, interferon (IFN)-γ, and tumor necrosis factor (TNF)-α were determined in tissues obtained 3 d after the transplantation, and the results are shown in Fig. 6 along the level of β-actin as a baseline.

The level of expression of the genes for IL-1α, -1β, and -6 were significantly up-regulated in the transplanted human RPE cells compared to that following the injection of vehicle only. Also, a statistically significant up-regulation was detected for IL-1α, -1β, and TNF-α 3 d after the transplantation of rat IPE cells compared to that of vehicle only. Thus, the strong expression of these cytokines may have influenced the rescue of photoreceptor cells and also the inflammatory reaction and graft rejection *(42)*.

Conclusions

Based on these experimental data, we concluded that the transplanted autologous IPE cells are functional in the subretinal space and may be the best alternative for transplantation to eliminate the problem of rejection.

Submacular Surgery and Cultured Autologous IPE Transplantation in AMD: Clinical Studies

Because of the results from the basic experiments, we decided to transplant autologous IPE cells at the time of the surgical removal of CNV in AMD patients. The procedures used conformed to the tenets of the Declaration of Helsinki, and informed consent was obtained from all subjects who participated. The nature of the surgical procedures, the possible consequences, and the materials to be used were carefully explained to each patient. This study was also approved by the Ethics Committee of Tohoku University on January 26, 1998.

Clinical Procedures

Preparation of autologous IPE cells and the surgical procedures have been reported in detail *(18,43,44)* and are also presented in the Methods of Culture and Proliferation Rates section of this article. In brief, iris tissue from the patients about to undergo transplantation was obtained under local anesthesia. A peripheral iridectomy (2 to 3 mm) was performed at the 11 o'clock position from the same eye to receive the transplantation. The IPE cells were mechanically detached from the stroma in balanced salt solution (BSS), and individual cells were isolated by pipetting through a glass pipet. By pipetting gently while viewing the procedure under a dissecting microscope, we were able to collect isolated IPE cells in Ham's F-12 medium supplemented with 15% autologous serum which had been obtained from venous blood of the same patient earlier. The medium was changed every three days. When the IPE cells attained confluency on the dishes approx 3 to 5 wk after iridectomy, the patient was admitted into the hospital *(28)*. Primary IPE cells at about 50% confluency were suspended in an EDTA solution (0.125% trypsin and 0.2 mM EDTA), and the cells were washed with BSS three times and resuspended in 15–20 µL of BSS for 30 min before transplantation. Approximately 2 to 20×10^4 cells were injected into the subretinal space.

For the surgery, a weak diathermal cauterization of the retina was performed, and a retinotomy site was created. Then a small retinal bleb was made by injecting BSS through a bent 30-gage needle and the CNV was detached from the neural retina and underlying Bruch's membrane. The CNV was then removed through the retinotomy site. The subretinal space was then gently flushed, and the cultured IPE cells were injected through the retinotomy site. In some of the eyes, a special double-barreled microinjection–aspiration needle was used *(44)*.

Results

Thirty-five patients with AMD received the IPE transplantation, and they were followed without any immunosuppressive drugs. One criterion used for the selection of patients was that their vision was worse than 0.1. They were followed pre- and postoperatively by ophthalmic examinations including visual acuity, perimetry, flicker, and single-flash electroretinograms, fundus photography, FA, and indocyanin green angiography with a scanning laser ophthalmoscope, and optical coherence tomography *(18,43–45)*.

There were no signs of rejection, no proliferative changes in the submacular region or vitreous cavity even without any immunosuppressive agents, and no recurrence of a submacular CNV. Although more than 60% of patients had an improvement in visual acuity after the transplantation, the best visual acuity was 0.3. This was not statistically better than that with a simple removal of CNV *(18, 4–45)*. No significant improvements were detected in the retinal functions determined by the other tests compared to the group with simple CNV removal. These results suggest that although the transplanted autologous IPE cells may have some of the functions of RPE cells, they had limited influence on the final visual acuity.

CHARACTERISTICS OF NEUROTROPHIC GENE TRANSFECTED IPE CELLS: BASIC STUDIES

The mutations of genes associated with the light-dependent phototransduction cascade or independent pathways induce apoptotic photoreceptor cell death. Prolonged light exposure and ischemia can also induce apoptotic photoreceptor cell death. In AMD, on the other hand, the photoreceptor cell death is caused by oxidative stress. The retinal degeneration induced by prolonged light exposure, aging or genetic mutations share a final common pathway, namely, apoptotic cell death. *(46–48)*.

Many experiments have shown that NT factors can slow the progression of photoreceptor cell death induced by these causes and preserve retinal function *(49–51)*. We have thus hypothesized that if transplanted autologous IPE cells could be made to supply the critical NT factors to the photoreceptors, the survival of photoreceptor cells would be enhanced *(52,53)*. We have not been able to examine the human eyes that have received transplanted autologous IPE cells, but the ineffectiveness in improving vision may be caused by their low rate of survival or differentiation on Bruch's membrane *(54)*. But if these transplanted autologous IPE cells could be made to produce sufficient amounts of NT factors to the neighboring cells, some positive survival effects may be provided to the photoreceptor cells.

The NT factor genes, such as nerve growth factor family members, have multiple functions on developing and mature neurons *(55)*. Among the neurotrophins, brain-derived NT factor (BDNF) has been reported to be the most abundant in the adult brain *(56)* and is well characterized on its effects on regeneration, synaptic modulation, and neuroprotection in the visual system in various retinal injury models in vivo and in vitro *(57)*, e.g., in ischemia-reperfusion injury *(19)*, light-induced photoreceptor damage *(52)*, and in animal models of inherited retinal degeneration *(50,51)*. It is possible that some NT factors could protect the retinal structure but not the function *(58)*. It is significant that BDNF is reported to have no negative effects.

Phagocytic Activity of Photoreceptor Outer Segments by NT Transfected IPE Cells

The phagocytic ability and capacity of cultured IPE cells were compared to that of RPE cells. The cultured IPE cells had 70% of the phagocytic activity of RPE cells, but IPE cells transduced with the bFGF gene had phagocytic activity levels comparable to

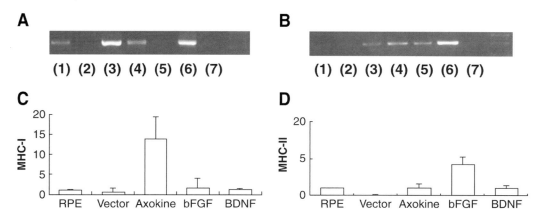

Fig. 7. Semiquantitative RT-PCR analysis of MHC-I and –II in gene-transduced RPE cells. Results of RT-PCR of MHC-I (**A**) and –II (**B**) are shown. Axokine enhanced the MHC-I expression (**C**). MHC-II expression was enhanced by bFGF-transfected cells (**D**).

that of RPE cells *(33,34)*. This improvement was significantly decreased by pretreatment of the IPE-bFGF cells with anti-bFGF antibody *(34)*.

Effects of Transplanting RPE Cells Transduced With NT Factor Genes (Axokine, bFGF, and BDNF) on MHC Class I and II Expression and Inflammatory Reactions

The expression of the MHC class I and II after transfection of three NT factor (Axokine, BDNF, and bFGF) genes was examined by real-time PCR. The ratio of CD4+ and CD8+ T cells and antibody production against the transplanted cells were also analyzed by flow cytometry *(59)*. The RPE cells transduced with Axokine induced 14 times more MHC class I expression than that of nontransduced RPE cells. The RPE cells transduced with bFGF increased the expression of MHC class II four times more than that of nontransfected RPE cells. In contrast, RPE cells transduced with BDNF induced a very weak expression of both MHC class I and II (Fig. 7). We found a cluster of genes in the RPE cells with transfection of bFGF.

Expression of BDNF and bFGF by Other NT Gene Transfection in RPE Cells and Subretinal Transplantation and Effects of Overexpression of NT Factors

It is clear that NT factors can enhance the survival of degenerating cells in vivo and in vitro, but it is important to determine the effects of the overexpression of the NT factors in the subretinal space or in the vitreous cavity. We examined the effects of injecting bFGF and BDNF on the surrounding tissues. Expressions in the normal, Axokine-, bFGF-, BDNF-, and vector-transduced RPE cells were quantified by RT-PCR. Axokine is not expressed in normal RPE cells, so the expression was not altered by the other NT gene transduction. bFDF and BDNF are endogenously expressed in normal cultured rat RPE cells, and the level of expression of these NT factors was affected by the other NT gene transduction. The expression in the normal and BDNF-transduced RPE cells was quantified by RT-PCR. The BDNF-transduced RPE cells expressed three times higher levels than non- or vector-transduced RPE cells. The expression of BDNF by Axokine- or bFGF-transduced cells was less than that of cultured cells or vector-transduced RPE cells (59).

Transduction Efficacy of BDNF Gene Into IPE Cells With AAV Vector

To be able to use IPE cells transduced with the BDNF gene (IPE-BDNF) clinically, we determined the conditions that will increase the efficiency of transducing BDNF into IPE cells using recombinant adeno-associated virus (rAAV). Of the six AAV serotypes, serotype 2 (rAAV2) is best characterized and therefore predominantly used in gene transfer studies (60,61). Therefore, we selected rAAV2 and determined the conditions that will increase the efficiency for transducing BDNF into IPE cells.

Human IPE cells had substantially lower levels of transfection than ARPE and HT1080 cells that are highly permissive cells for rAAV2. But the use of hydroxyurea-sodium butyrate (HU-SB) successfully increased the transduction efficiency in human IPE cells, and a mixture of human IPE cells with HU-SB-Tyr (tyrphostin-1) increased the level even more. The levels in the culture medium and cell lysates were measured by sandwich enzyme-linked immunosorbent assay (ELISA) and Western blot analysis for BDNF and the findings are shown in Fig. 8A,B. To determine the multiplicity of infection (MOI) of rAAV that affected the level of transgene expression, human IPE cells were infected with rAAV-hBDNF at various MOI, and the amount of vector that resulted in an increased gene expression was determined (Fig. 8C). If the AAV2-BDNF infectious concentration was $1 \chi 10^7$ capsids/ml or higher, photoreceptor protective effects were observed and 1×10^9 was the best concentration.

Systemic Dissemination

The eye is an ideal organ for gene therapy as it is separated from other systems in the body by the blood–ocular barrier, and it is also an immunologically privileged site. In many experiments, the targeting virus vector-DNA was injected into the subretinal space through the sclera and choroid. Unfortunately, the choroid is a vascular membrane and injection through it can lead to hemorrhages. In addition, the direct injection of naked AAV vector into the subretinal space allows an extraocular dissemination to the optic nerve, brain, lung, liver, kidney, and testis, and such widespread dissemination is supposed to be circulation induced. Once the NT genes were transfected into IPE cells, chance of dissemination is decreased (62; Yoshioka et al., in preparation).

Photoreceptor Cells Rescue Effects by Subretinal Transplantation of AAV-BDNF-Transfected Cells

Initially, we used the lipofection technique to transfect rat bFGF or BDNF genes into cultured rat IPE cells (bFGF-IPE and BDNF-IPE) (53,63). The BDNF gene was also inserted into AAV2 (AAV2-BDNF) and the recombinant AAV2 was transduced into rat IPE cells as described in Expression of BDNF and bFGF by Other NT Gene Transfection in RPE Cells and Subretinal Transplantation and Effects of Overexpression of NT Factors section. The effects of transplanting these IPE cells on photoreceptor protection were observed by procedures as described (60). The expression of each neurotrophic factor by the transfected cells was examined by RT-PCR and by sandwich ELISA. The neuroprotective ability of the BDNF-IPE cells against *N*-methyl-D-aspartate

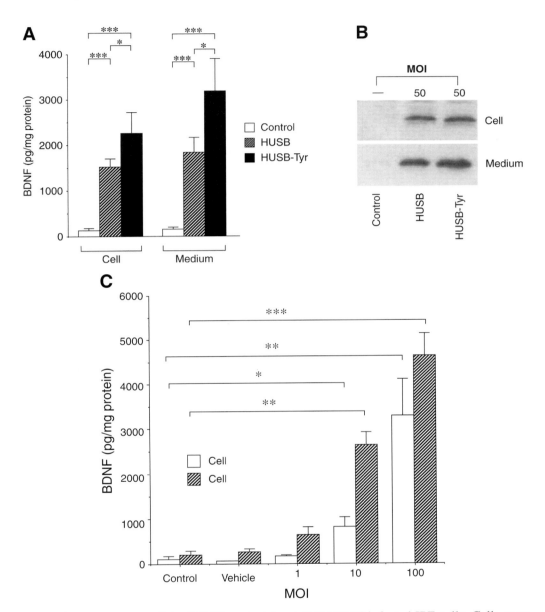

Fig. 8. Measurement of the BDNF content in rAAV-hBDNF-infected IPE cells. Cells were infected with the rAAV-hBDNF at a MOI of 50. The BDNF levels in the culture medium and cell lysates were measured by sandwich ELISA (**A**). Error bars represent standard deviation (*$p < 0.05$, **$p < 0.001$). Western blot analysis for BDNF is also shown in (**B**). Expression of BDNF in cells infected with various MOI (**C**). Human IPE cells were infected with the indicated amounts of rAAV-hBDNF following the pretreatment of a mixture of HU-SB plus Tyr (500 μM). Error bars represent standard deviation (*$p < 0.05$, **$p < 0.01$, ***$p < 0.001$).

(NMDA)-induced neurotoxicity was investigated in vitro and against an inherited mechanism in dystrophic RCS rats *(60)* or in rats with phototoxic damage *(63)* in vivo. The bFGF-IPE or BDNF-IPE cells were transplanted into the subretinal region in the superior

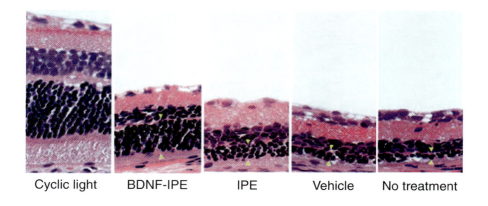

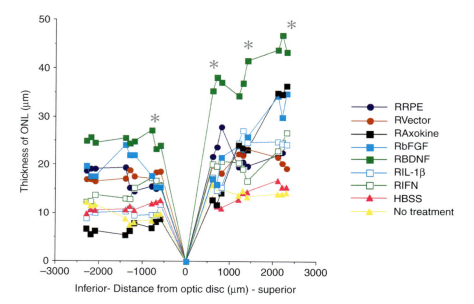

Fig. 9. Thickness of the outer nuclear layer of the retina of an eye that had received a transplantation of gene-transduced IPE cells. Rats were exposed to 3000 lux light 1 d after the transplantation of gene-transduced IPE cells into the subretinal space. Light microphotographs of rat retina taken 700 μm from the optic disc in the transplanted hemisphere of the eye.

half of the eye (Fig. 9). In the AAV2-BDNF-IPE cells, the rats were placed under constant light on days 1 and 90 after transplantation and examined.

The expression of the BDNF gene in the subretinal space was higher in AAV-BDNF-IPE transplantation than transplantation of IPE only. A statistically significant photoreceptor protection was observed on days 1 and 90 in eyes receiving the AAV2-BDNF-IPE transplant, in both the superior transplant site and the inferior hemispheres which did not receive the transplant (60).

bFGF-IPE and BDNF-IPE cells expressed higher levels of the mRNA and proteins of each NTfactor than nontransfected IPE cells (53,63). A significant increase in the protection of photoreceptor cells against NMDA neurotoxicity was observed in the neuroretinal cells cultured with BDNF-transfected IPE cells than in those cultured with nontransfected IPE cells ($p = 0.0029$) or with nontreated cells ($p = 0.0010$) (60). bFGF-IPE cells could protect photoreceptor outer segments (53). These results suggest IPE cells with transfection of the BDNF genes may be a useful tool for delivering these factors to the subretinal space and have protective effects against the mechanisms leading to the degeneration of photoreceptor cells.

OUR THERAPEUTIC STRATEGY FOR RETINAL DYSTROPHIES AND AMD

Gene therapy is now being used to treat a broad variety of diseases and many approaches for delivering the targeted genes to the appropriate sites have been attempted. Recombinant viral vectors have been the most extensively used, and the vectors can transport a gene to a cell to replace the defective gene, to suppress the expression of a mutant gene, or to deliver a protective gene to delay degeneration. The AAVs are members of the *Parvoviridae* family, and the AAV2 is one of the vectors most extensively studied and developed for clinical use (64,65). AAV2 transduction in animal models has progressed from rodents to nonhuman primates and is now being used in humans in phase I safety trials (66,67).

We are continuing our basic and clinical studies of autologous IPE cells proceeding logically from our observations described previously. Our results to date have demonstrated that IPE cells transduced with recombinant AAV2-mediated genes transplanted into the subretinal space may be the most promising treatment for protecting or slowing apoptotic photoreceptor cell death caused by retinal degeneration or AMD.

ACKNOWLEDGMENTS

I wish to thank many of my collaborators in the Department of Ophthalmology, Tohoku University School of Medicine since 1986 to the present. I wish to acknowledge especially Drs. S.-I. Ishiguro, T. Abe, K. Yamaguchi, Y. Wada, H. Tomita, and M. Yoshida for their hard work and cooperation in performing and analyzing the basic and clinical studies for many years. I also thank Dr. Duco Hamasaki for helpful advice. The main part of this chapter was presented as the special lecture at the 108th Japanese Ophthalmology Congress in April 17, 2004.

REFERENCES

1. Hewitt AT, Adler R. The retinal pigment epithelium and interphotoreceptor matrix: structure and specialized functions. In: Ryan S, ed. Retina. (St. Louis, MO: Mosby-Year Book, Inc., 1994;58–71.
2. Marshall J. The aging retina: physiology or pathology. Eye 1987;1:282–295.
3. Chang GQ, Hao Y, Wong F. Apoptosis: final common pathway of photoreceptor death in rd, rds, and rhodopsin mutant mice. Neuron 1993;11:595–605.
4. Thompson DA, Gal A. Vitamin A metabolism in the retinal pigment epithelium: genes, mutations, and diseases. Prog. Retinal Eye Res 2003;22:683–703.
5. Wada Y, Nakazawa T, Abe T, Fuse N, Tamai M. Clinical variability of patients associated with gene mutations of visual cycle protein, arrestin, RPE65 and RDH5. Invest Ophthalmol Vis Sci 2000;41(4):S617.

6. Li L, Turner JE. Optimal conditions for long-term photoreceptor cell rescue in RCS rats: The necessity for healthy RPE transplants. Exp Eye Res 1991;52:669–679.

7. Yamaguchi K, Gaur VP, Turner JE. Retinal pigment epithelial cell transplantation into aging retina: A possible approach to delay age-related cell death. Jpn J Ophthalmol 1993;37: 16–27.

8. Rezai KA, Kohen L, Wiedemann P, Heimann K. Iris pigment epithelium transplantation. Graefes Arch Clin Exp Ophthalmol 1997;235:558–562.

9. Heriot WJ, Machemer R. Pigment epithelial repair. Graefe's Arch Clin Exp Ophthalmol 1992;230:91–100.

10. Wallow IH. Repair of the pigment epithelial barrier following photocoagulation. Arch Ophthalmol 1984;102:126–135.

11. Bird A, Marshall J. Retinal pigment epithelial detachments in the elderly. Trans Ophthalmol Soc UK 1986;105:674–682.

12. Abe T, Tomita H, Kano T, et al. Autologous iris pigment epithelial cell transplantation in monkey subretinal region Curr Eye Res 2000;20:268–275.

13. Kimizuka Y, Yamada T, Tamai M. Quantitative study on regenerated retinal pigment epithelium and the effects of growth factor. Curr Eye Res 1997;16:1081–1087.

14. Valentino TL, Kaplan HJ, Del Priore LV, Fang SR, Berger A, Silverman MS. Retinal pigment epithelial repopulation in monkeys after submacular surgery. Arch Ophthalmol, 1995;113:932–938.

15. Sheedlo H, Li L, Turner J. Functional and structural characteristics of photoreceptor cells rescued in RPE-cell grafted retinas of RCS dystrophic rats. Exp Eye Res 1989;48:841–854.

16. Grossniklaus HE, Hutchinson AK, Capone A, Woolfson J, Lambert HM. Clinicopathologic features of surgically excised choroidal neovascular membranes. Ophthalmology 1994; 101:1099–1111.

17. Seregard S, Algvere PV, Berglin L. Immunohistochemical characterization of surgically removed subfoveal fibrovascular membranes. Graefe's Arch Clin Exp Ophthalmol 1994;232:325–329.

18. Abe T, Yoshida M, Kano T, Tamai M. Visual function after removal of suberetinal neovascular membranes in patients with age-related macular degeneration. Graefe's Arch Clin Exp Ophthalmol 2001;239:927–936.

19. Gass JDM. Biomicroscopic and histopathologic considerations regarding the feasibility of surgical excision of subfoveal neovascular membranes. Am J Ophthalmol 1994;118:285–298.

20. Algvere PV, Berglin L, Gouras P, Sheng Y. Transplantation of fetal retinal pigment epithelium in age-related macular degeneration with subfoveal neovascularization. Graefe's Arch Clin Exp Ophthalmol 1994;232:707–716.

21. Algvere PV, Berglin L, Gouras P, Sheng Y, Kopp ED. Transplantation of RPE in age-related macular degeneration: observations in disciform lesions and dry RPE atrophy. Graefe's Arch Clin Exp Ophthalmol 1997;235:149–158.

22. Weisz JM, Humayun MS, De Juan Jr., E, et al. Allogenic fetal retinal pigment epithelial cell transplant in a patient with geographic atrophy. Retina 1999;19:540–545.

23. Algvere PV, Gouras P, Dafgard KE, et al. Long-term outcome of RPE allografts in non-immunosuppressed patients with AMD. Eur J Ophthalmol 1997;9:217–230.

24. Peyman GA, Blinder KJ, Paris CL, Alturki W, Nelson Jr, NC, Billson FA. A technique for retinal pigment epithelium transplantation for age-related macular degeneration secondary to extensive subfoveal scarring. Ophthalmic Surgery 1991;22:102–108.

25. Binder S, Krebs I, Hilgers R-D, et al. Outcome of transplantation of autologous retinal pigment epithelium in age-related macular degeneration: a prospective trial. Invest Ophthalmol Vis Sci 2004;45:4151–4160.

26. Streilein JW. Anterior chamber associated immune deviation: the privilege of immunity in the eye. Surv Ophthalmol 1990;35:67–73.

27. Zhang X, Bok D. Transplantation of retinal pigment epithelial cells and immune response in the subretinal space. Invest Ophthalmol Vis Sci 1998;39:1021–1027.

28. Abe T, Tomita H, Ohashi T, et al. Characterization of Iris Pigment Epithelial Cell for Auto Cell Transplantation. Cell Transplant 1999;8:501–510.

29. Durlu YK, Tamai M. Transplantation of retinal pigment epithelium using viable cryopreserved cells. Cell Transplant 1997;6:149–162.

30. Abe T, Sato M, Tamai M. Dedifferentiation of the retinal pigment epithelium compared to the proliferative membranes of proliferative vitreoretinopaty. Curr Eye Res 1998;17:1103–1109.

31. Tamai M. Retinal pigment epithelial cell transplantation: Perspective. J Jpn Ophthalmol Soc 1996;100:982–1006.

32. Kociok N, Heppekausen H, Schraermeyer U, et al. The mRNA expression of cytokines and their receptors in cutured iris pigment epithelial cells: a comparison with retinal pigment epithelial cells. Exp Eye Res 1998;67:237–250.

33. Rezai KA, Lappas A, Farrokh-Siar L, Kohen L, Wiedeman P, Heimann K. Iris pigment epithelial cells of long evans rats demonstrate phagocytic activity. Exp Eye Res 1997;65:23–29.

34. Sakuragi M, Tomita H, Abe T, Tamai M. Changes of phagocytic capacity in basic fibroblast growth factor-transfected iris pigment epithelial cells in rats. Curr Eye Res 2001;23:185–191.

35. Sugano E, Tomita H, Abe T, Yamashita A, Tamai M. Comparative study of cathepsins D and S in rat IPE and RPE cells. Exp Eye Res 2003;77:203–209.

36. Li L, Turner JE. Optimal conditions for long-term photoreceptor cell rescue in RCS rats: The necessity for healthy RPE transplants. Exp Eye Res 1991;52:669–679.

37. Sheedlo HJ, Li L, Turner JE. Photoreceptor cell rescue at early and late RPE-cell transplantation periods during retinal disease in RCS dystrophic rats. J Neural Transplant Plast 1991;2:55–63.

38. Little CW, Castillo B, DiLoreto DA, et al. Transplantation of human fetal retinal pigment epithelium rescues photoreceptor cells from degeneration in the Royal College of Surgeons rat retina. Invest Ophthalmol Vis Sci 1996;37:204–211.

39. Bhatt NS, Newsome DA, Fenech T, et al. Experimental transplantation of human retinal pigment epitheliual cells on collagen substrates. Am J Ophthalmol 1994;117:214–221.

40. Akaishi K, Ishiguro S-I, Durlu YK, Tamai M. Quantitative analysis of major histocompatibility complex class II-positive cells in posterior segment of Royal College of Surgeons rat eyes. Jpn J Ophthalmol 1998;42:357–362.

41. Fealy MJ, Most D, Huie P, et al. Association of down-regulation of cytokine activity with rat hind limb allograft survival. Transplantation 1995;59:1475–1480.

42. Abe T, Takeda Y, Yamada K, et al. Cytokine gene expression after subretinal transplantation. Tohoku J Exp Med 1999;189:179–189.

43. Abe T, Yoshida M, Tomita H, et al. Functional analysis after auto iris pigment epithelial cell transplantation in patients with age-related macular degeneration. Tohoku J Exp Med 1999;189:295–305.

44. Abe T, Yoshida M, Tomita H, et al. Auto iris pigment epithelial cell transplantation in patients with age-related macular degeneration: short term results. Tohoku J exp Med 2000;191:7–20.

45. Tamai M. Progress in pathogenesis and therapeutic research in retinitis pigmentosa and age-related macular degeneration. J Jpn Ophthalmol Soc 2004;108:750–769.

46. Chang GQ, Hao Y, Wong F. Apoptosis: final common pathway of photoreceptor death in rd, rds, and rhodopsin mutant mice. Neuron 1993;11:595–605.

47. Wong P. Apoptosis, retinitis pigmentosa, and degeneration. Biochem Cell Biol 1994;72:489–498.

48. Ranganathan R. Cell biology: a matter of life or death. Science 2003;299:1677–1679.

49. LaVail MM, Yasumura D, Matthes MT, et al. Protection of mouse photoreceptors by survival factors in retinal degenerations. Invest Ophthalmol Vis Sci 1998;39:592–602.
50. Okoye G, Zimmer J, Sung J, et al. Increased expression of brain-derived neurotrophic factor preserves retinal function and slows cell death from rhodopsin mutation or oxidative damage. J Neurosci 2003;23:4164–4172.
51. Lawrence JM, Keegan DJ, Muir EM, et al. Transplantation of Schwann cell line clones secreting GDNF or BDNF into the retinas of dystrophic Royal College of Surgeons rat. Invest Ophthalmol Vis Sci 2004;45:267–274.
52. Unoki K, LaVail MM. Protection of the rat retina from ischemic injury by brain-derived neurotrophic factor, ciliary neurotrophic factor, and basic fibroblast growth factor. Invest Ophthalmol Vis Sci 1994;35:907–915.
53. Tamai M, Takeda Y, Yamada K, et al. bFGF transfected iris PE may rescue photoreceptor cell degeneration in RCS rat. In: LaVail MM, Anderson RE, Hollyfield JG, eds. Retinal Degeneration. New York: Plenum Press, 1997;323–328.
54. Itaya H, Gullapalli V, Sugino IK, Tamai M, Zarbin M. Iris pigment epithelium attachment to aged submacular human Bruch's membrane. Invest Ophthalmol Vis Sci 2004;45:4520–4528.
55. Bibel M, Barde YA. Neurotrophins: key regulators of cell fate and cell shape in the vertebrate nervous system. Genes Dev 2000;14:2919–2937.
56. Leibrock J, Lottspeich F, Hohn A, et al. Molecular cloning and expression of bain-derived neurotrophic factor. Nature 1989;341:149–152.
57. von Bartheld CS. Neurotrophins in the developing and regenerating visual system. Histol Histopathol 1998;13:437–459.
58. Liang FQ, Aleman TS, Dejneka NS, et al. Long-term protection of retinal structure but not function using RAAV.CNTF in animal models of retinitis pigmentosa. Mol Ther 2001;4:461–472.
59. Saigo Y, Abe T, Hojo M, Tomita H, Sugano E, Tamai M. Transplantation of Transduced Retinal Pigment Epithelium in Rats. Invest Ophthalmol Vis Sci 2004;45:1996–2004.
60. Hojo M, Abe T, Sugano E, et al. Photoreceptor Protection by Iris Pigment Epithelial Transplantation Transduced with AAV-Mediated Brain-Derived Neurotrophic Factor Gene. Invest Ophthalmol Vis Sci 2004;45:3721–3726.
61. Bennett J, Maguire AM, Cideciyan AV, et al. Stable transgene expression in rod photoreceptors after recombinant adeno-associated virus-mediated gene transfer to monkey retina. Proc Natl Acad Sci USA 1999;96:9920–9925.
62. Yoshioka Y, et al. (in preparation)
63. Kano T, Abe T, Tomita H, Sakata T, Ishiguro S-I, Tamai M. Protective effect against ischemia and light damage of iris pigment epithelial cells transfedted with the BDNF gene. Invest Ophthalmol Vis Sci 2002;43:3744–3753.
64. Rabinowitz JE, Samulski J. Adeno-associated virus expression systems for gene transfer. Curr Opin Biotechnol 1998;9:470–475.
65. Monahan PE, Samulski RJ. AAV vectors: Is clinical success on the horizon? Gene Ther 2000;7:24–30.
66. Kay MA, Manno CS, Ragni MV, et al. Evidence for gene transfer and expression of factor IX in haemophilia B patients treated with an AAV vector. Nat Genet 2000;24:257–261.
67. Rabinowitz JE, Rolling F, Li C, et al. Cross-packaging of a single adeno-associated virus (AAV) type 2 vector genome into multiple aav serotypes enables transduction with broad specificity. J Virol 2002;76:791–801.

19

Recent Results in Retinal Transplantation Give Hope for Restoring Vision

Robert B. Aramant, PhD, Norman D. Radtke, MD, and Magdalene J. Seiler, PhD

CONTENTS

ABSTRACT

Transplanting sheets of fetal retinal pigment epithelium (RPE) together with its neuronal retina offers potential as a viable technique to prevent blindness and restore vision. This chapter presents results of the first promising FDA-approved clinical trial with a limited number of patients, bringing cautious optimism to the evaluation of the results. The basic research with several rodent degeneration models shows that transplants restore visually evoked responses in the brain, in an area of the superior colliculus corresponding to the placement of the transplant in the retina. Retinal transplants can preserve vision in an optokinetic acuity test. The mechanism of functional restoration is still unclear, but research results indicate that likely both synaptic connectivity between transplant and host and rescue of host photoreceptors are involved. The conclusion can be drawn that co-transplants of RPE together with retinal sheets have a beneficial functional effect in several animal retinal degeneration models as well as in patients in the recent clinical trial.

INTRODUCTION

Retinal Transplantation: A Hope for Incurable Retinal Diseases

Retinal diseases such as age-related macular degeneration (AMD) *(1,2)*, or retinitis pigmentosa (RP), a group of inherited diseases with mutations in photoreceptor or retinal pigment epithelium (RPE) genes, affect a considerable part of the American population.

From: *Ophthalmology Research: Retinal Degenerations: Biology, Diagnostics, and Therapeutics*
Edited by: J. Tombran-Tink and C. J. Barnstable © Humana Press Inc., Totowa, NJ

In such diseases, photoreceptors and/or RPE become dysfunctional or degenerate and need to be replaced whereas the neural retina that connects to the brain can still remain functional *(3–6)* (reviewed in ref. *7*). If the diseased cells can be replaced and the new cells can make appropriate connections with the functional part of the host retina, a degenerated retina might be repaired and vision restored.

Vitamin supplements with zinc *(8)* and gene therapy to introduce trophic factors *(9)* or to correct mutated genes *(10,11)* may be helpful in the early stages of a disease, but once photoreceptors are lost, they must be replaced to restore vision (reviewed in refs. *12,13*). With the exception of microchip implantation (reviewed in refs. *14,15*), there are presently no realistic alternative techniques to retinal transplantation for the treatment of end-stage retinal diseases. In many retinal diseases, both photoreceptors and RPE are affected *(7)* and need to be replaced. To meet this need, our group has developed a procedure to transplant sheets of fetal RPE together with its neuroblastic retina *(16–18)*.

There is now evidence that transplanted fetal neurosensory retina can re-establish connections with the residual neural network *(19)*. In addition, the transplanted tissue might exert a positive rescue effect on the recipient's retina, as has been shown in animal experiments in vitro and in vivo *(20)*. The first FDA-approved clinical trials in retinal transplantation have shown very promising results *(21)*.

Retinal Remodeling

In pigmented Royal College of Surgeons (RCS) rat retina up to the age of 515 d, no ultrastructural abnormalities in synaptic counts and in ganglion cell characteristics were found *(22)*. However, subsequent studies showed that ganglion cells change their properties about 3 mo after photoreceptor loss in the RCS rat, owing to the abnormal in growth of blood vessels from the choroid *(23–25)*. Similar changes occur in the *rd* mouse retina *(26–28)*.

Remodeling of the inner retina is a major secondary effect of outer retinal degenerations *(29,30)*. It is thought that this process occurs as a result of denervation of the inner retinal neurons, and subsequent attempts by these neurons to find new synaptic input. This process involves cell death; rewiring, i.e., the formation of new circuits to replace lost innervation; and cell migration *(30)*.

Properties and Use of Fetal Donor Tissue

There are many reasons why all of our studies have used fetal donor cells and not adult donor cells. Fetal cells have a high capacity to sprout processes and to produce trophic substances that will aid host and transplant cells to establish contacts. They can multiply, so that the transplant can grow to cover a larger area, and transplants of retinal aggregates to nude rats can grow larger the younger the donor age *(31)*. Fetal retinal cells can also overcome the trauma of transplantation much easier than adult cells because they do not depend as heavily on oxygen *(32)*. Further, fetal retinal tissue is likely less immunogenic than adult tissue because it contains less microglia than older tissue *(33,34)*. Research has shown no rejection if the tissue is transplanted to the central nervous system (CNS) or the eye of the same species (*see* the Retinal Transplant Immunology section).

In January of 1993, President Clinton overturned the ban on federally funded fetal tissue research. Public Law 103-43 (also known as the National Institutes of Health [NIH] Revitalization Act of 1993) explicitly made funding of fetal tissue transplantation research legal when certain conditions are met. Strict ethical guidelines must be followed so as not to give any incentive for abortions. Donors must remain anonymous and informed consent procedures must be in place to explain the research purpose and potential risks. Our research has used private, nongovernmental sources, and we have adhered to all NIH guidelines regardless of our funding source.

Technical Challenges With Fetal Sheet Transplantation

Although fetal tissue has many advantages, one major weakness is that it is very fragile and presents several challenges during transplantation. It is of outmost importance not to damage the donor tissue or the host. The first challenge is to precisely dissect an intact monolayer RPE sheet attached to the neural retinal sheet. To perform this delicate task, our group uses custom-designed ultra-precision forceps. The second challenge is to implant the fragile donor sheets into the subretinal space without damage, and we have developed a proprietary instrument to perform this step. The implantation instrument is a hand-held tool (stainless-steel hand piece) with a flat, flexible disposable nozzle tube (many different sizes, according to the purpose) that fits over a stationary mandrel (not a movable plunger). This innovative device provides the surgeon with precise manual control and allows gentle placement of the transplant into the target area of the subretinal space with minimal trauma to the donor tissue and host eye. When the nozzle tip containing the tissue is on the target, the surgeon holds his hand completely still and releases a spring that retracts the nozzle. The surgeon has complete control over the speed of retraction of the nozzle tip and in this way exposes and "places" the tissue on target. The nozzle size is chosen according to the donor tissue so that very little fluid is delivered. All other methods *(35,36)* push or inject the tissue into a large subretinal bleb so that trauma is exerted on the host and donor tissue and the increased pressure can easily push out the donor tissue through the retinotomy site. In our earlier research *(16,37,38)*, we used a matrix coating to protect the donor tissue. However, our custom-made implantation instrument so exceeded expectations that the procedure could be performed without use of the matrix coating. The implantation tool provides the surgeon with the precise control required for very gentle delivery of the fragile fetal graft.

Retinal Transplant Immunology

The subretinal space is regarded as an immunological privileged site *(39)* so that there is a reduced probability of rejection of allografts of fetal tissue. The neural retina is non-immunogenic but the RPE and the microglial cells in the donor retina are immunogenic *(40,41)*. Dissociated RPE cells seem to initiate an immune response after transplantation to the subretinal space *(42,43)*. However, allografted sheets of RPE are not rejected when transplanted to the kidney capsule and thus are immunologically privileged *(41)*. Postnatal retinal tissue however was rejected. Despite the potential for rejection based on the microglia in the donor retina, our hypothesis is that rejection will probably not happen because of the immunological privileged site of the subretinal space. So far, our hypothesis has been confirmed in our results with patients *(21,44)*.

Most of the microglial cells are associated with blood vessels and migrate postnatally into the rat retina *(33)* and from 16-wk gestation into the human retina *(34)*. The number of immunogenic microglial cells in fetal rat retina is much lower than in postnatal retina *(33)*. Therefore, it is likely that fetal retina is less immunogenic than postnatal retina because fetal retina still lacks inner retinal vessels. However, no group has yet tested this hypothesis. In our model, we have seen stable transplants in rats 6 to 10 mo after surgery. This indicates that allogeneic retinal sheet transplants can be tolerated in the subretinal space of rats with retinal degeneration.

Use of Stem Cells in Our Model

Stem cell transplantation is considered to have a great potential. Stem cells are cells early in development that have the capacity to differentiate into different cell types and different tissues. Progenitor cells are still multipotential but restricted to a specific tissue. For example, fetal retina contains progenitor cells that have the capacity to differentiate into various retinal cell types, but not other cells. Our laboratory has worked with retinal progenitor cells derived from rat E17 retina in vitro or in vivo for more than 5 yr.

Retinal progenitor (stem) cells have been isolated from the ciliary margin of the adult retina of different ages *(45,46)*, or from fetal retina *(47–51)*. After transplantation, progenitor cells integrate and migrate into the retina depending on the age, the disease, or the injury status of the recipient retina. Previous studies showed that a limited percentage of progenitor cells can express opsin *(47)*, but most appeared to be limited to a glial lineage after transplantation to an adult host with retinal degeneration *(49)*. However, retinal progenitor cells that have been maintained in defined culture conditions *(51)* develop to mostly opsin expressing cells after transplantation to recipients with slow and fast retinal degeneration *(52)*. Initially, in our laboratory, we included "stem cells," progenitor cells derived from younger rat E17 retina, with our fetal sheet transplants with the hypothesis that the progenitor cells could help with the connections of transplant and host. These experiments did not produce the expected results. The experiments with retinal progenitor cells alone continued *(48,49)*, and showed some promise after switching to cells maintained and proliferating in serum-free medium *(51,52)*. It is interesting to compare the results of these "stem" retinal progenitor cells with the results of retinal sheet progenitor cells (rat E19) in our transplant model. The photoreceptors of the sheet transplants show a high degree of differentiation. By example, they can show migration of the phototransduction proteins dependent on the light cycle, indicating that they are able to transfer light into electrical signals *(38)*. In contrast, transplanted retinal progenitor cells have not yet been shown to contain photoreceptor outer segments (OS).

Ongoing studies in our laboratory are investigating various aspects of "stem" retinal progenitor cells to determine if these cells can do what has been shown for retinal sheet transplants: (1) develop to fully functional retinal cells, (2) contain all substances specific for each cell, (3) send out processes to host cells and establish synapses with the host, and (4) establish meaningful communications and restore vision in a host with retinal degeneration. Although stem/progenitor cells show great promise, there appears to be a long road of research ahead before any possible clinical application.

CLINICAL TRIALS

Clinical trials of retinal transplantation have been motivated by the lack of available treatment to recover or prevent vision loss from RP and other diseases of the outer retina. Although oral vitamin A therapy has been shown to slow the rate of electroretinogram (ERG) loss in RP, it has no effect on vision loss *(53)*. Gene therapy and pharmacological therapy are underway but are still under development, with clinical gene therapy trials to be started soon *(54,55)*. A clinical trial to deliver encapsulated ciliary neurotrophic factor-producing RPE cells to the eye of patients with RP has just begun (no published data available yet) *(56)*. These trials aim at delaying photoreceptor degeneration or correcting gene defects. Finally, development and use of a visual prosthesis is being actively pursued in many centers but the potential of existing devices is not known *(14,15,57–59)*.

With the exception of the visual prosthesis, gene and pharmacological therapies can only help when the retinal degeneration has not progressed too far. Once most of the photoreceptors are lost, replacement of degenerating cells by RPE together with neural retina may be the only means to restore the atrophying neural retina, the pigment epithelium, and part of the choroid.

RPE and IPE Transplantation

The success of RPE transplants in RCS rats *(60,61)* led to clinical trials in patients with AMD by a team in Sweden in collaboration with Columbia University, NY *(42)* and in the United States *(62)*. The results were mixed; rejection was observed depending on the status of retinal degeneration, the presence of an intact blood–brain barrier, and immunosuppression *(63)*. In summary, problems with RPE allografts, related to rejection, inflammation and/or changes in the RPE cells after tissue culture, prevented any long-term beneficial effects. Immunosuppressive treatment appeared to prevent graft failure *(63)*.

To avoid rejection, autologous transplants of adult RPE cells *(64)* and iris pigment epithelial (IPE) cells *(65,66)* have been performed, mostly to patients with "wet" AMD. Subjective improvements in visual acuity were reported.

Another strategy has been macular translocation to expose the macula to still healthy RPE cells *(67)*. However, several patients with non-exudative AMD developed clear evidence of new geographic atrophy of the RPE in the area of the translocated fovea after macular translocation *(68)*. This means that it is likely an intrinsic defect in the photoreceptors that negatively affects previously healthy RPE, and gives another argument for the need of combined transplants of retina together with its RPE.

Transplants of Neural Retina

The rationale for transplantation of photoreceptors or neural retina is photoreceptors cannot regenerate once they have undergone apoptosis (reviewed in ref. *69*). Fourteen patients with RP in India *(70)*, and eight patients with RP and one patient with AMD in the United States received aggregate fetal retinal transplants *(71)*. Ten patients received adult photoreceptor sheet transplants *(72,73)*. There was no improvement in vision. No clinical signs of rejection were observed, but a subtle, clinically not evident rejection cannot be excluded.

Our group performed retina-only transplants to four patients with RP. In one patient, transient functional improvement was observed as measured by multifocal ERG (mfERG) *(74)*. This patient also reported subjective improvements. However, no changes in Early Treatment Diabetic Retinopathy Study (ETDRS) vision occurred.

Recent Results of First FDA-Approved Clinical Trial: Transplants of Fetal RPE Sheets With Retina Can Improve Vision

The uniqueness of our approach is to co-transplant sheets of human fetal RPE together with its retina. In preclinical experiments, it was shown that in transplants of sheets of human fetal RPE with its retina to athymic nude rats, the RPE sheet can remain as a monolayer of co-transplanted RPE cells with junctional complexes at 9 mo after transplantation, supporting the development and maintenance of transplant photo-receptors *(17)*.

All of our studies met the local Institutional Review Board requirements and state and local laws for the handling of fetal tissue. Our patients' surgery, preoperative testing, and postoperative testing were funded by a private sponsor to make it possible for the patients to have the surgery without any financial outlay on their part. This private funding has been critical to the success of this clinical trial.

Patient Selection

Our clinical trial focuses on patients with RP and "dry" (non-exudative) AMD, particularly the central areolar choroidal pigment epithelial dystrophy because these diseases have an established pathology and are the most common diseases of the outer retina. Diseases that involve the formation of subretinal neovascular membranes have been excluded, such as presumed ocular histoplasmosis syndrome, and "wet" macular degeneration because of the possible bleeding as well as the excessive trauma to the foveal area that can occur when entering the subretinal space.

In both diseases, the retinal transplantation studies involving both RPE and neural retina in sheets have more potential for measurable success then transplanting either of these tissues alone. To determine the safety of the procedure, we started with patients with the most severe forms of visual problems with RP. As safety was shown, patients in a less advanced stage of the disease with better vision could be selected.

Preoperative Testing and Follow-Up

Complete assessments, ocular examinations, fluorescein angiography, ETDRS proto-col for visual acuity testing, photopic mfERG (VERIS science software, EDI Inc., San Mateo, CA), scanning laser ophthalmoscope (SLO), and optical coherent tomography were performed preoperatively and postoperatively and repeated to assess potentially corresponding physiological changes in the region of the transplant. Patients were tested three times preoperatively and at 2 wk and 1, 6, and 12 mo postoperatively. A great deal of consideration was given to what objective preoperative testing should be and could be done to definitively define improvement in vision. The ETDRS visual acuity testing was used as baseline. In addition to this, new techniques needed to be identified that might be able to show subtle improvements in visual acuity even though the patient might not see any subjective improvement.

The person measuring the ETDRS visual acuity was masked as to which eye had the surgery. The ETDRS protocol for visual acuity testing was followed as described *(75)*. mfERGs and SLO testing were done at different sites in the United States, in the same time frame. The examiners at these sites also did not know which eye had received the transplant. To compensate for the patients' eye movement after each measurement with the SLO, clearly defined vascular landmarks were used.

Results to Date

As of August 2005, a total of 11 patients received cografts of retina with its RPE, ten with RP and one with "dry" AMD. After FDA approval (BB-IND no. 8354 about clinical retinal transplantation, Principal Investigator Norman D. Radtke), cografted sheets of fetal retina together with its RPE were transplanted in five patients with RP with light perception or no light perception *(44)*. No adverse effects but also no vision improvement were noted. After these five patients, the food and drug administration (FDA) was satisfied with the safety of the procedure, and allowed us to proceed with transplants in patients with vision of 20/800 or worse in one eye. So far, only patients with preoperative vision in the range of 20/800 to hand motion or better have shown improvement.

The most encouraging results have been seen with a patient with RP who was transplanted in February 2002. The preoperative ETDRS vision was 20/800 in the left surgery eye, and 20/400 in the right nonsurgery eye. One year after transplantation, ETDRS vision in the surgery eye improved to 20/160 and remained stable at 20/200 at 2 yr and 2 mo postoperatively whereas the nonsurgery eye remained unchanged at 20/400 *(21)*. Presently, more than 4 yr after transplantation, the vision is still stable. An independent evaluation in another eye hospital by a SLO test showed visual improvement in the surgery eye from 20/270 at 9 mo to 20/84 at 2 yr and 3 mo after surgery. Interestingly, the right eye which had been 20/369 at 9 mo improved to 20/169 at 2 yr and 3 mo *(21)*. However, no changes were observed with mfERG. No clinical evidence of rejection was observed although the transplant sheet lost its pigmentation by 6 mo. The patient's quality of life has improved so she is now able to read and handle e-mail from a computer, using a magnification glass. She can now write checks, do some sewing, and has taken up her hobby of ceramic painting again. She now predominantly uses her surgery eye, whereas she previously relied on the other unoperated eye. After this patient, the FDA allowed us to use patients with 20/400 vision.

A second patient with RP (surgery in June 2003) showed improvement from hand motion to 20/400. In addition, one patient with AMD (surgery in March 2004) improved from 20/640 to 20/320 at 1.5 yr. With SLO testing at 6 mo, the fixation was over the pigmented transplant area, which means that the patient was using the transplant area. The SLO at 6 mo showed less scatter than preoperatively. Both patients also reported subjective vision improvements. These early findings are encouraging but confirmation from long-term follow-up studies is needed. One RP patient with surgery in March 2003 slightly improved from light perception to hand motion. Another RP patient with surgery February 2004 continued to worsen from 20/400 to 20/300 at 2 yr.

What has been gained from the results to date and where should we proceed? So far, our study has too few patients, so only tentative inferences can be made. The best recipient candidates may be in the range of 20/400 to 20/800 ETDRS vision. Our patient who exhibited the best results had autosomal dominant RP and also received oxygen therapy for 1 wk. Postoperative oxygen treatment is now part of the protocol in all patients on a voluntary basis. The rationale for oxygen treatment is based on studies in experimental animals showing that oxygen reduces the glial reactivity after retinal detachment *(76)* and preliminary transplantation studies that have shown a beneficial effect of oxygen *(77)*. Another avenue for improvement may be visual training by exposing the patient to visual stimulation patterns, especially during the first 6 months when the grafted cells are developing. This may enforce the use of the transplant for vision and encourage the formation of new appropriate connections.

ANIMAL RESEARCH WITH FETAL SHEET TRANSPLANTS

Retinal Degeneration Animal Models Used for Transplantation

The different retinal degeneration models used for retinal transplantation were extensively reviewed *(18,69)*. For the initial studies *(37,38)*, a light damage model was used in which albino rats were exposed to continuous moderate blue light for 2 to 4 d. This selectively damaged the rod photoreceptors while leaving the RPE intact *(78)*.

Since 1999, we have used three different inherited models of retinal degeneration: albino RCS rats *(11,16)*, and two different lines of transgenic rats with the mutant human rhodopsin S334ter, produced by Chysalis DNX, and kindly provided by M. LaVail, UC San Francisco: the fast degenerating line 3 *(79)*, and the slow degenerating line 5 *(80)*.

Morphological Repair of Damaged Retinas by Fetal Retinal Sheets With and Without RPE

The following sections will present the results from research performed to transplant fetal retinal sheets with and without its RPE in animal models of retinal degeneration, using our custom-made implantation instrument and procedure which gently places the tissue into the subretinal space (reviewed in refs. *18,69*; *see* Technical Challenges With Fetal Sheet Transplantation section).

Only 20 to 30% of fetal sheet transplant in rodents are laminated similar to a normal retina with photoreceptor OS in contact with grafted *(16)* or host RPE *(37,81)*. The rest of the transplants will develop rosettes sometimes with parallel inner retinal layers. The crucial problem is to insert the instrument nozzle at the correct angle to place the tissue into the subretinal space without damaging the host RPE or Bruch's membrane, or placing the tissue into the vitreous. This is very delicate in rodents because the transscleral approach does not allow the surgeon to see the operation area in the eye.

There is also a "time window" for successful transplantation because the retina and the RPE seal together in later stages of retinal degeneration, and this glial seal prevents the insertion of anything into the subretinal space.

The preparation of cografts poses an additional challenge: because fetal retina has not yet developed OS, the RPE can easily detach from its retinal sheet, and it requires very careful dissection to keep both together. An example of a human RPE cografted with retina

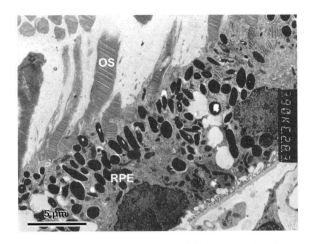

Fig. 1. Human fetal pigmented RPE cografted with its neural retina in subretinal space of athymic albino nude rat, in contact with photoreceptor outer segments of transplant photoreceptors. Part of Bruch's membrane and the choroid can be seen in the right lower corner of the micrograph. Donor 14 wk, 8.9 mo after transplantation. Bar = 5 μm. OS, outer segment; RPE, retinal pigment epithelial.

to an athymic albino nude rat is shown in Fig. 1; a rat fetal cograft to a S334ter line 5 rat is shown in Fig. 2A–D.

Connectivity

The integration of retinal aggregate transplants appears to depend on the status of host photoreceptors *(82)*. If the host photoreceptors are lost completely, bridging neuronal fibers can be seen between transplants and hosts. Similar results can be seen with fetal retinal sheet transplants in retinal degenerate rats. Connectivity and transplant "integration" is also related to the reaction of glial cells. It has recently been shown that retinal aggregate transplants integrate better in transgenic mice that lack both of the intermediate filaments vimentin and glial fibrillary acidic protein *(83)*.

Both the fetal retinal sheet transplant and the host have an inner nuclear layer with various retinal interneurons. If transplant and host neurons form synaptic connections, the wiring would be different from a normal retina. It is still unknown to what extent the host retina and brain can extract information from light stimulation of the transplant. However, our results indicate that the transplants have a direct effect on visual responses (*see* Section Visual Function of Transplants Demonstrated by Electrophysiology), and that transplants form synaptic connections, albeit to a limited extent, with the host retina (*see* next Section Indications for Transplant-Host Connectivity).

Indications for Transplant-Host Connectivity

Sometimes, transplants are too perfect and form a glial barrier towards the host (e.g., Fig. 2 in ref. *69*). In other cases, the transplants appear to merge with the host retina so that it is sometimes impossible to tell where the transplant ends and where the host retina begins. Examples for this are shown in Fig. 2. The current work is focused on

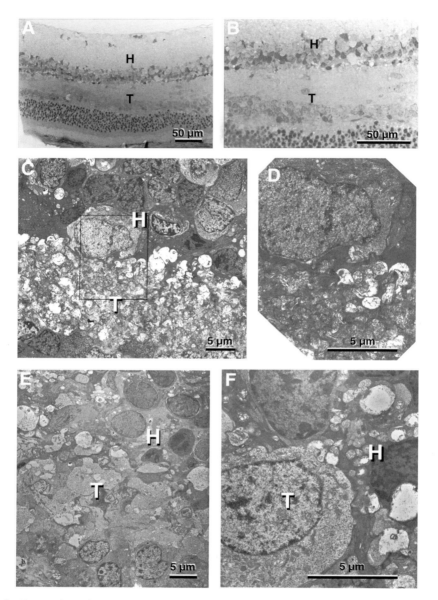

Fig. 2. Examples of transplant (T)/host (H) integration without glial barriers. Note the apparent fusion between transplant and host. **(A–D)** Cograft of E20 retina with RPE to S334-ter line 5 rat, 3 mo after surgery at the age of 1 mo. **(E,F)** Transplant of E17 retina-only to light-damaged rat, 6.3 mo after surgery. Magnification bars: **A,B:** 50 μm, **C–E:** 1 μm.

characterizing transplant-host synapses by using donor tissue derived from transgenic rats expressing human alkaline phosphatase (data not shown). The donor tissue derived from these rats can be detected on the light and electron microscopy level.

Pseudorabies Virus Tracing

To demonstrate synaptic connections between transplant and host retina, we have performed trans-synaptic tracing from the host brain to the transplant *(19)* using an

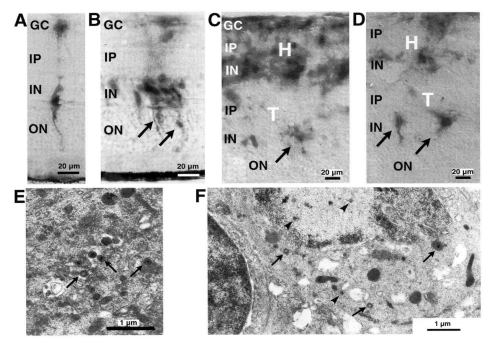

Fig. 3. Transsynaptic virus tracing—an indication for synaptic connections. Ba Blu virus expressing *Escherichia coli* β-galactosidase was injected into the visual brain center, the superior colliculus (SC), and detected by X-gal histochemistry (**A–D**), and electron microscopy (**E,F**) in the retina. (**A,B,E**) Normal retina; (**C,D,F**) transplants. (**A,B**) Normal retina, 48 h (**A**) and 67 h (**B**) after virus injection into the SC. (**A**) Müller cells are labeled with one ganglion cell. (**B**) With longer survival times, the virus is spreading through all retinal layers and also laterally. (**C,D**) Transplant to s334ter transgenic rat, 65 h after Bablu virus injection into SC. Arrows point to some virus-labeled cells in transplant with apparent neuronal morphology. E18 retinal transplant, 9.8 mo after transplantation. (**E**) Normal retina at 2 d (52 h) after virus injection: virus-labeled ganglion cell. Arrows point to enveloped virus. (**F**) Cograft of retina with RPE (E20) to RCS rat, 68 h after Bablu virus injection into SC, and 4.9 mo after transplantation. Arrowheads indicate nonenveloped virus in nucleus and cytoplasm. Arrows indicate enveloped virus in cytoplasm. Magnification bars: **A–D**: 20 μm; **E,F**: 1 μm. H, host; T, transplant; GC, ganglion cell layer; IP, inner plexiform layer; IN, inner nuclear layer; OP, outer plexiform layer; ON, outer nuclear layer; OS, outer segments; PRV, pseudorabies virus.

attenuated pseudorabies virus (PRV), which is specifically transferred from one neuron to the next at synaptic contact points, and transported exclusively in the retrograde direction *(84)*. This tracer, PRV Bartha, has been used for outlining multisynaptic circuitry in the CNS for many years *(85)*. The transfer of the virus between neurons depends on the development of functional synapses *(86)*. A somewhat confusing issue for scientists unfamiliar with the virus properties is that the virus can infect glial cells. Glial cells can however not produce a complete infectious virus so that no virus can leave the cell. This prevents nonspecific contamination *(84)*. Neuronal cells are labeled in retinal sheet transplants after injection of PRV into the superior colliculus into an area corresponding to the placement of the transplant in the retina *(19)*. Examples are shown in Fig. 3. However, Müller cells of the host retina inevitably are labeled shortly after label of the ganglion cells, but cannot produce infectious enveloped virus, similar to what has been shown in the brain.

Retinal Sheet Transplants Restore Visual Function

Light/Dark Shift of Phototransduction Proteins in Transplant Photoreceptors

Are transplant photoreceptors capable of responding to light?

Certain phototransduction proteins, such as arrestin (S-antigen) and rod α-transducin, are found in different compartments of the rod photoreceptor dependent on the light/dark cycle *(87)*. Arrestin is in the inner segments in the dark and in the OS in the light, whereas the distribution for rod α-transducin is reversed. The cytoskeleton is required for the translocation of these proteins *(88)*. The translocation of arrestin does not depend on transducin signaling *(89)*.

Photoreceptors in sheet transplant to light-damaged rats show translocation of arrestin and rod α-transducin dependent on the light cycle, indicating that the photo-transduction process takes place and that they respond to light *(38)*. This was shown by fixation of transplanted rat eyes either in light or dark.

Visual Function of Transplants Demonstrated by Electrophysiology

By recording from the primary target of retinal ganglion cells in rodents, the superior colliculus (SC), we showed in several rodent models of retinal degeneration that visually evoked multi-unit responses could be recorded in a small area of the SC corresponding to the retinal placement of the transplant *(11,79–81)*. The visual stimulus consisted of a bright light flash that specifically stimulated cone photoreceptors. A common phenomenon with all responses from transplanted rats was the longer latency (i.e., the response time) as compared with normal control rats.

In albino RCS rats, responses were found in 19 of 29 transplanted rats. Visual responses were also found in age-matched sham surgery rats; however, the responses from transplanted rats had higher amplitudes and shorter latencies *(11)*. The rats were transplanted at the age of 37 to 69 d, when photoreceptors are already committed to apoptosis, and it would be unlikely for transplants of RPE only to have a rescue effect at this age on host photoreceptors as shown previously *(90)*.

In transgenic S334ter-3 rats with rapid retinal degeneration, responses were found in 7 of 11 transplanted rats, but not in any sham surgery rats *(79)*. In transgenic S334ter line-5 rats with slow retinal degeneration, transplant-derived responses could be distinguished in 8 of 13 rats at 8 mo of age when a scotoma had developed in the transplant area *(80)*. In all three rat models, no rescue effect of the transplant on host cones was seen with qualitative immunohistochemistry for S-antigen (arrestin) which labels both rods and blue cones. The quality of responses appeared to depend on the quality of the transplant (better laminated transplants gave better responses). If a rescue effect was the case, disorganized transplants should also have given a response.

This was different in *rd* mice, where responses were found in three of seven mice transplanted at the age of 31 to 38 d *(81)*. The three light-responsive eyes had rosetted or disorganized grafts, but no responses were found in eyes with well-laminated grafts. On the other hand, the immunoreactivity for recoverin in the host retina overlying the graft was much higher in the responsive eyes compared with untreated age-matched *rd/rd* mice, or nonresponsive grafted eyes. Thus, in *rd* mice, the restoration of visual responses by retinal sheet transplants appeared to depend on a rescue of host cones. A previous study with retinal aggregate transplants found ganglion cell responses in 3 of 10 mice transplanted at the age of 2 wk, but no responses in mice transplanted at an older age *(91)*. This was in contrast to the recent study which showed an effect in mice transplanted at the age of 30–38 d *(81)*, demonstrating the advantage of sheet transplantation.

Transplant Effect on Visual Acuity

Optokinetic head tracking has been used to test visual acuity in pigmented rats to show the time course of retinal degeneration in different models *(92–94)* and photoreceptor rescue after RPE or Schwann cell transplantation *(92,95,96)*. Using an improved design which made it possible to test each eye independently, it could be demonstrated that the deterioration of the optokinetic response in S334ter-3 rats with rapid retinal degeneration could be delayed by retinal sheet transplants *(97)*.

Mechanism of Transplant Effect on Vision

Two mechanisms appear to be involved: a trophic effect by the transplanted fetal tissue, resulting in rescue of host cones (in refs. *20,98* for reviews); and a direct effect from transplant photoreceptors mediated by synaptic connections between transplant and host *(19)*. So far, visual responses from transplanted animals in the SC could only be recorded at mesopic or photopic light levels where cones can be stimulated, and not at scotopic light levels less than -3 log cd/m^2. Normal rats show responses less than -5 log cd/m^2 *(99)*. The mechanism of visual restoration appears to be different in *rd* mice (because there was no correlation between transplant organization and visual responses) *(81)* and the different rat degeneration models where a correlation between transplant organization and visual responses was found *(11,79,80)*. Irrespective of the mechanism involved, our results clearly show that transplants have a beneficial effect on vision.

SUMMARY OF RESEARCH ACCOMPLISHMENTS BUILDING THE BASIS FOR CLINICAL TRIALS

The research presented in this chapter has demonstrated many accomplishments indicating the viability of retinal transplantation as a remedy for retinal diseases. These include:

1. Use of a gentle procedure and instrument.
2. Maintenance of the primordial circuitry between the donor retinal neuroblastic cells by sheet transplantation.
3. Reconstruction of an area of a damaged retina by a laminated transplant resembling a normal retina.
4. Ability of transplant photoreceptors to transform light into electrical signals as normal photoreceptors, as indicated by light/dark shift of phototransduction proteins.
5. Synaptic connections between transplant and host retina are indicated by transsynaptic tracing from the host brain.
6. Light sensitivity can be restored or preserved in brain area that corresponds to the placement of the transplant in the retina.
7. Retinal degeneration rats can preserve functional vision after transplantation in a visual acuity optokinetic test.

8. Many retinal diseases require the replacement of both photoreceptors and RPE. Fetal retinal epithelium can be gently co-transplanted with the retina and remain as a monolayer sheet after ten months, apparently supporting exchange of nutrition between the choroid and the photoreceptors.

9. The donor tissue is well tolerated in the subretinal space, can survive without immunosuppression, and is safe to use in patients.

10. In several retinal degeneration models, intact-sheet transplants unequivocally have beneficial effects on vision, as shown also in recent clinical trials.

FUTURE DIRECTIONS

There are indications that synaptic connections between host and graft are involved in the beneficial effects from retinal transplantation. Although now only limited connections may be involved, the positive sign is that there might be a large margin for improvement. The increase of appropriate graft-host connections might be helped with several different treatments: (1) Treatment with brain-derived neurotrophic factor has shown promising effects in preliminary studies, as well as introduction of other trophic factors. (2) Use of a laser to shave off the superficial layer of the donor tissue may improve integration. (3) A noninvasive treatment with "visual stimulation" of patients in the first 6 mo after surgery when the immature graft cells are developing has a good probability to improve results.

After many years of dedicated research, retinal transplantation appears very promising as a viable treatment to bring hope to patients with "incurable" retinal diseases.

ACKNOWLEDGMENTS

This work was supported by funds from an anonymous sponsor; The Vitreoretinal Research Foundation, Louisville KY; The Foundation Fighting Blindness; the Murray Foundation Inc., New York; an unrestricted grant from the Research to Prevent Blindness; NIH EY08519; NIH EY03040; Foundation for Retinal Research; and Fletcher Jones Foundation.

The authors want to thank Zhenhai Chen, Xiaoji Xu, Betty Nunn, and Lilibeth Lanceta for their technical assistance.

The authors can acknowledge only the most significant collaborators involved owing to space limitations. Biju Thomas, Guanting Qiu, Srinivas Sadda (Doheny Eye Institute, Los Angeles, CA); Shinichi Arai (now Niigata University, Niigata, Japan); Botir T. Sagdullaev (now Washington University, St. Louis); Norman D. Radtke, Heywood M. Petry, Maureen A. McCall; Peng Yang (University of Louisville); Gustaw Woch (Hershey Medical School, Hershey, PA); and Sherry L. Ball (Cleveland University, Cleveland, OH) contributed to the work presented in this chapter. We thank J.P. Card, University of Pittsburg, and Lynn Enquist, Princeton University, for their gift of the pseudorabies virus strains, and their helpful advice; Matthew M. LaVail, UCSF, for the founder breeding pairs of transgenic S334ter rats; and Eric Sandgren, University of Wisconsin, for founder breeders of transgenic hPAP rats.

Robert B. Aramant, Norman D. Radtke, and Magdalene J. Seiler have a proprietary interest in the implantation instrument and method.

REFERENCES

1. Ambati J, Ambati BK, Yoo SH, Ianchulev S, Adamis AP. Age–related macular degeneration: etiology, pathogenesis, and therapeutic strategies. Surv Ophthalmol 2003;48: 257–293.
2. Zarbin MA. Current concepts in the pathogenesis of age-related macular degeneration. Arch Ophthalmol 2004;122:598–614.
3. Papermaster D, Windle J. Death at an early age. Apoptosis in inherited retinal degenerations. Invest Ophthalmol Vis Sci 1995;36:977–983.
4. Santos A, Humayun MS, de Juan E Jr, et al. Preservation of the inner retina in retinitis pigmentosa. A morphometric analysis. Arch Ophthalmol 1997;115:511–515.
5. Milam AH, Li ZY, Fariss RN. Histopathology of the human retina in retinitis pigmentosa. Prog Retin Eye Res 1998;17:175–205.
6. Humayun MS, Prince M, de Juan E Jr, et al. Morphometric analysis of the extramacular retina from postmortem eyes with retinitis pigmentosa. Invest Ophthalmol Vis Sci 1999;40:143–148.
7. Pacione LR, Szego MJ, Ikeda S, Nishina PM, McInnes RR. Progress toward understanding the genetic and biochemical mechanisms of inherited photoreceptor degenerations. Annu Rev Neurosci 2003;26:657–700.
8. AREDS Report No 8. A randomized, placebo-controlled, clinical trial of high-dose supplementation with vitamins C and E, beta carotene, and zinc for age-related macular degeneration and vision loss: AREDS report no. 8. Arch Ophthalmol 2001;119: 1417–1436.
9. Liang FQ, Dejneka NS, Cohen DR, et al. AAV-mediated delivery of ciliary neurotrophic factor prolongs photoreceptor survival in the rhodopsin knockout mouse. Mol Ther 2001;3:241–248.
10. Acland GM, Aguirre GD, Ray J, et al. Gene therapy restores vision in a canine model of childhood blindness. Nat Genet 2001;28:92–95.
11. Woch G, Aramant RB, Seiler MJ, Sagdullaev BT, McCall MA. Retinal transplants restore visually evoked responses in rats with photoreceptor degeneration. Invest Ophthalmol Vis Sci 2001;42:1669–1676.
12. Dejneka NS, Bennett J. Gene therapy and retinitis pigmentosa: advances and future challenges. Bioessays 2001;23:662–668.
13. Weleber RG, Kurz DE, Trzupek KM. Treatment of retinal and choroidal degenerations and dystrophies: current status and prospects for gene-based therapy. Ophthalmol Clin North Am 2003;16:583–593, vii.
14. Hetling JR, Baig-Silva MS. Neural prostheses for vision: designing a functional interface with retinal neurons. Neurol Res 2004;26:21–34.
15. Loewenstein JI, Montezuma SR, Rizzo JF 3rd. Outer retinal degeneration: an electronic retinal prosthesis as a treatment strategy. Arch Ophthalmol 2004;122:587–596.
16. Aramant RB, Seiler MJ, Ball SL. Successful cotransplantation of intact sheets of fetal retinal pigment epithelium with retina. Invest Ophthalmol Vis Sci 1999;40:1557–1564.
17. Aramant RB, Seiler MJ. Transplanted sheets of human retina and retinal pigment epithelium develop normally in nude rats. Exp Eye Res 2002;75:115–125.
18. Aramant RB, Seiler MJ. Progress in retinal sheet transplantation. Prog Retin Eye Res 2004;23:475–494.
19. Seiler MJ, Sagdullaev BT, Woch G, Thomas BB, Aramant RB. Transsynaptic virus tracing from host brain to subretinal transplants. Eur J Neurosci 2005;21:161–172.
20. Mohand-Said S, Hicks D, Leveillard T, Picaud S, Porto F, Sahel JA. Rod-cone interactions: developmental and clinical significance. Prog Retin Eye Res 2001;20:451–467.

21. Radtke ND, Aramant RB, Seiler MJ, Petry HM, Pidwell DJ. Vision change after sheet transplant of fetal retina with RPE to a retinitis pigmentosa patient. Arch Ophthalmol 2004;122:1159–1165.

22. Eisenfeld AJ, LaVail MM, LaVail JH. Assessment of possible transneuronal changes in the retina of rats with inherited retinal dystrophy: cell size, number, synapses, and axonal transport by retinal ganglion cells. J Comp Neurol 1984;223:22–34.

23. Caldwell RB, Slapnick SM, Roque RS. RPE-associated extracellular matrix changes accompany retinal vascular proliferation and retino-vitreal membranes in a new model for proliferative retinopathy: the dystrophic rat. Prog Clin Biol Res 1989;314:393–407.

24. Villegas-Perez MP, Vidal-Sanz M, Lund RD. Mechanism of retinal ganglion cell loss in inherited retinal dystrophy. Neuroreport 1996;7:1995–1999.

25. Wang S, Villegas-Perez MP, Holmes T, et al. Evolving neurovascular relationships in the RCS rat with age. Curr Eye Res 2003;27:183–196.

26. Liu LO, Li G, McCall MA, Cooper NG. Photoreceptor regulated expression of Ca(2+)/calmodulin-dependent protein kinase II in the mouse retina. Brain Res Mol Brain Res 2000;82:150–166.

27. Wang S, Villegas-Perez MP, Vidal-Sanz M, Lund RD. Progressive optic axon dystrophy and vascular changes in rd mice. Invest Ophthalmol Vis Sci 2000;41:537–545.

28. Liu LO, Laabich A, Hardison A, Cooper NG. Expression of ionotropic glutamate receptors in the retina of the rdta transgenic mouse. BMC Neurosci 2001;2:7.

29. Jones BW, Watt CB, Frederick JM, et al. Retinal remodeling triggered by photoreceptor degenerations. J Comp Neurol 2003;464:1–16.

30. Marc RE, Jones BW, Watt CB, Strettoi E. Neural remodeling in retinal degeneration. Prog Retin Eye Res 2003;22:607–655.

31. Aramant RB, Seiler MJ. Human embryonic retinal cell transplants in athymic immunodeficient rat hosts. Cell Transplant 1994;3:461–474.

32. Wasselius J, Ghosh F. Adult rabbit retinal transplants. Invest Ophthalmol Vis Sci 2001; 42:2632–2638.

33. Ashwell KW, Hollander H, Streit W, Stone J. The appearance and distribution of microglia in the developing retina of the rat. Vis Neurosci 1989;2:437–448.

34. Provis JM, Leech J, Diaz CM, Penfold PL, Stone J, Keshet E. Development of the human retinal vasculature: cellular relations and VEGF expression. Exp Eye Res 1997;65: 555–568.

35. Ghosh F, Arnér K, Ehinger B. Transplant of full-thickness embryonic rabbit retina using pars plana vitrectomy. Retina 1998;18:136–142.

36. Ghosh F, Ehinger B. Full-Thickness Retinal Transplants: A Review. Ophthalmologica 2000;214:54–69.

37. Seiler MJ, Aramant RB. Intact sheets of fetal retina transplanted to restore damaged rat retinas. Invest Ophthalmol Vis Sci 1998;39:2121–2131.

38. Seiler MJ, Aramant RB, Ball SL. Photoreceptor function of retinal transplants implicated by light-dark shift of S-antigen and rod transducin. Vision Res 1999;39: 2589–2596.

39. Streilein JW, Okamoto S, Sano Y, Taylor AW. Neural control of ocular immune privilege. Ann N Y Acad Sci 2000;917:297–306.

40. Ma N, Streilein JW. T cell immunity induced by allogeneic microglia in relation to neuronal retina transplantation. J Immunol 1999;162:4482–4489.

41. Wenkel H, Streilein JW. Evidence that retinal pigment epithelium functions as an immune-privileged tissue. Invest Ophthalmol Vis Sci 2000;41:3467–3473.

42. Algvere PV, Gouras P, Dafgard Kopp E. Long-term outcome of RPE allografts in non-immunosuppressed patients with AMD. Eur J Ophthalmol 1999;9:217–230.

43. Zhang X, Bok D. Transplantation of retinal pigment epithelial cells and immune response in the subretinal space. Invest Ophthalmol Vis Sci 1998;39:1021–1027.

44. Radtke ND, Seiler MJ, Aramant RB, Petry HM, Pidwell DJ. Transplantation of intact sheets of fetal neural retina with its retinal pigment epithelium in retinitis pigmentosa patients. Am J Ophthalmol 2002;133:544–550.
45. Tropepe V, Coles BL, Chiasson BJ, et al. Retinal stem cells in the adult mammalian eye. Science 2000;287:2032–2036.
46. Coles BL, Angenieux B, Inoue T, et al. Facile isolation and the characterization of human retinal stem cells. Proc Natl Acad Sci USA 2004;101:15,772–15,777.
47. Chacko DM, Rogers JA, Turner JE, Ahmad I. Survival and differentiation of cultured retinal progenitors transplanted in the subretinal space of the rat. Biochem Biophys Res Commun 2000;268:842–846.
48. Yang P, Seiler MJ, Aramant RB, Whittemore SR. In vitro isolation and expansion of human retinal progenitor cells. Exp Neurol 2002;177:326–331.
49. Yang P, Seiler MJ, Aramant RB, Whittemore SR. Differential lineage restriction of rat retinal progenitor cells in vitro and in vivo. J Neurosci Res 2002;69:466–476.
50. Lu B, Kwan T, Kurimoto Y, Shatos M, Lund RD, Young MJ. Transplantation of EGF-responsive neurospheres from GFP transgenic mice into the eyes of rd mice. Brain Res 2002;943:292–300.
51. Qiu G, Seiler MJ, Arai S, Aramant RB, Sadda SR. Alternative culture conditions for isolation and expansion of retinal progenitor cells. Curr Eye Res 2004;28:327–336.
52. Qiu G, Seiler MJ, Mui C, et al. Photoreceptor differentiation and integration of retinal progenitor cells transplanted into transgenic rats. Exp Eye Res 2005;80:515–525.
53. Sibulesky L, Hayes KC, Pronczuk A, Weigel-DiFranco C, Rosner B, Berson EL. Safety of <7500 RE (<25000 IU) vitamin A daily in adults with retinitis pigmentosa. Am J Clin Nutr 1999;69:656–663.
54. Chong NH, Bird AC. Management of inherited outer retinal dystrophies: present and future. Br J Ophthalmol 1999;83:120–122.
55. Dejneka NS, Rex TS, Bennett J. Gene therapy and animal models for retinal disease. Dev Ophthalmol 2003;37:188–198.
56. Burnham CM. Encapsulated cell technology could prevent blindness. Drug Discov Today 2003;8:146–147.
57. Chow AY, Pardue MT, Perlman JI, et al. Subretinal implantation of semiconductor-based photodiodes: durability of novel implant designs. J Rehabil Res Dev 2002;39:313–321.
58. Margalit E, Maia M, Weiland JD, et al. Retinal prosthesis for the blind. Surv Ophthalmol 2002;47:335–356.
59. Rizzo JF 3rd, Wyatt J, Loewenstein J, Kelly S, Shire D. Methods and perceptual thresholds for short-term electrical stimulation of human retina with microelectrode arrays. Invest Ophthalmol Vis Sci 2003;44:5355–5361.
60. Li L, Turner JE. Inherited retinal dystrophy in the RCS rat: prevention of photoreceptor degeneration by pigment epithelial cell transplantation. Exp Eye Res 1988;47:911–917.
61. Lopez R, Gouras P, Kjeldbye H, et al. Transplanted retinal pigment epithelium modifies the retinal degeneration in the RCS rat. Invest Ophthalmol Vis Sci 1989;30:586–588.
62. Weisz JM, Humayun MS, De Juan E Jr, et al. Allogenic fetal retinal pigment epithelial cell transplant in a patient with geographic atrophy. Retina 1999;19:540–545.
63. Del Priore LV, Kaplan HJ, Tezel TH, Hayashi N, Berger AS, Green WR. Retinal pigment epithelial cell transplantation after subfoveal membranectomy in age-related macular degeneration: clinicopathologic correlation. Am J Ophthalmol 2001;131:472–480.
64. Binder S, Stolba U, Krebs I, et al. Transplantation of autologous retinal pigment epithelium in eyes with foveal neovascularization resulting from age-related macular degeneration: a pilot study. Am J Ophthalmol 2002;133:215–225.
65. Abe T, Yoshida M, Tomita H, et al. Auto iris pigment epithelial cell transplantation in patients with age-related macular degeneration: short-term results. Tohoku J Exp Med 2000;191:7–20.

66. Thumann G, Aisenbrey S, Schraermeyer U, et al. Transplantation of autologous iris pigment epithelium after removal of choroidal neovascular membranes. Arch Ophthalmol 2000; 118:1350–1355.

67. Fujii GY, Au Eong KG, Humayun MS, de Juan E Jr. Limited macular translocation: current concepts. Ophthalmol Clin North Am 2002;15:425–436.

68. Eckardt C, Eckardt U. Macular translocation in nonexudative age-related macular degeneration. Retina 2002;22:786–794.

69. Aramant RB, Seiler MJ. Retinal Transplantation - Advantages of Intact Fetal Sheets. Prog Retin Eye Res 2002;21:57–73.

70. Das T, del Cerro M, Jalali S, et al. The transplantation of human fetal neuroretinal cells in advanced retinitis pigmentosa patients: results of a long-term safety study. Exp Neurol 1999;157:58–68.

71. Humayun MS, de Juan E, del Cerro M, et al. Human neural retinal transplantation. Invest Ophthalmol Vis Sci 2000;41:3100–3106.

72. Berger AS, Tezel TH, Del Priore LV, Kaplan HJ. Photoreceptor transplantation in retinitis pigmentosa: short-term follow-up. Ophthalmology 2003;110:383–391.

73. Kaplan HJ, Tezel TH, Berger AS, Del Priore LV. Retinal transplantation. Chem Immunol 1999;73:207–219.

74. Radtke ND, Aramant RB, Seiler MJ, Petry HM. Preliminary report: indications of improved visual function following retina sheet transplantation to retinitis pigmentosa patients. Am J Ophthalmol 1999;128:384–387.

75. Ferris FL 3rd, Kassoff A, Bresnick GH, Bailey I. New visual acuity charts for clinical research. Am J Ophthalmol 1982;94:91–96.

76. Sakai T, Lewis GP, Linberg KA, Fisher SK. The ability of hyperoxia to limit the effects of experimental detachment in cone-dominated retina. Invest Ophthalmol Vis Sci 2001; 42:3264–3273.

77. Aramant RB, Thomas BB, Arai S, Chen Z, Sadda SR, Seiler MJ. Oxygen treatment improves visual responses to low light in retinal degenerate rats after retinal sheet transplantation. Invest Ophthalmol Vis Sci 2004;44:ARVO E-abstract 5183.

78. Seiler MJ, Liu OL, Cooper NG, Callahan TL, Petry HM, Aramant RB. Selective photoreceptor damage in albino rats using continuous blue light. A protocol useful for retinal degeneration and transplantation research. Graefes Arch Clin Exp Ophthalmol 2000;238: 599–607.

79. Sagdullaev BT, Aramant RB, Seiler MJ, Woch G, McCall MA. Retinal transplantation-induced recovery of retinotectal visual function in a rodent model of retinitis pigmentosa. Invest Ophthalmol Vis Sci 2003;44:1686–1695.

80. Thomas BB, Seiler MJ, Sadda SR, Aramant RB. Superior colliculus responses to light - preserved by transplantation in a slow degeneration rat model. Exp Eye Res 2004;79:29–39.

81. Arai S, Thomas BB, Seiler MJ, et al. Restoration of visual responses following transplantation of intact retinal sheets in rd mice. Exp Eye Res 2004;79:331–341.

82. Zhang Y, Arner K, Ehinger B, Perez MT. Limitation of anatomical integration between subretinal transplants and the host retina. Invest Ophthalmol Vis Sci 2003;44:324–331.

83. Kinouchi R, Takeda M, Yang L, et al. Robust neural integration from retinal transplants in mice deficient in GFAP and vimentin. Nat Neurosci 2003;6:863–868.

84. Card JP, Rinaman L, Lynn RB, et al. Pseudorabies virus infection of the rat central nervous system: ultrastructural characterization of viral replication, transport, and pathogenesis. J Neurosci 1993;13:2515–2539.

85. Card JP, Rinaman L, Schwaber JS, et al. Neurotropic properties of pseudorabies virus: uptake and transneuronal passage in the rat central nervous system. J Neurosci 1990; 10:1974–1994.

86. Rinaman L, Levitt P, Card JP. Progressive postnatal assembly of limbic-autonomic circuits revealed by central transneuronal transport of pseudorabies virus. J Neurosci 2000; 20:2731–2741.
87. Whelan JP, McGinnis JF. Light dependent subcellular movement of photoreceptor proteins. J Neurosci Res 1988;20:263–270.
88. McGinnis JF, Matsumoto B, Whelan JP, Cao W. Cytoskeleton participation in subcellular trafficking of signal transduction proteins in rod photoreceptor cells. J Neurosci Res 2002;67:290–297.
89. Mendez A, Lem J, Simon M, Chen J. Light-dependent translocation of arrestin in the absence of rhodopsin phosphorylation and transducin signaling. J Neurosci 2003;23:3124–3129.
90. Li L, Turner JE. Optimal conditions for long-term photoreceptor cell rescue in RCS rats: the necessity for healthy RPE transplants. Exp Eye Res 1991;52:669–679.
91. Radner W, Sadda SR, Humayun MS, et al. Light-driven retinal ganglion cell responses in blind rd mice after neural retinal transplantation. Invest Ophthalmol Vis Sci 2001;42:1057–1065.
92. Lund RD, Kwan AS, Keegan DJ, Sauve Y, Coffey PJ, Lawrence JM. Cell transplantation as a treatment for retinal disease. Prog Retin Eye Res 2001;20:415–449.
93. Hetherington L, Benn M, Coffey PJ, Lund RD. Sensory capacity of the Royal College of Surgeons rat. Invest Ophthalmol Vis Sci 2000;41:3979–3983.
94. Thaung C, Arnold K, Jackson IJ, Coffey PJ. Presence of visual head tracking differentiates normal sighted from retinal degenerate mice. Neurosci Lett 2002;325:21–24.
95. Coffey PJ, Girman S, Wang SM, et al. Long-term preservation of cortically dependent visual function in RCS rats by transplantation. Nat Neurosci 2002;5:53–56.
96. Lund RD, Adamson P, Sauve Y, et al. Subretinal transplantation of genetically modified human cell lines attenuates loss of visual function in dystrophic rats. Proc Natl Acad Sci USA 2001;98:9942–9947.
97. Thomas BB, Seiler M, Sadda SR, Coffey PJ, Aramant RB. Optokinetic test to evaluate visual acuity of each eye independently. J Neurosci Methods 2004;138:7–13.
98. Leveillard T, Mohand-Said S, Fintz AC, Lambrou G, Sahel JA. The search for rod-dependent cone viability factors, secreted factors promoting cone viability. Novartis Found Symp 2004;255:117–127; discussion 127–130, 177–178.
99. Thomas BB, Aramant RB, Sadda SR, Seiler MJ. Retinal transplantation-A treatment strategy for retinal degenerative diseases. In: Retinal Degenerative diseases, Hollyfield JG, Anderson RE, Medicine and Biology, Vol. 572, Springer, New York, NY, 2006, pp. 367–376.

20

Stem Cells and Retinal Transplantation

Joanne Yau, MB, Henry Klassen, MD, PhD, Tasneem Zahir, PhD, and Michael Young, PhD

INTRODUCTION

Stem cells are characterized by their potential to self renew and generate different cell types. However, their ability to give rise to various cell types, also referred to as phenotypic plasticity, is contingent on the source from which these cells are derived. At present, stem cells are being isolated and expanded in vitro from early developing embryos or specific tissues such as blood, brain, and retina.

The advent of stem cell biology has called into question the long-held view that diseases of the central nervous system, particularly those involving cell loss, are effectively untreatable. Indeed, an explicit aim of the burgeoning field of regenerative medicine is the use of stem cells to replace dead cells in the brain, spinal cord, and retina. In this chapter, we will focus on retinal degenerations and the potential role of stem cells in cell replacement strategies for these conditions. Because retinal degenerative diseases frequently involve the loss of photoreceptors, along with the connections of these cells to retinal interneurons, photoreceptor replacement has the potential to restore sight to the blinded eye.

Most retinal degenerative diseases originate in the sensory retina or adjacent supporting tissues, such as the retinal pigment epithelium (RPE) and choroid, and result in the common final pathway of photoreceptor death. The initial trigger for this process can often be traced back to specific genetic defects, primarily affecting rods, cones, or RPE cells, ultimately resulting in photoreceptor apoptosis. Despite the commonalities, retinal degenerative diseases exhibit a range of clinical phenotypes and progress at varying rates.

From: *Ophthalmology Research: Retinal Degenerations: Biology, Diagnostics, and Therapeutics*
Edited by: J. Tombran-Tink and C. J. Barnstable © Humana Press Inc., Totowa, NJ

For instance, the loss of photoreceptor cells can be limited to the macula or begin in the periphery and eventually extend to the entire retina. The most common type of retinal degeneration is age-related macular degeneration, which is limited to the central retina and occurs predominantly in older individuals. Hereditary forms of central retinal degeneration also exist and typically have an earlier age of onset. Some examples include Stargardt's Disease, Best's Disease, cone dystrophy, pattern dystrophy, and Malattia Leventinese. Macular degenerations are characterized by a loss of central vision and a decreased ability to discriminate fine details. The peripheral visual field and scotopic vision usually remain intact. Symptoms, however, vary greatly in severity between conditions and from one individual to another. As a general rule, these diseases cause significant visual impairment but rarely progress to complete blindness.

Hereditary retinal degenerations that involve the entire retina tend to be more severe. The most common types of these diseases are retinitis pigmentosa (RP) and Usher syndrome. RP is the name given to a group of retinal degenerative diseases with a gradual progression of scotopic symptoms, often occurring over many years or decades. The rods are the first cells to be affected in RP; thus, one of the earliest symptoms is night blindness followed by a progressive loss of peripheral vision leading to "tunnel vision." Although the majority of patients with RP do not suffer from other neurological deficits, hearing loss can occur in some patients. The combination of RP and congenital hearing impairment is known as Usher syndrome.

Whatever the subtype, retinal degenerative diseases share a common trait—they cannot be cured with existing therapies. In this context, the transplantation of stem cells to the degenerating retina provides a promising strategy for replacing dead cells with new ones, including the potential to integrate with remaining host circuitry and restore the transmission of visual information to the brain.

RETINAL TRANSPLANTATION: FROM PAST TO PRESENT

Retinal transplantation was first performed in 1946 by Tansley who demonstrated features of retinal differentiation in embryonic ocular tissue when transplanted into the brains of young rat *(1)*. The fact that the graft was transplanted into the brain and not the eye did not detract from the momentous discovery that embryonic ocular tissue had the capacity to undergo retinal differentiation even in an ectopic site. Later in 1959, Royo and Quay described the first intraocular retinal transplantation procedure which demonstrated that fetal retina could survive in the anterior chamber of the maternal parent *(2)*.

However, significant interest in the field of retinal transplantation was not generated until the mid-1980s when it was shown that retinal grafts could survive when transplanted into various sites within the eye such as the anterior chamber *(3)* or the subretinal space *(4)*. It was discovered that graft survival could be enhanced by using a younger donor *(5)*. However, it was also shown that very young donors (early in embryogenesis) yielded less well-organized transplants, implying that there could be a critical period for retinal development *(6)*. Thus, the late fetal period was suggested as the optimal age of tissue for retinal transplantation.

All of these earlier experiments involved transplantation of full-thickness retina. Although transplants remained ordered and viable, they showed a limited ability to integrate with the host retina. Dissociated retinal microaggregates and retinal cell suspensions were subsequently tried by several investigators *(7–9)*. These microaggregate suspensions were easily introduced into the subretinal space with minimal trauma but were shown to form rudimentarily differentiated rosettes rather than well-organized layers *(9,10)*.

Following this pioneering work, intact photoreceptor sheets were proposed as a more suitable tissue for transplantation. Photoreceptor sheets were first successfully harvested using a vibratome to section gelatin-embedded retinal tissue *(11)*. Subsequently, excimer laser ablation was used by several investigators for the preparation of photoreceptor sheets *(12)*. Although the harvesting of photoreceptor sheets is technically more challenging, it does appear to minimize the problem of rosette formation.

The motivation for using photoreceptor grafts is that the visual pathway downstream of the photoreceptor layer often remains functional, even in advanced retinal degenerations. However, it is now clear that after photoreceptor cells are lost, the remaining outer aspect of the retina reorganizes and eventually resembles a glial scar *(13)*. Nonetheless, the neural retina remains viable, the optic nerve and secondary connections to the visual cortex remain functional, and the visual cortex too, remains intact. This suggests that photoreceptor replacements alone might be sufficient to restore vision in the diseased eye, at least prior to the late-occurring cellular reorganization of the retina.

Several animal studies have suggested that transplanted retinal cells may be able to form synapses and integrate anatomically with host retinal tissue *(14–19)*. The functional capability of the grafted tissues has also been demonstrated via ectopic electrophysiological responses *(20)* and light-dependent shifts in phototransduction proteins in photoreceptor cells *(21)*. However, functional integration and host visual improvement after transplantation into the eye have been less well established.

In contrast to intraocular grafts, intracranial transplantations have proven to be successful in terms of functional integration with the host. McLoon et al. (1982) showed that embryonic retinal grafts transplanted into the brain of newborn rats established projections to the superior colliculus, demonstrating that a degree of neural plasticity is present in the mammalian visual system beyond the normal period of development of the visual pathways *(22)*. Researchers in the Lund laboratory later demonstrated that retinal transplants were able to establish functional connectivity in the brain by driving a pupillary reflex via the pretectum and that the magnitude of this response reflected the strength of graft-host integration *(23–25)*.

Although transplantation of retinal cells has shown some encouraging results in animals, it is not yet a treatment available for use in humans. In human studies, transplantation of intact sheets of photoreceptor cells *(26)* or retinal microaggregates *(27)* into the subretinal space of adults with RP has not yielded reproducible evidence of improved visual function. Retinal cell transplantation is still in its preliminary stages of investigation in the laboratory. The good news, however, is that studies have thus far found that when photoreceptor cells are transplanted to the retina of animals, some features of normal photoreceptors are either maintained within the graft or develop after transplantation. Such studies have shown that transplanted photoreceptors, either in the form of cell suspensions *(28)* or intact retinal sheets *(11,29)*, display good survival and morphologic preservation in the host. Transplants of embryonic retinal sheets showed apparent fusion of the graft's inner plexiform layer (which tends to lose ganglion cells) and the host's remnant inner nuclear layer (INL). In addition, grafted retinas are able to maintain an organized photoreceptor layer with outer segments that contact the host RPE. Photoreceptors of intact retinal sheets were also shown to have functional phototransduction properties after transplantation into the host. Unfortunately, the grafted photoreceptors and retinal sheets showed limited integration with the host neural retina, precluding downstream transmission of neural signals to the INL.

The lack of anatomical or functional integration of grafted photoreceptors with the host has been attributed to an inherent lack of plasticity in these "mature" donor neurons. Recent studies have shown that although retinal grafts derived from postnatal mice express many markers characteristic of mature retina (e.g., rhodopsin, conventional protein kinase Cα), very few of the grafted cells migrate into, or extend neurites into, the host retina when transplanted into the eyes of mature *rd* mice *(30)*. Conversely, brain progenitor cells (BPCs) transplanted into the subretinal space of *rd* mice readily migrate into, and integrate with, the host retina but showed very limited ability to differentiate into mature retinal neurons *(30)*. Therefore, although mature retinal neurons have the ability to potentially restore function to the diseased retina because of their expression of retinal-specific markers, they appear to be resistant to integration within the host. On the other hand, whereas progenitor cells possess the plasticity to migrate and integrate with the host, they do not necessarily differentiate into retinal neurons following transplantation.

This dichotomy of graft-host integration exhibited by mature neurons and neural progenitor cells (NPCs) has led to research aimed at achieving three goals: (1) improving the integration of whole retinal grafts or photoreceptor sheets, (2) enhancing the ability of progenitor cells to differentiate into functional mature neurons, and (3) using progenitor cells to enhance graft-host integration.

Although there is not yet conclusive evidence that retinal progenitor cell (RPC) transplants result in long-term improved or restored vision in animals or humans with a retinal degeneration, research done thus far demonstrates that the intrinsic multipotentiality and plasticity of progenitor cells offers great potential because RPCs are proliferative, multipotent cells that give rise to postmitotic progeny that ultimately differentiate into the various cell types that comprise the retina, these cells may be the ideal candidates for replacing cells lost to disease. Although NPCs from the retina or brain (RPCs or BPCs) present a promising future approach for treating retinal degenerative diseases, there are still a number of significant obstacles to be surmounted in the coming years.

STEM CELL CHARACTERISTICS

Stem cells are multipotent, self-renewing cells that sit at the top of a lineage hierarchy and proliferate to form differentiated cell types of a given tissue in vivo. In adult organisms, stem cells can divide repeatedly to replenish a tissue or may be quiescent, as in the mammalian retina. Within their niche, stem cells can divide symmetrically to expand their numbers (self-renewing) or undergo asymmetric division to yield both stem cells and committed precursors.

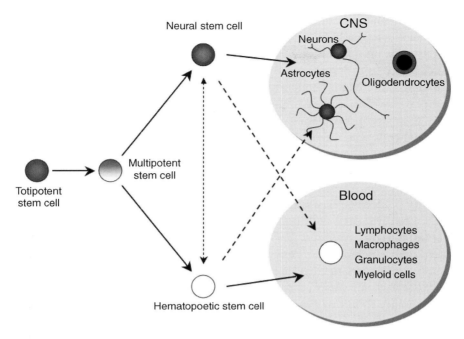

Fig. 1. Schematic of stem cell fate and differentiation.

Not all stem cells are created equal (Fig. 1). The potential for stem cells to differentiate into mature, specialized cells changes as an embryo develops. Stem cells can be:

- **Totipotent:** In the early stages after fertilization, the stem cells in a dividing zygote are considered totipotent. These cells can form any type of cell, and by definition must be able to generate a complete, viable organism with placenta.
- **Pluripotent:** Several days after fertilization, the cells of the developing embryo lose "totipotency" and are then considered pluripotent. These cells give rise to the three embryonic germ cell layers and can form any kind of cell found in adults, but can no longer give rise to a complete organism.
- **Multipotent:** Eventually, the progeny of pluripotent stem cells become multipotent stem cells—the type found in adults. Multipotent stem cells can no longer develop into all or most cell types. Instead, they are limited in terms of cell lineages and can typically develop into certain cell types within a specific tissue, organ, or system. Multipotent stem cells play an important role later in development and replenish the senescent or damaged cells in adult tissue.

Multipotent stem cells exist in many tissues and organs of the body, such as the brain, retina, liver, muscles, cornea, and bone marrow. In recent years, these multipotent stem cells have been isolated and cultured in vitro from different parts of the mammalian retina *(31–34)*. However, in most adult mammals these retinal stem cells (RSCs) remain quiescent and do not mobilize in response to injury. Hence, significant injuries are not repaired and functions lost through neuronal cell death are not regained. In contrast, lower vertebrates such as amphibians, birds, and fish respond to injury by neuronal regeneration. The cellular source for regeneration differs for each class: the RPE in amphibians and embryonic birds, Müller glia in post-hatch birds and intrinsic stem cells in fish as described in a recent review by Hitchcock et al. *(35)*.

In mammals, multipotent stem cells are generally more restricted in their developmental capacity and are largely programmed to develop into certain cell types within a specific tissue or organ. However, recent evidence suggests that these stem cells also have the capacity to differentiate into cell types specific for the tissue into which they are transplanted. For example, neural stem cells (NSCs) have been shown to differentiate into bone marrow derived cells when grafted into the bloodstream of irradiated hosts *(36)*. Conversely, bone marrow-derived stem cells have been shown to differentiate into retinal cells when transplanted into injured rat retina *(37)*. This capability to display more potential phenotypes in alternate niches could potentially allow stem cells derived from one body part to form cells of other body parts, a quality called plasticity. This experimental data remains controversial and it will likely be many years before the true plasticity of stem cells is fully understood.

Over the last few years, multipotent stem cells have been isolated from different parts of the central nervous system (CNS) including the brain, retina, and spinal cord. These "CNS stem cells" include NSCs that are brain derived and have the capability to generate all cells of the CNS lineage, namely neurons, astrocytes, and oligodendrocytes, as well as RSCs that can generate retinal-specific neurons, but not oligodendrocytes, as none are normally found within the retina. The isolation of stem/progenitor cells from various regions of the CNS has raised the possibility of using them as a donor cell source for retinal transplantation, where they offer great promise for repair of the diseased retina.

The mammalian retina presents formidable challenges to medical therapeutics because of its restricted capacity for endogenous self repair and regeneration. In addition, attempts at exogenous repair/restoration are severely constrained by its delicate structure and cytoarchitectural complexity. The lack of retinal regeneration and the paucity of effective strategies for repair invariably result in partial or complete visual deficits in patients suffering from retinal degenerations. Tissue engineering using progenitor cells isolated from the CNS offers a potential therapy for retinal degenerations because CNS stem cells are not only multipotent and capable of self-renewal, but also, more importantly, they satisfy the immunogenic requirement for CNS transplantation by having an intrinsic immune privilege status, as will be discussed shortly.

RETINAL TRANSPLANTATION OF CNS STEM CELLS

A major technical challenge facing attempts at intraocular grafting of fetal retina is not so much survival or differentiation, but instead a lack of widespread functional integration of the graft with the remaining circuitry of the host retina. Thus, a fundamental prerequisite to functional success is widespread integration of graft-derived neurons within the mature degenerating mammalian retina. Furthermore, the grafted cells must exhibit this capacity in the face of active retinal disease.

These criteria were first met by hippocampal progenitor cells isolated and grown from the brain of adult rats *(38)*. Hippocampal progenitor cells are a type of NSC derived from a region of active neurogenesis in adult mammals. In the first report on grafting this type of cell to the retina, Takahashi et al. demonstrated that intravitreal injections of hippocampal progenitors resulted in a spectacular degree of morphological integration within the neural retina of neonatal rats *(39)*. In a subsequent article, we reported that these rat hippocampal progenitor cells were capable of widespread migration and

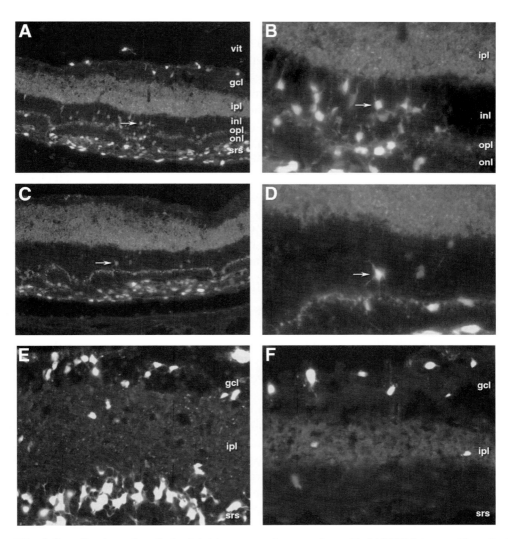

Fig. 2. Localization of grafted adult hippocampal progenitor cells (AHPCs) to specific retinal layers. Cells were grafted into the vitreous of 4 (**A–D**), 10 (**E**), and 18 (**F**) -wk-old rats, and examined 4 wk later. Sections were stained with antisynaptophysin/Cy3 antibody and viewed under flouroscein isothyocyanate (FITC) and Cy3 fluorescent illumination. Arrow in **A** indicates cell seen in **B** at higher power; arrow in **C** indicates cell seen in **D** at higher power. vit, vitreous; gcl, ganglion cell layer; ipl, inner plexiform layer; inl, inner nuclear layer; opl, outer plexiform layer; onl, outer nuclear layer; srs, subretinal debris and degenerating photoreceptor elements. (Reprinted from ref. *46*, with permission.)

morphological integration into the degenerating retina of mature Royal College of Surgeons rats during the phase of active retinal degeneration *(40)*. In this case, grafted cells differentiated into neurons based on marker expression and morphologically showed indications of an ability to respond to the host retinal cytoarchitecture (Fig. 2). For instance, the somata of grafted cells were located predominantly within the cellular layers of the retina, whereas their processes extended into the plexiform layers, frequently at an appropriate orientation, and with finer processes branching off within specific sublaminae (Fig. 2). In addition, GFP+ donor cells within the ganglion cell layer expressed the neuronal marker NF-200, known to be expressed by retinal ganglion cells, and extended large numbers of growth cone-tipped processes into the host optic nerve. We have also found that xenografts of neural progenitors can integrate with the host retina, although there may well be limits to this capacity *(41)*.

Indeed, the factors controlling the integration of stem and progenitor cells within the CNS are likely to be many. In the case of xenografts, basic metabolic and size considerations will necessarily play a role. In addition, the degree of genetic disparity between graft and host will, to some extent, limit the efficacy of intercellular signaling. Furthermore, the relative plasticity of the donor cell line will influence outcome, as will the developmental state of the host and any ongoing degeneration or inflammation. Finally, there is the important factor of immune tolerance. Recent studies provide a baseline for predicting the immunological consequences of transplanting stem and progenitor cells to various sites within the CNS, as will be discussed after further considering the relationship between stem cell plasticity and graft-host integration.

Plasticity vs Commitment: A Stem Cell Conundrum

Plasticity, the very property that endows stem cells with the ability to engraft in the mature CNS, also carries with it a significant burden. The more plasticity a cell possesses, the less committed it is to a specific lineage. Although the lack of commitment of an embryonic stem (ES) cell can be exploited to generate a number of cell types from a single cell source, it is also clear that ES cells cannot, at least at present, be induced to differentiate into all types of cells. In those cell types that can be generated, it remains unclear how "complete" the differentiation truly is, i.e., it may look like a specific cell, or even express markers of that cell, but does it become a fully mature and functional replica of that cell? The dichotomy between plasticity and commitment is perhaps the most important issue to be addressed in the field of stem cell biology. This relationship is especially true in the context of retinal transplantation.

Both embryonic and mature retinal tissue has been used for transplantation studies in animals and humans for decades. Although these grafts invariably differentiate into mature retinal tissue, they have shown a very limited capacity to integrate (e.g., migrate or send neurites into the host retina). Our early work with hippocampal stem cells showed the opposite result: engraftment, but only limited differentiation into retinal cell types. We further investigated this issue in a controlled study by grafting neural stem cells, or developing retina, into the same hosts (*rd* mice). Our results reinforced the hypothesis that plasticity and commitment are in some ways mutually exclusive. Conventional tissue grafts have the intrinsic potential to differentiate into the lineages from which they were obtained, but lack the plasticity to fully engraft in mature hosts. Conversely, stem cells have the intrinsic ability to engraft, but often differentiate incompletely or in unpredictable ways. At present, we do not possess the tools to take the reins of stem cell development and induce them to make the cell types we need or desire. We describe this in terms of the experimenter lacking the tools, rather than an intrinsic lack of developmental potential of the stem cell, because it is clear that ES cells, for example, have the "potential" to make all cells in the body. Harnessing the power of stem cells for therapeutic purposes requires that we unravel the secrets of development, thereby unleashing the potential for stem cells to repair the diseased or injured body.

Immunological Aspects of Stem Cell Transplantation

A series of in vitro and in vivo experiments have revealed mouse NPCs to be immune privileged cells. These cells survive allografting without the need for immune suppression and are more likely to be immunologically tolerated than solid tissue grafts of brain or retina, which contain major histocompatibility complex (MHC) class II-expressing microglia. On a cautionary note, the immunological situation in large mammals, including humans, may be more complex than in the mouse. For instance, we already know that human NPCs express abundant MHC class I, although, like their murine counterparts, they do not express class II *(42)*. Because NPCs appear to exhibit relative immune privilege as a donor cell type, and because transplantation of these cells is directed towards recipient cites that are themselves immune privileged, such as the retina, brain, or spinal cord, it seems reasonable to conclude that transplantation of these cells will prove to be less challenging from an immunological perspective than has been the case with hematopoietic stem cells (that express class II MHC antigens) or solid organ transplants (that contain class II-expressing passenger leukocytes). It can be hoped that systemic immunosuppression will not be necessary when grafting NPCs to the retina clinically, although it remains to be seen whether this is in fact the case.

Transplantation of Retinal Stem Cells

A profound degree of engraftment can be achieved through the use of brain-derived progenitor cells for retinal transplantation. It has become apparent, however, that brain-derived cells do not differentiate into authentic retinal neurons in the microenvironment of the mature, diseased retina. Several possible strategies can be employed in an effort to overcome this obstacle. For example, one could either attempt to modify a brain-derived cell to induce transdifferentiation along a retinal lineage, or induce the differentiation of a more plastic, less differentiated cell type such as an ES cell into such a fate. We and several other groups have chosen a third strategy, namely, the isolation of progenitor cells from the neural retina of the developing mammalian eye (Fig. 3) *(34)*. Although these studies are in the early stages, one important discovery is that these cells, upon transplantation to the retina of adult mammals with retinal disease, possess both the integrative plasticity that is a hallmark of CNS stem cells, as well as the ability to differentiate into retinal neurons, including photoreceptors (Fig. 4). We are hopeful that further studies of retinal progenitor cell grafts will point the way forward to the development of clinical strategies aimed at restoring vision to the blinded eye.

Functional Repair of the Diseased Retina

The results of the previous studies make a strong case for the need for RSCs. If one could isolate stem or progenitor cells from the retina, the properties of self renewal and multipotentiality could allow for a large supply of donor cells for use in retinal repair. Researchers have identified and isolated RSCs from both the mature ciliary marginal zone and the developing neurosensory retina.

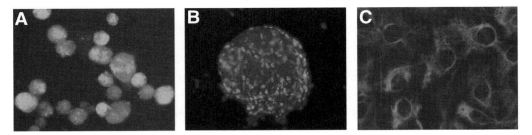

Fig. 3. Expression of phenotypic markers by GFP+ retinal progenitor cells. Cultured under proliferation conditions, RPC neurospheres exhibit endobgenous GFP **(A)** and widespread immunolabeling for Ki-67 **(B)** and nestin **(C)**. (Reprinted from ref. *34*, with permission.)

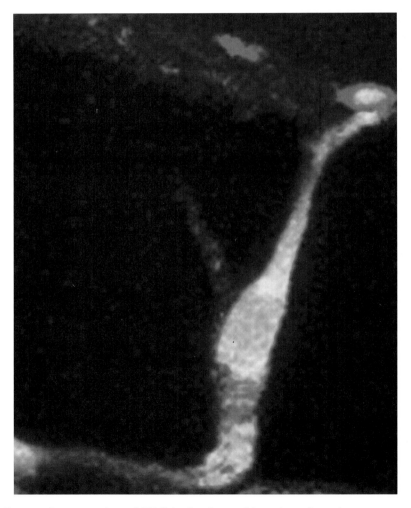

Fig. 4. Image of a tranasplanted RPC in the dystrophic retina of an rd mouse expressing the retinal photoreceptor marker rhodopsin (red) and GFP (green), with co-localization of these markers shown as yellow in the merged image. (Reprinted from ref. *34*, with permission.)

The ciliary marginal zone stem cells have an impressive ability to replicate in the absence of proteins and have been shown to differentiate into cells expressing retinal markers in culture. On a cautionary note, however, that same lack of growth factor dependence, together with limited differentiation potential in the mature, diseased retina, makes them a questionable choice for transplantation studies.

The work of Ahmad's group has provided a wealth of information on the intracellular signaling pathways of RSCs derived from developing retinal tissue. We have also been studying a variety of sources for isolation of RSCs. Neural retina from the period of late neurogenesis (postnatal day 1 [P1] for mice) appears to be optimal for generating cells that both expand through repeated passaging in culture and generate retinal cells upon transplantation to the mature host.

We recently described the utilization of this source of RSCs in detail *(34)*. In culture, we demonstrated that RPCs isolated from P1 GFP mice could be greatly expanded while maintaining their multipotentiality, which included the expression of rod, bipolar, and glial markers by distinct subsets of differentiating cells. Under proliferation conditions, RPCs expressed a number of neurodevelopmental genes and surface markers. Reverse transcriptase-polymerase chain reaction demonstrated expression of nestin and Sox2, as well as other neurodevelopmental genes including Notch1, Hes1, Hes5, Sox2, Prox1, Mash1, numb, and NeuroD. Analysis by flow cytometry showed surface expression of GD_2 ganglioside, CD15 (LeX), and the tetraspanins CD9 and CD81.

After grafting to the degenerating retina of mature mice, a subset of the retinal progenitor cells developed into mature neurons, including cells expressing the photoreceptor markers recoverin, rhodopsin, or cone opsin. Importantly, cells could be observed differentiating into photoreceptors that had both morphological and cytochemical hallmarks of mature rods. When grafted into rho–/– hosts, we found rescue of host cells in the outer nuclear layer (ONL), along with widespread integration of donor cells into the inner retina. A subset of grafted cells expressed cone markers in this model. Moreover, recipient mice showed improved light-mediated behavior compared to controls. Greater thickness of the host ONL was seen in stem cell-grafted eyes, but not in sham-operated controls and photoreceptor density was higher in the treated eye than the untreated eye. The increase in photoreceptor density correlated with graft location. Graft-associated rescue likely reflects an indirect neuroprotective effect, similar to that reported previously *(9,42,44)*. The behavioral results suggest that grafted RPCs decrease the tempo of luminance detection loss in dystrophic (rho–/–) mice, especially at low light levels. Preservation of visual function was detected over a 25-wk period, extending into a period with limited photoreceptor survival. As there are a number of technical difficulties associated with functional assessment in mice, expanding this approach to larger animals would be useful for accurate determination of graft efficacy, as well as development of surgical approaches for potential clinical application of transplantation research.

Human Retinal Progenitor Cells

RPCs can also be obtained from cadaveric human retinal tissue. We have recently reported that retinas obtained from postmortem premature infants can be enzymatically dissociated, and viable proliferative cells obtained and grown in the presence of epithelial growth factor and basic fibroblast growth factor. Such cultures grow to confluence repeatedly for up to 3 mo. Again, the cells can be grown as suspended spheres or adherent monolayers, depending on how they are cultured. Cultured human RPCs (hRPCs) express a range of markers consistent with CNS progenitor cells and similar to those found in human BPCs (hBPCs) from the same donors, including nestin, nucleostemin, vimentin, Sox2, and the proliferation marker Ki-67. Also expressed are the surface markers GD2 ganglioside, CD15 (Lewis X), the tetraspanins CD9 and CD81, the CD95 "death receptor" (Fas), CD133, and MHC class I antigens, however, no MHC class II expression was detected, nor was expression of Pou5f1 (Oct4) or Nanog (Schwartz and Klassen, unpublished data), two genes expressed in ES cells. hRPCs, but not hBPCs, expressed the genes for Dach1, Pax6, Six3, Six6, and recoverin. hBPCs, but not hRPCs, expressed the genes for Dlx2, Dlx5, Gad67, and Olig2 (Schwartz and Klassen, unpublished data). Minority subpopulations of both hRPCs and hBPCs expressed the protoneuronal genes doublecortin and β-III tubulin, as well as the glial gene GFAP (glial fibrillary acid protein), consistent with increased lineage restriction in subsets of cultured cells. These data suggest that although immature neuroepithelial cells taken from different regions of the immature CNS express a number of gene products common to CNS progenitor cells, they also retain genes indicative of fate specification events occurring in the region from which they were harvested.

Based on these findings, it is evident that hRPCs, derived from retinal tissue obtained postmortem from premature infants at just past mid-gestation, represent a rather heterogeneous population of progenitors. Estimating the neurodevelopmental age of the 7-d-old rodent to be roughly equivalent to the newborn human, a mid-gestation human approximately corresponds to the E13 rodent. Studies of the E13 rodent retina have shown that it contains precursors of rods, amacrine cells, cones, ganglion cells, and horizontal cells, with little evidence of bipolar or Müller cell development *(45)*. Our immunocytochemical studies of progenitors from the developing human retina are entirely consistent with a heterogeneous population, based on morphology and marker expression, suggesting that in vitro cultures represent to some extent the heterogeneity found in vivo. The presence of subpopulations expressing doublecortin (DCX), recoverin, β-III tubulin, and GFAP (Fig. 5) among the majority of cells expressing the immature markers nestin, Sox2, and vimentin, suggests a tendency toward continuous differentiation in these cultures, even under proliferation conditions (Fig. 5). DCX- and GFAP-expressing subpopulations were also present in brain-derived human progenitors, however, further studies will be necessary to determine whether these findings are specific to cells harvested at this particular developmental time point or can be generalized to a wider range of CNS progenitors, either human or from other species. Another point of considerable interest is the potential of hRPCs to integrate and differentiate into retinal neurons following transplantation, particularly in the setting of photoreceptor loss.

CONCLUSIONS

Stem cell biology is providing new insight into the development and pathophysiology of the mammalian retina. Retinal specificity of cell fate is now known to relate to changes in transcription factor expression during lineage choices made by stem and progenitor cell populations. Furthermore, the responses of grafted stem cells to retinal

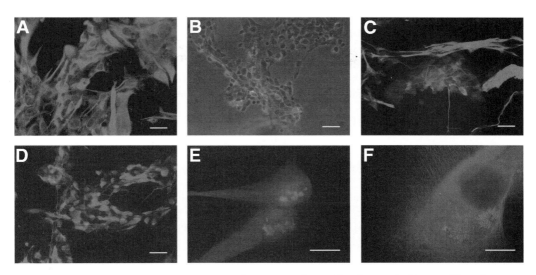

Fig. 5. Phenotypic markers in human retinal progenitor cell cultures. **A–E** = proliferation conditions; **F** = differentiation conditions. (**A**) Nestin (green) staining showed a cytoplasmic pattern, Sox2 (red) a nuclear pattern, and GD_2 ganglioside (blue) a surface pattern consisting of discrete punctate foci of variable size and number. (**B**) CD15 immunoreactivity (red) was variable and most evident on a subset of cultured cells with small, rounded profiles (shown against phase contrast). (**C**) Distinct subpopulations within hRPC cultures expressed either the neuronal marker b-III tubulin (red) or the glial marker GFAP (green). (**D**) Other subpopulations could be distinguished by expression of the neuroblast marker DCX (red) as compared to the photoreceptor marker recoverin (green). (**E**) Nestin (blue) and Sox2 (red) co-localized with Ki-67 (green) in hRPCs grown under proliferation conditions. (**F**) Under proneuronal differentiation conditions, there was an absence of Ki-67 staining, whereas Sox2 (red) was still detectable but now assumed a perinuclear distribution together with cytoplasmic expression of FRRI (blue). Original magnification X40, except **F** = X100. Bars = 50 μm, except **E,F** = 10 μm. (Reprinted with permission from ref. *47*, with permission.)

injury cues, such as photoreceptor degeneration, provide a means of evaluating local homeostatic mechanisms in the diseased microenvironment. Although much work remains to be done, especially with respect to the investigation of host visual benefits after transplantation, it is already apparent that RSC transplantation provides an important new strategy for altering the course of retinal degenerations.

REFERENCES

1. Tansley K. The development of the rat eye in the graft. J Exp Bio 1946;22:221–223.
2. Royo PE, Quay WB. Retinal transplantation from fetal to maternal mammalian eye. Growth 1959;23:313–336.
3. del Cerro M, Gash DM, Rao GN, Notter MF, Wiegand SJ, Gupta M. Intraocular retinal transplants. Invest Ophthalmol Vis Sci 1985;26:1182–1185.
4. Turner JE, Blair JR. Newborn rat retinal cells transplanted into a retinal lesion site in adult host eyes. Brain Res 1986;391:91–104.

5. Aramant R, Seiler M, Turner JE. Donor age influences on the success of retinal grafts to adult rat retina. Invest Ophthalmol Vis Sci 1988;29:498–503.

6. Ghosh F, Arner K, Ehinger B. Transplant of full-thickness embryonic rabbit retina using pars plana vitrectomy. Retina 1998;18:136–142.

7. Gouras P, Du J, Gelanze M, Kwun R, Kjeldbye H, Lopez R. Transplantation of photoreceptors labeled with tritiated thymidine into RCS rats. Invest Ophthalmol Vis Sci 1991; 32:1704–1707.

8. Du J, Gouras P, Kjeldbye H, Kwun R, Lopez R. Monitoring photoreceptor transplants with nuclear and cytoplasmic markers. Exp Neurol 1992;115:79–86.

9. Juliusson B, Bergstrom A, van Veen T, Ehinger B. Cellular organization in retinal transplants using cell suspensions or fragments of embryonic retinal tissue. Cell Transplant 1993;2:411–418.

10. Gouras P, Algvere P. Retinal cell transplantation in the macula: new techniques. Vision Res 1996;36:4121–4125.

11. Silverman MS, Hughes SE. Transplantation of photoreceptors to light-damaged retina. Invest Ophthalmol Vis Sci 1989;30:1684–1890.

12. Huang JC, Ishida M, Hersh P, Sugino IK, Zarbin MA. Preparation and transplantation of photoreceptor sheets. Curr Eye Res 1998;17:573–585.

13. Marc RE, Jones BW, Watt CB, Strettoi E. Neural remodeling in retinal degeneration. Prog Retin Eye Res 2003;22:607–655.

14. del Cerro M, Notter MF, del Cerro C, Wiegand SJ, Grover DA, Lazar E. Intraretinal transplantation for rod-cell replacement in light-damaged retinas. J Neural Transplant 1989; 1:1–10.

15. Gouras P, Du J, Gelanze M, et al. Survival and synapse formation of transplanted rat rods. J Neural Transplant Plast 1991;2:91–100.

16. Silverman MS, Hughes SE, Valentino TL, Liu Y. Photoreceptor transplantation: anatomic, electrophysiologic, and behavioral evidence for the functional reconstruction of retinas lacking photoreceptors. Exp Neurol 1992;115:87–94.

17. Zucker CL, Ehinger B, Seiler M, Aramant RB, Adolph AR. Ultrastructural circuitry in retinal cell transplants to rat retina. J Neural Transplant Plast 1994;5:17–29.

18. Aramant RB, Seiler MJ. Fiber and synaptic connections between embryonic retinal transplants and host retina. Exp Neurol 1995;133:244–255.

19. Ghosh F, Bruun A, Ehinger B. Graft-host connections in long-term full-thickness embryonic rabbit retinal transplants. Invest Ophthalmol Vis Sci 1999;40:126–132.

20. Adolph AR, Zucker CL, Ehinger B, Bergstrom A. Function and structure in retinal transplants. J Neural Transplant Plast 1994;5:147–161.

21. Seiler MJ, Aramant RB, Ball SL. Photoreceptor function of retinal transplants implicated by light-dark shift of S-antigen and rod transducin. Vision Res 1999;39:2589–2596.

22. McLoon LK, Lund RD, McLoon SC. Transplantation of reaggregates of embryonic neural retinae to neonatal rat brain: differentiation and formation of connections. J Comp Neurol 1982;205:179–189.

23. Klassen H, Lund RD. Retinal transplants can drive a pupillary reflex in host rat brains. Proc Natl Acad Sci USA 1987;84:6958–6960.

24. Klassen H, Lund RD. Parameters of retinal graft-mediated responses are related to underlying target innervation. Brain Res 1990;533:181–191.

25. Klassen H, Lund RD. Retinal graft-mediated pupillary responses in rats: restoration of a reflex function in the mature mammalian brain. J Neurosci 1990;10:578–587.

26. Kaplan HJ, Tezel TH, Berger AS, Wolf ML, Del Priore LV. Human photoreceptor transplantation in retinitis pigmentosa. A safety study. Arch Ophthalmol 1997;115:1168–1172.

27. Humayun MS, de Juan E Jr, del Cerro M, et al. Human neural retinal transplantation. Invest Ophthalmol Vis Sci 2000;41:3100–3106.
28. Gouras P, Du J, Kjeldbye H, Yamamoto S, Zack DJ. Long-term photoreceptor transplants in dystrophic and normal mouse retina. Invest Ophthalmol Vis Sci 1994;35:3145–153.
29. Aramant RB, Seiler MJ. Retinal transplantation—advantages of intact fetal sheets. Prog Retin Eye Res 2002;21:57–73.
30. Lu B, Kwan T, Kurimoto Y, Shatos M, Lund RD, Young MJ. Transplantation of EGF-responsive neurospheres from GFP transgenic mice into the eyes of rd mice. Brain Res 2002;943:292–300.
31. Tropepe V, Coles BL, Chiasson BJ, et al. Retinal stem cells in the adult mammalian eye. Science 2000;287:2032–2036.
32. Ahmad I, Dooley CM, Thoreson WB, Rogers JA, Afiat S. In vitro analysis of a mammalian retinal progenitor that gives rise to neurons and glia. Brain Res 1999;831:1–10.
33. Ahmad I, Tang L, Pham H. Identification of neural progenitors in the adult mammalian eye. Biochem Biophys Res Commun 2000;270:517–521.
34. Klassen HJ, Ng TF, Kurimoto Y, et al. Multipotent retinal progenitors express developmental markers, differentiate into retinal neurons, and preserve light-mediated behavior. Invest Ophthalmol Vis Sci 2004;45:4167–4173.
35. Hitchcock P, Ochocinska M, Sieh A, Otteson D. Persistent and injury-induced neurogenesis in the vertebrate retina. Prog Retin Eye Res 2004;23:183–194.
36. Bjornson CR, Rietze RL, Reynolds BA, Magli MC, Vescovi AL. Turning brain into blood: a hematopoietic fate adopted by adult neural stem cells in vivo. Science 1999;283:534–537.
37. Tomita M, Adachi Y, Yamada H, et al. Bone marrow-derived stem cells can differentiate into retinal cells in injured rat retina. Stem Cells 2002;20:279–283.
38. Gage FH, Ray J, Fisher LJ. Isolation, characterization, and use of stem cells from the CNS. Annu Rev Neurosci 1995;18:159–192.
39. Takahashi M, Palmer TD, Takahashi J, Gage FH. Widespread integration and survival of adult-derived neural progenitor cells in the developing optic retina. Mol Cell Neurosci 1998;12:340–348.
40. Whiteley SJ, Klassen H, Coffey PJ, Young MJ. Photoreceptor rescue after low-dose intravitreal IL-1beta injection in the RCS rat. Exp Eye Res 2001;73:557–568.
41. Mizumoto H, Mizumoto K, Whiteley SJ, Shatos M, Klassen H, Young MJ. Transplantation of human neural progenitor cells to the vitreous cavity of the Royal College of Surgeons rat. Cell Transplant 2001;10:223–233.
42. Klassen H, Schwartz MR, Bailey AH, Young MJ. Surface markers expressed by multipotent human and mouse neural progenitor cells include tetraspanins and non-protein epitopes. Neurosci Lett 2001;312:180–182.
43. Reh TA. Cellular interactions determine neuronal phenotypes in rodent retinal cultures. J Neurobiol 1992;23:1067–1083.
44. Klassen H, Imfeld KL, Ray J, Young MJ, Gage FH, Berman MA. The immunological properties of adult hippocampal progenitor cells. Vision Res 2003;43:947–956.
45. Livesey FJ, Cepko CL. Vertebrate neural cell-fate determination: lessons from the retina. Nat Rev Neurosci 2001;2:109–118.
46. Young MJ, Ray J, Whiteley SJ, Klassen H, Gage FH. Neuronal differentiation and morphological integration of hippocampal progenitor cells transplanted to the retina of immature and mature dystrophic rats. Mol Cell Neurosci 2000;16:197–205.
47. Klassen H, Ziaeian B, Kirov II, Young MJ, Schwartz PH. Isolation of retinal progenitor cells from post-mortem human tissue and comparison with autologous brain progenitors. J Neurosci Res 2004;77:334–343.

21

Application of Encapsulated Cell Technology for Retinal Degenerative Diseases

Weng Tao, MD, PhD, and Rong Wen, MD, PhD

Contents

THE ENCAPSULATED CELL TECHNOLOGY

Encapsulated cell technology (ECT) is essentially a cell-based delivery system that can be used to deliver therapeutic agents to the target tissue, including the central nervous system (CNS) and the eye to treat chronic disorders. In this chapter, we focus on its application in retinal degenerative diseases.

Advances in molecular biology over the last two decades have led to the discovery of many protein molecules with promising therapeutic potentials, including cytokines and neurotrophic (NT) factors. However, the value of these new molecules has not been fully realized for clinical use, mainly due to the lack of an effective delivery system. The blood–brain barrier or blood–retinal barrier prevent large molecules in the blood stream from entering the brain or the retina. Circumventing these barriers is one of the major challenges for long-term sustained delivery of proteins to the CNS and retina.

For protein delivery to the CNS or the retina, the traditional approaches are quite limited. This is exemplified by the failure of a clinical trial to systemically administer ciliary NT factor (CNTF 24 kD), a member of the interleukin (IL)-6 family of cytokines, for amyotrophic lateral sclerosis (sponsored by Regeneron). In this trial, systemic administered CNTF (subcutaneous injection) resulted in no detectable CNTF in the CNS despite the high doses used. Consequently, no therapeutic benefit was demonstrated. In fact, the high peripheral CNTF levels were associated with major side effects, such as fever, fatigue, and blood chemistry changes that are consistent with activation of the acute phase

From: *Ophthalmology Research: Retinal Degenerations: Biology, Diagnostics, and Therapeutics*
Edited by: J. Tombran-Tink and C. J. Barnstable © Humana Press Inc., Totowa, NJ

response *(1,2)*. Thus, systemic administration of large molecules, such as CNTF and other NT factors, is simply not an effective approach for CNS and ocular disorders.

There are two other options for delivering proteins to the CNS or the retina: bolus injection of purified recombinant proteins and gene therapy. Bolus injection is clinically impractical because it requires repetitive injections for long-term therapy. Gene therapy, on the other hand, can achieve sustained expression of a given protein. However, the doses of therapeutic protein are difficult to control due to the fact that no reliable means is available to regulate the expression levels of the transgene. Furthermore, it is impossible to reverse the treatment once the gene is delivered.

The ECT is a delivery system that uses live cells to secrete a therapeutic agent. This is usually achieved by genetically engineering a specific type of cells to overexpress a particular agent. The engineered cells are then encapsulated in semipermeable polymer capsules. The capsule is then implanted into the target sites. There are several advantages of ECT. First, it allows potentially any gene encoding for a therapeutic protein to be engineered into the cells and therefore has a broad range of applications. Also, the therapeutic protein is freshly synthesized and released *in situ* so that a relatively small amount of the protein is needed to achieve a therapeutic effect. The long-lasting output assures that the availability of the protein at the target site is not only continuous, but also long-term. Furthermore, the output of an ECT implant can be controlled to achieve the optimal dose for treatment. Finally, the treatment by ECT can be terminated if it becomes desirable by simply retrieving the implant. Thus, ECT is a very effective means of long-term delivery of biologically active proteins and polypeptides to the CNS and the retina. In fact, ECT is now proven to be an excellent choice for retinal degenerative diseases, especially considering the limited distribution volume, easy access to the eye, and the chronic nature of the diseases.

The therapeutic efficacy of growth factors delivered by ECT has been demonstrated in a number of animal models of neurodegenerative diseases, including CNTF in the rodent and primate models of the Huntington's disease *(3,4)*, Glial cell-line derived neurotrophic factor (GDNF) in rat model of Parkinson's disease *(5,6)*, and nerve growth factor (NGF) in rodent and primate models of Alzheimer's disease *(7–9)*. Furthermore, studies have shown that growth factors produced by mammalian cells, synthesized *de novo*, are more potent than purified recombinant ones expressed in *Escherichia coli (10,11)*. And transplantation of encapsulated mammalian cells delivers therapeutic agents to the target site in the CNS to produce therapeutic effects at lower dosage than are required with other means of delivery *(10)*.

An intraocular implantable device prototype of ECT has been developed by Neurotech for long-term delivery of therapeutic agents to treat ophthalmic disorders (Fig. 1) *(12)*. The device consists of genetically modified cells packaged in a hollow tube of semipermeable membrane that prevents immune molecules, e.g., antibodies and host immune cells, from entering the device, while it allows nutrients and therapeutic molecules to diffuse freely across the membrane. The encapsulated cells secrete therapeutic agents continuously, and derive nourishment from the host milieu. The device is designed for implantation through a small incision in pars plana and a small titanium wire loop on the device allows it to be anchored to the sclera. The current device is about 6.0 mm in length (including the titanium loop) and approx 1 mm in diameter. These dimensions assure that the device is outside the visual axis in the human eye.

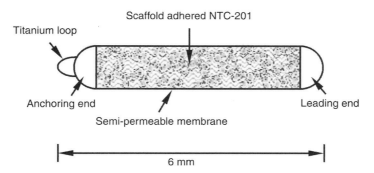

Fig. 1. Schematic illustration of the ECT device. The device is constructed with a section of semipermeable polymer membrane and supportive matrices to accommodate live cells. The two ends of the polymer section are sealed and a titanium loop is placed on the anchoring end. The loop allows the device to be anchored to the sclera. The total length of the device is 6 mm. (Reproduced from ref. *12*, with permission.)

THERAPEUTIC EFFICACY OF THE NT-501 DEVICE FOR PHOTORECEPTOR PROTECTION

Ophthalmic disorders represent a rapidly growing disease area that is associated with an increase in aging population *(13–18)*. Patients suffering from potentially blinding diseases have become one of the largest segments of the health care field with more than 50 million patients in the United States alone. Their sight is threatened by age-related macular degeneration (AMD) *(19–21)*, diabetic retinopathy *(22–26)*, glaucoma *(27–30)*, or retinitis pigmentosa (RP) *(31–34)*. Apart from AMD and RP, glaucoma is now considered to be a retinal degenerative disease because control of intraocular pressure (IOP) alone does not prevent ganglion cell degeneration.

Few effective treatments for retinal degenerative disorders are available to date. Newly discovered NT factors provide a great promise for treating these diseases. However, without a practical delivery system, the realization of this promise would be very difficult. The ECT device for intraocular implantation is specifically designed to overcome this obstacle. The first such device is NT-501, an ECT-CNTF product that consists of encapsulated cells that secrete recombinant human CNTF. NT-501 is manufactured to be sterile, nonpyrogenic, and retrievable. It is intended to deliver CNTF intraocularly for treating photoreceptor degenerations. NT-501 has been tested for preclinical efficacy, pharmacokinetics, and toxicology in dogs, pigs, and rabbits.

CNTF and RP

The promise of growth factors as potential therapeutics for photoreceptor degeneration was first demonstrated in 1990 *(35)*. Since then, many growth factors, NT factors, and cytokines have been tested in a variety of photoreceptor degeneration models, mainly by intravitreal injection of purified recombinant proteins in short-term experiments *(35–38)*. Among them, CNTF has been shown to be the most effective one in almost every model *(38)*. However, the chronic nature of RP (years) makes repetitive intraocular injection of purified recombinant CNTF (which is only effective for short duration) impractical. In fact, the obstacles of intraocular delivery have prevented the initiation of clinical trials and its further development. Other factors that have shown protective effect in animal models of retinal degeneration include brain-derived NT factor (BDNF), NT-4, Axokine, basic fibroblast growth factor, insulin-like growth factor II, transforming growth factor-β2, IL-1β, tumor necrosis factor, NGF *(36)*, pigment epithelium-derived factor (PEDF) *(39)*, GDNF *(40)*, lens epithelium-derived growth factor *(41,42)*, and cardiotrophin-1 (CT-1) *(43)*.

RP affects approx 100,000 Americans. It is a group of retinal degenerative diseases that have a complex molecular etiology. More than 100 mutations in several genes, including rhodopsin, peripherin, and phosphodiesterase (PDE)β, are believed to be responsible for RP, although the genotypes of the majority of RP patients are unknown. Despite the genetic heterogeneity, the phenotypes are very similar. Typically, a patient experiences a decrease in night vision early in life as a result of the loss of rod photoreceptor. Although the genetic defects affect only rods, cone photoreceptors eventually degenerate, leading to a progressive decrease in patient's visual field and eventually to total blindness. The similarity in phenotypes and perhaps also in pathogenesis pathways enables medical intervention with a common approach without the need to identify the genotype of a patient.

The NT-501 device has been developed for intraocular delivery of CNTF for RP. Proof-of-principle experiments were conducted in two animal models of photoreceptor degeneration, the S334ter-3 transgenic rat and the *rcd1* mutant dog. Pharmacokinetics studies were performed in the rabbits. The data from these studies indicate that CNTF delivered by NT-501 is not only effective in protecting photoreceptors, but also long-lasting. These preclinical studies of NT-501 paved the way for human clinical trial, which is currently ongoing.

Protective Effect of NTC-201 in a Rat Model of Photoreceptor Degeneration

NTC-201 cells are genetically engineered human retinal pigment epithelial (RPE) cells to overexpress human CNTF. In culture, these cells secret CNTF at a rate of 100 ng/million cells/d. We first assessed the efficacy of these cells in the heterozygous S334ter-3 rats carrying the rhodopsin mutation S334ter. Photoreceptor degeneration in these animals begins soon after birth (postnatal day 8 [P8]) and progresses rapidly. By P20, more than 90% photoreceptors are degenerated *(44)*. The size of rat eyes makes it impossible to accommodate the NT-501 device so that only unencapsulated CNTF secreting cells were used. NTC-201 cells (approx 10^5 in 2 μL phosphate buffered saline [PBS]) were injected into the left eyes intravitrealy at P9. Control animals were injected with untransfected parental cells (NTC-200). The contralateral eyes (right eyes) were untreated. In addition, a group of animals were treated with 1 μg of purified recombinant CNTF protein (in 1 μL of PBS, intravitreal injection to the left eye at P9) for comparison. Eyes were collected at P20, and processed for histological evaluation.

In untreated eyes of S334ter-3 transgenic rats, severe photoreceptor degeneration was observed by P20. The outer nuclear layer (ONL) contained only 1 row of nuclei (Fig. 2A), reduced from 10 to 12 rows of a normal animal. The NTC-201 injected eyes had five to six rows of nuclei in the ONL (Fig. 2C), whereas in the control eyes that were injected with untransfected cells (NTC-200), only one to two rows of nuclei remained (Fig. 2B). No evidence of retinal inflammation was observed in any of the treated or control eyes. In animals treated with a single intravitreal injection of purified

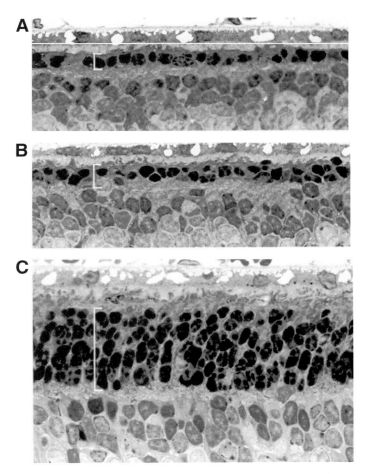

Fig. 2. Photoreceptor protection by CNTF secreting NTC-201 cells. Sections of retina from transgenic rats carrying the rhodopsin mutation S334ter were examined at PD 20 by light microscopy. **(A)** Untreated eye, **(B)** NTC-200 parental (control) cell treated eye, and **(C)** NTC-201 cell (CNTF secreting) treated eye. Eyes were injected with cells on P9. The ONL of untreated eye contained only one row of photoreceptor nuclei **(A)**. In the retina treated with control cells, the ONL had one to two rows of nuclei, whereas in the retina treated with CNTF secreting cells, the ONL contained five to six rows, indicating significant protection by CNTF released from those cells. Brackets denote ONL. Plastic embedded sections stained with toland blue. (Reproduced from ref. *12*, with permission.)

human recombinant CNTF, the ONL had two to three rows of nuclei (data not shown). These results clearly demonstrate that continuous delivery of CNTF delivered via mammalian cells protected against retinal degeneration in this model.

NT-501 Device Protects Photoreceptors in the **rcd1** Dog RP Model

The efficacy of NT-501 devices was investigated in the *rcd1* dog model. These dogs carry a mutation on the PDE6B gene encoding the β-subunit of the rod cGMP PDEβ (kindly provided by the Retinal Disease Studies Facility, Kennett Square, PA). The retinal degeneration of this model is well characterized *(45,46)*. Photoreceptor degeneration begins 3.5 wk after birth in these animals and continues for 1 yr, with 50% photoreceptor loss at 7 wk of age. NT-501 devices secreting 1 to 2 ng/d of CNTF were surgically implanted into the left eye of each *rcd1* dog at 7 wk of age, the earliest time point that the surgical procedure can be performed without disruption of the retina (the eyes of younger dogs would be too small to accommodate an earlier version of NT-501 of 10 mm in length). In a normal dog the ONL contains 10 to 12 layers of photoreceptor nuclei, but in these animals at 7 wk of age, only 5 to 6 layers of nuclei in the remain. The contralateral eye was not treated.

At the endpoint of the experiment (14 wk of age), the devices were explanted and assayed for CNTF output and cell viability, and the eyes were collected and processed for histological evaluation. The ONL in untreated eyes contained only two to three rows of photoreceptor nuclei. In contrast, the ONL in the NT-501 treated eyes still had five to six rows remaining, similar to the number of nuclei rows present at the time when the treatment began (Fig. 3). The protection of photoreceptors was evenly distributed throughout the retina and not localized near the implant site. No apparent adverse effects were found in the retina. All explanted devices contained viable cells.

Photoreceptor Protection by NT-501 Devices is Dose Dependent

To determine the minimum effective dose and the optimal therapeutic dose of CNTF, a dose-ranging study was conducted. Thirty-one *rcd1* dogs were included in this study. Devices that released different levels of CNTF were implanted into one eye of an animal at 7 wk of age. The contralateral eye was not treated. The level of device CNTF output (ng/d) was defined as follows: <0.1 ($n = 4$), 0.2–1 ($n = 8$), 1–2 ($n = 7$), 2–4 ($n = 9$), 5–15 ($n = 3$). The devices were explanted at 14 wk of age and assayed for CNTF output and viability, and the eyes were processed for histological evaluation. As shown in Fig. 4, photoreceptor protection by the NT-501 devices in the *rcd1* dog model was dose dependent. Complete protection was achieved at the highest dose tested (5–15 ng/d of CNTF), and minimal, but statistically significant, protection was observed at levels as low as 0.2–1 ng/d of CNTF. CNTF delivered less than 0.1 ng/d had no protective effect. No cellular evidence of an immune reaction, inflammation or damage to the retina was observed. Evaluation indicated that all devices contained healthy, viable cells.

PHARMACOKINETICS OF NT-501 DELIVERED CNTF

To evaluate the pharmacokinetics of CNTF in the vitreous humor and the long-term function in vivo, NT-501 devices were implanted into rabbit eyes and explanted at different time points and vitreous samples harvested. The CNTF output from the explanted devices and CNTF levels in vitreous samples were determined by enzyme-linked immunosorbent assay (ELISA).

As shown in Fig. 5, the explanted NT-501 devices produced a consistent amount of CNTF up to 12 mo in vivo. CNTF was readily detectable in the vitreous. Data from these pharmacokinetic and long-term device function studies indicate that the CNTF secreting function of the NT-501 device last at least for one year when implanted into

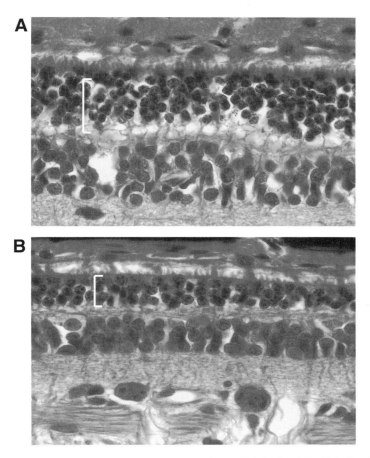

Fig. 3. Photoreceptor protection by CNTF secreting ECT device NT-501. Sections of retina from *rcd1* dog model of retinitis pigmentosa were examined by light microscopy. A device was implanted into one eye at 7 wk of age and explanted at 14 wk of age. The contralateral eye was not treated. **(A)** Treated eye, **(B)** untreated eye. The ONL of treated retina contained five to six rows of nuclei. In contrast, the untreated retina had only two to three rows. Thus the NT-501 device provided significant protection to photoreceptors in the *rcd1* dogs. Brackets denote ONL. (Reproduced from ref. *12*, with permission.)

the eye. The released CNTF was throughout the vitreous, readily available for retinal cells. The functional results are confirmed by the histological evaluation showing that all devices contained healthy, viable cells.

POTENTIAL APPLICATION OF ECT FOR OTHER RETINAL DISEASES

Neuroprotection in Glaucoma

Glaucoma is a leading cause of blindness worldwide and the second cause of irreversible blindness in the United States. Approximately 2 million people in the United States have glaucoma, although roughly half are not even aware of it. Glaucoma is a group of diseases characterized by abnormal IOP and progressive death of retinal ganglion

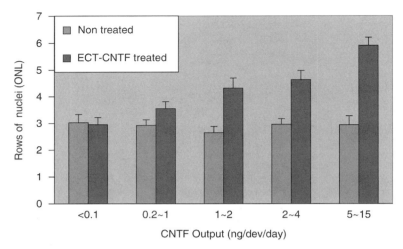

Fig. 4. Dose–response protection of photoreceptors in *rcd1* dogs. Data are presented as rows of nuclei in the ONL in ECT-CNTF treated eyes vs nontreated eyes (Mean ± SEM). The levels of CNTF output (ng/dev/d) were: <0.1 ($n = 4$, $p = 0.744$), 0.2–1 ($n = 8$, $p = 0.0009$), 1–2 ($n = 7$, $p = 0.0001$), 2–4 ($n = 9$, $p = 0.0004$), and 5–15 ($n = 3$, $p = 0.043$). The preservation of photoreceptor nuclei in the ONL depends on the amount of CNTF out put of the device. (Reproduced from ref. *12*, with permission.)

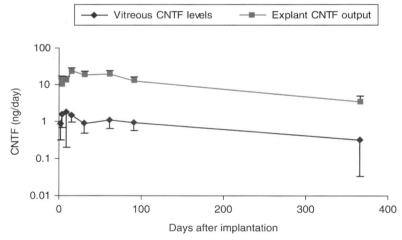

Fig. 5. Time courses of CNTF output and vitreous CNTF levels in rabbit eyes after NT-501 implantation. The NT-501 devices were implanted into rabbit eyes. At indicated time points, the devices were explanted and vitreous samples were collected. The CNTF output of devices (ng/device/d) and CNTF vitreous levels (ng/mL) were assayed by ELISA. Data are presented as Mean ± SEM.

cells. The pathologic hallmark of glaucomatous optic neuropathy is the selective death of retinal ganglion cells associated with structural changes in the optic nerve head. Glaucoma is still considered to be a disease associated with abnormal IOP, and most current approaches for treating glaucoma are directed toward pressure control. However, in some cases the progress of retinal ganglion cell death continues even when

IOP is under control, indicating IOP independent mechanisms associated with the development of glaucomatous optic neuropathy. In fact, it is now recognized that glaucoma is also a retinal degenerative disease. Investigators in the glaucoma research field are actively seeking new approaches aimed at protecting retinal ganglion cells in patients with glaucoma.

Many neurotrophic factors have been found to protect retinal ganglion cells, including GDNF *(47,48)*, BDNF *(49,50)*, and CNTF *(49,51–53)*. All these factors are deliverable by the intraocular ECT device. The fact that CNTF also protects retinal ganglion cells indicates that NT-501 could be used for glaucoma as well.

Neuroprotection in AMD

AMD is the leading cause of irreversible blindness in people over the age of 50. About 15 million people in the United States alone and 25–30 million people worldwide are affected by AMD.

The pathogenesis of AMD is not clearly understood. The disease affects the central region of the retina, the macular, hence the name. Cone photoreceptor and RPE cell degeneration is evident in patients with AMD. In some cases, newly formed blood vessels from the choroid (choroidal neovascularization [CNV]) invade the affected retinal area. This usually results in severe complications and loss of central vision.

Approximately 85–90% of the cases of AMD are of the "dry" form, characterized by soft drusen, retinal pigmentary disturbance, and/or focal retinal atrophy. Vision loss is generally not severe in patients with the dry form of AMD. Delivery of CNTF by NT-501 or other NT factors may help in photoreceptor protection in these cases.

Another form of AMD, the exudative or "wet" form, occurs in only 10–15% of all AMD cases. The wet form is characterized by CNV as new blood vessels originating from the choroid penetrate Bruch's membrane to enter the retina. As a result, the normal architecture of the retina is destroyed with devastating complications, including hemorrhage, retinal detachment, disciform scar formation, and loss of central vision.

Currently, there are two Food and Drug Administration-approved therapies (*Visudyne*® and *Macugen*) for CNV in wet AMD aimed at inhibiting neovascularization. *Visudyne* (Novatis) is a photosensitizer used in photodynamic therapy to block new blood vessels with a laser beam. The other approved therapy, *Macugen* (Eyetech) is an aptamer that binds and inhibits vascular endothelial growth factor. Although the most important goal for treating wet AMD is to inhibit neovascularization, photoreceptor degeneration is still a problem that needs to be addressed. Again, ECT delivery of NT factors could be helpful in protecting the visual function for patients with wet AMD.

ECT Delivery of Antiangiogenic Factors for Ocular Neovascularization

There are two major neovascular diseases that affect the retina, exudates AMD and diabetic retinopathy. As mentioned previously, the existence of CNV is a characteristic of exudates AMD. CNV originates from choriocapillaris and invades the retina, causing blinding complications. In diabetic retinopathy, neovascularization in the retinal vasculature leads to hemorrhage and other complications, leading to blindness.

Neovascularization research, pioneered by Dr. Jodah Fokman of Harvard University, has lead to the discovery of many polypeptide anti-angiogenic factors, including endostatin, angiostatin, and PEDF. These factors could be delivered by ECT. In addition, these anti-angiogenic factors could be combined with neurotrophic factors in a single device.

ECT-Based Delivery of Anti-Inflammatory Factors in Uveitis

Uveitis is inflammation of the uvea. Approximately 80% of the cases in humans are autoimmune related and chronic, whereas the remaining 20% are caused by infectious processes. Topical and systemic corticosteroids and immunomodulators are the current standard of care. Studies have demonstrated the possible use of anti-inflammatory cytokines for treating uveitis. For example, IL-10, a 35-kD homodimeric cytokine synthesized by monocytes, B cells, and Th2 lymphocytes, induces mainly immunosuppressive effects through the down regulation of macrophage functions and the inhibition of the synthesis of pro-inflammatory cytokines, such as IL-1β, IL-6, IL-8, interferon-γ, and granulocyte/macrophage colony-stimulating factors, produced by Th1 lymphocytes or monocytes. Because of its short half-life, daily injections of recombinant IL-10 are required to produce therapeutic effects. Repetitive injections of IL-10 inhibit experimental autoimmune uveoretinitis (EAU) *(54)* and endotoxin-induced uveitis *(55)*. Inhibition of EAU by systemic and subconjunctival adenovirus-mediated transfer of the IL-10 gene has been reported *(56)*, indicating sustained availability of IL-10 would be beneficial. Obviously, ECT would be a good choice for intraocular delivery of IL-10 for uveitis.

SUMMARY

Preclinical development of ECT has demonstrated the therapeutic efficacy, long-term delivery, and relative safety in the animal eyes. Based on this data, a clinical phase I safety study of NT-501 has been initiated at the National Eye Institute to treat RP. If safety and consistent delivery are demonstrated in clinical trials, ECT could potentially serve as a delivery system not only for RP, but also for a number of ophthalmic diseases for which no effective therapies are currently available.

REFERENCES

1. Cedarbaum JM. The pharmacokinetics of subcutaneously administered recombinant human ciliary neurotrophic factor (rHCNTF) in patients with amyotrophic lateral sclerosis: relation to parameters of the acute-phase response. The ALS CNTF Treatment Study (ACTS) Phase I-II Study Group. Clin Neuropharmacol 1995;18:500–514.
2. Cedarbaum JM. A phase I study of recombinant human ciliary neurotrophic factor (rHCNTF) in patients with amyotrophic lateral sclerosis. The ALS CNTF Treatment Study (ACTS) Phase I-II Study Group. Clin Neuropharmacol 1995;18:515–532.
3. Emerich DF, Lindner MD, Winn SR, Chen EY, Frydel BR, Kordower JH. Implants of encapsulated human CNTF-producing fibroblasts prevent behavioral deficits and striatal degeneration in a rodent model of Huntington's disease. J Neurosci 1996;16:5168–5181.
4. Emerich DF, Winn SR, Hantraye PM, et al. Protective effect of encapsulated cells producing neurotrophic factor CNTF in a monkey model of Huntington's disease. Nature 1997; 386:395–399.

5. Emerich DF, Plone M, Francis J, Frydel BR, Winn SR, Lindner MD. Alleviation of behavioral deficits in aged rodents following implantation of encapsulated GDNF-producing fibroblasts. Brain Res 1996;736:99–110.

6. Tseng JL, Baetge EE, Zurn AD, Aebischer P. GDNF reduces drug–induced rotational behavior after medial forebrain bundle transection by a mechanism not involving striatal dopamine. J Neurosci 1997;17:325–333.

7. Emerich DF, Hammang JP, Baetge EE, Winn SR. Implantation of polymer-encapsulated human nerve growth factor- secreting fibroblasts attenuates the behavioral and neuropathological consequences of quinolinic acid injections into rodent striatum. Exp Neurol 1994;130:141–150.

8. Emerich DF, Winn SR, Harper J, Hammang JP, Baetge EE, Kordower JH. Implants of polymer-encapsulated human NGF-secreting cells in the nonhuman primate: rescue and sprouting of degenerating cholinergic basal forebrain neurons. J Comp Neurol 1994;349:148–64.

9. Kordower JH, Winn SR, Liu YT, et al. The aged monkey basal forebrain: rescue and sprouting of axotomized basal forebrain neurons after grafts of encapsulated cells secreting human nerve growth factor. Proc Natl Acad Sci USA 1994;91:10,898–10,902.

10. Lindner MD, Kearns CE, Winn SR, Frydel B, Emerich DF. Effects of intraventricular encapsulated hNGF-secreting fibroblasts in aged rats. Cell Transplant 1996;5:205–223.

11. Hoane MR, Puri KD, Xu L, et al. Mammalian-cell-produced neurturin (NTN) is more potent than purified escherichia coli-produced NTN [In Process Citation]. Exp Neurol 2000;162:189–193.

12. Tao W, Wen R, Goddard MB, et al. Encapsulated cell-based delivery of CNTF reduces photoreceptor degeneration in animal models of retinitis pigmentosa. Invest Ophthalmol Vis Sci 2002;43:3292–3298.

13. Leibowitz HM, Krueger DE, Maunder LR, et al. The Framingham Eye Study monograph: an ophthalmological and epidemiological study of cataract, glaucoma, diabetic retinopathy, macular degeneration, and visual acuity in a general population of 2631 adults, 1973–1975. Surv Ophthalmol 1980;24:335–610.

14. Klein BE, Klein R. Cataracts and macular degeneration in older Americans. Arch Ophthalmol 1982;100:571–573.

15. Martinez GS, Campbell AJ, Reinken J, Allan BC. Prevalence of ocular disease in a population study of subjects 65 years old and older. Am J Ophthalmol 1982;94:181–189.

16. Gibson JM, Rosenthal AR, Lavery J. A study of the prevalence of eye disease in the elderly in an English community. Trans Ophthalmol Soc UK 1985;104:196–203.

17. Vinding T. Age-related macular degeneration. Macular changes, prevalence and sex ratio. An epidemiological study of 1000 aged individuals. Acta Ophthalmol (Copenh) 1989;67:609–616.

18. Bressler NM, Bressler SB, West SK, Fine SL, Taylor HR. The grading and prevalence of macular degeneration in Chesapeake Bay watermen. Arch Ophthalmol 1989;107:847–852.

19. Klein R, Klein BE, Linton KL. Prevalence of age-related maculopathy. The Beaver Dam Eye Study. Ophthalmology 1992;99:933–943.

20. Vingerling JR, Dielemans I, Hofman A, et al. The prevalence of age-related maculopathy in the Rotterdam Study. Ophthalmology 1995;102:205–210.

21. Hawkins BS, Bird A, Klein R, West SK. Epidemiology of age-related macular degeneration. Mol Vis 1999;5:26.

22. Davis MD, Hiller R, Magli YL, et al. Prognosis for life in patients with diabetes: relation to severity of retinopathy. Trans Am Ophthalmol Soc 1979;77:144–170.

23. Podgor MJ, Cassel GH, Kannel WB. Lens changes and survival in a population-based study. N Engl J Med 1985;313:1438–1444.

24. Klein R, Klein BE, Moss SE, Davis MD, DeMets DL. The Wisconsin epidemiologic study of diabetic retinopathy. II. Prevalence and risk of diabetic retinopathy when age at diagnosis is less than 30 years. Arch Ophthalmol 1984;102:520–526.
25. Klein R, Moss SE, Klein BE, DeMets DL. Relation of ocular and systemic factors to survival in diabetes. Arch Intern Med 1989;149:266–272.
26. Klein R, Klein BE, Moss SE. Age-related eye disease and survival. The Beaver Dam Eye Study. Arch Ophthalmol 1995;113:333–339.
27. Coleman AL. Glaucoma. Lancet 1999;354:1803–1810.
28. West SK. Looking forward to 20/20: a focus on the epidemiology of eye diseases. Epidemiol Rev 2000;22:64–70.
29. Hoyng PF, van Beek LM. Pharmacological therapy for glaucoma: a review. Drugs 2000;59:411–434.
30. Weih LM, VanNewkirk MR, McCarty CA, Taylor HR. Age-specific causes of bilateral visual impairment. Arch Ophthalmol 2000:118:264–269.
31. Dryja TP, Li T. Molecular genetics of retinitis pigmentosa. Hum Mol Genet 1995;4:1739–1743.
32. Holt IJ, Harding AE, Petty RK, Morgan-Hughes JA. A new mitochondrial disease associated with mitochondrial DNA heteroplasmy. Am J Hum Genet 1990;46:428–433.
33. Narcisi TM, Shoulders CC, Chester SA, et al. Mutations of the microsomal triglyceride-transfer-protein gene in abetalipoproteinemia. Am J Hum Genet 1995;57:1298–1310.
34. Meindl A, Dry K, Herrmann K, et al. A gene (RPGR) with homology to the RCC1 guanine nucleotide exchange factor is mutated in X-linked retinitis pigmentosa (RP3). Nat Genet 1996;13:35–42.
35. Faktorovich EG, Steinberg RH, Yasumura D, Matthes MT, LaVail MM. Photoreceptor degeneration in inherited retinal dystrophy delayed by basic fibroblast growth factor. Nature 1990;347:83–86.
36. LaVail MM, Unoki K, Yasumura D, Matthes MT, Yancopoulos GD, Steinberg RH. Multiple growth factors, cytokines, and neurotrophins rescue photoreceptors from the damaging effects of constant light. Proc Natl Acad Sci USA 1992;89:11,249–11,253.
37. Unoki K, Ohba N, Arimura H, Muramatsu H, Muramatsu T. Rescue of photoreceptors from the damaging effects of constant light by midkine, a retinoic acid-responsive gene product. Invest Ophthalmol Vis Sci 1994;35:4063–4068.
38. LaVail MM, Yasumura D, Matthes MT, et al. Protection of mouse photoreceptors by survival factors in retinal degenerations. Invest Ophthalmol Vis Sci 1998;39:592–602.
39. Hauswirth WW LQ, Raisler B, Timmers AM, et al. Range of retinal diseases potentially treatable by AAV-vectored gene therapy. Novartis Found Symp 2004:255:179–194.
40. Lawrence JM KD, Muir EM, Coffey PJ, et al. Transplantation of Schwann cell line clones secreting GDNF or BDNF into the retinas of dystrophic Royal College of Surgeons rats. Invest Ophthalmol Vis Sci. 2004;45:267–274.
41. Machida S CP, Shinohara T, Singh DP, et al. Lens epithelium-derived growth factor promotes photoreceptor survival in light-damaged and RCS rats. Invest Ophthalmol Vis Sci 2001;42:1087–1095.
42. Nakamura M SD, Kubo E, Chylack LT Jr, Shinohara T. LEDGF: survival of embryonic chick retinal photoreceptor cells. Invest Ophthalmol Vis Sci 2000;41:1168–1175.
43. Song Y, Zhao L, Tao W, Laties AM, Luo Z, Wen R. Photoreceptor protection by cardiotrophin-1 in transgenic rats with the rhodopsin mutation s334ter. Invest Ophthalmol Vis Sci 2003;44:4069–4075.
44. Liu C, Li Y, Peng M, Laties AM, Wen R. Activation of caspase-3 in the retina of transgenic rats with the rhodopsin mutation s334ter during photoreceptor degeneration. J Neurosci 1999;19:4778–4785.

45. Aguirre G, Farber D, Lolley R, et al. Retinal degenerations in the dog III abnormal cyclic nucleotide metabolism in rod-cone dysplasia. Exp Eye Res 1982;35:625–642.
46. Ray K, Baldwin VJ, Acland GM, Blanton SH, Aguirre GD. Cosegregation of codon 807 mutation of the canine rod cGMP phosphodiesterase beta gene and rcd1. Invest Ophthalmol Vis Sci 1994;35:4291–4299.
47. Wu WC, Lai CC, Chen SL, et al. GDNF gene therapy attenuates retinal ischemic injuries in rats. Mol Vis 2004;10:93–102.
48. Lindqvist N, Peinado-Ramonn P, Vidal-Sanz M, Hallbook F. GDNF, Ret, GFRalpha1 and 2 in the adult rat retino-tectal system after optic nerve transection. Exp Neurol 2004; 187:487–499.
49. Mey J, Thanos S. Intravitreal injections of neurotrophic factors support the survival of axotomized retinal ganglion cells in adult rats in vivo. Brain Res 1993;602:304–317.
50. Ko ML, Hu DN, Ritch R, Sharma SC, Chen CF. Patterns of retinal ganglion cell survival after brain-derived neurotrophic factor administration in hypertensive eyes of rats. Neurosci Lett 2001;305:139–142.
51. Watanabe M, Fukuda Y. Survival and axonal regeneration of retinal ganglion cells in adult cats. Prog Retin Eye Res 2002;21:529–553.
52. Van Adel BA, Kostic C, Deglon N, Ball AK, Arsenijevic Y. Delivery of ciliary neurotrophic factor via lentiviral-mediated transfer protects axotomized retinal ganglion cells for an extended period of time. Hum Gene Ther 2003;14:103–115.
53. Ji JZ EW, Yip HK, Lee VW, Yick LW, Hugon J, So KF. CNTF promotes survival of retinal ganglion cells after induction of ocular hypertension in rats: the possible involvement of STAT3 pathway. Eur J Neurosci 2004;19:265–272.
54. Rizzo LV XH, Chan CC, Wiggert B, Caspi RR. IL-10 has a protective role in experimental autoimmune uveoretinitis. Int Immunol 1998;10:807–814.
55. Rosenbaum JT, Angell E. Paradoxical effects of IL-10 in endotoxin-induced uveitis. J Immunol 1995;155:4090–4094.
56. De Kozak Y, Thillaye-Goldenberg B, Naud MC, Da Costa AV, Auriault C, Verwaerde C. Inhibition of experimental autoimmune uveoretinitis by systemic and subconjunctival adenovirus-mediated transfer of the viral IL-10 gene. Clin Exp Immunol 2002;130:212–223.

22

Effective Treatment for the Canine RPE65 Null Mutation, a Hereditary Retinal Dystrophy Comparable to Human Leber's Congenital Amaurosis

Kristina Narfström, DVM, PhD, Gregory E. Tullis, PhD, and Mathias Seeliger, MD, PhD

CONTENTS

INTRODUCTION

Mutations in the human retinal pigment epithelial (RPE)65 gene underlie some forms of early childhood blindness, including a form of Leber's congenital amaurosis (LCA), early-onset severe retinal dystrophy, and juvenile retinitis pigmentosa (RP) *(1–4)*. LCA is an autosomal recessively inherited disease *(5)*, although a few families with autosomal dominant inheritance have also been reported *(6)*. In general, LCA is diagnosed when there is marked visual impairment from birth, whereas the disease is considered juvenile RP if vision is lost during the first 2 yr of life. Mutations in several genes other than RPE65 have also been identified in other ocular phenotypes designated as LCA *(7)*.

The disease is characterized by profound visual loss or total blindness, searching nystagmus, and hyperopia, usually recognized the first few months of life. Fundus appearance is usually normal in young patients, but with increasing age vascular attenuation and rarefaction are seen, and later in life heterogeneous retinal changes appear, such as whitish specks and/or pigmentation of the fundus. Electroretinograms (ERGs) are usually nonrecordable or severely reduced from infancy *(8)*. In cases deficient of RPE65, histological studies of human fetuses have confirmed the hypothesis of prenatal onset of the disease as well as the presence of lipid and vesicular inclusions in the RPE *(9,10)*. The incidence of LCA is low, about 1.9% of all retinal dystrophies in one

From: *Ophthalmology Research: Retinal Degenerations: Biology, Diagnostics, and Therapeutics*
Edited by: J. Tombran-Tink and C. J. Barnstable © Humana Press Inc., Totowa, NJ

University Hospital (Zurich, Switzerland) *(11)*. Currently, there is no effective treatment for this congenital and further progressive blinding disease.

RPE65 is a 61-kDa microsomal protein expressed almost exclusively in the RPE *(12,13)*. The gene spans 20 kb and is divided into 14 exons *(14)* and is localized to chromosome 1p31 *(14,15)*. The transcript consists of about 0.1% of the total messenger RNA isolated from the RPE and is approx 2.9 kb. The 533-amino acid protein is highly conserved between various species. RPE65 is associated with the smooth endoplasmic reticulum and is necessary for the synthesis of the 11-*cis*-retinal chromophore of the photoreceptor cell visual pigments *(16)*. In an RPE65 knockout mouse, 11-*cis*-retinoids cannot be generated and there is an accumulation of all-*trans*-retinyl esters in the RPE, forming lipoid inclusion bodies *(16)*.

LCA-Like Disease of Dogs

The Briard dog is affected by a hereditary retinal dystrophy *(17)* and has been shown to be a very suitable animal model for LCA *(18,19)*. The molecular defect that underlies this canine disorder is a 4-bp deletion in the RPE65 gene *(20,21)*. Dogs homozygous for the RPE65 null mutation are congenitally night blind, and show severe visual deficits in daylight. Most, but not all, cases show a fast-quivering nystagmus. Their resting pupillary size is slightly larger than in normal dogs in accordance with their reduced sensitivity to light *(22)*. ERG studies are already diagnostic at the age of 5 wk *(19)*. There are no, or barely recordable, scotopic responses, and photopic ERGs are usually of low amplitudes, most clearly observed in 30 Hz flicker recordings. Ophthalmoscopic examination reveals normal fundus appearance until about 3 yr of age, then there is a generalized vascular attenuation and a slight paling of the fundus. In most affected older animals, grayish to white spots, mainly in the central tapetal and nontapetal fundus, appear. These spots increase slowly with age and spread peripherally with time (Fig. 1). Thus, the disease is slowly progressive clinically. Large lipoid-like inclusions accumulate primarily in the RPE of the central fundus, but they become more generalized at later ages. There is also disorganization of photoreceptor outer segments (OS) at an early age, followed by degenerative changes and later by rod, and then cone loss, with a gradient going from the peripheral retina to the central parts with increasing age *(18,23)*. There are obviously close similarities between the clinical characteristics of the diseases resulting from RPE65 gene mutations in dogs and in humans *(24)*.

Retinal Gene Therapy

Gene therapy is a promising technology for the treatment of inherited genetic disorders in which the genetic defect is known. The retina is an especially good target for gene therapy because it is easily accessible and because it is an immunoprivileged site. The gene therapy vector can be injected either into the vitreous or the subretinal space that lies between the neuroretina and RPE with only minimal local side effects. Additionally, in the advent of an unexpected adverse reaction to the therapy, the eye can be removed surgically.

Immune rejection of the transgene is a potential problem in gene replacement therapy, because many patients express a truncated form of the protein. Therefore, the new transgene may contain epitopes that are recognized as foreign by the patient's immune

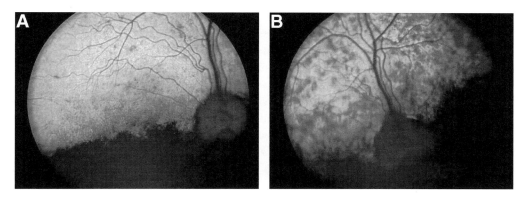

Fig. 1. (A) Minor ophthalmoscopic signs of the RPE65 null mutation are illustrated in this 3.5-yr-old affected dog, treated by subretinal gene transfer 1 yr previously (area of injection not shown in this fundus photograph). Some grayish spots are seen in the fundus and there is a slight generalized vascular attenuation. **(B)** Shows the fundus in the same dog 3 yr later. The aberrant spots have increased markedly as well as the vascular attenuation.

system. However, like other internal structures in the eye, the subretinal space displays two important immune privilege features: (1) it will tolerate tissue grafts without immune rejection, and (2) it promotes the acquisition of systemic immune deviation to antigens placed within it *(25)*. Therefore, long-term transgene expression is possible in the retina without clearance of transduced cells by the immune system.

Viruses are naturally occurring gene therapy vectors and can be highly efficient in delivering their DNA or RNA to the target cells. Vectors based on a parvovirus called adeno-associated virus (AAV) have been highly successful for transduction of the retina *(26)*. AAV virions are comprised of 60 protein molecules surrounding a 4.7-kb single-stranded DNA. Of this, approx 4.4 kb can be used to express the transgene. The only AAV DNA in the recombinant AAV is the 145-bp terminal repeats located at each end. These terminal repeats contain the AAV origin of replication and the packaging signals. When AAV infects a cell, the single-stranded DNA becomes converted to double-stranded DNA and the transgene is then expressed. AAV vector DNA can persist for many years in cells as either long concatemers or circular intermediates. Vector DNA can also integrate randomly over time and can potentially express the transgene over the lifetime of the patient. Expression of therapeutic levels of RPE65 has been observed in treated animals for over 3 yr.

Epithelial cells such as in the RPE are excellent targets for AAV vectors. By injecting the recombinant AAV into the subretinal space it is possible to place the virus in close contact with the potential target. This allows for extremely efficient transduction of the RPE layer within the bleb. AAV vectors can also transduce the RPE layer when injected into the vitreous. This suggests that the vector can migrate through the neural retina and reach the RPE. Currently, eight different serotypes of AAV have been used for gene therapy (AAV1-8). We have used the most widely studied strain, AAV2, in all experiments described here. AAV4 and AAV6 vectors appear to transduce the RPE more specifically than AAV2 *(27,28)*. The AAV6 result is somewhat surprising because AAV1 vectors transduce ganglion cells and photoreceptors in addition to RPE cells *(29)*. AAV6 derives from a recombination of AAV2 and AAV1 such that the capsid proteins of AAV1 and AAV6 are highly homologous.

Because the effects of loss of RPE65 function appear to be restricted to the retina, localized gene therapy appears to be a promising approach to treat LCA. Acland et al. *(30)* observed a significant improvement in vision in three 4-mo-old dogs treated with the AAV vector that expressed RPE65 from the β-actin promoter. In-depth experiments to assess the efficacy of AAV-mediated gene therapy in reversing the effects of the RPE65 mutation in RPE65–/– dogs were conducted in our laboratories *(31–34)*. A canine RPE65 complementary DNA (cDNA) was cloned into a recombinant AAV2 vector. We have used a strong, constitutive promoter from cytomegalovirus (CMV) to drive RPE65 expression. Ancestors of the group of dogs used in these experiments were originally discovered in Sweden and transported to the University of Missouri where gene transfer surgeries were initiated in 2001. During the past 4 yr, 20 RPE65–/– dogs have undergone intraocular gene therapy. Treated animals have been studied using various clinical and experimental techniques. Some of the affected animals have been followed for more than 3 yr postoperatively with some very dramatic and exciting results.

MATERIALS AND METHODS

Preparation of the Gene Construct

Development of our gene construct has been described *(31)* and, in short, was performed as follows: Normal RPE65 dog cDNA was cloned into a pCI vector (Promega, Madison, WI) carrying the human CMV promoter and the late SV40 polyadenylation signal (poly A). Following transfection into Cos-7 cells (ATCC, Rockville, MD), the expression of RPE65 protein was analyzed by Western blotting. Subsequently, the RPE65 expression cassette was cloned into pSSV9 *(35)* in between the AAV2 terminal repeats. A control vector with green fluorescent protein (GFP) cDNA in place of the RPE65 cDNA was also constructed. Human embryonic kidney (HEK 293) cells (ATCC, Rockville, MD) were transfected with pAAV.CMV.GFP or pAAV.CMV.RPE65 and assessed for the expression of the transgenes by fluorescence microscopy and Western blot analysis, respectively. The excision and replication of the rAAV.RPE65 was assessed by Hirt analysis *(36)*. Large-scale production of recombinant viruses was done at the Vector Core Facility of the University of North Carolina Gene Therapy Center (Chapel Hill, NC) *(37)*. AAV vectors are typically assessed for their ability to transduce target cells (e.g., HeLa) and for the number of DNA-containing virions by quantitative polymerase chain reaction (PCR). The titer of the rAAV.GFP preparations were approx 2×10^{10} transducing units per milliliter and ranged from 5.5×10^{11} to 8.6×10^{11} single-stranded vector genomes per milliliter. The four rAAV.RPE65 preparations were approx 10^{12} vector genomes per milliliter. Purity of the virus preparations were evaluated by silver-staining sodium dodecyl sulfate (SDS)-polyacrylamide gel electrophoresis gels. Unfortunately, two out of four rAAV.RPE65 preparations were contaminated with nonvector proteins, which resulted in an inflammatory reaction in some treated animals (*see* Immediate Postoperative Results section). The purity of more recent batches was approx 95% and we have not observed any inflammatory reactions using these batches.

Methods for Dog Experiments

A majority of dogs that underwent gene transfer were treated surgically at the age of 4 mo, whereas two of the RPE65–/– dogs were 2.5 and 4 yr old, respectively, at the time of gene transfer. From age 5 to 7, wk baseline clinical studies were performed. Initially, their visual behavior was tested by merely watching the dogs walk in unknown surroundings in the dark and in daylight conditions. Their ability to follow a strong beam of light in both lighting conditions and see falling cotton balls was also studied. Direct and indirect pupillary light reflexes were tested in both conditions and indirect ophthalmoscopy, slit-lamp biomicroscopy, and fundus photography were performed regularly *(31)*. Baseline ERGs were obtained at least twice before surgery *(31,32,34)*. Follow-up studies after surgery were performed in similar ways, but extended with maze testing for objective visual behavior *(34)*, multifocal ERG (mfERG) studies *(38,39)*, and imaging of the treated and untreated fundi using Fluorescein (FL) and Indocyanine Green (ICG) angiography. Fundus images were obtained by confocal scanning laser ophthalmoscopy (SLO; Heidelberg, Engineering Retina Angiograph, Heidelberg, Germany), equipped with an argon blue laser (488 nm), a green laser (514 nm), and an infrared laser (835 nm) for visualization of fluorescence and fundus during the angiography. All electrophysiological studies were performed under general anesthesia using propofol (Diprivan, 1%, Astra Zeneca Pharmaceutical LP, Wilmington, DE, 6 mg/kg iv) and isoflurane (Isoflurande USP, Abbott Laboratories, North Chicago, IL). Imaging studies were performed in deep sedation using medetomidine (Domitor vet., Vetpharma, Lund, Sweden, 0.01 mg/kg IM) and ketamine (Ketalar, Park-Davies, Morris Plains, NJ, 2.5 mg/kg IM).

Gene transfer was performed under microscopic visualization using routine aseptic methods by two trained vitreo-retinal surgeons. In most cases, both eyes were treated with subretinal injections: one eye with rAAV.RPE65 and the contralateral eye with rAAV.GFP *(31)*. One hundred microliters of each construct was injected subretinally in most of the dogs.

A lateral canthotomy was performed and conjunctiva and Tenon's capsule were dissected so that access to the sclera was obtained. Two sclerotomies were performed in the temporal sclera about 4 mm apart and 6–8 mm from the limbus. A fiberoptic light was inserted into the vitreous cavity through one sclerotomy and a custom-made glass micropipet through the other opening and guided toward the fundus, under direct visualization through an operating microscope and a Machemer flat lens (Ocular Instruments, Bellevue, WA) on the cornea. For subretinal injections, the inferior-nasal central part of the neuroretina was perforated and a bleb was obtained with the injection of one of the constructs. The neuro-retinal detachment encompassed approximately 30% of the total fundus area. After the subretinal deposition, the micropipette and light guide were withdrawn and the sclerotomies sutured using 7/0 Vicryl. The conjunctiva, Tenon's capsule and subcutaneous tissues were sutured in a routine manner. A subconjunctival injection in each eye of decadrone phosphate (Dexamethasone, Butler, Columbus, OH, 1 mg in 0.25 mL) was given postoperatively.

RESULTS

Immediate Postoperative Results

Approximately 85% of the surgeries were completely successful (in a total of 38 eyes), with the construct injected clearly subretinally without complications. Otherwise, the most common complication observed was slight hemorrhage from a sclerotomy site in conjunction with surgery. However, the bleeding did not affect the final outcome of the surgery in any of the treated dogs. Further, in several cases the subretinal bleb did not form perfectly well, and most of the gene construct leaked out into the vitreous. The most serious immediate complication to the subretinal injection was a large partial retinal detachment that occurred in one case.

In nine of the treated dogs, postoperative intraocular inflammatory reactions (uveitis) were found to develop on the 2nd to 6th day in the rAAV.RPE65 treated eyes only *(31,40,41)*. All were low grade except one, which was complicated with choroiretinitis and vitritis, refractory to treatment. The other eight cases of uveitis subsided after 4–12 wk of systemic and/or topical anti-inflammatory treatment.

Further investigation of the uveitis showed that these were most probably due to contaminating proteins in two of the four different batches of the rAAV.RPE65 gene construct used. Testing of the AAV showed that purity was approx 95% as estimated from both high-performance liquid chromatograms and silver-staining SDS polyacrylamide gels. Owing to problems with uveitis, batches used lately have been further tested using SDS-polyacrylamide gels and silver staining. Also filtering has been performed through a 0.45-μm polyethersulfane (PES) membrane as well as screening for numerous viral and microbial contaminants by PCR. Because these rigorous measures have been taken, no further development of aberrant reactions following the gene transfer has occurred.

Early Postoperative Results

Clinical Findings and Visual Behavior

All dogs were systemically healthy postoperatively. Excellent functional improvement in visual behavior could be demonstrated as early as 4 wk following surgery in the previously blind affected dogs. The behavioral studies showed definite improvement of both day and night vision in dogs treated with subretinal injections of the rAAV.RPE65 gene construct. Vision appeared better in daylight than under dim light conditions. Preferential looking was observed from the side of the AAV.RPE65 treated eye only. Approximately 10 wk following the subretinal treatment with rAAV.RPE65, the nystagmus disappeared in both eyes in all affected dogs treated by unilateral subretinal gene transfer *(31)*.

Early ERG Findings

Dark-adapted low- and high-intensity ERGs were mainly nonrecordable prior to treatment in the RPE65–/– dogs. Only a few of the affected dogs showed recordable scotopic low amplitude b-wave responses at the high intensity stimulus. In the light-adapted state, however, small amplitude responses, including single flash b-wave and 30 Hz flicker responses were obtained in many of the affected dogs. Fifty Hz flicker recordings were, however, not recorded in any of the affected dogs prior to the gene transfer treatment.

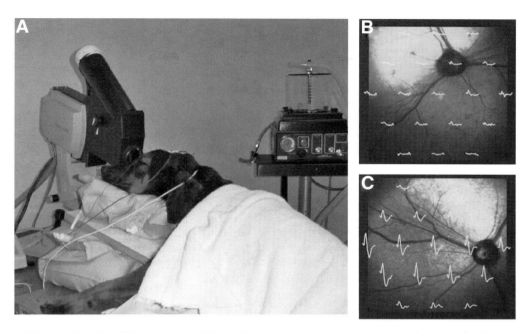

Fig. 2. (A) mfERG in progress with the dog under general anesthesia and mechanically ventilated (for maximum relaxation and no muscular movement). A SLO is used, equipped with an infrared laser for fundus visualization. **(B)** Fundus image with mfERG recordings from the left, nontreated eye and **(C)** similar recordings from the gene transfer treated eye. The mfERG stimulator is a prototype developed by Dr. Eric Sutter.

In all rAAV.RPE65 eyes treated with subretinal injections, the ERG amplitude responses improved 4–6 wk following surgery and statistically significant differences between AAV.RPE65 treated eyes and the contralateral control eyes were found. Also, ERG b-wave thresholds were close to normal in the treated dogs *(42)*. ERG responses from low- and high-intensity scotopic light stimuli were obtained in treated dogs and photopic single flash, 30 and 50 Hz flicker responses were recorded in all gene transfer treated dogs performed successfully. The dark-adapted b-wave amplitudes had recovered to an average of 25% of normal, and the light adapted b-wave amplitudes to 20% of normal.

mf ERGs were performed in seven of the gene transfer treated RPE65–/– dogs 3–6 mo after surgery with continuous monitoring of the retinal position of the stimulus *(38)*, to show the topographical restoration of visual function. A marked difference in response amplitudes in treated and untreated areas of eyes that had undergone gene transfer were found. An area of increased responses could be clearly correlated to the treatment area (Fig. 2). It was possible to obtain a focal response-intensity series in treated eyes, whereas the untreated eyes did not show such an increase in response amplitudes with increasing stimulus intensity at this time after surgery (3 mo) (Fig. 3). Although response amplitudes were greatest in the treated region, ERG responses were also detected when areas of the retina outside of the treated region was stimulated. This is consistent with a spreading of the treatment effect extending beyond the borders of the injection site.

Intensity series of mfERG responses

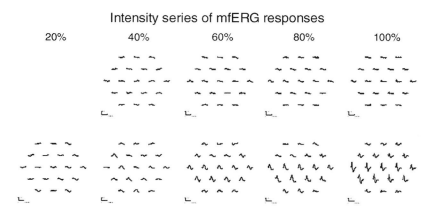

Fig. 3. The topography of functional rescue was evaluated using bilateral mfERGs. An intensity series was performed in both eyes of unilaterally treated RPE65–/– dogs. This showed wave forms of increasing amplitudes in relation to increase in light intensity in the area of the gene transfer treatment in the right eye, depicted in the lower row of mf ERG responses.

Long-Term Postoperative Results

Visual Behavior

Six of the treated dogs were used for long-term follow-up. Using a maze test and observers blinded to the study up to 2 yr following gene transfer, objective visual testing was performed which demonstrated that the treated dogs had significantly improved vision over presurgical observations *(34)*. Bright light vision was better than dim light vision, as evidenced by a significantly greater number of collisions with objects occurring under dim light conditions. Vision in control dogs (RPE65+/+) was not significantly different in dim- and bright-light conditions. Further, although a significantly higher number of collisions were noted in the treated affected dogs than in normal control animals in dim light conditions, in bright light conditions the affected dogs that had undergone gene therapy performed as well as normal control dogs. Although vision in treated dogs was not completely normal, especially in dim light, these results represent a potentially dramatic improvement in the general quality of life as a result of the gene therapy.

ERG

The maximum improvement of rod ERG responses was found at 3 mo after gene transfer. The dark-adapted b-wave maximum amplitudes recovered to an average of 28% of normal, and the light-adapted b-wave maximum amplitudes to 47% of normal *(31,32)*. After 3 mo, there was a slight reduction in rod responses over time, and at 18 mo postsurgery, scotopic high-intensity b-wave amplitudes, photopic single-flash and 30 Hz flicker b-wave amplitudes were still significantly increased above baseline values. For the cone system, there was a slow, long-term improvement, which was either sustained or continued to increase up to 18 mo following the gene transfer treatment. By 21 to 24 mo following treatment, rod responses and maximum b-wave amplitude values in the treated eye had declined further, and continued to decrease in amplitudes at 24 and 33

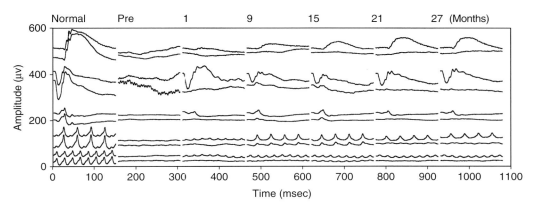

Fig. 4. The actual ERG recordings from a normal control dog, far left, and an affected gene transfer treated littermate before surgery (pre) and at 1, 9, 15, 21, and 27 mo postoperatively. The affected dog was injected with the gene construct (rAAV.RPE65) subretinally into the right eye and, in a similar way, with rAAV.GFP into the left eye. The latter construct was used in the right eye (only) of the control dog. Upper recording in each set of two traces are ERG responses from the right eye and lower recording from the left eye. The simultaneous bilateral full-field ERG responses from each animal were obtained using scotopic low-light intensity stimuli (-2 log cd·s/m^2); 1st row of responses, and scotopic high-light stimuli levels (4 cd·s/m^2); 2nd row. After 10 min of light adaptation photopic single-flash stimulation was used (5 Hz) at 1 cd·s/m^2; 3rd row, then flicker stimulation at 30 and 50 Hz, using 1cd·s/m^2 of light stimulation; 4th and 5th rows, respectively.

mo after surgery. Cone responses, indicated by photopic single-flash b-wave responses were still significantly elevated over preoperative values up to 30 mo following surgery. At 33 mo following surgery, however, no significant differences were found for either rod or cone response amplitudes in comparison to preoperative ERG recordings *(43)* (Fig. 4).

An Unexpected But Positive Finding

An unexpected finding was encountered at long-term follow-up in five of the dogs used for long-term studies: there was a surprising appearance of low-amplitude ERG responses (at high-intensity scotopic and photopic recordings) also in the fellow eyes starting approx 6–9 mo after surgery. This contralateral eye effect was sustained (for 30 Hz flicker responses) up to 27–30 mo after treatment. Although these positive ERG responses were of low amplitude in the fellow eyes, there was a definite qualitative improvement of these responses compared to preoperative responses, most of which were nonrecordable.

Through careful control studies of the contralateral eye effect we can conclude that the positive ERG responses seen on the contralateral (control) eye at long-term follow-up was real and not an artifact, and appears as a result of the gene transfer treatment. These results could indicate transmission of RPE65 protein, of the gene transfer construct, or most likely, of 11-*cis*-retinal to the fellow eye, and suggest that cells outside of the treated area may also benefit from the rAAV.RPE65 gene therapy. This is of particular importance in adapting this therapeutic approach to humans because the most important region of the retina in which to restore function is the macula. However, it would be detrimental to inject the construct in the macular area owing to the risk of causing permanent damage from the temporary retinal detachment. If the treatment effect can be detected in the untreated eye, surely it would extend beyond the injection site in the treated eye. Thus, it is likely that a subretinal injection using a gene therapy construct peripheral to the macula would also restore macular function.

Wide Time Window for Surgery

Subretinal gene transfer treatments were given to two older dogs, the oldest being 4 yr old at the time of surgery, without complications. Both dogs showed a markedly improved visual behavior 4–6 wk postoperatively and increased ERG sensitivity and amplitudes as previously described *(42)*. There was no obvious difference in visual behavior upon follow-up of these older dogs (now 39 and 7 mo after surgery, respectively), compared to younger treated animals. It appears that components of the photo-transduction cascade and the retinal circuitry remain functional, despite the absence of normal photoreceptor activity for several years in untreated dogs. As has been shown for Rpe65–/– mice, injections of 11-*cis*-retinal result in regeneration of rhodopsin and improved photoreceptor function—irrespective of age *(44)*. Thus, there is a wide time window for surgery in conjunction with the RPE65 null mutation, a fact that has positive implications, also for human patients.

Effect of Subretinal Gene Therapy on In Vivo Retinal Morphology

In order to determine the magnitude and distribution of GFP transduced retinal cells (Fig. 5) retinal fluorescence was examined in vivo and detailed fundus images obtained using the SLO, 2 yr following the subretinal injection. Transduction was still obvious in the left eye, as observed by the GFP fluorescence in the area of the subretinal (control) injection. Further, retinal morphology following gene transfer studies were performed in treated eyes using the SLO and FL and ICG angiography, 2 to 2.5 yr after treatment. Surprisingly, good SLO images were obtained with no major alterations observed in the injection area in five long-term studied dogs. However, when using FL and ICG angiography, there were marked changes. Specifically at the injection site, in a region smaller than the original bleb area, there was a distinct part with successively increasing hyperfluorescence, possibly indicating leakage from retinal vessels and/or atrophic focal changes in the RPE (Fig. 5). These observations are potentially of major significance for application of this treatment to human subjects. Damage to the retina at the injection site would indicate that injecting the construct too near the macula should be avoided. The fact that abnormalities were restricted to the injection site indicates that there is no geographic spreading of this negative effect, however.

Light and Electron Microscopy of Treated Dogs

So far six RPE65–/– dogs given subretinal injections of the AAV-RPE65 gene therapy vector have been analyzed for the effects of treatment on retinal morphology. The first dog was euthanized 3 mo after the gene therapy treatment. Modest recovery of visual sensitivity was observed in this dog in the treated eye, as measured with ERG *(31)*. The dog had severe uveitis in the AAV-RPE65 treated eye that did not respond to

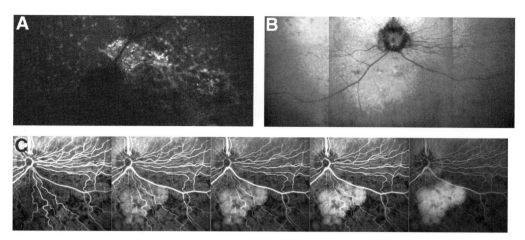

Fig. 5. SLO images of a gene transfer treated dog 2 yr after rAAV.RPE65 into the right eye and rAAV.GFP into the left eye, both injections given subretinally. (**A**) Transduction is still obvious in the left eye, as observed by AAV.GFP fluorescence in an area close to the optic disc, although significantly less than at 3 mo following surgery (for 3-mo follow-up, *see* ref. *31*). (**B**) In the subretinal rAAVRPE65 bleb area (inferior-nasal to the disc) in the right eye, 2 yr after treatment, there are only minor changes and severe scarring was not observed. (**C**) Fluorescein angiography, however, in this gene transfer treated eye shows a late hyperfluorescence in the neuroretinal bleb area, indicating distinct morphologic changes specifically at the injection site.

intensive topical and systemic anti-inflammatory medication. The clinically observed uveitis was accompanied by lymphocytic infiltration along the inner retinal surface. In the AAV-RPE65-treated eye, RPE65 immunolabeling indicated that RPE65 protein was being produced by the RPE only in the treated region of the eye *(31,33)*. In the fellow eye, GFP expression was pronounced in the RPE throughout the region where the subretinal bleb had been made. Electron microscopic analysis demonstrated that in the fellow eye the RPE contained numerous large lipid droplets in both injected and un-injected regions of the retina. In the eye treated with the AAV-RPE65 vector, the lipid droplets had almost completely disappeared from the RPE in the treated region. However, outside of the treated region, the RPE still contained large numbers of inclusions *(31)*. Despite reversal of the RPE lipid droplet accumulation in the treated region, there was no apparent recovery of photoreceptor OS morphology 3 mo after surgery. Thus, the observed functional recovery, measured by ERG, must have been mediated by the remaining OS fragments in the treated region or by photoreceptor cells outside of the treated region.

A littermate to the first dog was euthanized 6 mo after gene transfer surgery and a third dog was euthanized 5 mo after the treatment. None of the eyes of these dogs had been affected by uveitis. The eyes of the latter dog were used primarily for immunocytochemistry studies (performed this time at the Wallenberg Retina Center at the University of Lund, Sweden). At 5 mo postsurgery, a significant amount of RPE65 immunolabeling was observed in the RPE of the treated eye, but none in the contralateral eye *(33)*. There was also a marked reduction of lipoid inclusions in the RPE in the treated eye, and further, an obvious difference between the structure of cone OS in the

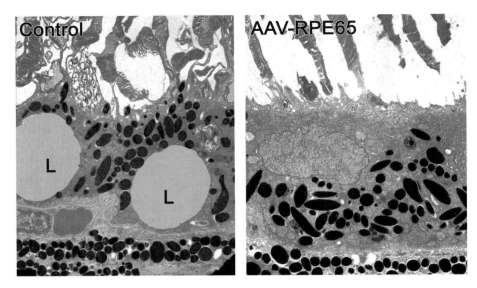

Fig. 6. Ultrastructural studies showed large lipoid inclusions (L) in the RPE of an untreated RPE65–/– dog (left image), littermate of a treated dog (right image) that had undergone gene transfer 10 mo earlier. In the latter dog there was a lack of lipoid inclusions in the area of the gene construct injection. Note also the normal appearing photoreceptor outer segments in this dog in comparison to those of the untreated littermate. Courtesy of Dr. Martin Natz.

treated eye compared to those of the untreated eye. In the gene transfer treated eye, the cones seemed to be aligned more orderly and appeared more robust than in the contralateral, untouched eye.

Morphological analysis was performed also on the eyes of two RPE65–/– littermates that were euthanized 10 mo after administration of the AAV vectors into one of the dogs. The untreated littermate was used as a control. The dog that had undergone gene transfer had shown a low-grade, transient uveitis in the AAV-RPE65 treated eye. Long-term follow-up ERG studies showed restoration of photoreceptor function, not only in the treated eye, but also slight recovery of photopic responses in the contralateral untreated eye, starting 6 mo postsurgery. Ultrastructural studies were undertaken to determine whether there were morphological changes that correlated with the functional recovery of the eyes. In the rAAV.RPE65 treated eye, almost no RPE lipoid inclusions characteristic of the RPE65 null mutation were observed within the treated region *(45)*. In the opposite, untreated eye, the numbers and sizes of lipid droplets were diminished relative to those present in the eyes of the untreated littermate. Rod and cone OS morphology appeared orderly and elongated in the treated eye compared to that of the untreated dog and the fellow eye (Fig. 6). These preliminary results for the gene therapy treated eye establish a basis for the functional recovery of scotopic and photopic ERG responses obtained upon clinical follow-up in the gene transfer treated eye.

These results were further verified when another dog used for long-term follow-up was euthanized 2.5 yr following the gene transfer *(46)*. This dog had exceptionally high-amplitude ERG responses in the gene transfer treated eye, as well as increased mainly photopic ERG responses in the fellow eye. Following light and electron microscopic studies of the retina, several exciting results were obtained: the photoreceptor outer and inner segments were orderly aligned in areas of the retina peripheral to the immediate injection site. The latter area was, however, morphologically disrupted. Outside of the immediate injection area the outer segments were normally elongated toward the RPE, and contained stacks of normally oriented lamellar discs. Lipoid inclusions were abolished in the former bleb area although still prevalent in more peripheral regions. Further studies are in progress with morphometric analysis of tissue from both eyes in order to evaluate the spatial distribution of treatment effects.

DISCUSSION AND CONCLUSION

Gene therapy in dogs affected with a hereditary retinal dystrophy caused by the RPE65 null mutation resulted in a remarkable improvement in visually mediated behavior and in retinal function as assessed by ERG. Rescue of visual function was observed in all dogs in which the vector was successfully delivered subretinally. Immunocytochemical analysis detected RPE65 expression only in the area of the RPE where the gene vector was applied and it was in this specific area that the lipoid inclusions were markedly diminished after treatment. The immediate injection site showed several alterations in retinal morphology as visualized at long-term follow-up by FL and ICG angiography. On the other hand, photoreceptor OS morphology appeared normalized by the gene transfer both within and outside of the treated region. The only adverse effects in the gene therapy treatment observed were transient inflammatory responses in the eyes of some of the injected animals as a result of impurities in the gene therapy vector preparations.

The therapeutic expression of RPE65 from an rAAV vector injected subretinally in RPE65–/– dogs was effectively achieved up to approx 30 mo following gene therapy. The effect of treatment was successively reduced to levels where ERG amplitudes were insignificantly increased in comparison to preoperatively. The reason for this diminished functional effect is not clear, but may be because of several factors such as promoter shut-off or transgene silencing *(47–50)*. Another reason for the reduced effect could be the loss of transgene expressing cells since long-term follow-up studies have clearly shown that there is a progression of clinical disease in the 3 yr following the initial gene transfer treatments (Fig. 1).

The contralateral eye effect shows up as a low-grade qualitative improvement of the ERG responses observed additionally in the fellow eyes of gene transfer treated dogs. The phenomenon appears to be most prominent on cone function and the effect is transient. The precise mechanism for this effect in the presently treated group of RPE65–/– dogs has not been elucidated, but there may be several possible explanations. It has been shown that unilateral cell injury causes multiple cellular responses in the contralateral eye *(51)*. Further, unilateral pseudorabies virus injections with GFP into the vitreous of one eye in hamsters infected the intergeniculate leaflet of the thalamus. It was later found that the retinal ganglion cells in the contralateral eye also expressed GFP, becoming infected after transsynaptic uptake and retrograde transport through infected retinorecipient neurons *(52)*. Our findings and those of others, thus clearly indicate that the contralateral or fellow eye, even if untreated, should not be used as a "control" eye. Gene transfer in the presently described unilaterally treated dogs resulted in the production of 11-*cis*-retinal, which long-term could be disseminated to the contralateral eye in small amounts, the mechanism of which is still obscure.

AAV-mediated gene therapy is an effective means of rescuing visual function in inherited diseases that result in retinal dysfunction. Modification of the gene construct may be necessary to achieve more long-term results of the treatment. However, the subretinal approach of treating outer retinal disease appears safe as long as the vector preparations are of high purity and are not given directly into the macular region.

ACKNOWLEDGMENTS

We would like to acknowledge the following coworkers, all of which have been involved in some parts of this project: Ragnheidur Bragadottir, MD, PhD, Norway; Anitha Bruun, PhD, Sweden; Lynnette Caro, DVM, USA; Marnie Ford, DVM, PhD, Canada; Martin Katz, PhD, USA; Helmut M. Mayser, MD, Germany; Piroska E. Rakoczy, PhD, Australia; T. Michael Redmond, PhD, USA; Ernst-Otto Ropstad, DVM, Norway; Vaegan, PhD, Australia.

For superb technical assistance, we would like to thank Jenny Garland, Ginny Dodam, Leilani Castaner, Deborah Becker, and Howard Wilson. We are grateful to the University of North Carolina Gene Therapy Vector Core facility of Dr. R. Jude Samulski, PhD, for providing the SSV9 plasmid.

This work was supported by the Foundation Fighting Blindness, Research to Prevent Blindness, Inc., and the University of Missouri Research Board.

REFERENCES

1. Gu SM, Thompson DA, Srikumari CR, et al. Mutations in the RPE65 cause autosomal recessive childhood-onset severe retinal dystrophy. Nat Genet 1997;17:194–197.
2. Marlhens F, Bareil C, Griffoin JM, et al. Mutations in RPE65 cause Leber's congenital amaurosis. Nat Genet 1997;17:139–141.
3. Morimura H, Fishman GA, Grover SA, Fulton AB, Berson EL, Dryja TP. Mutations in the RPE65 gene in patients with autosomal recessive retinitis pigmentosa or Leber congenital amaurosis. Proc Natl Acad Sci USA 1998;95:3088–3093.
4. Thompson DA, Gyurus P, Fleischer LL, et al. Genetics and phenotypes of RPE65 mutations in inherited retinal degeneration. Invest Ophthalmol Vis Sci 2000;41:4293–4299.
5. Lambert SA, Sherman S, Taylor D, Kriss R, Coffey R, Pembrey M. Concordance and recessive inheritance of Leber congenital amaurosis. Am J Med Gen 1993;46:275–277.
6. Heckenlively JR. Retinitis pigmentosa. In: Heckenlively JR, ed. Retinitis Pigmentosa Philadelphia: J.B. Lippincott Co., 1988;1–5.
7. Lorenz B, Preising M, Bremser D, et al. Early-onset severe rod-cone dystrophy in young children with RPE65 mutations. Invest Ophthalmol Vis Sci 2000;41:2735–2742.
8. Grieshaber MC, Boltshauser E, Niemeyer G. Leber's congenital amaurosis. Clinical heterogeneity and electroretinography in 27 patients. In: Hollyfield JG, Anderson RE, LaVail MM, eds. Retinal Degenerative Diseases and Experimental Therapy. New York: Kluwer Academic/ Plenum Publishers, 1999:95–104.
9. Perrault I, Rozet JM, Calvas P, et al. Histological study and in situ hybridization of the retGC gene in a human fetus affected with Leber's congenital amaurosis (LCA1). Invest Ophthalmol Vis Sci 1998;39:104–102.
10. Porto FBO, Perrault I, Hicks D, et al. Prenatal human ocular degeneration occurs in Leber's congenital amaurosis (LCA2). J Gene Med 2002;4:390–396.
11. Niemeyer G, Stahli P. Electroretinographic diagnoses and differential diagnoses: results over 6 years. Klin Monatsblat Augenheilkd 1996;208:306–310.

12. Hamel CP, Tsilou E, Harris E, et al. A developmentally regulated microsomal protein specific for the pigment epithelium of the vertebrate retina. J Neurosci Res 1993a;34:414–425.

13. Hamel CP, Tsilou E, Pfeffer BA, Hooks JJ, Detrick B, Redmond TM. Molecular cloning and expression of RPE65, a novel retinal pigment epithelium-specific microsomal protein that is post-transcriptionally regulated in vitro. J Biol Chem 1993b;268: 15,751–15,757.

14. Nicoletti A, Wong DJ, Kawase K, et al. Molecular characterization of the human gene encoding an abundant 61 dDA protein specific to the retinal pigment epithelium. Hum Mol Genet 1995;4:641–649.

15. Hamel CP, Jenkins NA, Gilbert DJ, Copeland NG, Redmond TM. The gene for the retinal pigment epithelium-specific protein RPE65 is localized to human 1p31 and mouse 3. Genomics 1994;20:509–512.

16. Redmond TM, Yu S, Lee E, et al. Rpe65 is necessary for the production of 11-cis-vitamin A in the retinal visual cycle. Nat Genet 1998;20:44–51.

17. Narfström K, Wrigstad A, Nilsson SEG. The Briard dog: a new animal model of congenital stationary night blindness. Br J Ophthalmol 1989;73:750–756.

18. Wrigstad A, Nilsson SEG, Narfström K. Ultrastructural changes of the retina in Briard dogs with hereditary congenital night blindness and partial day blindness. Exp. Eye Res 1992;55:805–818.

19. Narfström K, Wrigstad A, Ekesten B, Nilsson SE. Hereditary retinal dystrophy in the briard dog: clinical and hereditary characteristics. Prog Vet Comp Ophthalmol 1994;4:85–92.

20. Veske A, Nilsson SE, Narfström K, Gal A. Retinal dystrophy of Swedish briard/briard-beagle dogs is due to a 4-bp deletion in RPE65. Genomics 1999;57:57–61.

21. Aguirre GD, Baldwin V, Pearce-Kelling S, Narfström K, Ray K, Acland GM. Congenital stationary night blindness in the dog: common mutation in the RPE65 gene indicates founder effect. Mol Vis 1998;4:23.

22. Seeliger MW, Grimm C, Ståhlberg F, et al. New views on RPE65 deficiency: the rod system is the source of vision in a mouse model of Leber congenital amaurosis. Nat Genet 2001;29:70–74.

23. Wrigstad A, Narfström K, Nilsson SEG. Slowly progressive changes of the retina and the retinal pigment epithelium in Briard dogs with hereditary retinal dystrophy: a morphological study. Doc Ophthalmol 1994a;87:337–354.

24. Wrigstad A. Hereditary dystrophy of the retina and the retinal pigment epithelium in a strain of Briard dogs: a clinical, morphological and electrophysiological study. Linkoping, Sweden: Linkoping Univ Med Dissertations, 1994b, No. 423.

25. Streilein JW, Ma N, Wenkel H, Ng TF, Zamiri P. Immunobiology and privilege of the neuronal retina and pigment epithelium transplants. Vision Res 2002;42:487–495.

26. Rolling F. Recombinant AAV-mediated gene transfer to the retina: gene therapy perspectives. Gene Ther 2004;Suppl 1:S26–S32.

27. Weber M, Rabinowitz J, Provost N, et al. Recombinant adeno-associated virus serotype 4 mediates unique and exclusive long-term transduction of retinal pigmented epithelium in rat, dog, nonhuman primate after subretinal delivery. Mol Ther 2003;7:774–781.

28. Yang GS, Schmidt M, Yan Z, et al. Virus-mediated transduction of murine retina with adeno-associated virus: effects of viral capsid and genome size. J Virol 2002;76: 7651–7660.

29. Auricchio A, Kobinger G, Anand V, et al. Exchange of surface proteins impacts on viral vector cellular specificity and transduction characteristics: the retina as a model. Hum Mol Genet 2001;10:3075–3081.

30. Acland GM, Aguirre GD, Ray J, et al. Gene therapy restores vision in a canine model of childhood blindness. Nat Genet 2001;28:92–95.

31. Narfström K, Katz ML, Bragadottir R, et al. Functional and structural recovery of the retina after gene therapy in the RPE65 mutation dog. Invest Ophthalmol Vis Sci 2003a; 44:1663–1672.

32. Narfström K, Katz ML, Ford M, Redmond TM, Rakoczy PE, Bragadottir R. In vivo gene therapy in young and adult RPE65–/– dogs produces long-term visual improvement. J Hered 2003b;94:31–37.

33. Narfström K, Bragadottir R, Redmond TM, Rakoczy PE, van Veen T, Bruun A. Functional and structural evaluation after AAV.RPE65 gene transfer in the canine model of Leber's congenital amaurosis, In: Hollyfield JG, Anderson RE, LaVail MM, eds. Retinal Degeneration Mechanisms and Experimental Therapy. New York: Kluwer Academic/Plenum Publishers, 2003c.

34. Ford M, Bragadottir R, Rakoczy PE, Narfström K. Gene transfer in the RPE65 null mutation dog: relationship between construct volume, visual behavior and electroretinographic (ERG) results. Doc Ophthalmol 2003;107:79–86.

35. Samulski RJ, Chang LS, Shenk T. Helper-free stocks of recombinant adeno-associated viruses: normal integration does not require viral gene expression. J Virol 1989;63: 3822–3828.

36. Skulimowski AW, Samulski RJ. Adeno-associated virus: integrating vectors for human gene therapy. Method Mol Genet 1995;7:3–12.

37. Zolotukhin S, Byrne BJ, Mason E, et al. Recombinant adeno-associated virus purification using novel methods improves infectious titer and yield. Gene Ther 1999;6:973–985.

38. Seeliger MW, Narfström K. Functional assessment of the regional distribution of disease in a cat model of hereditary retinal degeneration. Invest Ophthalmol Vis Sci 2000;41: 1998–2005.

39. Mayser HM, Narfström K, Bragadottir R, Rakoczy E, Redmond YM, Seeliger MW. Assessment of local functional improvement following gene therapy in the RPE65 null mutation Dog model. ARVO 2003; Abstract no. 3592.

40. Caro ML, Estes DM, Cohn LA, Narfström K. The systemic and ocular immune response after recombinant adeno-associated virus gene transfer of RPE65 in 6 dogs affected with the RPE65 null mutation. Invest Ophthalmol Vis Sci 2002;43:E-Abstract 4594.

41. Caro ML, Tullis G, Estes DM, Narfström K. Long-term expression of RPE65 in the canine model for LCA fails to elicit a humoral response to the transgene following gene therapy using rAAV.RPE65. Invest Ophthalmol Vis Sci 2004;45:E-Abstract 3693.

42. Bragadottir R, Lei B, Narfström K. Electroretinographic monitoring of gene therapy treated RPE65 null mutation dog, a model for human Leber's Congenital Amaurosis. Invest Ophthalmol Vis Sci 2002;43:E-Abstract 4596.

43. Narfström K, Vaegan, Katz M, Bragadottir R, Seeliger M. Assessment of structure and function over a 3-year period after gene transfer in RPE65–/– dogs. International Society for Clinical Electrophysiology of Vision (ISCEV) 42nd Annual Symposium, Dorado, Puerto Rico, 2004:50.

44. Rohrer B, Goletz P, Znoiko S, et al. Correlation of regenerable opsin with rod ERG signal in Rpe65–/– mice during development and aging. Invest Ophthalmol Vis Sci 2003;44: 310–315.

45. Narfström K, Vaegan, Ropstad E-O, Ford MM, Katz ML. Sustained long-term bilateral cone function in the degenerate retina of monocular gene therapy treated RPE65–/– dogs. ARVO 2004; Abstract no. 3487.

46. Katz ML, Seeliger MW, Vaegan, et al. Rescue of retinal function and structure via AAV-mediated gene therapy in RPE65 null mutation dogs. Presented at the 1st International Symposium on Translational Clinical Research for Inherited and Orphan Retinal Diseases, Washington, DC, 2005:423–430.

47. Scharfmann R, Axelrod JH, Verma IM. Long-term in vivo expression of retrovirus-mediated gene transfer in mouse fibroblast implants. Proc Natl Acad Sci USA 1991; 88:4626–4630.
48. Loser P, Jennings GS, Strauss M, Sandig V. Reactivation of the previously silenced cytomegalovirus major immediate-early promoter in the mouse liver: involvement of NFkappaB. J Virol 1998;72:180–190.
49. Gaetano C, Catalano A, Palumbo R, et al. Transcriptionally active drugs improve adenovirus vector performance in vitro and in vivo. Gene Ther 2000;7:1624–1630.
50. Chen WY, Townes TM. Molecular mechanism for silencing virally transduced genes involves histone deacetylation and chromatin condensation. Proc Natl Acad Sci USA 2000;97:377–382.
51. Bodeutsch N, Siebert H, Dermon C, Thanos S. Unilateral injury to the adult rat optic nerve causes multiple cellular responses in the contralateral site. J Neurobiol 1999;38:116–128.
52. Smith BN, Banfield BW, Smeraski CA, et al. Pseudorabies virus expressing enhanced green fluorescent protein: a tool for in vitro electrophysiological analysis of transsynaptically labeled neurons in identified central nervous system circuits. Proc Natl Acad Sci USA 2000;97:9264–9269.

23

Neuroprotective Factors and Retinal Degenerations

Joyce Tombran-Tink, PhD and Colin J. Barnstable, DPhil

INTRODUCTION

Mammalian neurons are postmitotic and, in general, are nonrenewable, so an individual cell must be capable of surviving a wide range of environmental conditions for many decades, if not a lifetime. Retinal neurons are isolated to some extent from fluctuations in levels of many circulating molecules by a specific blood–retinal barrier, a structure analogous to the blood–brain barrier elsewhere in the central nervous system (CNS). This barrier provides protection to the retina by selectively transporting those molecules required for normal metabolism and filtering out components that could be detrimental to the health of the tissue.

Alterations in nutrient level and composition, light levels, synaptic activity, and physical stresses can all fluctuate in the eye around a mean or normal level and a variety of endogenous homeostatic mechanisms compensate for most of these variations from this mean. However, if the transient stress goes outside the boundaries of the homeostatic capability of the cell, it is likely to result in death of the neuron, an event seen in most retinal degenerative diseases. For example, if the level of oxygen gets too low, the level of light too high, or the level of intraocular pressure too great then intrinsic safeguard mechanisms can no longer maintain normal cell function and a cascade of cell death signals is initiated.

From: *Ophthalmology Research: Retinal Degenerations: Biology, Diagnostics, and Therapeutics*
Edited by: J. Tombran-Tink and C. J. Barnstable © Humana Press Inc., Totowa, NJ

If we were able to extend the homeostatic capacity of retinal cells then it would be possible to substantially reduce cell death and lessen the risk of blindness in many retinal degenerative diseases. Two ways of doing so are to identify and potentiate the inherent neuroprotective mechanisms in the eye and/or to use factors that will allow the cells to withstand larger extremes of transient stress.

In this chapter, we will consider several classes of neuroprotective molecules that are strong candidates to increase the capacity of the retina to tolerate stress. These include neurotransmitters, steroids, fatty acids and their derivatives, and polypeptide neurotrophic (NT) factors, which can exert neuroprotective actions on a range of different target molecules. For example, some neurotransmitters and polypeptides activate membrane receptors, which can in turn trigger intracellular neuroprotective signals, whereas fatty acids and Co-enzyme Q can activate mitochondrial uncoupling proteins to alter abnormal amounts of reactive oxygen produced within a cell.

USING NEUROTRANSMITTERS AND THEIR RECEPTORS TO PROTECT RETINAL GANGLION CELLS FROM GLAUCOMA-ASSOCIATED DEGENERATION

Neurotransmitters are molecules whose primary function is to facilitate the transmission of signals between neurons or from neurons to a target tissue. In some cases, however, excessive activity in neuronal tissues can lead to dangerously toxic concentrations of a specific neurotransmitter. This phenomenon was first reported in 1957 by Lucas and Newhouse, who showed that increased amounts of glutamate, above physiological levels, are toxic to the mammalian eye *(1)*. The term "excitotoxicity" was coined from the observation that systemic injection of glutamate into neonatal mice led to destruction of the inner retinal layers, most notably the retinal ganglion cell (RGC) layer, which was evident by ultrastructural examination *(2)*. Later, it was determined that glutamate caused depolarization of RGCs and that glutamate toxicity was associated with prolonged cation influx *(3–6)*. Cation influxes are known to activate many intracellular pathways including excess transport of calcium into the mitochondria, with subsequent induction of permeability pores, release of cytochrome c, and initiation of an apoptotic cascade *(7)*, as well as the production of nitric oxide and peroxide, both of which are highly toxic to cells *(8,9)*. In addition, many intracellular signal transduction cascades are under the control of glutamate-induced cation influxes. Some of these, for example activation of p38 and ERK (extracellular-signal-regulated kinase) MAPKs, have been directly linked to the initiation of cell death signals *(10–12)*.

Attempts to protect retinal ganglion cells from excitotoxicity by blocking glutamate receptors with specific antagonists has generated much interest. The problem with this approach, however, is that although general blocking of glutamate receptors may prevent cell death, it also prevents normal cell function including the transmission of visual signals through the ganglion cells. As an alternative to using strong glutamate receptor antagonists, several groups have explored the use of the weaker compound memantine, an *N*-methyl-D-aspartate (NMDA) receptor noncompetitive antagonist *(13,14)*. Memantine blocks excessive activation of the NMDA receptor but does not block normal signaling. It is sufficiently antagonistic that it reduces calcium influx through the NMDA receptor to levels that are not toxic. The success of this approach in a variety of model systems has led to clinical trials of memantine as a therapeutic agent for glaucoma, and its widespread use is awaiting Food and Drug Administration approval *(15)*.

There is, however, clear evidence that some neurotransmitters can function as neuroprotective agents in the retina and other parts of the CNS. This is especially true for acetylcholine (ACh) where activation of the α-7 nicotinic ACh receptor (α-7 nAChR) by ACh has been linked to protection against glutamate-induced excitotoxicity in the brain *(16–19)*. A similar effect has been found in the retina where ACh protects ganglion cells from high concentrations of glutamate *(20)*. One possible mechanism by which it protects these cells is through activation of the α-7 nAChR followed by a calcium-dependent activation of its intracellular signaling partners that are associated with the PI3-kinase and MAPK pathways *(21,22)*.

Norepinephrine is another neurotransmitter linked to neuroprotection throughout the CNS *(23,24)*. Many of the protective effects of this transmitter appear to be mediated by the α-2-adrenergic receptor *(25,26)*. Activation of α-2-adrenergic receptors, which are expressed on RGCs and other cells in the inner nuclear layer, has been associated with neuroprotective outcomes in the retina *(27)*. When the norepinephrine agonist, brimonidine, engages with the α-2-adrenergic receptors, it induces strong protection of RGCs in many models of glaucoma and in ischemia-reperfusion models of general retinal damage *(28,29)*. Brimonidine is effective at preserving retinal ganglion cells after optic nerve crush when given either at the time of or 14 h before injury. Pharmacologically active concentrations of brimonidine can be achieved in the retina (>2 nM) when this compound is applied to the ocular surface. One response of the retina to brimonidine is increased expression of the brain-derived NT factor (BDNF) by RGCs *(30)*. Because brimonidine also reduces intraocular pressure, it is being studied as a potentially useful agent to treat glaucoma.

The approach of modulating the actions of specific neurotransmitters or the actions of their cognate receptors for neuroprotective ends is less appropriate for macular degeneration and other photoreceptor-associated retinal degenerations because these cells are primary sensory neurons that do not receive synaptic input. There are other classes of factors, however, that can promote survival of photoreceptors, which are more relevant for these conditions.

STEROIDS AND LIPIDS AS NEUROPROTECTIVE FACTORS FOR RETINAL DEGENERATIONS

Two steroid hormones most typically associated with reproductive function are estrogen and progesterone. Estrogen affects differentiation and neurite outgrowth of neurons and is protective against several toxic insults including oxidative stress, glutamate excitotoxicity, hypoglycemia, and ischemia *(31–37)*. In addition, an observed increase in risk for degenerative eye diseases in postmenopausal women is reduced with estrogen treatment *(38)*. The classical method by which estrogen works upon entry into a cell is that it binds to its cognate receptor, this receptor–ligand complex then dimerizes, translocates to the nucleus, and interacts with an estrogen receptor binding motif on the promoter regions of many genes to activate gene expression. One mechanism by which estrogen may exert neuroprotective effects in the retina is by promoting expression of antiapoptotic and neuroprotective molecules, such as BCL-2 and thioredoxin *(39,40)*.

Estrogen may also act upstream at the level of the plasma membrane to induce activation of cell survival signals associated with the ERK/MAPK and PI3 kinase/Akt cascades *(41,42)*.

Although progesterone can also reduce the death of retinal neurons resulting from global ischemia and glutamate excitotoxicity, there is less extensive evidence for the molecular mechanisms of the neuroprotective actions of this hormone *(43)*. Larger scale studies to elucidate the effects of these hormones on eye diseases have been limited by the use of synthetic steroids. For example, hormone replacement therapies using synthetic progestins have complicated the studies because these molecules are not neuroprotective and may even be toxic *(44)*.

There is now mounting evidence that fatty acids may also contribute to the survival of neurons in conditions of stress. This is not entirely surprising because omega-3 fatty acids are natural body constituents and found in high concentrations in the retina, particularly in photoreceptors. Two of the most common polyunsaturated fatty acids in the retina are arachidonic acid (AA) and docosahexanoic acid (DHA) *(45)*. These molecules serve as building blocks for the synthesis of eicosanoids and docosanoids respectively, when released from phospholipids by the action of phospholipases *(46)*. AA protects RGCs from cell death induced by high levels of glutamate but can be toxic when used above concentrations of 10 μM *(47)*. Whether AA itself is the active neuroprotective compound or whether it needs to be converted into prostaglandins or leukotrienes for activity is still being determined.

This neuroprotective compound NPD1 (10,17S-docosatriene) is formed from DHA by the action of a 15-lipoxygenase-like enzyme *(46,48)*. NPD1 is neuroprotective in a wide range of in vivo and in vitro systems including ischemia-reperfusion or chemically induced oxidative stress. In patients suffering from Alzheimer's disease, the concentration of NPD1 is significantly reduced in brain regions undergoing degeneration suggesting that it may play a role in augmenting neurodegenerative processes *(49)*. NPD1 triggers increased expression of a number of key anti-apoptotic molecules including Bcl-2, Bcl-xl, and Bfl-1(A1). It also reduces the expression of pro-inflammatory molecules suggesting that its neuroprotective role may be mediated by these actions *(50)*. Therefore, strategies that can modulate in vivo levels of NPD1 has the potential to be effective in reducing neuronal degeneration although, so far, it has been difficult to regulate ocular levels of its precursor, DHA, by dietary supplementation.

ANTIOXIDANTS OFFER PROTECTION TO PHOTORECEPTORS: MITOCHONDRIAL UNCOUPLING PROTEINS

Elevated levels of reactive oxygen species (ROS) are toxic and their accumulation in a tissue causes damage to membranes, proteins, and DNA eventually leading to death of the cell *(51,52)*. The retina is particularly sensitive to the increased ROS generated through the high metabolic activity of photoreceptors and RGCs *(53)*. In accord with this, there is evidence that antioxidants provide protection for photoreceptors by reducing intracellular levels of ROS generated by oxidative stress *(54)*.

Intracellular generation of reactive oxygen species is a natural function of mitochondria because the proton gradient that drives production of adenosine triphosphate (ATP) produces reactive oxygen species as a byproduct. The greater the rate at which mitochondria

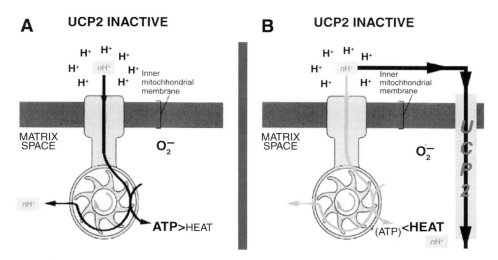

Fig. 1. Schematic illustration of ATP synthase functioning in the absence of UCP2 activity (**A**), and in the presence of active UCP2 (**B**). The proton gradient drives ATP synthase and ATP production. If the proton gradient decreases as a result of the UCP2 activity, ATP production decreases, energy is dissipated in the form of heat (thermogenesis), superoxide (O_2^-) production is decreased, and calcium efflux is increased. However, despite decreased ATP production by individual mitochondria, overall the neurons will have more ATP available, because uncoupling is accompanied by mitochondrial proliferation. (Adapted from ref. *66*.)

produce ATP, the greater is the endogenous production of ROS. Increased intracellular levels of ROS can lessen the ability of a cell to tolerate additional ROS generated extracellularly in tissues exposed to oxidative stress.

The intracellular levels of ROS however, can be reduced by uncoupling proteins (UCPs), located in the inner membrane of the mitochondria, which serve as intracellular antioxidants (Fig. 1). The primary function of these proteins is to allow hydrogen ions to leak from the intermembrane space into the matrix of the mitochondria and in this way dissipate the energy in the form of heat *(55–60)*. By decreasing the driving force of ATP synthase, the enzyme that catalyzes ATP synthesis, UCPs reduce the amount of ATP and ROS produced *(61,62)*.

The most well-characterized UCP, is UCP1, which is expressed solely in brown adipose tissue and is mainly responsible for thermogenesis in small rodents *(55,63)*. Brown adipose tissue is virtually insignificant for normal physiology in primates and, until recently, little attention was paid to the action of uncoupling proteins in other tissues of the body. In the last few years, however, several other members of the UCP family have been identified and found to promote partial uncoupling of oxidation from phosphorylation in vitro. The five putative UCPs differ greatly in tissue distribution and regulation and may have distinct physiological roles. UCP2, UCP4, and BMCP1 are predominantly expressed in the central nervous system, including the retina, but are also detected in muscle, spleen, and adipose tissue. UCP1 and UCP3 are expressed only in peripheral tissues *(56–60)*.

The relevance of these uncoupling proteins to neurodegenerative processes has been shown in studies that linked increased activity of UCP2 with protection of cells from seizure-induced excitotoxicity or injury induced by MPTP (1-methyl-4-phenyl-1,2,5,6 tetrahydropyridine) in several models of neurodegenerative diseases *(64,65).* Further support for a role of UCPs in cell death is that UCP2 overexpressing mice show an increased number of RGCs owing to decreased programmed cell death in the early post-natal stages of development *(66).*

UCPs can be regulated by a number of factors, most importantly by their co-factors co-enzyme Q (CoQ) and fatty acids. Although studies have not yet separated the general antioxidant function of CoQ from its specific effects on mitochondrial UCPs there is a general view that most antioxidants including the CoQ offer some protection for photoreceptors.

POLYPEPTIDES THAT PROMOTE SURVIVAL
OF NEURONS IN THE CNS

The other major class of factors that promote survival of neurons in the CNS is a group of polypeptides that also have trophic influence and other functions during normal development. Four of these are currently considered to have the most impact on neurons in the eye. These are BDNF, ciliary NT factor (CNTF), glial-derived NT factor (GDNF), and pigment epithelium-derived factor (PEDF).

All four polypeptides are produced at multiple sites in the brain and in other non-neural tissues. All four are also expressed in the normal retina, though by different cell types. Immunocytochemical studies show that BDNF is localized to ganglion cells and other cells of the inner retina *(67).* CNTF, on the other hand, is found primarily in Müller glia and astrocytes *(68).* Similarly, the major sites of synthesis of GDNF in the retina are glial cells *(69,70).* PEDF is synthesized in the retinal pigment epithelial (RPE), Müller glia, and ganglion cells of the retina, as well as cells in the ciliary body *(71).* These patterns of expression have been defined in the normal retina and may differ in conditions of retinal injury or disease where all four polypeptides are upregulated as part of a homeostatic response. As discussed later in this chapter, expression of these factors by various retinal cell types including microglia and vascular endothelial cells in injury may be an important component in limiting damage to the retina.

All four polypeptides have been tested extensively for their neuroprotective properties in a wide range of in vitro and in vivo models. Some of these, such as the axotomy model of neurodegeneration, have been studied extensively but are outside the scope of this chapter. Here we will only review the actions of these polypeptides on RGCs injured by the excito-toxin glutamate, to all retinal neurons damaged by oxidative stress, and to photoreceptors induced to degenerate by mutations or by excessive light exposure.

Neuroprotective Polypeptides Impede Glutamate Excitotoxicity

As discussed in a previous section, excessive amounts of glutamate in the nervous system can kill neurons. Cell death is proportional to the concentration of the excito-toxin present as is clearly shown in studies in which dissociated cultures of neurons are treated with micromolar concentrations of glutamate *(72).*

Both BDNF and GDNF counteract glutamate excitotoxicity by reducing NMDA-receptor-mediated Ca^{2+} influx through an ERK-dependent pathway, as shown in a number of different culture models *(73,74).* GDNF has additional autocrine actions in the retina

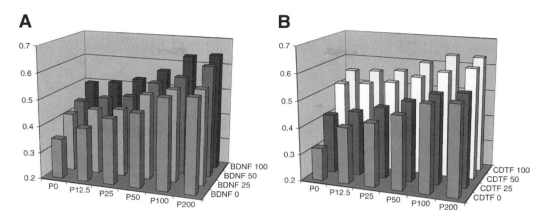

Fig. 2. Neuroprotective polypeptides protect RPE cells from H_2O_2-induced cell death. In each histogram, combinations of two neuroprotective factors were used for each point. The increased survival given by increasing doses of PEDF is show from left to right. The increased survival induced by BDNF (**A**) or CNTF (**B**) is shown from front to back. Even at saturating doses of one factor, additional protection can be given by the addition of a second factor.

where it increases expression of the GLAST glutamate transporter in glia and in so doing may help reduce glutamate levels and glutamate excitotoxity *(75)*. A number of studies have shown that nanomolar concentrations of PEDF allow cells to withstand the toxic influence of glutamate, which would otherwise induce apoptotic cell death in many neurons. (reviewed in ref. *71*.)

Neuroprotective Polypeptides Reduce Oxidative Stress-Related Damage

As we have already discussed, one of the most common causes of neuronal death and a possible contributing factor to many forms of retinal degeneration is oxidative stress. One way of experimentally inducing oxidative stress to test the neuroprotective efficacy of a compound is to treat cells with low concentrations of hydrogen peroxide, a naturally occurring toxic byproduct of visual transduction. Many of the NT polypeptides that reduce the death of neurons that are challenged with toxic levels of glutamate also shield against oxidative stress (Fig. 2). In one study, we observed that when retinal neurons were pretreated with PEDF, they develop resistance to moderate concentrations of hydrogen peroxide *(76)*. It is important to note that protection only occurred when cells were pretreated for at least an hour with PEDF and even after pretreatment, high concentrations of hydrogen peroxide were still toxic.

Similarly, others have shown that BDNF offers protection to photoreceptors and RGCs from oxidative stress *(77–79)*. It has been suggested that BDNF does so by reducing endogenous production of ROS, possibly by its actions at the level of the mitochondria *(80,81)*. It is interesting to speculate that uncoupling proteins may be direct or indirect targets for BDNF actions in reducing oxidative damage. Both GDNF and CNTF can prevent neuronal injury caused by oxidative stress as shown in many models of CNS injury *(76,82,83)*. Such findings support the idea that the damage caused by oxidative stress can be reduced by the endogenous neuroprotective mechanisms controlled by these four polypeptides.

Neuroprotective Polypeptides Block Light-Induced and Genetic Forms of Retinal Degeneration

The efficacy of polypeptide factors in promoting the survival of retinal neurons has been established in genetic models of retinal degenerations and experimental models in which damage is induced by exposing animals to constant bright light. Fibroblast growth factor 2 (FGF2) was one of the first factors to be tested for neuroprotective actions in such an in vivo model. FGF2 showed a significant delay in photoreceptor degeneration following intravitreal or subretinal injections of the soluble, purified protein, in the Royal College of Surgeons strain of rats carrying an inherited retinal dystrophy *(84)*. Later, it was shown that PEDF can also slow the progression of photoreceptor degeneration in *rd* mutant mice and reduce apoptosis in photoreceptor cells in *rds* mutant mice *(85)*, two mouse models that have human counterparts in forms of retinitis pigmentosa (RP). Similarly, GDNF increases survival of rod photoreceptors in both *rd/rd* mouse mutants and the TgN S334ter-4 rhodopsin line of transgenic rats *(86,87)*.

The effects of FGF2 on the survival of neurons were also observed in rat photoreceptors in a constant light damage model of retinal degeneration *(88)*. However, further interest was diminished in the use of this polypeptide as a suitable candidate for retinal degenerations because the risks outweighed the potential benefits. FGF2 causes an increase in the number of retinal microglia, promotes the formation of cataracts, induces angiogenesis, and is a potent mitogen for RPE and Müller glia cells. Nevertheless, these observations were some of the earliest to provide support for the idea that neuroprotective factors can reduce the effects of injurious insults to photoreceptors.

Several other endogenous polypeptides had similar biological actions on photoreceptors that are exposed to constant bright light. In a larger study, intravitreal injections of FGF1, FGF2, BDNF, CNTF, and interleukin (IL)-1b, 2 d before rats were exposed for 1 wk to constant bright light, were able to block the damage to photoreceptors under this condition *(89)*. Like FGF, IL-1b had undesirable side effects. After IL-1b treatment, a large increase in the infiltration of macrophages and the induction of folds and rosettes were seen in the retina. BDNF and CNTF, on the other hand, showed promise as good candidates to protect photoreceptor without the side effects observed for FGF and IL-1b. In separate experiments, it was also found that PEDF, like the other polypeptides, was as effective in reducing the degeneration of rat photoreceptors that have been exposed to constant light (Fig. 3) *(90)*.

The trials conducted with CNTF, although successful in promoting survival of photoreceptors, resulted in diminished general cellular activity in the retina, although this effect was dependent on the concentration of CNTF used *(91)*. In addition, the narrow range of effective dose of CNTF and changes in body weight after administration of the protein systemically are still some concerns that need to be addressed *(92)*.

The data, so far, obtained from studies using neuroprotective polypeptides to slow down retinal degenerations are encouraging and suggest that the normal retina contains more than one factor that engage in keeping neurons alive in the presence of hazardous conditions and that exogenous application of these molecules can also reduce pathological damage to the cells. Multiple treatments of single factors, application of a neuroprotective cocktail of these factors, and delivery of the polypeptides using

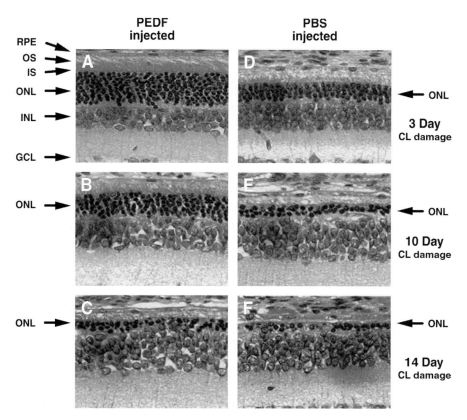

Fig. 3. Rescue of photoreceptors by PEDF as a function of constant light-exposure. Animals were injected intravitreally with PEDF or PBS 2 d before light exposure. PEDF-injected eyes exposed to constant light for **(A)** 3, **(B)** 10, and **(C)** 14 d. PBS-injected eyes exposed to constant light for **(D)** 3, **(E)** 10, and **(F)** 14 d. GCL, ganglion cell layer; INL, inner nuclear layer. (Adapted from ref. *76*.)

intraocular sustained release systems, could maximize protection in the retina over a longer time frame. This subject is discussed more fully in the section on delivery of neuroprotective factors.

NON-NEURONAL RETINAL TARGETS FOR NEUROPROTECTIVE FACTORS

The in vitro studies described earlier suggest that the bioactive polypeptides act directly on neurons to regulate molecular pathways essential to promote cell survival. This assumption is generally accepted; however, there is evidence that in vivo, their actions could be mediated, in part, through other adjacent non-neuronal cells. Support for this hypothesis comes from the finding that receptors for BDNF are not detected on normal rod photoreceptors yet rods are protected from toxic damage by BDNF *(93)*. Similarly, CNTF protects rod photoreceptors and may be involved in their early differentiation although there is controversy over the presence of receptors and signaling cascades for CNTF in rods *(94,95)*.

On the other hand, there is clear evidence that receptors for both BDNF and CNTF are found on Müller cells and that CNTF activates both MAP kinase and STAT3 intracellular transduction pathways in these cells *(93,96,97)*. Therefore, the possibility exists that the primary effects of these polypeptides are on Müller cells and that they, in turn, release molecules that control survival of photoreceptors.

There is less information available about the direct target cells, signaling mechanisms, and receptor expression for PEDF and GDNF. PEDF acts on several types of cells including neurons and glia, its receptor has not yet been identified and characterized, and it is expressed by many cells in the eye including the RPE on which it has autocrine actions *(98)*. We still do not know what the primary cell target for GDNF are in the retina although, because its receptors have been found on Müller glial cells and other neuronal cell types, there may very well be multiple cell types in the retina that respond to this factor *(99–101)*.

In addition to Müller cells, other targets in the retina for neuroprotective factors include RPE, vascular endothelial cells, and microglia. The responses of these cells to various pathological insults may be an important component of a number of retinal degenerative diseases including macular degeneration and diabetic retinopathy in which activation of vascular endothelial cells is the cornerstone for neovascularization in the eye. This hypothesis is partially supported by the observations that BDNF is mitogenic for more than one microglial cell line and its receptors are found on these cells in brain tissue sections *(102,103)* suggesting that its actions on neurons may be mediated through the microglia. Of course, we cannot rule out the possibility that such factors may have pleiotropic effects and that their actions on adjacent non-neural cells are unrelated to neuronal cell survival.

Another example of possible non-neuronal cell intervention that ultimately results in neuroprotection is evident in studies showing that RPE cells, which are closely juxtaposed to photoreceptors and exert trophic influence on these cells, express both PEDF and BDNF as well as its cognate receptor, TrkB. Both factors have been shown to promote differentiation, survival, and function of RPE cells obtained from several species. In vivo, this action could possibly occur by BDNF and PEDF autocrine feedback loops in the RPE or by a paracrine pathway *(104–106)*. On the other hand, endothelial cells not only synthesize PEDF, but also respond to exogenous PEDF to decrease their cell proliferation activity. There is general agreement that endothelial cells also secrete BDNF and express BDNF receptors as well, but there is disagreement about whether this factor can regulate endothelial cell function *(107,108)*.

A similar situation exists for CNTF. This polypeptide is produced by RPE and microglial cells although there is no clear demonstration of its actions on these cells, or on endothelial cells in the retina or choroid. In contrast, GDNF is synthesized by microglial cells and can enhance their survival as well as regulate the permeability of endothelial cell junctions *(109,110)*.

We have already mentioned that many types of retinal cells contribute to the pathogenesis of macular degeneration and other retinal degenerations. There is evidence that one of the later identified NT factors, PEDF is intimately involved in the pathophysiology of these cells. PEDF is not only synthesized and secreted by RPE and vascular endothelial cells, but also it can alter the morphology and properties of these cells, as well as Müller glia. It also causes the production of other factors by microglial cells and has multiple inhibitory effects on endothelial cells during neovascularization *(111,112)*.

As is true for any neuron, the health of the photoreceptors is tightly linked to the normal function of supporting cells. These findings are important because they imply that the survival of the neural retina is governed by more than one cell type and more than one endogenous factor. This generates a broader range of therapeutic targets that can be scrutinized for developing retinal degeneration treatments. Consequently, although it is prudent to focus on developing exogeneous strategies that influence the survival of photoreceptors, it may also be worthwhile to promote endogenous neuroprotective mechanisms and augment normal function of supporting cells to optimize treatment for retinal pathologies associated with photoreceptor cell death.

DELIVERY OF NEUROPROTECTIVE FACTORS

Systemic Delivery

It is one thing to show that neuroprotective factors are effective inhibitors of photoreceptor cell death; it is quite another to translate this finding into therapeutics for retinal degenerative diseases. The indications from experimental models used so far suggest that the amounts of neuroprotective factors required for activity are relatively small— they work well in the low nanomolar to picomolar ranges.

The simplest way to provide a neuroprotective factor to the retina is by systemic delivery. There is some evidence that this route will allow therapeutic doses of some factors to reach the retina. For example, in a mouse model of ischemia-induced retinopathy, daily systemic injections of microgram quantities of PEDF can prevent retinal neovascularization *(113)*. Systemic injections of 5 to 11 µg/d were sufficient to generate a therapeutic response in the eye. Despite the restriction of PEDF movement by the blood–retinal barrier, the binding of PEDF to other sites in the body and the possible degradation of PEDF, it is clear that sufficient PEDF reaches the eye to be of therapeutic value.

There may also be less rapid degradation of PEDF in systemic routes because this protein is secreted by the liver, and significant amounts of it is found in circulating plasma *(114,115)*. Thus systemic injections of PEDF and other NT molecules normally found in circulating plasma could be an effective way of treating degenerative diseases if the blood–retina barrier would permit, or could be manipulated to allow, their access to the retina in nanomolar doses. Although this is a practical approach, a major concern that arises is that of unwanted side effects in other organs with systemic injections of therapeutic molecules. Two such examples of side effects were observed in the clinical trials of CNTF as a therapy for amyotrophic lateral sclerosis and GDNF for Parkinson's disease where there was substantial weight loss among participants *(92,116)*.

Site-Targeted Delivery

A more effective and direct approach to deliver neuroprotective factors is to inject them at the target site of action. This strategy is somewhat more invasive but shows promising results in several animal models. Injection of 1 µL of a 1 µg/µL solution of BDNF, CNTF, or PEDF into the vitreous of the superior hemisphere of rat eyes provides protection to photoreceptors against the damaging effects of 1 wk of continuous light exposure to the eye *(90)*. Although a number of diseases routinely require patients to carry out multiple systemic injections at home, intraocular injections will require more careful administration by highly skilled personnel and will most likely have greater risks of retinal damage and ocular infections.

Slow Release Polymers

A variation of this strategy currently under investigation is the use of safe and effective long-term slow release polymers that contain neuroprotective factors. Biodegradable polymers such as poly(lactide-co-glycolide) have been used to encapsulate proteins and peptides and can release therapeutic doses of the molecules for extended periods *(117,118)*. Encapsulated NT factors have already been shown to augment the survival of transplants in the brain *(119,120)*. Although not eliminating the need for intraocular injections, such an approach would lessen the frequency of injections and possibly make it a more acceptable form of therapy.

Eyedrop Formulations

Finally, we should not exclude the application of NT molecules to the ocular surface to target the retina. For many years, eyedrops containing drugs at pharmacologically active doses that diffused to the ciliary epithelium were used in the medical management of glaucoma. There is some evidence that drugs applied in this way can affect retinal physiology. Likewise, therapies for retinal degenerations could be formulated for topical application. It may even be possible to have factors diffuse through the sclera overlying the retina. Experimental measurements of molecular diffusion across the sclera have shown that the rate of diffusion is inversely proportional to size of the molecule. Proteins and nucleic acids large enough to be therapeutic agents can be induced to cross the sclera by application of a mild electric current *(121)*. Although such methods have not yet been tried in animal models of retinal degeneration, it is easy to see how such treatments could be self administered or provided during simple office visits.

Unfortunately, the diffusion of large neuroprotective polypeptides poses a problem for such delivery because they are slower to cross the cornea or sclera than the small molecules currently used in eyedrops. Therefore, it is less likely that therapeutic doses of such factors could reach the retina by this route. One way of circumventing this problem would be to find small peptide fragments that are biologically active and that are able to diffuse more easily through to the retina. So far, only one such fragment, a 44 amino acid region of PEDF, shows neuroprotective actions consistent with those of the parent protein *(122,123)*. As we have shown, phylogenetic analysis of the NT polypeptides is a useful method to identify highly conserved surface epitopes that are unique to a protein and which may contain the biological activity *(123)*. Such studies may identify small peptides from BDNF, CNTF and GDNF that have neuroprotective activity and are more likely to diffuse rapidly through the cornea and sclera.

Biological Vectors

The other two methods of delivering neuroprotective polypeptides to the retina rely on the use of biological vectors to carry the factor or engineering stem cells and mammalian cells to secrete the therapeutic protein of interest. Adenovirus and, more recently, adeno-associated viral vectors have been used to deliver therapeutic genes to localized target tissues. In the eye, virally delivered CNTF improves the survival of RGCs and rod photoreceptors, as indicated by physiological and anatomical measurements *(124–128)* and virally vectored PEDF into the vitreous or subretinal compartment can block retinal neovascularization and protect photoreceptor from further degeneration *(129–133)*. These studies, though preliminary, show that viral mediated delivery of NT factors is effective in attenuating cell death and blood vessel growth.

There are also some problems and risks to patients encountered when using current DNA-mediated gene-transfer technologies. These include (1) obtaining clinically effective viral titers, (2) toxicity and immunogenicity as a result of the expression of viral genes, (3) stable transgene expression in individuals requiring long-term treatment, and (4) insertional mutagenesis by random viral integration into the host genome. These concerns are being addressed and whether they prevent future development of this approach for retinal degenerations remains to be seen.

However, in efforts to side step some of the current difficulties associated with gene transfer using viral vectors, mammalian cells engineered to produce neuroprotective factors have been generated. Engineered iris epithelial cells represent a good example of such a technology. These are usually obtained from the patients needing treatment, grown in culture, and then manipulated to produce high levels of the specific neuroprotective factors. This would provide autologous transplants of cells to repopulate the diseased eyes, avoiding issues of tissue rejection and thereby decreasing side effects. Whether or not there is appropriate integration into the target tissue and secretion of therapeutic amounts of the protein over the period necessary to prevent degeneration are questions still being addressed.

Success of the strategy, however, was shown in a rat model, in which autologous iris epithelial cells, stably transfected to produce high levels of PEDF, were transplanted subretinally in the eye *(134)*. The transplanted cells secreted PEDF, blocked blood vessel growth in models of retinal neovacularization, and delayed photoreceptor death in a model of retinal degeneration. In both cases, higher levels of PEDF were expressed in the retina. It is possible that other types of cells, such as RPE cells from fetal or donor eye tissues, could also be exploited for this form of neuroprotective strategy in which autologous cells are not readily available. Such an approach may not only generate elevated levels of the neuroprotective factors but may also provide functional RPE cells that would correct some of the pathologies resulting from the defective RPE cells associated with both macular degeneration and RP.

An approach, still in its infancy, is the use of stem cells genetically engineered to produce a target neuroprotective factor. The use of stem cells to prevent additional loss of photoreceptors is also a good alternate approach to autologous iris pigment epithelial transfer and is currently under investigation for macular degeneration therapy. In theory at least, after transplantation in the subretinal space, these cells would have the additional advantage of being given the appropriate retinal cues to undergo controlled differentiation and integration into the retina to replace lost photoreceptors. Stem cells altered to produce specific neuroprotective factors are less susceptible to the microenvironment that led to loss of the photoreceptors in the first place. These cells may provide implantable reservoirs of neuroprotective factors and are not restricted to autologous cells. Engineered stem cell populations could also be encapsulated to lessen the probability that they will disperse from the site of injection or mount an immune response. Such an approach would correct the visual deficits caused by macular degeneration or RP, while, at the same time, allows adequate amounts of neuroprotective factor to become available to prevent more cells from dying.

SUMMARY AND PROSPECTS

At the beginning of this chapter, we suggested that in many forms of degenerative diseases retinal neurons die because transient stresses overwhelm the cellular homeostatic mechanisms. As we learn more about the wide range of molecules that can protect retinal neurons in experimental models of neurodegenerative diseases, we can look ahead to the day when most can be prevented, or their progress at least dramatically slowed down. To reach this goal we first need better diagnostic tools and earlier diagnosis. Although much is known about the genetics of many forms of RP, there is still a dearth of genetic markers for these disorders; although we can now account for much of the genetic risk for macular degeneration, we still need a better understanding of the environmental triggers for the disease. We also need to define the molecular pathways by which the various environmental and genetic insults lead to retinal degeneration.

What we now know emphasizes three important concepts when considering the use of neuroprotective molecules to treat some forms of retinal degenerations: (1) protection by these factors is finite, (2) strong toxic insults will still cause death in the presence of NT molecules, and (3) neuroprotective pathways must be activated before a rapidly acting toxic insults like hydrogen peroxide or glutamate can trigger cell death cascades.

What we do not know is precisely how these factors work and whether there are synergistic actions among them and the intracellular pathways they control. There is clearly cross talk between these signaling pathways as shown diagramatically in Fig. 4. We have discussed the possibility that common degenerative signals and common neuroprotective pathways may be regulated by multiple factors in the retina. However, what need to be explored more carefully are the questions of whether different types of insult trigger a common degenerative cascade in all neurons or whether initial damage caused by a specific injury is unique to that insult but converges on a key set of downstream molecules that execute final degenerative decisions.

For therapeutic purposes, it is essential to determine if these factors act at the initial stages of the disease or only on the downstream common pathways, or whether they all activate the same intracellular signaling mechanisms. If the factors work on different pathways and have synergistic actions, then we need to consider using a cocktail of these agents to achieve maximum therapeutic efficacy.

Finally, we discussed the need for safer and convenient mechanisms of drug delivery. Experimental models using single injections show only transient protection, probably resulting from clearance of the molecule. Frequent intraocular injections carry risks and are not convenient for the patient. We proposed that injections of drug formulations that have long-term effectiveness would be clinically more attractive alternatives. An even

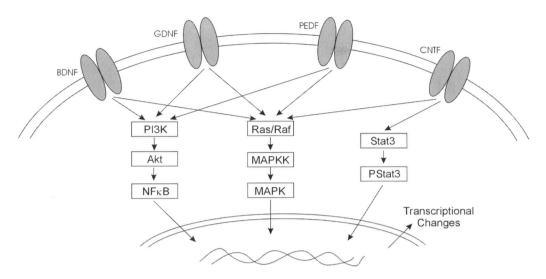

Fig. 4. Diagrammatic representation of the major transduction pathways used by the neuroprotective factors discussed in this chapter. Although novel transduction pathways may yet be defined for these factors, it is clear that the factors act in very similar ways and all that four can influence the phosphorylation of the MAPK pathway.

better approach would be to develop therapeutic small molecules that can cross the sclera or are effective after application to the cornea.

With the increasing pace of work in each of these areas, we are confident that effective long-term therapies for retinal degenerative diseases will soon become a reality.

REFERENCES

1. Lucas DR, Newhouse JP. Toxic effect of sodium L-glutamate on the inner layers of the retina. Arch Ophthalmol 1957;58:193–201.
2. Olney JW. Glutamate-induced retinal degeneration in neonatal mice. Electron microscopy of the acute developing lesion. J Neuropathol Exp Neurol 1969;28:455–474.
3. Aizenman E, Frosch MP, Lipton SA. Responses mediated by excitatory amino acid receptors in solitary retinal ganglion cells from rat. J Physiol 1988;396:75–91.
4. Sucher NJ, Lipton SA, Dreyer EB. Molecular basis of glutamate toxicity in retinal ganglion cells. Vision Res 1997;37:3483–3493.
5. Quigley HA. Neuronal death in glaucoma. Prog Ret Eye Res 1998;18:39–57.
6. Lam TT, Abler AS, Kwong JMK, Tso MOM. N-methyl-D-aspartate (NMDA)-induced apoptosis in rat retina. Invest Ophthalmol Vis Sci 1999;40:2391–2397.
7. Budd SL, Tenneti L, Lishnak T, Lipton SA. Mitochondrial and extramitochondrial apoptotic signaling pathways in cerebrocortical neurons. Proc Natl Acad Sci USA 2000;97:6161–6166.
8. Lipton SA, Choi YB, Pan ZH, et al. A redox-based mechanism for the neuroprotective and neurodestructive effects of nitric oxide and related nitroso-compounds. Nature 1993;364:626–632.
9. Bonfoco E, Krainc D, Ankarcrona M, Nicotera P, Lipton SA. Apoptosis and necrosis: two distinct events induced, respectively, by mild and intense insults with N-methyl-D-aspartate or nitric oxide/superoxide in cortical cell cultures. Proc Natl Acad Sci USA 1995;92:7162–7166.

10. Kikuchi M, Tenneti L, Lipton SA. Role of p38 mitogen-activated protein kinase in axotomy-induced apoptosis of rat retinal ganglion cells. J Neurosci 2000;20:5037–5044.

11. Manabe S, Lipton SA. NMDA signals leading to proapoptotic and antiapoptotic pathways in the rat retina. Invest Ophthalmol Vis Sci 2003;44:385–392.

12. Munemasa Y, Ohtani-Kaneko R, Kitaoka Y, et al. Contribution of mitogen-activated protein kinases to NMDA-induced neurotoxicity in the rat retina. Brain Res 2005;1044:227–240.

13. Osborne NN, Quack G. Memantine stimulates inositol phosphates production in neurones and nullifies N-methyl-D-aspartate-induced destruction of retinal neurones. Neurochem Int 1992;21:329–336.

14. Chen HS, Lipton SA. Mechanism of memantine block of NMDA-activated channels in rat retinal ganglion cells: uncompetitive antagonism. J Physiol 1997;499 (Pt 1):27–46.

15. Lipton SA. Possible role for memantine in protecting retinal ganglion cells from glaucomatous damage. Surv Ophthalmol 2003;48 Suppl 1:S38–S46.

16. Marin P, Maus M, Desagher S, Glowinski J, Premont J. Nicotine protects cultured striatal neurones against N-methyl-D-aspartate receptor-mediated neurotoxicity. Neuroreport 1994;5:1851–1855.

17. Shimohama S, Akaike A, Kimura J. Nicotine-induced protection against glutamate cytotoxicity. Nicotinic cholinergic receptor-mediated inhibition of nitric oxide formation. Ann N YAcad Sci 1996;17:356–361.

18. Kaneko S, Maeda T, Kume T, et al. Nicotine protects cultured cortical neurons against glutamate-induced cytotoxicity via alpha7-neuronal receptors and neuronal CNS receptors. Brain Res 1997;765:135–140.

19. Dajas-Bailador FA, Lima PA, Wonnacott S. The alpha/nicotinic receptor subtype mediates nicotine protection against NMDA excitotoxicity in primary hippocampal cultures through a Ca2+ dependent mechanism. Neuropharm 2000;39:2799–2807.

20. Wehrwein E, Thompson SA, Coulibaly SF, Linn DM, Linn CL. Acetylcholine protects isolated adult pig retinal ganglion cells from glutamate-induced excitotoxicity. Invest Ophthalmol Vis Sci 2004;45:1531–1543.

21. Kihara T, Shimohama S, Sawada H, et al. Alpha 7 nicotinic receptor transduces signals to phosphatidylinositol 3-kinase to block A beta-amyloid-induced neurotoxicity. J Biol Chem 2001;276:13,541–13,546.

22. Dineley KT, Westerman M, Bui D, Bell K, Ashe KH, Sweatt JD. Beta-amyloid activates the mitogen-activated protein kinase cascade via hippocampal alpha7 nicotinic acetylcholine receptors: in vitro and in vivo mechanisms related to Alzheimer's disease. J Neurosci 2001;21:4125–4133.

23. Troadec JD, Marien M, Darios F, et al. Noradrenaline provides long-term protection to dopaminergic neurons by reducing oxidative stress. J Neurochem 2001;79:200–210.

24. Marien MR, Colpaert FC, Rosenquist AC. Noradrenergic mechanisms in neurodegenerative diseases: a theory. Brain Res Rev 2004;45:38–78.

25. Veyrac A, Didier A, Colpaert F, Jourdan F, Marien M. Activation of noradrenergic transmission by alpha2-adrenoceptor antagonists counteracts deafferentation-induced neuronal death and cell proliferation in the adult mouse olfactory bulb. Exp Neurol 2005;194:444–456.

26. Srinivasan J, Schmidt WJ. Treatment with alpha2-adrenoceptor antagonist, 2-methoxy idazoxan, protects 6-hydroxydopamine-induced Parkinsonian symptoms in rats: neurochemical and behavioral evidence. Behav Brain Res 2004;154:353–363.

27. Wheeler LA, Gil DW, WoldeMussie E. Role of alpha-2 adrenergic receptors in neuroprotection and glaucoma. Surv Ophthalmol 2001;45 Suppl 3:S290–S294.

28. Donello JE, Padillo EU, Webster ML, Wheeler LA, Gil DW. Alpha(2)-adrenoceptor agonists inhibit vitreal glutamate and aspartate accumulation and preserve retinal function after transient ischemia. J Pharmacol Exp Ther 2001;296:216–223.

29. Wilensky JT. The role of brimonidine in the treatment of open-angle glaucoma. Surv Ophthalmol 1996;41 Suppl 1:S3–S7.

30. Gao H, Qiao X, Cantor LB, WuDunn D. Up-regulation of brain-derived neurotrophic factor expression by brimonidine in rat retinal ganglion cells. Arch Ophthalmol 2002;120:797–803.

31. Toran-Allerand CD. Sex steroids and the development of the newborn mouse hypothalamus and preoptic area in vitro: implications for sexual differentiation. Brain Res 1976;106:407–412.

32. Toran-Allerand CD. Sex steroids and the development of the newborn mouse hypothalamus and preoptic area in vitro II. Morphological correlates and hormonal specificity. Brain Res 1980;189:413–427.

33. Behl C, Widmann M, Trapp T, Holsboer F. 17-beta estradiol protects neurons from oxidative stress-induced cell death in vitro. Biochem Biophys Res Commun 1995;216:473–482.

34. Simpkins JW, Green PS, Gridley KE, Singh M, de Fiebre NC, Rajakumar G. Role of estrogen replacement therapy in memory enhancement and the prevention of neuronal loss associated with Alzheimer's disease. Am J Med 1997;103:19S–25S.

35. Singer CA, Rogers KL, Strickland TM, Dorsa DM. Estrogen protects primary cortical neurons from glutamate toxicity. Neurosci Lett 1996;212:13–16.

36. Dubai DB, Kashon ML, Pettigrew LC, et al. Estradiol protects against ischemic injury. J Cereb Blood Flow Metab 1998;18:1253–1258.

37. Simpkins JW, Rajakumar G, Zhang YQ, et al. Estrogens may reduce mortality and ischemic damage caused by middle cerebral artery occlusion in the female rat. J. Neurosurg 1997;87:724–730.

38. Snow KK, Seddon JM. Age-related eye diseases: impact of hormone replacement therapy, and reproductive and other risk factors. Int J Fertil Womens Med 2000;45:301–313.

39. Alkayed NJ, Goto S, Sugo N, et al. Estrogen and Bcl-2: gene induction and effect of transgene in experimental stroke. J Neurosci 2001;21:7543–7550.

40. Chiueh C, Lee S, Andoh T, Murphy D. Induction of antioxidative and antiapoptotic thioredoxin supports neuroprotective hypothesis of estrogen. Endocrine 2003;21:27–31.

41. Bryant DN, Bosch MA, Ronnekleiv OK, Dorsa DM. 17-Beta estradiol rapidly enhances extracellular signal-regulated kinase 2 phosphorylation in the rat brain. Neuroscience 2005;133:343–352.

42. Dhandapani KM, Wade FM, Mahesh VB, Brann DW. Astrocyte-derived transforming growth factor-{beta} mediates the neuroprotective effects of 17{beta}-estradiol: involvement of nonclassical genomic signaling pathways. Endocrinology 2005;146:2749–2759.

43. Singh M. Mechanisms of progesterone-induced neuroprotection. Ann N Y Acad Sci 2005;1052:145–151.

44. Nilsen J, Brinton RD. Impact of progestins on estrogen-induced neuroprotection: synergy by progesterone and 19-norprogesterone and antagonism by medroxyprogesterone acetate. Endocrinology 2002;143:205–212.

45. Chen H, Anderson RE. Comparison of uptake and incorporation of docosahexaenoic and arachidonic acids by frog retinas. Curr Eye Res 1993;12:851–860.

46. Bazan NG. Neuroprotectin D1(NPD1): a DHA-derived mediator that protects brain and retina against cell injury-induced oxidative stress. Brain Pathol 2005;15:159–166.

47. Kawasaki A, Han M-H, Wei J-Y, Hirata K, Otori Y, Barnstable CJ. Arachidonic acid protects rat retinal ganglion cells from glutamate neurotoxicity. Invest Ophthalmol Vis Sci 2002;43:1835–1842.

48. Mukherjee PK, Marcheselli VL, Serhan CN, Bazan NG. Neuroprotectin D1: a docosahexaenoic acid-derived docosatriene protects human retinal pigment epithelial cells from oxidative stress. Proc Natl Acad Sci USA 2004;101:8491–8496.

49. Lukiw WJ, Cui JG, Marcheselli VL, et al. A role for docosahexaenoic acid-derived neuro-protection D1 in neural cell survival and Alzheimer disease. J Clin Invest 2005;115: 2774–2783.

50. Marcheselli VL, Hong S, Lukiw WJ, et al. Novel docosanoids inhibit brain ischemia-reperfusion-mediated leukocyte infiltration and pro-inflammatory gene expression. J Biol Chem 2003;278:43,807–43,817.

51. Geller HM, Cheng KY, Goldsmith NK, et al. Oxidative stress mediates neuronal DNA damage and apoptosis in response to cytosine arabinoside. J Neurochem 2001;78:265–275.

52. See V, Loeffler JP. Oxidative stress induces neuronal death by recruiting a protease and phosphatase-gated mechanism. J Biol Chem 2001;276:35,049–35,059.

53. Winkler BS, Boulton ME, Gottsch JD, Sternberg P. Oxidative damage and age-related macular degeneration. Mol Vis 1999;5:32.

54. Beatty S, Koh H, Phil M, Henson D, Boulton M. The role of oxidative stress in the pathogenesis of age-related macular degeneration. Surv Ophthalmol 2000;45:115–134.

55. Bouillaud F, Ricquier D, Thibault J, Weissenbach J. Molecular approach to thermogenesis in brown adipose tissue: cDNA cloning of the mitochondrial uncoupling protein. Proc Natl Acad Sci USA 1985;82:445–448.

56. Fleury C, Neverova M, Collins S, et al. Uncoupling protein-2: a novel gene linked to obesity and hyperinsulinemia. Nat Genet 1997;15:269–272.

57. Boss O, Samec S, Paoloni-Giacobino A, et al. Uncoupling protein-3: a new member of the mitochondrial carrier family with tissue-specific expression. FEBS Lett 1997;408:39–42.

58. Vidal-Puig A, Solanes G, Grujic D, Flier JS, Lowell BB. UCP3: an uncoupling protein homologue expressed preferentially and abundantly in skeletal muscle and brown adipose tissue. Biochem Biophys Res Commun 1997;235:79–82.

59. Mao W, Yu XX, Zhong A, et al. UCP4, a novel brain-specific mitochondrial protein that reduces membrane potential in mammalian cells. FEBS Lett 1999;443:326–330.

60. Sanchis D, Fleury C, Chomiki N, et al. BMCP1, a novel mitochondrial carrier with high expression in the central nervous system of humans and rodents, and respiration uncoupling activity in recombinant yeast. J Biol Chem 1998;273:34,611–34,615.

61. Lowell BB, Spiegelman BM. Towards a molecular understanding of adaptive thermogenesis. Nature 2000;404:652–660.

62. Negre-Salvayre A, Hirtz C, Carrera G, et al. A role for uncoupling protein-2 as a regulator of mitochondrial hydrogen peroxide generation. FASEB J 1997;11:809–815.

63. Nicholls DG, Locke RM. Thermogenic mechanism in brown fat. Physiol Rev 1984;64:1–64.

64. Diano S, Matthews RT, Patrylo P, et al. Uncoupling protein 2 prevents neuronal death including that occurring during seizures: a mechanism for preconditioning. Endocrinology. 2003;144:5014–5021.

65. Horvath TL, Diano S, Leranth C, et al. Coenzyme Q induces nigral mitochondrial uncoupling and prevents dopamine cell loss in a primate model of Parkinson's disease. Endocrinology 2003;144:2757–2760.

66. Barnstable CJ, Li M, Reddy R, Horvath TL. Mitochondrial uncoupling proteins: regulators of retinal cell death. In: LaVail MM, Hollyfield JG, Anderson RE, eds. Retinal Degeneration 2002. Kluwer Academic/Plenum Press, 2003:269–275.

67. Garcia M, Forster V, Hicks D, Vecino E. In vivo expression of neurotrophins and neuro-trophin receptors is conserved in adult porcine retina in vitro. Invest Ophthalmol Vis Sci 2003;44:4532–4541.

68. Walsh N, Valter K, Stone J. Cellular and subcellular patterns of expression of bFGF and CNTF in the normal and light stressed adult rat retina. Exp Eye Res 2001;72:495–501.

69. Karlsson M, Lindqvist N, Mayordomo R, Hallbook F. Overlapping and specific patterns of GDNF, c-ret and GFR alpha mRNA expression in the developing chicken retina. Mech Dev 2002;114:161–165.

70. Jomary C, Darrow RM, Wong P, Organisciak DT, Jones SE. Expression of neurturin, glial cell line-derived neurotrophic factor, and their receptor components in light-induced retinal degeneration. Invest Ophthalmol Vis Sci 2004;45:1240–1246.
71. Tombran-Tink J, Barnstable CJ. PEDF: a multifaceted neurotrophic factor. Nat Rev Neurosci 2003;4:628–636.
72. Otori Y, Wei JY, Barnstable CJ. Neurotoxic effects of low doses of glutamate on purified rat retinal ganglion cells. Invest Ophthalmol Vis Sci 1998;39:972–981.
73. Munemasa Y, Ohtani-Kaneko R, Kitaoka Y, et al. Contribution of mitogen-activated protein kinases to NMDA-induced neurotoxicity in the rat retina. Brain Res 2005;1044: 227–240.
74. Nicole O, Ali C, Docagne F, et al. Neuroprotection mediated by glial cell line derived neurotrophic factor: involvement of a reduction of NMDA-induced calcium influx by the mitogen-activated protein kinase pathway. J Neurosci 2001;21:3024–3033.
75. Delyfer MN, Simonutti M, Neveux N, Leveillard T, Sahel JA. Does GDNF exert its neuro-protective effects on photoreceptors in the rd1 retina through the glial glutamate transporter GLAST? Mol Vis 2005;11:677–687.
76. Cao W, Tombran-Tink J, Chen W, Mrazek D, Elias R, McGinnis JF. Pigment epithelium-derived factor protects cultured retinal neurons against hydrogen peroxide-induced cell death. J Neurosci Res 1999;57:789–800.
77. Okoye G, Zimmer J, Sung J, et al. Increased expression of brain-derived neurotrophic factor preserves retinal function and slows cell death from rhodopsin mutation or oxidative damage. J Neurosci 2003;23:4164–4172.
78. Unoki K, LaVail MM. Protection of the rat retina from ischemic injury by brain-derived neurotrophic factor, ciliary neurotrophic factor, and basic fibroblast growth factor. Invest Ophthalmol Vis Sci 1994;35:907–915.
79. Castillo B Jr, del Cerro M, Breakefield XO, et al. Retinal ganglion cell survival is promoted by genetically modified astrocytes designed to secrete brain-derived neurotrophic factor (BDNF). Brain Res 1994;647:30–36.
80. Gabaizadeh R, Staecker H, Liu W, Van De Water TR. BDNF protection of auditory neurons from cisplatin involves changes in intracellular levels of both reactive oxygen species and glutathione. Mol Brain Res 1997;50:71–78.
81. Yamagata T, Satoh T, Ishikawa Y, et al. Brain-derived neurotropic factor prevents superox-ide anion-induced death of PC12h cells stably expressing TrkB receptor via modulation of reactive oxygen species. Neurosci Res 1999;35:9–17.
82. Cheng H, Fu YS, Guo JW. Ability of GDNF to diminish free radical production leads to protection against kainate-induced excitotoxicity in hippocampus. Hippocampus 2004; 14:77–86.
83. Iwata E, Asanuma M, Nishibayashi S, et al. Different effects of oxidative stress on activa-tion of transcription factors in primary cultured rat neuronal and glial cells. Brain Res Mol Brain Res 1997;50:213–220.
84. Faktorovich EG, Steinberg RH, Yasumura D. Photoreceptor degeneration in inherited reti-nal dystrophy delayed by basic fibroblast growth factor. Nature 1990;347:83–86.
85. Cayouette M, Smith SB, Becerra SP, Gravel C. Pigment epithelium-derived factor delays the death of photoreceptors in mouse models of inherited retinal degenerations. Neurobiol Dis 1999;6:523–532.
86. Frasson M, Picaud S, Leveillard T, et al. Glial cell line-derived neurotrophic factor induces histologic and functional protection of rod photoreceptors in the rd/rd mouse. Invest Ophthalmol Vis Sci 1999;40:2724–2734.
87. McGee Sanftner LH, Abel H, Hauswirth WW, Flannery JG. Glial cell line derived neurotrophic factor delays photoreceptor degeneration in a transgenic rat model of retinitis pigmentosa. Mol Ther 2001;4:622–629.

88. Faktorovich EG, Steinberg RH, Yasumura D, Matthes MT, LaVail MM. Basic fibroblast growth factor and local injury protect photoreceptors from light damage in the rat. J Neurosci 1992;12:3554–3567.

89. LaVail MM, Unoki K, Yasumura D, Matthes MT, Yancopoulos GD, Steinberg RH. Multiple growth factors, cytokines, and neurotrophins rescue photoreceptors from the damaging effects of constant light. Proc Natl Acad Sci USA 1992;89:11,249–11,253.

90. Cao W, Tombran-Tink J, Elias R, Sezate S, Mrazek D, McGinnis JF. In vivo protection of photoreceptors from light damage by pigment epithelium-derived factor. Invest Ophthalmol Vis Sci 2001;42:1646–1652.

91. Zeiss CJ, Allore HG, Towle V, Tao W. CNTF induces dose-dependent alterations in retinal morphology in normal and rcd-1 canine retina. Exp Eye Res 2006;82:395–404.

92. Mattson MP. Lose weight STAT: CNTF tops leptin. Trends Neurosci 2001;24:313–314.

93. Rohrer B, Korenbrot JI, LaVail MM, Reichardt LF, Xu B. Role of neurotrophin receptor TrkB in the maturation of rod photoreceptors and establishment of synaptic transmission to the inner retina. J Neurosci 1999;19:8919–8930.

94. Rhee KD, Yang XJ. Expression of cytokine signal transduction components in the postnatal mouse retina. Mol Vis 2003;9:715–722.

95. Valter K, Bisti S, Stone J. Location of CNTFR alpha on outer segments: evidence of the site of action of CNTF in rat retina. Brain Res 2003;985:169–175.

96. Sarup V, Patil K, Sharma SC. Ciliary neurotrophic factor and its receptors are differentially expressed in the optic nerve transected adult rat retina. Brain Res 2004;1013:152–158.

97. Zhang SSM, Wei JY, Kano R, et al. Stat3 but not MAPK signaling controls neural precursor fate in mouse retina. Invest Ophthalmol Vis Sci 2004;45:2407–2412.

98. Barnstable CJ, Tombran-Tink J. Neuroprotective and antiangiogenic actions of PEDF in the eye: molecular targets and therapeutic potential. Prog Ret Eye Res 2004;23:561–577.

99. Karlsson M, Lindqvist N, Mayordomo R, Hallbook F. Overlapping and specific patterns of GDNF, c-ret and GFR alpha mRNA expression in the developing chicken retina. Mech Dev 2002;114:161–165.

100. Harada C, Harada T, Quah HM, et al. Potential role of glial cell line-derived neurotrophic factor receptors in Muller glial cells during light-induced retinal degeneration. Neuroscience 2003;122:229–235.

101. Jomary C, Darrow RM, Wong P, Organisciak DT, Jones SE. Expression of neurturin, glial cell line-derived neurotrophic factor, and their receptor components in light-induced retinal degeneration. Invest Ophthalmol Vis Sci 2004;45:1240–1246.

102. Zhang J, Geula C, Lu C, Koziel H, Hatcher LM, Roisen FJ. Neurotrophins regulate proliferation and survival of two microglial cell lines in vitro. Exp Neurol 2003;183:469–481.

103. Knott C, Stern G, Kingsbury A, Welcher AA, Wilkin GP. Elevated glial brain-derived neurotrophic factor in Parkinson's diseased nigra. Parkinsonism Relat Disord 2002;8:329–341.

104. Hackett SF, Friedman Z, Freund J, et al. A splice variant of trkB and brain-derived neurotrophic factor are co-expressed in retinal pigmented epithelial cells and promote differentiated characteristics. Brain Res 1998;789:201–212.

105. Liu ZZ, Zhu LQ, Eide FF. Critical role of TrkB and brain-derived neurotrophic factor in the differentiation and survival of retinal pigment epithelium. J Neurosci 1997;17:8749–8755.

106. Malchiodi-Albedi F, Feher J, Caiazza S, et al. PEDF (pigment epithelium-derived factor) promotes increase and maturation of pigment granules in pigment epithelial cells in neonatal albino rat retinal cultures. Int J Dev Neurosci 1998;16:423–432.

107. Donovan MJ, Lin MI, Wiegn P, et al. Brain derived neurotrophic factor is an endothelial cell survival factor required for intramyocardial vessel stabilization. Development 2000;127:4531–4540.

108. Kim H, Li Q, Hempstead BL, Madri JA. Paracrine and autocrine functions of brain-derived neurotrophic factor (BDNF) and nerve growth factor (NGF) in brain-derived endothelial cells. J Biol Chem 2004;279:33,538–33,546.

109. Salimi K, Moser K, Zassler B, et al. Glial cell line-derived neurotrophic factor enhances survival of GM-CSF dependent rat GMIR1-microglial cells. Neurosci Res 2002;43:221–229.

110. Igarashi Y, Chiba H, Utsumi H, et al. Expression of receptors for glial cell line-derived neurotrophic factor (GDNF) and neurturin in the inner blood-retinal barrier of rats. Cell Struct Funct 2000;25:237–241.

111. Bouck N. PEDF: anti-angiogenic guardian of ocular function. Trends Mol Med 2002;8:330–334.

112. Tombran-Tink J, Barnstable CJ. Therapeutic prospects for PEDF: more than a promising angiogenesis inhibitor. Trends Mol Med 2003;9:244–250.

113. Stellmach V, Crawford SE, Zhou W, Bouck N. Prevention of ischemia-induced retinopathy by the natural ocular antiangiogenic agent pigment epithelium-derived factor. Proc Natl Acad Sci USA 98:2593–2597.

114. Sawant S, Aparicio S, Tink AR, Lara N, Barnstable CJ, Tombran-Tink J. Regulation of factors controlling angiogenesis in liver development: a role for PEDF in the formation and maintenance of normal vasculature. Biochem. Biophys. Res. Comm 2004;325:408–413.

115. Petersen SV, Valnickova Z, Enghild JJ. Pigment-epithelium-derived factor (PEDF) occurs at a physiologically relevant concentration in human blood: purification and characterization. Biochem J 2003;374(Pt 1):199–206.

116. Nutt JG, Burchiel KJ, Comella CL, et al. Randomized, double-blind trial of glial cell line-derived neurotrophic factor (GDNF) in PD. Neurology 2003;60:69–73.

117. Saltzman WM, Olbricht WL. Building drug delivery into tissue engineering. Nat Rev Drug Discov 2002;1:177–186.

118. Cypes SH, Saltzman WM, Giannelis EP. Organosilicate-polymer drug delivery systems: controlled release and enhanced mechanical properties. J Control Release 2003;90:163–169.

119. Haller MF, Saltzman WM. Nerve growth factor delivery systems. J Control Release 1998;53:1–6.

120. Mahoney MJ, Saltzman WM. Millimeter-scale positioning of a nerve-growth-factor source and biological activity in the brain. Proc Natl Acad Sci USA 1999;96:4536–4539.

121. Davies JB, Ciavatta VT, Boatright JH, Nickerson JM. Delivery of several forms of DNA, DNA-RNA hybrids, and dyes across human sclera by electrical fields. Mol Vis 2003;9:569–578.

122. Bilak MM, Becerra SP, Vincent AM, Moss BH, Aymerich MS, Kuncl RW. Identification of the neuroprotective molecular region of pigment epithelium-derived factor and its binding sites on motor neurons. J Neurosci. 2002;22:9378–9386.

123. Tombran-Tink J, Aparicio S, Xu X, et al. PEDF and the Serpins: phylogeny, sequence conservation, and functional domains. J Struct Biol 2005;151:130–150.

124. Cayouette M, Gravel C. Adenovirus-mediated gene transfer of ciliary neurotrophic factor can prevent photoreceptor degeneration in the retinal degeneration (*rd*) mouse. Hum Gene Ther 1997;8:423–430.

125. Ng TF, Streilein JW. Light-induced migration of retinal microglia into the subretinal space. Invest Ophthalmol Vis Sci 2001;42:3301–3310.

126. Cayouette M, Behn D, Sendtner M, Lachapelle P, Gravel C. Intraocular gene transfer of ciliary neurotrophic factor prevents death and increases responsiveness of rod photoreceptors in the retinal degeneration slow mouse. J Neurosci 1998;18:9282–9293.

127. Liang FQ, Dejneka NS, Cohen DR, et al. AAV-mediated delivery of ciliary neurotrophic factor prolongs photoreceptor survival in the rhodopsin knockout mouse. Mol Ther 2001;3:241–248.

128. van Adel BA, Kostic C, Deglon N, Ball AK, Arsenijevic Y. Delivery of ciliary neurotrophic factor via lentiviral-mediated transfer protects axotomized retinal ganglion cells for an extended period of time. Hum Gene Ther 2003;14:103–115.

129. Auricchio A, Behling KC, Maguire AM, et al. Inhibition of retinal neovascularization by intraocular viral-mediated delivery of anti-angiogenic agents.Mol Ther 2002;6:490–494.

130. Duh EJ, Yang HS, Suzuma I. Pigment epithelium-derived factor suppresses ischemia-induced retinal neovascularization and VEGF-induced migration and growth.Invest Ophthalmol Vis Sci 2002;43:821–829.

131. Mori K, Gehlbach P, Ando A, McVey D, Wei L, Campochiaro PA. Regression of ocular neovascularization in response to increased expression of pigment epithelium-derived factor. Invest Ophthalmol Vis Sci 2002;43:2428–2434.

132. Raisler BJ, Berns KI, Grant MB, Beliaev D, Hauswirth WW. Adeno-associated virus type-2 expression of pigmented epithelium-derived factor or Kringles 1-3 of angiostatin reduce retinal neovascularization. Proc Natl Acad Sci USA 2002;99:8909–8914.

133. Takita H, Yoneya S, Gehlbach PL, Duh EJ, Wei LL, Mori K. Retinal neuroprotection against ischemic injury mediated by intraocular gene transfer of pigment epithelium-derived factor. Invest Ophthalmol Vis Sci 2003;44:4497–4504.

134. Semkova I, Kreppel F, Welsandt G, et al. Autologous transplantation of genetically modified iris pigment epithelial cells: a promising concept for the treatment of age-related macular degeneration and other disorders of the eye. Proc Natl Acad Sci USA 2002;99: 13,090–13,095.

24

Carbonic Anhydrase Inhibitors as a Possible Therapy for RP17, an Autosomal Dominant Retinitis Pigmentosa Associated With the R14W Mutation, Apoptosis, and the Unfolded Protein Response

George Rebello, PhD, Jacquie Greenberg, PhD, and Raj Ramesar, PhD

CONTENTS

INTRODUCTION

Of the 14 loci so far associated with autosomal dominant retinitis pigmentosa (RP), the most recently cloned gene is the one responsible for RP17 which occurs in six South African families comprising 187 individuals. Of these, 60 individuals are currently affected with RP and 16 are at risk by virtue of their relatedness to a known mutation carrier. The gene responsible for RP17 is carbonic anhydrase IV *(CA4)*, one of a family of carbonic anhydrases which are involved in the interconversion of carbon dioxide and carbonic acid *(1)*. The proposed mechanism of disease suggests that a drug-based treatment may be possible.

DISCUSSION

Of the six South African families linked to the RP17 focus genealogical studies have revealed the links between five of these families, whereas the sixth family cannot be linked at present. The average age of onset of RP in the affected individuals is 25 yr, and the average age of all affected individuals in the cohort is 49 yr.

Ophthalmological findings in RP17 individuals demonstrate the same high degree of variability usually associated with retinal degenerative disorders. On average though, the following description will fit most patients:

From: *Ophthalmology Research: Retinal Degenerations: Biology, Diagnostics, and Therapeutics*
Edited by: J. Tombran-Tink and C. J. Barnstable © Humana Press Inc., Totowa, NJ

- **Electroretinogram:** Nonrecordable.
- **Dilated fundal examination:** Typical changes of RP, including waxy pallor of the optic nerve heads, attenuated retinal vasculature, marked depigmentation, and choroidal atrophy in the macular area and peripheral fundus.

Bone spicule pigmentary deposits were distributed through all four segments of the retinas. No cataracts were present, but marked cellularity of the vitreous was observed.

The macular showed cystoid changes with marked retinal epithelial thinning.

The RP17 locus was initially mapped to the 17q22 chromosomal region by Bardien et al. in 1995 *(2)*. This region was refined and a number of candidate genes were excluded by Bardien et al. in 1997 *(3)*. The locus was fine mapped into a 1-cM region by Bardien-Kruger in 1999 *(4)*, whereafter the intensive hunt for a candidate gene in the region was started. Also in 1999, den Hollander et al. *(5)* published mapping results based on a Dutch family which mapped into the RP17 region. Because the original South African family was of Dutch extraction, it was assumed that the Dutch family and the South African families were related, albeit only by the RP17 mutation.

The hunt for the RP17 gene after 1999 was assisted by intensive bioinformatic effort to characterize candidates in the region and to construct sequence-based maps across the candidate region. Initially sequence-based maps were limited by the paucity of completed sequence in the region, although this situation quickly changed and it was possible to create a physical map to compare to the linkage map that had formed the backbone of the gene hunting effort up until that time.

In order to focus the effort of gene screening by DNA sequencing, which was at that time a time-limiting step, we needed to further refine the candidate region and this was performed by creating new STS markers based on the published sequence in the region. A number of steps allowed us to reduce the region from 3 Mb to a much more manageable 410 kb. Within this region, a number of genes were screened by sequencing in a couple of affected South African RP17 individuals from each of the families. Finally, a previously undescribed sequence change was detected in exon 1 of the *CA4* gene. The change is at base 40 of the complementary DNA sequence has been detected in all affected individuals; this change has not been detected in 36 unaffected relatives and 100 unrelated individuals from the same population. The C to T transition mutation leads to a change from an arginine to a tryptophan in the signal sequence at position –5 relative to the signal peptidase cleavage site (R14W). This signal sequence variant is predicted not to alter the sequence of the mature *CA4*. The R14W mutation creates an *MscI* restriction endonuclease site which was used to screen for the mutation in the extended family.

The gene, *CA4*, is a member of the carbonic anhydrase family of genes the products of which catalyse the reversible reaction between carbon dioxide and carbonic acid, according to the following reaction:

$$H_2O + CO_2 \leftrightarrow HCO_3^- + H^+.$$

The CAIV enzyme is membrane bound and functions in the transport of CO_2 across membranes and into, or out of, solution in various tissues of the body. *CA4* is expressed in the luminal surfaces of vessels in a number of tissues including; the proximal renal tubules *(6)*, the lungs *(7)*, and the choriocapillaris of the eye *(8)*. Hageman et al. *(8)* demonstrated expression of *CA4* in only two tissues in the eye, namely the vessels of the choriocapillaris, and in the lens.

Functional studies on the R14W *CA4*, and comparisons with wild-type *CA4* in Cos7 cells have shown that the R14W mutant leads to upregulation of markers of the unfolded protein response (UPR), namely, upregulation of the Endoplasmic Reticulum (ER) chaperone BiP, upregulation and activation of the ER kinase, PERK, and induction of CHOP (GADD153) *(1)*. Secondarily, the cells expressing the mutant *CA4* have been shown to induce apoptosis as measured by TUNEL staining and annexin V binding.

This evidence leads us to speculate that the R14W mutation is leading to cell death by the UPR and apoptosis. This will, in turn, lead to defective CO_2 and HCO_3 transport and eventual retinal destruction.

Subsequent work by Bonapace et al. *(9)* demonstrated in cell models that a number of nonspecific chemical chaperones influence the processing of the mutant (R14W) CAIV enzyme, and reduce the level of apoptotic loss associated with the mutation. Interestingly, other proteins that bind to the defective molecule also assist with rescuing the biological defect/process. In this respect, Bonapace et al. *(9)* investigated various carbonic anhydrase inhibitors, which bind to carbonic anhydrase. These drugs/reagents are a proven treatment for glaucoma. Bonapace et al. *(9)* showed that immunohistochemical staining for BiP, PERK, and CHOP were positive in 82.5, 69.0, and 85.7% of the R14W mutant-expressing cells, respectively, (these markers were detected in fewer than 5% of cells expressing wild-type *CA4*). Expression of each of these markers in R14W expressing cells was reduced to around 20% by 10 μM acetazolamide. Similarly, the markers of apoptosis, TUNEL staining, and annexin V binding were reduced from 65 and 85% of R14W expressing cells respectively to 25 and 20%, respectively, in the presence of acetazolamide.

This work demonstrates that, at least in vitro, the phenotype of the R14W *CA4* mutant may be rescued to a large degree. Before these results an be applied in a possible treatment trial in R14W mutant RP sufferers, three issues remain to be addressed:

1. Route of administration.
2. Dosage.
3. Duration of treatment.

Route of Administration

Clearly, the best approach will be to try to evaluate the treatment that enables the lowest dose of active ingredient to be used. This limits the choice of carbonic anhydrase inhibitors because acetazolamide is systemically administered, and dorzolamide and brinzolamide are topically delivered, thereby requiring much lower doses.

Dosage

The concentration of active ingredient evaluated by Bonapace et al. *(9)* was 10 μM and the only evidence for tissue concentration post treatment is available for dorzolamide in which Sugrue examined the tissue layers in rabbit eyes after 14 ds of twice daily administration and showed that the concentration of dorzolamide in the various layers were: retina, – approx 3.6 μM; choroid, approx 4.0 μM; and Sclera, approx 3.6 μM *(10)*. These concentrations are less that half the desired 10 μM and clinical efficacy, on this evidence alone, would be unsure. When this reservation was communicated to W. Sly (the group leader who carried out the work reported in the Bonapace et al. article *[9]*), they performed a second analysis using Dorzolamide at different concentrations and measuring the percentage of TUNEL positive cells.

This work showed that at concentrations from 0.1 μM and up, the percentage of TUNEL positive cells was 5% or less. This is strongly suggestive, assuming that rabbit and human eyes behave similarly, that treatment with dorzolamide drops will deliver an effective dose of active ingredient to the cell in the choriocapillaris.

Duration of Treatment

An equally important aspect of the design of a potential trial is that the metrics used in the assessment of vision will be able to yield detectable changes that may then be attributed to the treatment. In order to evaluate this we are in the process of starting a 2-yr evaluation of our patient cohort to determine the baseline values for measures of vision, and to determine the rate of change in these parameters over time.

CONCLUSION

It is obvious that the most trustworthy drug-response experiments will be derived from work on an animal model of the disease, and this is in progress. We are working to set the stage so that when the animal work has been concluded we will be in a position to move to a treatment trial without further delay. Thereby, we are fulfilling our promise to our subjects, and the implicit promise of molecular genetics as a whole, to use the genetic discoveries to bring some meaningful change to the lives of sufferers.

REFERENCES

1. Rebello G, Ramesar R, Vorster A, et al. Apoptosis-inducing signal sequence mutation in carbonic anhydrase IV identified in patients with the RP17 form of retinitis pigmentosa. Proc Natl Acad Sci USA 2004;101:6617–6622.
2. Bardien S, Ebenezer N, Greenberg J, et al. An eighth locus for autosomal dominant retinitis pigmentosa is linked to chromosome 17q. Hum Mol Genet 1995;4:1459–1462.
3. Bardien S, Ramesar R, Bhattacharya S, Greenberg J. Retinitis pigmentosa locus on 17q (RP17): fine localization to 17q22 and exclusion of the PDEG and TIMP2 genes. Hum Genet 1997;101:13–17.
4. Bardien-Kruger S, Greenberg J, Tubb B, et al. Refinement of the RP17 locus for autosomal dominant retinitis pigmentosa, construction of a YAC contig and investigation of the candidate gene retinal fascin. Eur J Hum Genet 1999;7:332–338.
5. den Hollander AI, van der Velde-Visser SD, Pinckers AJ, Hoyng CB, Brunner HG, Cremers FP. Refined mapping of the gene for autosomal dominant retinitis pigmentosa (RP17) on chromosome 17q22. Hum Genet 1999;104:73–76.
6. Wistrand PJ, Knuuttila KG. Renal membrane-bound carbonic anhydrase. Purification and properties. Kidney Int 1989;35:851–859.
7. Zhu XL, Sly WS. Carbonic anhydrase IV from human lung. Purification, characterization, and comparison with membrane carbonic anhydrase from human kidney. J Biol Chem 1990;265:8795–8801.
8. Hageman GS, Zhu XL, Waheed A, Sly WS. Localization of carbonic anhydrase IV in a specific capillary bed of the human eye. Proc Natl Acad Sci USA 1991;88:2716–2720.
9. Bonapace G, Waheed A, Shah GN, Sly WS. Chemical chaperones protect from effects of apoptosis-inducing mutation in carbonic anhydrase IV identified in retinitis pigmentosa 17. Proc Natl Acad Sci USA 2004;101:12,300–12,305.
10. Sugrue MF. Pharmacological and ocular hypotensive properties of topical carbonic anhydrase inhibitors. Prog Retin Eye Res 2000;19:87–112.

25

Macular Degeneration—An Addendum

Colin J. Barnstable and Joyce Tombran-Tink

In the last decade there has been an increasing focus on developing new treatments for age-related macular degeneration because of the burgeoning aging population in the Western world. Many attempts have been made to find genes important to the onset of macular degeneration in the hope that we will understand the biological mechanisms and pathways that trigger the disease so that we can develop the most effective intervention strategies. Linkage and candidate gene studies have indicated that macular degeneration is a multigenic disease. In chapter 2, Wang et al. describe the evidence for a possible role of the ABCA-4, a retina-specific ATP-binding cassette transporter protein and the Apo-E gene, which encodes for a lipoprotein that maintains normal levels of cholesterol, as two genes that are risk factors for AMDR when they are dysfunctional. A recent review provides more details of the linkage studies that suggest a role for these genes and lists others that have been proposed (1). ABCA-4 is clearly involved in the early onset Stargardt's disease but has a less clearly defined role in adult AMD. Different alleles of Apo-E can confer risk for or protection from AMD but the effect is relatively minor. Hemicentin-1, a gene on chromosome 1 encoding an extracellular matrix protein, was previously thought to be important in AMD but is now viewed as a marker for real AMD risk genes in the same region of chromosome 1. The linkage and candidate gene studies have resulted in our understanding of how the dysfunction of some genes may be risk factors for the pathogenesis of AMD. However, these do not account for a significant number of the diseased cases or provide a clear indication of the causes of AMD.

Recently, a new generation of high-density genetic analysis tools to exploit SNP genotyping has been applied to AMD studies. SNPs are single-nucleotide polymorphisms that are scattered throughout the genome and have been used widely as polymorphic markers for over two decades to study point mutations in cancers and congenital disease. Previously, the laborious gel-based analyses of single-stranded conformational polymorphisms (SSCP) have been the method for these genetic studies of polygenic traits in a given population. However, a worldwide-based effort to identify SNPs has now resulted in the identification of millions of these markers in public databases with precisely matched chromosomal localization using information from the human genome project. The overwhelming number of SNPs to screen for genetic diseases has therefore resulted in the generation of arrays of 100,000 or more SNPs spaced evenly along the genome, a much more efficient method to test an association of any

From: *Ophthalmology Research: Retinal Degenerations: Biology, Diagnostics, and Therapeutics*
Edited by: J. Tombran-Tink and C. J. Barnstable © Humana Press Inc., Totowa, NJ

disease or trait with a specific gene. Although the analysis and interpretation of these whole genome studies are very complex, a number of studies of AMD have resulted in important new information.

We have the spent the last year trying to put together this volume to include cutting edge research and therapies for retinal degenerations and feel it is important to include this brief essay as an update of the recent exciting findings of two other genes reported to be associated with macular degeneration. The first report was based on a whole-genome case-control association study for genes involved in the dry form of AMD carried out by Klein et al. in 2005 *(2)*. To maximize the chance of success, this group chose clearly defined phenotypes for disease cases and controls. Case individuals all exhibited at least some large drusen, the most prominent phenotypic marker for AMD, combined with evidence of sight-threatening AMD (geographic atrophy or neovascular AMD). Control individuals had either no or only a few small drusen. Data were analyzed using a statistically conservative approach to correct for the large number of SNPs tested, thereby guaranteeing that the probability of a false positive was no greater than the reported *p* values. The study used a subset of individuals who participated in the Age-Related Eye Disease Study (AREDS) sponsored by the National Eye Institute. Among the 116,204 single-nucleotide polymorphisms genotyped, an intronic and common variant in the complement factor H gene (CFH), a gene associated with complement inactivation, was strongly associated with AMD (nominal *p* value $<10^{-7}$). In individuals homozygous for the risk allele, the likelihood of AMD was increased by a factor of 7.4 (95% confidence interval 2.9–19). In subsequent studies, polymorphism in linkage disequilibrium with the risk allele representing a tyrosine–histidine change at amino acid 402 was identified. This polymorphism is in a region of CFH that binds heparin and C-reactive protein. The CFH gene is located on chromosome 1 in a region that has repeatedly been linked to AMD in family-based studies. These findings were confirmed and reported at the same time by two other groups *(3,4)* and by numerous later studies of AMD patient populations *(5–9)*. Some of these reports have shown that other polymorphisms in the factor H gene are less prevalent in AMD and may play a protective role in this disease *(10,11)*.

The significance of the association of factor H and AMD is still being investigated; however, we know that the function of this protein is to shut off complement activation. It is therefore possible that the amino acid change at position 402 alters the efficacy of this action, leading to enhanced inflammation that in turn leads to AMD. It is unclear whether factor H polymorphisms are causative for AMD but it is likely that they govern the response to events triggered by other genes or environmental factors. The strong association between factor H and AMD has rekindled interest in the idea that this disease has an important inflammatory component *(12)* and in the future we may see increasing attempts to slow down the progression or block AMD by a number of anti-inflammatory approaches in susceptible individuals.

There is also new evidence that complement component, factor B, is associated in populations with the dry form of AMD *(13)*. Factor B acts upstream of factor H in the alternate complement pathway and solidifies the important role of this pathway in AMD.

While the factor H polymorphism is associated with a significant percentage of AMD patients, it still represents a genetic risk factor for only a portion of the AMD population. A more recent pair of studies presented evidence for a gene that confers substantial risk for the wet form of the disease *(14,15)*. This study focused on an Asian population where the wet form of the disease is more prevalent and the factor H polymorphism is less common. After examining almost 100,000 SNPs, only one showed a significant association with the patient group. This SNP was located on chromosome 10q26, a region previously associated with AMD by linkage studies. More detailed analysis suggests that the polymorphism associated with AMD resides in the promoter region of the HTRA1 gene, a gene encoding a heat-shock serine protease. The prediction from this study is that AMD patients transcribe this gene at a different rate than controls. In support of this is the finding that in AMD patients, the levels of HTRA1 RNA and protein are elevated *(15)*. Whether the HTRA1 enzyme is responsible for promoting the vessel growth that is a hallmark of the wet form of AMD is not known. These genetic investigations have, however, allowed the formulation of specific hypotheses about the role of this protein.

The findings of very different genes as major risk factors for the dry and wet forms of AMD imply that these are two distinct diseases, a hypothesis which is also supported by the clinical presentation of the disorders. Such a conclusion, however, may be premature because immunocytochemical studies indicate elevated levels of HTRA1 in drusen, the hallmark of the dry form of AMD *(15)*, suggesting that HTRA1 could play a role in the pathogenesis of both forms of the disease.

We have clearly moved into a new era of studying AMD and ongoing genetic studies are almost certain to define more genes conferring significant risk for the disease.

If the products of these genes interact with each other, then we may be able to identify a set of polymorphisms, or a haplotype, strong enough to be used as a diagnostic predictor of AMD. From the biological point of view, the genetic findings are telling us how the pathology of AMD develops. A better understanding of these pathways is a necessary prerequisite to developing the next generation of therapies to combat this prevalent and devastating disease.

REFERENCES

1. Haddad S, Chen CA, Santangelo SL, Seddon JM. The genetics of age-related macular degeneration: a review of progress to date. Surv Ophthalmol 2006;51:316–363.
2. Klein RJ, Zeiss C, Chew EY, et al. Complement factor H polymorphism in age-related macular degeneration. Science 2005;308:385–389.
3. Haines JL, Hauser MA, Schmidt S, et al. Complement factor H variant increases the risk of age-related macular degeneration. Science 2005;308:419–421.
4. Edwards AO, Ritter R 3rd, Abel KJ, Manning A, Panhuysen C, Farrer LA. Complement factor H polymorphism and age-related macular degeneration. Science 2005;308:421–424.
5. Zareparsi S, Branham KE, Li M, et al. Strong association of the Y402H variant in complement factor H at 1q32 with susceptibility to age-related macular degeneration Am J Hum Genet 2005;77:149–153.
6. Souied EH, Leveziel N, Richard F, et al. Y402H complement factor H polymorphism associated with exudative age-related macular degeneration in the French population. Mol Vis 2005;11:1135–1140.
7. Sepp T, Khan JC, Thurlby DA, et al. Complement factor H variant Y402H is a major risk determinant for geographic atrophy and choroidal neovascularization in smokers and nonsmokers. Invest Ophthalmol Vis Sci 2006;47:536–540.

8. Okamoto H, Umeda S, Obazawa M, et al. Complement factor H polymorphisms in Japanese population with age-related macular degeneration. Mol Vis 2006;12:156–158.

9. Despriet DD, Klaver CC, Witteman JC, et al. Complement factor H polymorphism, complement activators, and risk of age-related macular degeneration. JAMA 2006;296:301–309.

10. Hageman GS, Anderson DH, Johnson LV, et al. A common haplotype in the complement regulatory gene factor H (HF1/CFH) predisposes individuals to age-related macular degeneration. Proc Natl Acad Sci USA 2005;102:7227–7232.

11. Hughes AE, Orr N, Esfandiary H, Diaz-Torres M, Goodship T, Chakravarthy U. A common CFH haplotype, with deletion of CFHR1 and CFHR3, is associated with lower risk of age-related macular degeneration. Nat Genet 2006;38:1173–1177.

12. Anderson DH, Mullins RF, Hageman GS, Johnson LV. A role for local inflammation in the formation of drusen in the aging eye. Am J Ophthalmol 2002;134:411–431.

13. Gold B, Merriam JE, Zernant J, et al. AMD Genetics Clinical Study Group; Hageman GS, Dean M, Allikmets R. Variation in factor B (BF) and complement component 2 (C2) genes is associated with age-related macular degeneration. Nat Genet 2006;38:458–462.

14. Dewan A, Liu M, Hartman S, et al. HTRA1 promoter polymorphism in wet age-related macular degeneration. Science 2006;314:989–992.

15. Yang Z, Camp NJ, Sun H, et al. A variant of the HTRA1 gene increases susceptibility to age-related macular degeneration. Science 2006;314:992–993.

INDEX